J. Allen Hill

Spri

Heidelberg
Science
Library

Heidelberg
Science
Library

Albrecht Unsöld

The New Cosmos

2nd Revised and Enlarged Edition

Translated by R. C. Smith
based on the translation by
W. H. McCrea of the 1st Edition

Springer-Verlag
New York
Heidelberg
Berlin

A. Unsöld
Institut für Theoretische Physik
und Sternwarte der Universität
23 Kiel,
Olshausenstrasse, Neue Universität
Haus 13
Federal Republic of Germany

R. C. Smith
Physics Building
Astronomy Centre
University of Sussex
Falmer, Brighton BN1 9QH
England

The photo on the cover shows the Crab nebula.

Library of Congress Cataloging in Publication Data
Unsöld, Albrecht, 1905–
The new cosmos.
(Heidelberg science library)
Translation of Der neue Kosmos.
Bibliography: p.
Includes index.
1. Astronomy. I. Title. II. Series.
QB43.2.U5713 1977 520 76-48061

First published in 1967: Heidelberger Taschenbücher, Band 16/17
Der neue Kosmos

Printed in the United States of America.

ISBN 0-387-90223-6 Springer-Verlag New York
ISBN 3-540-90223-6 Springer-Verlag Berlin Heidelberg

In Memoriam M. G. J. Minnaert

(12.II.1893-26.X.1970)

From the Author's Preface to the First Edition

This book may serve to present the modern view of the universe to a large number of readers whose over-full professional commitments leave them no time for the study of larger monographs . . . Such a book should not be too compendious. Accordingly, the author has been at pains to allow the leading ideas of the various domains of astronomical investigation to stand out plainly in their scientific and historical settings; the introductory chapters of the three parts of the book, in the framework of historical surveys, should assist the general review. With that in mind, the title was chosen following Alexander von Humboldt's well known book "*Kosmos, Entwurf einer physischen Weltbeschreibung*" (1827–1859). On the other hand, particular results—which admittedly first lend color to the picture—are often simply stated without attempting any thorough justification.

The reader seeking further information will find guidance in the Bibliography. This makes no pretensions to completeness or historical balance. References in the text or in captions for the figures, by quoting authors and years, make it possible for the reader to trace the relevant publications through the standard abstracting journals.

I wish to thank my colleagues V. Weidemann, E. Richter and B. Baschek for their critical reading of the book and for much helpful counsel, and H. Holweger for his tireless collaboration with the proofs. Similarly, my thanks are due to Miss Antje Wagner for the careful preparation of the typescript.

April 1966

ALBRECHT UNSÖLD

Kiel
Institute for Theoretical Physics
and University Observatory

Translator's Foreword to the First Edition

Many graduates in mathematics and physics turn to research in optical or radio astronomy, in astrophysics, in space science, or in cosmology. Professor Unsöld has provided the concise but comprehensive introduction to modern astronomy that all such students need at the outset of their work. Scientists in other fields who follow current advances in astronomy will find in it a compact work of reference to provide the background for their reading. Professor Unsöld has had in mind the widest possible circle of readers who for any reason want to know what modern astronomy is about and how it works. It is a privilege to help to extend this circle of readers to include those who prefer to read the book in English.

I have sought to put the work into serviceable English while losing as little as possible of the force and economy of Professor Unsöld's own masterly writing. It has not been my concern to create any illusion that the work was originally written in English.

I thank Professor Unsöld for reading my translation and for all the helpful comments which have, indeed, convinced me that he could himself have written his book in English better than anyone else. He has also supplied corrections of a few minor errors in the German text. I am grateful to Dr John Hazlehurst for preparing all the diagrams and their legends for the English edition. I thank Miss Shirley Ansell for taking infinite trouble with the typescript and the proofs.

Winter 1968 WILLIAM H. McCREA

Falmer, Sussex
Astronomy Centre
of the University of Sussex

Author's Preface to the Second Edition

The development of astronomy in the last ten years has been nothing short of explosive. This second edition of The New Cosmos, considerably revised and enlarged, tries to share this development with its readers. Let us mention a few key words: from moon landings, planetary probes, and continental drift through pulsars, X-ray and γ-ray sources, interstellar molecules, quasars, and the structure and evolution of stars and stellar systems right up to cosmological models.

As before, the most important task of this book is to give a not too difficult introduction to present-day astronomy and astrophysics, both to the student of astronomy and to the specialist from a neighboring discipline. We therefore draw to the attention of the reader, as an essential part of our description, the numerous illustrations—many of them new—and their detailed captions. As far as possible we link a description of important observations with basic features of the theory. On the other hand, when it comes to detail we often content ourselves with a brief description, leaving the detailed explanation to the specialist literature. The transition to the specialist literature should be eased by the Bibliography at the end of the book. Important new investigations are noted in the text by their year, not so much for historical reasons as to enable the original work to be found in the Astronomy and Astrophysics Abstracts (1969 on).

The amateur astronomer should not let himself be frightened by a few formulas. Rather, it is best for him in the meantime to accept their numerical results in good faith and to pursue his reading in the spirit of "divine curiosity" so prized by Einstein.

My colleagues V. Weidemann, H. Holweger, D. Reimers, and T. Gehren have upheld me in the friendliest way by reading the manuscript, by a variety of helpful advice, and by reading the proofs. Mrs G. Mangelsen and Mrs G. Hebeler have helped untiringly with the production of the manuscript. To all these are due my heartiest thanks.

August 1974 ALBRECHT UNSÖLD

Kiel
Institute for Theoretical Physics
and University Observatory

Translator's Foreword to the Second Edition

This second edition of The New Cosmos is a worthy successor to the now well-known first edition. Professor Unsöld has taken the opportunity not only to bring his book up to date but also to expand on some topics which previously received only a brief mention. The task of translation has been greatly lightened by the high quality of the book.

Like Professor McCrea, whose excellent translation of the first edition I have used where possible, I have tried mainly to produce a faithful translation in clear English. Despite that, there are inevitable differences in style between the two translations, and I can only hope that these are not too glaringly obvious.

I am grateful to Professor Unsöld for patiently answering my queries and for his percipient comments on my translation. I also wish to thank Drs Brian and Deborah Charlesworth for comments on Section 31. I am grateful to Ms Esther Salve, to Ms Elizabeth Barnes, and in particular to Mrs Hazel Freeman for producing the typescript.

July 1977

ROBERT C. SMITH

Falmer, Sussex
Astronomy Centre
of the University of Sussex

Contents

I. Classical Astronomy

II. Sun and Stars
Astrophysics of Individual Stars

III. Stellar Systems
Milky Way and Galaxies; Cosmogony and Cosmology

I. Classical Astronomy

1. Stars and Men: Observing and Thinking

Historical introduction to classical astronomy

Through all the ages the stars have run their courses uninfluenced by man. So the starry heavens have ever symbolized the "Other"—Nature, Deity—the antithesis of the "Self" with its world of inner consciousness, desires, and activities. The history of astronomy forms one of the most stirring chapters in the history of the human spirit. All along the emergence of new modes of thought has interlocked with the discovery of new phenomena, often made with the use of new-fashioned instruments.

Here we cannot recount the great contributions of the peoples of the ancient orient, the Sumerians, Babylonians, Assyrians and Egyptians. Also we must forbear to describe what was in its own way the highly developed astronomy of the peoples of the far east, the Chinese, Japanese, and Indians.

The concept of the cosmos and of its investigation in our sense of the word goes back to the Greeks. Discarding all notions of magic and aided by an immensely serviceable language, they set about constructing forms of thought that made it possible step by step to "understand" the cosmical manifestations.

How daring were the ideas of the Presocratics! Six centuries before Christ, Thales of Miletus was already clear about the Earth being round, he knew that the Moon is illumined by the Sun, and he had predicted the solar eclipse of the year 585 B.C. Is it not just as important, however, that he sought to refer the entire universe to a single basic principle, namely "water?"

The little that we know about Pythagoras (mid-sixth century B.C.) and his school has an astonishingly modern ring. Here already is talk of the sphericity of the Earth, Moon, and Sun, of the rotation of the Earth and of the revolution of at least the two inferior planets Mercury and Venus round the Sun.

Because after the dissolution of the Greek states science had found a new home in Alexandria, there the quantitative investigation of the heavens based upon systematic measurements made rapid advances. Rather than dwell here upon the numerical results, it is of particular interest to note how the great Greek astronomers above all dared to apply *geo*metrical laws to the cosmos. Aristarchus of Samos, who lived in the first half of the third century B.C., sought to compare quantitatively the Sun-Earth and Moon-Earth distances as well as the diameters of these three heavenly bodies. His starting point was that at the Moon's first and third quarters the Sun-Moon-Earth triangle is right-angled at the Moon. Following this first measurement in outer space, Aristarchus was the first to teach the heliocentric world-system. He appreciated its portentous consequence that the distances of the fixed stars must be stupendously greater than the distance of the Sun from the Earth. How far he was ahead of his time is best shown by the fact that the succeeding generation proceeded to forget this great discovery of his. Soon after these notable achievements of Aristarchus, Eratosthenes carried out between Alexandria and Syene the first measurement of a degree of arc on the Earth's surface. He compared the difference in latitude between the two places with their separation along a well-used caravan route, and he inferred thus early fairly exact values of the circumference and diameter of the Earth. The greatest observer of antiquity was Hipparchus (about 150 B.C.), whose star catalog was scarcely surpassed in accuracy even in the sixteenth century. Even though his equipment naturally did not suffice for him significantly to improve the determination of the fundamental parameters of the planetary system, he nevertheless succeeded in making the important discovery of precession, that is, of the advance of the equinoxes and so of the difference between the tropical and sidereal year.

Within the compass of Greek astronomy, the theory of planetary motion, which we must now discuss, had to remain a problem in geometry and kinematics. Gradual improvement of observations on the one hand, and the development of new mathematical approaches on the other hand, provided the material from which Philolaus, Eudoxus, Heracleides, Apollonius, and others strove to construct a representation of planetary motions by means of ever more complicated interlacings of circular motions. It was much later that the astronomy and planetary theory of antiquity first attained its definitive form through Claudius Ptolemy who about 150 A.D. in Alexandria wrote the thirteen books of his Handbook of Astronomy (Mathematics) *Μαθηματικῆσ Συντάξεωσ βιβλία ιγ*. The "Syntax" later earned the nickname *μεγίστη* (greatest), from which the arabic title *Almagest* ultimately emerged. Although the contents of the Almagest rested extensively upon the observations and investigations of Hipparchus, Ptolemy did nevertheless contribute something new, especially to the theory of planetary motion. Here we need give only a brief sketch of Ptolemy's geocentric world system: The Earth rests at the center of the cosmos. The motions of the Moon and the Sun around the sky may be fairly well represented simply by circular paths. Ptolemy described the motions of the planets using the theory of epicycles. A planet moves round a circle, the so-called epicycle, the immaterial center of which moves around the Earth in a second circle, the deferent. Here we need not go into the elaboration of the system using further circles, some of them eccentric, and so on. In Section 6 we shall consider the connexion between this system and the heliocentric world system of Copernicus, and the way they differ. In its intellectual attitude, the Almagest shows clearly the influence of Aristotelian philosophy, or rather of *Aristotelianism*. Its scheme of thought, which had developed from the tools of vital investigations into what eventually became the dogmas of a rigid doctrine, contributed not a little to the astonishing historical durability of the Ptolemaic system.

We cannot here trace in detail the way in which, after the decline of the Academy at Alexandria, first the nestorian Christians in Syria and then the Arabs in Baghdad took over and extended the work of Ptolemy.

Translations and commentaries upon the Almagest formed the essential sources for the first western textbook of astronomy, the *Tractatus de Sphaera* of Johannes de Sacrobosco, an Englishman by birth who taught in the University of Paris until his death in 1256. The *Sphaera* was time and again reissued and commented upon; it was the text for academic instruction even to the time of Galileo.

Suddenly an altogether new spirit in thought and life showed itself in the fifteenth century, first in Italy, and soon afterwards in the north. Today for the first time we are beginning to appreciate the penetrating meditations of Cardinal Nicolaus Cusanus (1401–1464). It is of the greatest interest to see how with Cusanus ideas about the infinity of the universe and about the quantitative investigation of nature sprang from religious and theological reflections. Toward the end of the century (1492) came the discovery of America by Christopher Columbus, who expressed the new feeling about the world in his classic remark, "il mondo e poco." A few years later Nicholas Copernicus (1473–1543) initiated the heliocentric world-system.

Part of the intellectual background of the new thinking came from the fact that after the sack of Constantinople by the Turks in 1453 many learned works of antiquity were made available in the west by Byzantine scholars. Certain very fragmentary traditions about the heliocentric systems of the ancients had obviously made a deep impression upon Copernicus. Furthermore, we notice a trend away from the rigid doctrine of Aristotle toward the much more vital mode of thought of the Pythagoreans and the Platonists. The "platonic" concept is that the advance of knowledge consists in a progressive adjustment of our inner world of ideas and thought-forms to the ever more fully investigated external world of phenomena. It has been shared by all the important investigators in modern times from Cusanus through Kepler to Niels Bohr. Finally, with the rise of industry, the question was no longer, "What does Aristotle say about so and so?" but, "How does one do so and so?"

About 1510 Copernicus sent to several notable astronomers in the form of a letter a communication, first rediscovered in 1877, *Nicolai Copernici de hypothesibus motuum caelestium a se constitutis commentariolus,* which already contained most of the results of his masterpiece, *De revolutionibus orbium coelestium libri VI,* first published in Nuremberg in 1543, the year of Copernicus's death.

Throughout his life Copernicus held to the idea of the perfection of circular motion that was inescapable throughout ancient and mediaeval times, and he never contemplated any other motions.

Following the traditions of the Pythagoreans and Platonists, Johannes Kepler (1571–1630) was the first to succeed in attaining a more general standpoint in "mathematico-physical aesthetics." Starting from the observations of Tycho Brahe (1546–1601), which far excelled all earlier ones in accuracy, he discovered his three laws of planetary motion (see Section 3). Kepler found the first two as a result of an incredibly burdensome trigonometric computation, using Tycho's observations of Mars, in his *Astronomia nova, seu physica coelestis tradita commentariis de motibus stellae Martis ex observationibus G. V. Tychonis Brahe* (Prague, 1609). The third of Kepler's laws was announced in the *Harmonices mundi libri V* (1619). Kepler's fundamental work on optics, his astronomical telescope, his Rudolphine Tables (1627), and much else, can be only barely mentioned here.

About the same time in Italy, Galileo Galilei (1564–1642) directed the telescope, which he had constructed in 1609, on the heavens and discovered in rapid succession: the maria, the craters, and other mountain-formations on the Moon, the many stars in the Pleiades and the Hyades, the four moons of Jupiter and their free revolution round the planet, the first indication of Saturn's rings and sunspots. Galileo's *Sidereus nuncius* (1610), in which he described his discoveries with the telescope, the *Dialogo delli due massimi sistemi del mondo, Tolemaico e Copernicano* (1632), and the book he produced after his condemnation by the Inquisition, the *Discorsi e dimostrazioni matematiche intorno a due nuove scienze attenenti alla meccanica ed ai movimenti locali* (1638), with its beginning of theoretical mechanics, are all masterpieces not only scientifically but also artistically in the way they are presented. The observations with the telescope, the observations of the supernovae of 1572 by Tycho Brahe and of 1604 by Kepler and Galileo, and finally the appearance of several comets, all promoted what was perhaps the most important perception of the times. Contrary to the Aristotelian view, it was that there is no basic difference between celestial and terrestrial matter, and that the same physical laws hold good in the realm of astronomy as in the realm of terrestrial physics. The Greeks had perceived this already so far as geometry was concerned. Recalling the case of Copernicus makes the difficulty of the concept evident. But it was this that gave wings to the enormous upsurge of natural science at the beginning of the seventeenth century. Also W. Gilbert's investigations of magnetism and electricity, Otto von Guericke's experiments with the air pump and the electrical machine, and much else sprang from the transformation of the astronomical outlook.

We cannot here pay tribute to the many observers and theorists who built up the new astronomy among whom such notabilities as Hevelius, Huygens, and Halley were outstanding.

A whole new epoch of natural philosophy began with Isaac Newton (1642–1727). His masterpiece, *Philosophiae naturalis principia mathematica* (1687), with the help of his newly created infinitesmal calculus (fluxions), placed theoretical mechanics for the first time upon a sure foundation. In combination with his law of gravitation, it accounted for Kepler's laws, and at a single stroke it founded the whole of terrestrial and celestial mechanics.

In the domain of optics, Newton invented the reflecting telescope and discussed the interference phenomena of "Newton's rings." Almost by the way, Newton developed the basic ideas for many branches of theoretical physics.

We may compare with him only the "princeps mathematicorum" Carl Friedrich Gauss (1777–1855) to whom astronomy owes the theory of determination of orbits, important contributions to celestial mechanics and advanced geodesy, as well as the method of least squares. Never has any other mathematician combined sure judgment in opening up new fields of investigation with such preeminent skill in solving special problems.

Again this is not the place to commemorate the great workers in celestial mechanics from Euler through Lagrange and Laplace to Henri Poincaré. Also we shall be able to mention the great observers like F. W. Herschel and J. F. W. Herschel, F. W. Bessel, F. G. W. Struve, and O. W. Struve only in connexion with their discoveries. As the conclusion of this review let a single historic date be recorded—that of the measurement of the first trigonometric stellar parallaxes and so of the distances of stars by F. W. Bessel (61 Cygni), F. G. W. Struve (Vega) and T. Henderson (α Centauri) in the year 1838. This conspicuous achievement of the technique of astronomical measurement forms basically the starting point for the modern advance into cosmic space.

(We shall preface Part II with some historical remarks on astrophysics and Part III with some on galactic research as well as cosmogony and cosmology.)

2. Celestial Sphere: Astronomical Coordinates: Geographic Latitude and Longitude

Man's imagination has from ancient times made star pictures out of easily recognizable groupings of stars (Figure 2.1). In the northern sky we easily recognize the Great Bear (Plough). We find the Polestar if we produce the line joining the two brightest stars of the Great Bear by about five times its length. The Polestar is the brightest star in the Little Bear; if we extend the line past it to about the same distance on the other side, we come to the "W" of Cassiopeia. With the help of a celestial globe or a star map, other

Figure 2.1. Circumpolar stars for a place of geographical latitude $\varphi = +50°$ (for example, approximately Frankfurt or Winnipeg). The coordinates are right ascension RA and declination (+ 40° to + 90°). The hour hand turning with the stars (see above), whose extension passes through the first point of Aries, shows *sidereal time* on the outer dial. *Precession:* the celestial pole turns about the pole of the ecliptic ENP once in 25,800 years. The position of the celestial north pole is denoted for various past and future dates.

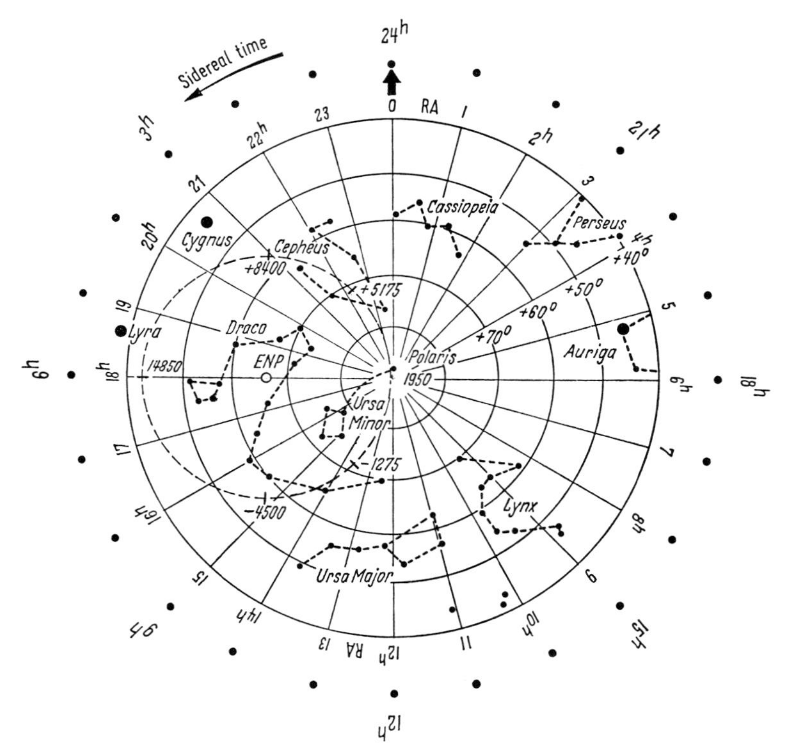

constellations are easily found. In 1603, J. Bayer in his *Uranometria nova* denoted the stars in each constellation in a generally decreasing order of brightness as $\alpha, \beta, \gamma, \ldots$. Nowadays these Greek letters are supplemented by the numbering introduced by the first Astronomer Royal, J. Flamsteed, in his *Historia Coelestis Britannica* (1725). The Latin names of the constellations are usually abbreviated to three letters.

As an example, the second brightest star in the Great Bear (Ursa Major) is known as β U Ma or 48 U Ma (read as 48 Ursae Majoris).

The *celestial sphere* is, mathematically speaking, the infinitely distant sphere upon which we see the stars to be projected. On this sphere we distinguish (Figure 2.2):

(1) The *horizon* with the directions north, west, south, east.
(2) Vertically above us the *zenith* and below us the *nadir*.
(3) The *meridian* through the celestial pole, the zenith, the south point, the nadir, and the north point.
(4) The *prime vertical* through the zenith, the west point, and the east point, at right angles to the horizon and to the meridian.

In the coordinate system so determined we describe the instantaneous position of a star by specifying two angles (Figure 2.2): (1) The *azimuth* is reckoned along the horizon from 0° to 180° (W or E) and is measured either from the south point or the north point. (2) The *altitude* = 90° − *zenith distance*.

Figure 2.2 Celestial sphere. Horizon with north, south, east and west points. The (celestial) meridian passes through the north point, the (celestial) pole, the zenith, the south point and the nadir. Coordinates are altitude and azimuth.

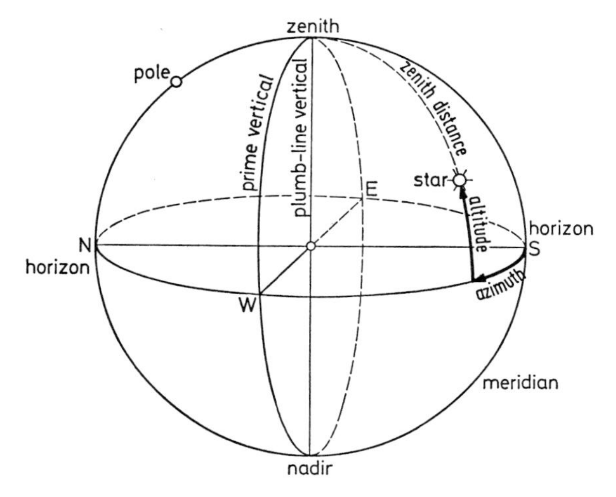

The celestial sphere with all the stars appears to rotate each day around the *axis* of the heavens, through the north pole and south pole of the sky. The *celestial equator* is in the plane perpendicular to this axis. At a particular instant we describe the *place* (position) *of a star* upon the celestial sphere (Figure 2.3), thought of as infinitely remote, by the *declination* δ, reckoned positive from the equator toward the north pole and negative toward the south pole, and the *hour angle* HA, reckoned from the meridian in the sense of the daily motion, that is, westward from the meridian.

In the course of a day, a star traverses a *parallel circle* on the sphere; it reaches its greatest altitude on the meridian at *upper culmination* and its least altitude at *lower culmination*.

Figure 2.3. Celestial coordinates. Right ascension RA and declination δ. Hour angle HA = sidereal time minus right ascension RA. Lower right: the Earth (flattening exaggerated). Polar altitude = geographical latitude φ.

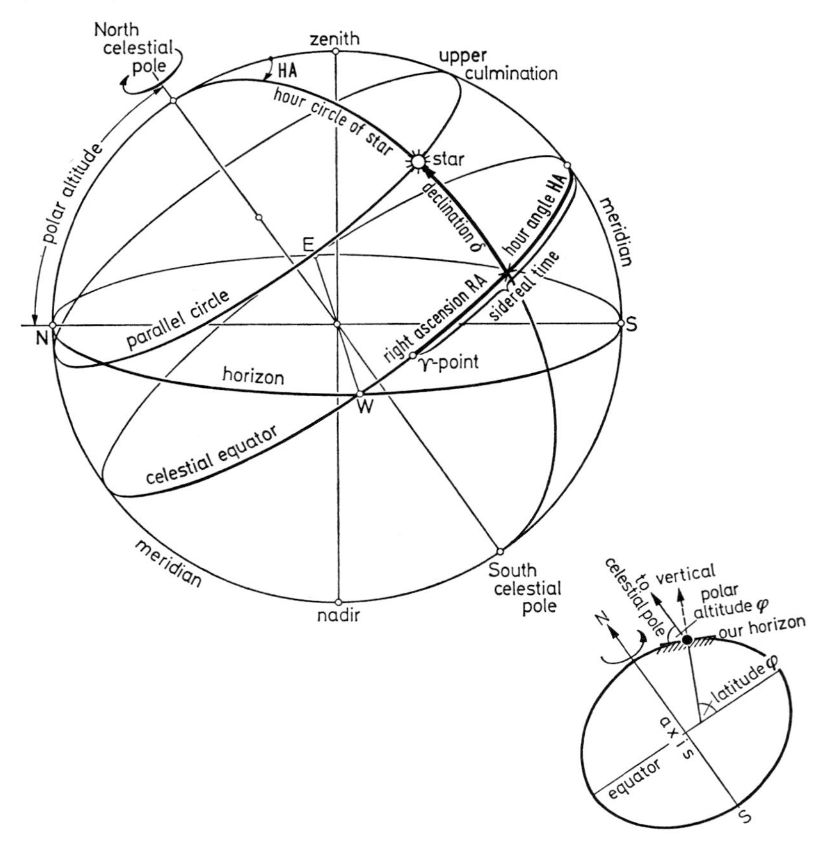

On the celestial equator, we now mark the *first point of Aries* ♈, which we shall explain in the next section as being the position of the Sun at the time of the spring equinox (March 21). Its hour angle gives *sidereal time* τ. In Figure 2.1, if we think of the arrow (which indicates the point ♈ on the celestial equator) as moving round with the stars, it will indicate sidereal time on the "clock dial" shown round the outside of the diagram.

Finally, we can define the position of a star on the celestial sphere independently of the time of day. We call the arc of the equator from ♈ to the hour circle of the star its *right ascension* RA. It is reckoned in hours, minutes, and seconds. Thus 24^h corresponds to 360° so that

$$\begin{aligned} 1^h &= 15^\circ \quad 1^\circ = \frac{1^h}{15} = 4^m \\ 1^m &= 15' \quad 1' = 4^s \\ 1^s &= 15'' \end{aligned}$$

We see at once from Figure 2.3 that

$$\text{hour angle HA} = \text{sidereal time } \tau \text{ minus right ascension RA.} \tag{2.1}$$

We have already introduced our second coordinate of the star, the *declination* δ.

If the astronomer wants to point his telescope at some particular star, planet, or other celestial object, he extracts its right ascension RA and declination δ from a star catalog, he reads the sidereal time τ on his sidereal clock, and then he sets on the graduated circles of his instrument the hour angle HA in hours, minutes, and seconds as derived from equation 2.1 and the declination δ in degrees, positive to the north, negative to the south. Specially accurately determined positions of the so-called *fundamental stars,* particularly for time determination (see below), together with positions of the Sun, Moon, planets, and so on, are to be found in the astronomical almanacs or *ephemerides,* the most important of which is "The Astronomical Ephemeris."

The Copernican system interpreted the apparent rotation of the celestial sphere by asserting that the Earth rotates about its axis once in 24 hours. The *horizon* is in the tangent plane of the Earth, or more precisely that of a water level, at our position. The *zenith* corresponds to the direction of the plumb line perpendicular to this plane, which is that of local gravity (including the effect of centrifugal force arising from the Earth's rotation). The pole height (= altitude of the pole above the horizon) is seen from Figure 2.3 to be equal to the *geographical latitude* φ (= angle between the plumb line and the plane of the equator). The pole height is easily measured as the mean value of the altitudes of the pole star, or any circumpolar star, at upper and lower culmination.

Geographical longitude l corresponds to the hour angle. If the hour angle HA of one and the same star is observed simultaneously at Greenwich (prime meridian, $l_G = 0°$) and, say, Kiel, then the difference is the geographical longitude of Kiel l_K. While the determination of geographical *latitude* requires only the measurement of angles, the measurement of *longitude* demands precise time transfer. In former days, time as indicated by the motion of the Moon or of Jupiter's satellites was employed. The invention of the mariners' chronometer by John Harrison (about 1760–1765), was a big advance, as was later the transmission of time signals first by telegraph and then by radio.

We remark further: At a place of (north) latitude φ a star of declination δ reaches altitude $h_{max} = \delta + 90° - \varphi$ at *upper culmination*, and reaches $h_{min} = \delta - (90° - \varphi)$ at *lower culmination*. *Circumpolar stars* with $\delta > 90° - \varphi$ are always above the horizon; stars with $\delta < -(90° - \varphi)$ are never above the horizon.

In measuring star altitudes h the refractive effect of the Earth's atmosphere has to be taken into account. The apparent raising of a star (apparent altitude − true altitude) is called the *refraction*. For average conditions of pressure and temperature in the atmosphere the refraction is as shown:

Star altitude h	0°	5°	10°	20°	40°	60°	90°
Refraction Δh	34′50″	9′45″	5′16″	2′37″	1′09″	33″	0″

The refraction decreases a little with increasing atmospheric temperature and with decreasing atmospheric pressure, for example in a region covered by a depression or on mountains.

3. Motion of the Earth: Seasons and the Zodiac: Day, Year, and Calendar

We now consider the orbital motion, or revolution, of the Earth around the Sun, in the sense of Copernicus, and also the rotation of the Earth around its axis as well as the motion of the axis itself. Here we adopt the observer's standpoint. We shall develop Newton's theory of the motions of the Earth and the planets in accordance with his theories of mechanics and gravitation in Section 6.

The apparent annual motion of the Sun across the heavens Copernicus ascribed to the description by the Earth of an almost circular path round the Sun. The plane of this path meets the celestial sphere in a great circle called the *ecliptic* (Figure 3.1). This cuts the celestial equator at an angle 23°27′, the *obliquity of the ecliptic*. Thus in its annual motion round the Sun, the axis of the Earth maintains a fixed direction in space, making an angle 90° − 23°27′ = 66°33′ with the plane of the Earth's orbit.

Here is a short summary of the circumstances of the onset of the *seasons* for the northern hemisphere of the Earth:

Beginning of seasons			Coordinates of the Sun		Zodiacal constellation entered by the Sun
			RA	δ	
Spring	March 21	Vernal equinox	0^h	$0°$	Aries (Ram) ♈
Summer	June 22	Summer solstice	6^h	$23°27'$	Cancer (Crab) ♋
Autumn	September 23	Autumnal equinox	12^h	$0°$	Libra (Scales) ♎
Winter	December 22	Winter solstice	18^h	$-23°27'$	Capricorn (Sea Goat) ♑

At the equinoxes the day and night segments of the Sun's diurnal motion are each 12 hours. The Sun reaches its greatest altitude at a place north of the tropics of latitude φ at noon on the summer solstice on June 22 with $h = 90° - \varphi + 23°27'$, and on December 22 it reaches its lowest noon altitude with $h = 90° - \varphi - 23°27'$. The Sun can reach the zenith in geographical latitudes up to $\varphi = 23°27'$, the tropic of Cancer. On the other hand, north of the arctic circle $\varphi \geqslant 90° - 23°27' = 66°33'$ around the time of the winter solstice the Sun remains below the horizon; around the time of the summer solstice, the "midnight sun" behaves as a circumpolar star.

In the southern hemisphere summer occurs when winter occurs in the north, the tropic of Capricorn corresponds to the tropic of Cancer, and so on.

Figure 3.1. Apparent annual motion of the Sun among the stars.
Ecliptic. Seasons.

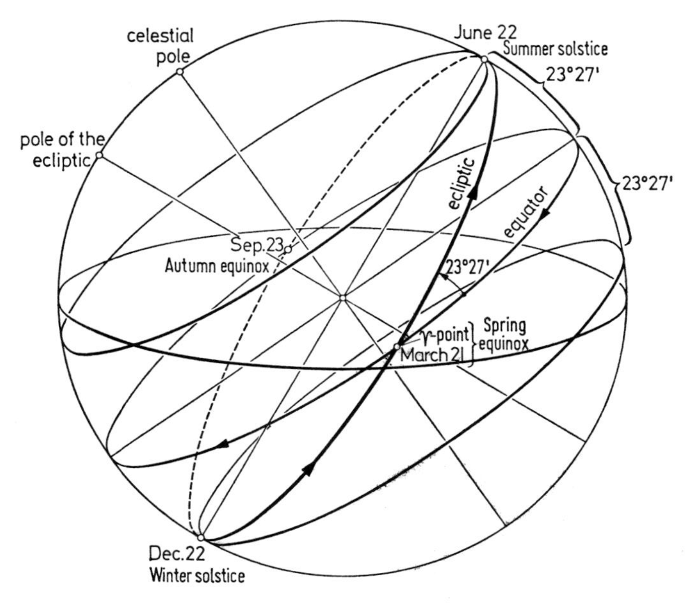

The *Zodiac* is an 18° wide belt of the sky, centred on the ecliptic. From ancient times, mankind has divided it up into twelve equal parts denoted by the *Signs of the Zodiac* (Figure 3.2).

For describing the motion of the Earth and the planets it is convenient to employ a coordinate system determined by the ecliptic and its poles. Ecliptic (or astronomical) *longitude* is measured from the first point of Aries along the ecliptic, analogously to right ascension, in the sense of the annual motion of the Sun. Ecliptic (or astronomical) *latitude* is measured analogously to declination, at right angles to the ecliptic. Astronomical latitude and longitude must not be confused with geographic latitude and longitude!

Astronomers of antiquity knew of the nonuniformity of the apparent annual motion of the Sun, and J. Kepler (1571–1630) recognized this as a consequence of the first two of his laws of planetary motion, to which we return more fully in Section 5.

Kepler's first law. Each planet moves in an ellipse with the Sun at one focus.

Kepler's second law. The radius vector from the Sun to any one planet describes equal areas in equal times.

Kepler's third law. For any two planets the squares of the periods are proportional to the cubes of the semimajor axes of their orbits.

The geometrical characteristics of the orbit of the Earth or of any other planet round the Sun are shown in Figure 3.3. The semimajor axis we denote by a; the distance from the center to a focus we denote by ae and we call the fraction e the *eccentricity* of the orbit. At *perihelion* the distance of the Earth from the Sun is $r_{\min} = a(1 - e)$; at aphelion, the distance is $r_{\max} = a(1 + e)$. The daily motion of the Sun in the heavens, (the angle turned through in one day by the radius vector to the Earth) at

Figure 3.2. Path of the Earth around the Sun. Seasons. Zodiac ("zone of animals") and its signs. The Earth is at perihelion (closest to the Sun) on January 2 and at aphelion (furthest from the Sun) on July 2.

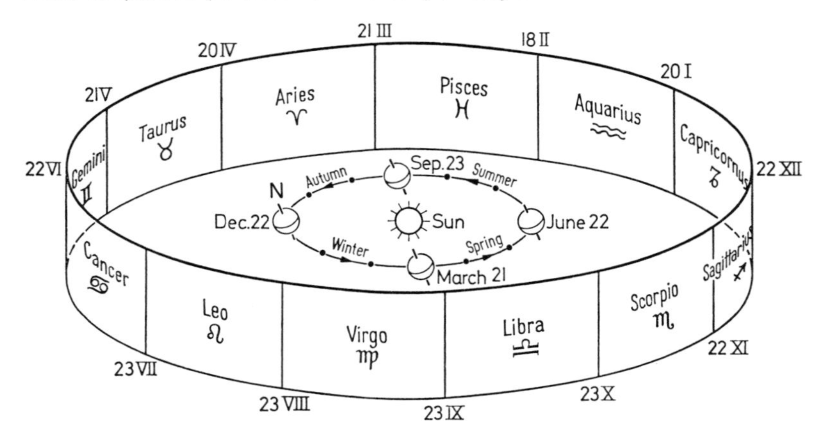

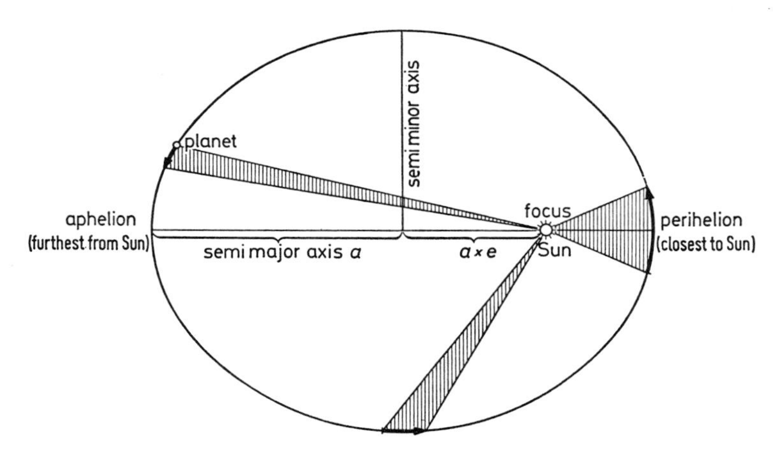

Figure 3.3. Elliptical orbit of a planet. Semimajor axis *a;* eccentricity *e;* distance from center to focus (Sun) *ae.* (The eccentricity of planetary orbits is much smaller than shown here)

perihelion and aphelion, according to Kepler's second law, are in the ratio $(r_{\max}/r_{\min})^2 = ((1 + e)/(1 - e))^2$; the corresponding values of apparent diameter of the Sun's disk are in the ratio $(1 + e)/(1 - e)$. Measurements of both of these ratios agree in giving for the eccentricity of the Earth's orbit $e = 0.01674$. At the present time, the Earth goes through perihelion about January 2. The approximate coincidence of this date with the beginning of the calendar year is fortuitous.

Hipparchus discovered that the point of the vernal equinox is not fixed in the sky but moves back along the ecliptic by some 50″ a year with respect to the stars. This means that, since ancient times, the point has moved out of the constellation Aries into the constellation Pisces. This precession of the equinoxes depends on the fact that the celestial pole moves round the sky with a period of 25 800 years on a circle of radius 23°27′ having its center at the pole of the ecliptic, this being fixed among the stars. In other words, the Earth's axis describes a cone of semiangle 23°27′ about the axis of the Earth's orbit once in 25 800 years.

Since the precession displaces the coordinate frame of right ascension RA and declination δ relative to the stars, star positions and star catalogs must be expressed with reference to the equinox to which the stated values of RA and δ refer. Since the star places change on account of the proper motions (Section 23), the *epoch* of the observations is also reduced to the same point of time. Table 3.1 gives the corrections for right ascension RA (for different values of RA and δ) and for declination (for different values of RA) consequent upon ten years' precession.

Precession with the 25 800-year period is superimposed upon a somewhat similar motion of smaller amplitude having a 19-year period; this is called *nutation*. Finally, the rotation axis of the Earth fluctuates by amounts of the order ±0″.2 relative to the body of the Earth, which can be

Table 3.1. Precession for Ten-Year Interval

a. ΔRA in minutes of time (+increase, −decrease)

Hours RA for northern objects	6	7 5	8 4	9 3	10 2	11 1
Declination |δ| in degrees	m	m	m	m	m	m
80°	+1.77	+1.73	+1.60	+1.40	+1.14	+0.84
70°	1.12	1.10	1.04	0.94	0.82	0.67
60°	0.898	0.885	0.846	0.785	0.705	0.612
50°	0.778	0.768	0.742	0.700	0.645	0.581
40°	0.699	0.693	0.674	0.644	0.606	0.560
30°	0.641	0.636	0.624	0.603	0.576	0.546
20°	0.593	0.590	0.582	0.570	0.553	0.533
10°	0.552	0.550	0.546	0.540	0.532	0.522
0°	+0.512	+0.512	+0.512	+0.512	+0.512	+0.512
Hours RA for southern objects	18	19 17	20 16	21 15	22 14	23 13

12 0	13 23	14 22	15 21	16 20	17 19	18	
m	m	m	m	m	m	m	
+0.51	+0.19	−0.12	−0.38	−0.58	−0.70	−0.75	80°
0.51	0.35	+0.21	+0.08	−0.02	−0.08	−0.10	70°
0.512	0.412	+0.319	+0.240	+0.178	+0.140	+0.126	60°
0.512	0.444	+0.380	+0.324	+0.282	+0.256	+0.247	50°
0.512	0.464	+0.419	+0.380	+0.350	+0.332	+0.335	40°
0.512	0.479	+0.448	+0.421	+0.401	+0.388	+0.384	30°
0.512	0.491	+0.472	+0.455	+0.442	+0.434	+0.431	20°
0.512	0.502	+0.492	+0.484	+0.478	+0.476	+0.473	10°
+0.512	+0.512	+0.512	+0.512	+0.512	+0.512	+0.512	0°
0 12	1 11	2 10	3 9	4 8	5 7	6	

b. Δδ in minutes of arc (increase, which implies a decrease of | δ | in the southern sky)

Hours RA	0 24	1 23	2 22	3 21	4 20	5 19
Δδ Minutes of arc	+3′.34	+3′.23	+2′.89	+2′.36	+1′.67	+0′.86

6 18	7 17	8 16	9 15	10 14	11 13	12
0′.0	−0′. 86	−1′. 67	−2′. 36	−2′.89	−3′.23	−3′.34

analysed into an irregular part and a periodic part for which the so-called Chandler period of 433 days is recognized. The corresponding *polar wandering* is kept under observation at a series of stations. We return in Section 6 to the explanation of the various motions of the Earth's axis.

We first give further consideration to the problem of time keeping. Our daily life is determined by the position of the Sun. So there was introduced

$$\text{true solar time} = \text{hour angle of the Sun} + 12^{\mathrm{h}}.$$

This is the time that a simple sundial shows, 12^{h} corresponding to the upper transit of the Sun. On account of the nonuniform orbital motion of the Earth (Kepler's second law) and of the obliquity of the ecliptic, true solar time does not proceed uniformly. Consequently we have recourse to *mean solar time*. We imagine a *mean sun* that traverses the equator uniformly in the same time that the true Sun takes for its annual passage round the ecliptic. The hour angle of this imagined mean sun defines mean solar time. The difference

$$\text{true solar time} - \text{mean solar time} = \text{equation of time}$$

is composed of two terms depending upon the eccentricity of the Earth's orbit and on the obliquity of the ecliptic. Its extreme values are

	February 12	May 14	July 26	November 4
equation of time	$-14^{\mathrm{m}}20^{\mathrm{s}}$	$3^{\mathrm{m}}45^{\mathrm{s}}$	$-6^{\mathrm{m}}23^{\mathrm{s}}$	$16^{\mathrm{m}}23^{\mathrm{s}}$.

Mean solar time is different for each meridian. For the purpose of ordinary intercourse, people have therefore agreed that throughout certain zones they will employ the time appropriate to one selected meridian. In Germany and central Europe they use Central European Time (Mitteleuropäische Zeit MEZ) which is local mean solar time on the meridian 15°E. In western Europe they use the time of the Greenwich Prime Meridian.

For scientific purposes, for instance for astronomical and geophysical measurements at stations that are often distributed all over the globe, scientists everywhere use

$$\text{universal time (UT)} = \text{mean solar time of the Greenwich meridian.}$$

We reckon in 24 hours starting with 0^{h} at midnight. For example 12^{h} UT is 13^{h} MEZ.

Using a small variable correction which is published in the *Astronomical Ephemeris* (for example, for 1965 it is $+35^{\mathrm{s}}$) we pass from UT to

$$\text{ephemeris time (ET)}$$

which we shall describe below.

For astronomical observation we need also the relation between mean solar time and sidereal time. Relative to the equinox, the "mean sun" moves in a year (=365 days) through 360° ($=24^{\mathrm{h}}$) from west to east. The mean solar day is therefore $24^{\mathrm{h}}/365$ or $3^{\mathrm{m}}56^{\mathrm{s}}$ longer than the sidereal day. In a month the sidereal clock gains about 2^h on the "ordinary" UT or MEZ clock. To amplify this, we quote the sidereal time for 0^h local time on certain dates. As is known, this is equal to the hour angle of the point ♈ and equal to the right ascension RA of stars passing the meridian at midnight (at the longitude where the observations are made).

0^h local time (midnight)	January 1	April 1	July 1	October 1
RA on meridian and sidereal time	6^h42^m	12^h37^m	18^h35^m	0^h38^m

The unit of time for longer intervals is the *year*. We define a

$$\textit{sidereal year} = 365.25636 \text{ mean solar days}$$

which is the time between two successive passages of the Sun through the same point in the sky. It is also the true period of revolution of the Earth relative to the fixed stars ("sidereal" from Latin *sidus*, a star).

The time between two successive passages of the Sun through the vernal equinox is a

$$\textit{tropical year} = 365.24220 \text{ mean solar days.}$$

Since this advances westwards by $50''\!.3$ in a year, the tropical year is correspondingly shorter than the sidereal year. The seasons and the calendar depend upon the tropical year ("tropic" from Greek *τρόπεῖν*, to turn).

Since for practical reasons every year should consist of an integral number of days, in ordinary life we use the

$$\textit{civil year} = 365.2425 = 365 + \frac{1}{4} - \frac{3}{400} \text{ mean solar days}$$

corresponding to the intercalary prescription of the *Gregorian calendar* introduced in 1582 by Pope Gregory XIII. After 3 years of 365 days follows a leap year (date divisible by 4) of 366 days except for the years of the centuries *not* divisible by 400. Here we cannot discuss the older *Julian calendar* of Julius Caesar 45 B.C., or other interesting questions of chronology. Modern proposals for calendar reform seek to ensure that the beginning of the year and the first days of the month shall always fall on the same days of the week. Also they would fix the present movable feasts, particularly Easter (the first Sunday after the first full moon after the spring equinox) and Whitsunday (50 days after Easter).

For ease of chronological reckoning over long periods of time, especially for such applications as to observations and ephemerides of variable stars, we seek to avoid the irregularities in the length of years and months. Following the proposal of J. Scaliger (1582) we therefore simply count the so-called *Julian days* consecutively. The Julian day begins at 12^h UT (mean Greenwich noon). The beginning of the Julian day 0 is put at 12^h UT on January 1 of the year 4713 B.C. On 1970 January 1 at 12^h UT began Julian day 2440588.

Astronomical time measurement depended for a long time on the (assumed) uniformity of the Earth's rotation. The physical basis of terrestrial time measurement was already known to Huygens (*Horologium oscillatorium,* published 1673): a clock consists of an oscillatory mechanism (pendulum, balance wheel etc.), which is isolated as far as possible from its surroundings and kept in motion by a drive (weight, spring, etc.) with the least possible feedback. The pendulum clock—always being further improved—was for over three centuries one of the most important instruments in every observatory. The quartz clock, much less sensitive to disturbances, uses an oscillating piezoquartz block or ring, which is kept in motion by a loosely coupled electronic oscillator. However the pinnacle of precision measurement was reached more recently in the atomic clock, in which the oscillation frequency of the atoms in caesium vapour (^{133}Cs) is used; the frequency corresponds to the transition between the two lowest hyperfine structure levels. In physical terms, the frequency in question is that due to the alteration of the direction of the nuclear spin with respect to the angular momentum vector of the rest of the atom. The excitation, frequency division (in order to obtain a lower frequency) and display are again controlled electronically. The enormous precision of the atomic clock, whose (relative) frequency precision reaches $\sim 10^{-13}$, is the basis for many fundamental measurements and observations in physics and astronomy.

The comparison of astronomical time measurements with sets of quartz clocks, and even more with atomic clocks, shows that the rotation period of the Earth is not constant, but exhibits fluctuations (of the order of a millisecond ms), which are partly irregular and partly periodic and depend on alterations in the mass distribution on the Earth.

Since 1967, one second is defined as the length of 9192631770 oscillation periods of the radiation which corresponds to the transition between the two hyperfine structure levels of the ground state of the ^{133}Cs atom.

From the atomic time data of a standard institute one obtains Coordinated Universal Time (=UTC). World Time or Universal Time, more exactly designated as UT1, which is corrected only for the fluctuations of the Earth's axis, is not uniform and differs from UTC by the correction

$$\Delta UT = UT1 - UTC$$

which is published regularly by the Bureau International de l'Heure (BIH) and other institutes. Since in daily life, just as in navigation, geophysics, etc., time reckoning must as before adjust itself to the Earth's rotation, UTC is "jumped" forward or backward by a whole second, from time to time, as soon as the amount of ΔUT approaches 1 s.

Independent of the progress of physical time measurement, it is found that over long periods of time the motions of the planets and of the Sun (or alternatively, of the Earth) and especially of the Moon show small, common deviations from the ephemerides calculated by Newtonian mechanics and gravitation theory. This has to do with a secular (progressive) increase in the length of the day, which must be caused by the braking of the Earth's rotation as a result of tidal friction (Section 6). A further part of the deviations exhibits no such evident regularity. The comparison of the deviations for different heavenly bodies compels us to ascribe them to deviations of "astronomical" measures of time from the "physical" time concerned in Newton's laws. Because of these experiences it was decided in 1950 to base all astronomical ephemerides on a time reckoning founded upon the fundamental laws of physics, the so-called *ephemeris time* ET. The small corrections, ET minus UT1, are discovered essentially from very exact observations of the motion of the Moon. They can be determined only retrospectively; however, for most purposes predictions can be made with sufficient accuracy by extrapolation. The unit of ephemeris time, the ephemeris second, was defined in 1956 as the 31556925.974 part of the tropical year 1900.

Ten years later it was decided also to tie the ephemeris second to the unit of atomic time. However, this violates the internal consistency of the system of ET. The question of a complete amalgamation of the time systems of atomic time UTC and ephemeris time ET and further the consideration of the position (that is, the gravitational potential) of the clock in the sense of general relativity theory (Section 30) requires further investigations and international discussions.

4. Moon: Lunar and Solar Eclipses

The Moon appears to us as a disk with average diameter 31′ and so with just about the same apparent size as the Sun. Its distance from the Earth can be got by triangulation from two observatories sufficiently far apart (say along the same meridian). Astronomers call the angle subtended at the Moon by the Earth's equatorial radius the equatorial horizontal parallax of the Moon. Its value in the mean is $3422''.6$. Since the Earth's radius is known to be 6378 km, we derive for the mean distance of the Moon from the center of the Earth

$$60.3 \text{ Earth radii} = 384\,400 \text{ km}$$

and hence for the radius of the Moon

$$0.272 \text{ Earth radius} = 1738 \text{ km}.$$

We shall concern ourselves with the physical structure of the Earth and the Moon in Section 7. Here we consider the path and motion of the Moon from the observational standpoint.

The Moon circles the Earth, in the same sense as the Earth goes round the Sun, in a *sidereal month* = 27.32 days. That is to say, after this time the Moon returns to the same place among the stars.

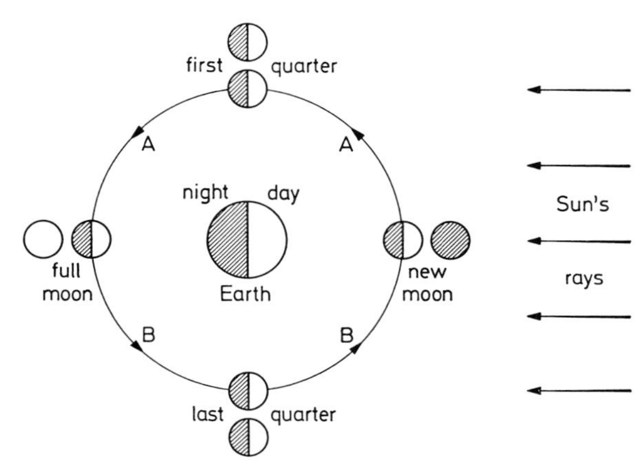

Figure 4.1. Phases of the Moon. The Sun is on the right. The outer pictures show the phases as seen from the Earth (i.e. from the center of the circle). Waxing moon A; waning moon B.

The origin of the *phases*[1] *of the Moon* is illustrated in Figure 4.1. Their period, the *synodic month* = 29.53 days (1 to 3 in Figure 4.2), after which the Moon returns to the same position relative to the Sun, is longer than the sidereal month (1 to 2 in Figure 4.2). Each day the Moon moves eastward $360°/29.53 = 12\overset{\circ}{.}2$, relative to the Sun, and $360°/27.32 = 13\overset{\circ}{.}2$ relative to the stars.

[1]The connection between the phases of the Moon and the weather has often been discussed. The famous astrologer Stupi Dass discovered that 95 percent of all sudden changes in weather occur within a week before or after full moon or new moon.

Figure 4.2. The synodic month (1–3) is longer than the sidereal month (1–2) since the Earth has meanwhile moved further along its orbit.

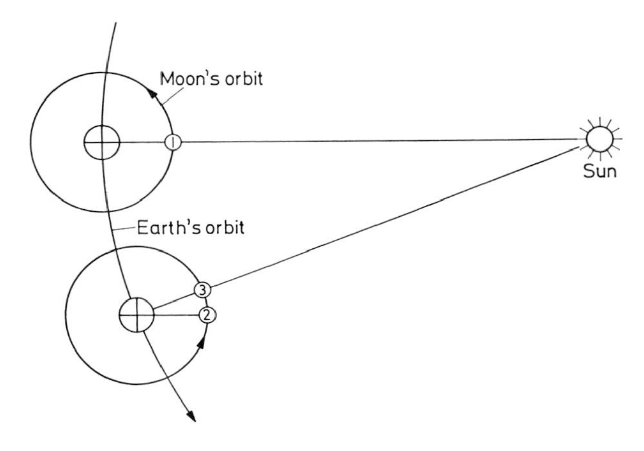

The difference between the sidereal and synodic daily motions of the Moon is equal to the daily motion of the Sun and so is $360°/365 \approx 1^{\circ}\!.0$. We can express this as

$$\frac{1}{\text{sidereal month}} - \frac{1}{\text{sidereal year}} = \frac{1}{\text{synodic month}}.$$

Actually the path of the Moon round the Earth is an ellipse of eccentricity $e = 0.055$. The point in the path where the Moon approaches nearest to the Earth is called *perigee* (analogous to perihelion in the motion of the Earth); where it is furthest from the Earth is called *apogee*. The plane of the Moon's orbit has an inclination $i \approx 5°$ to the plane of the Earth's orbit (ecliptic). The Moon rises "above" the ecliptic from south to north at the *ascending node,* and passes "below" the ecliptic (in the sense of dwellers in the northern hemisphere) at the *descending* node.

In consequence of the perturbations produced by the attraction of the Sun and the planets, the Moon's orbit further performs the following motions:

(1) The perigee moves around the Earth, in the plane of the Moon's orbit, with "direct" motion, that is, in the sense of the motion of the Earth, with a period of 8.85 years.
(2) The line of nodes, in which the planes of the orbits of the Moon and the Earth intersect, has "retrograde" motion in the ecliptic, that is, in the opposite sense to the motion of the Earth, with a period of 18.61 years, the so-called *nutation period.*

This regression of the nodes causes among other things a corresponding "nodding" of the Earth up to a maximum of 9″, the *nutation* of the Earth's axis already mentioned.

The mean time between successive passages of the Moon through the same node is called the *draconitic* month = 27.2122 days. It is important for the computation of eclipses (see below).

If we could view the paths of the Moon and the Earth around the Sun from a space vehicle, we should confirm that, in agreement with a simple calculation, the path of the Moon is always concave toward the Sun (Figure 4.3).

Here we shall now consider the *rotation* of the Moon and its motion relative to its mass center. This can be measured very accurately from observations of a sharply defined crater, or the like, on the Moon's disk.

The fact that, by and large, the Moon always presents the same face to us depends upon its rotation period being equal to its period of revolution, that is, to the sidereal month. The equalization of the two periods has clearly been brought about by the tidal interaction of the Moon and the Earth (Section 6).

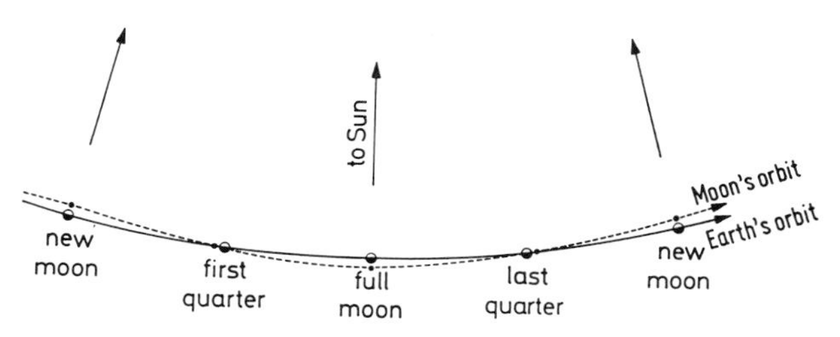

Figure 4.3. The orbits of the Earth and Moon around the Sun.

Precise observation shows nevertheless that the face of the Moon wobbles somewhat. The so-called *geometrical librations* of the Moon arise from the following causes:

(1) The equator of the Moon and the plane of the Moon's orbit are inclined at an angle of about 6°.7: the consequent *libration in latitude* is thus about ±6°.7.
(2) In agreement with the law of inertia, the Moon's rotation is uniform. In accordance with Kepler's second law, taking account of the eccentricity of the orbit, the Moon's revolution is not uniform. Hence arises the *libration in longitude* of about ±7°.6.
(3) The equatorial radius of the Earth subtends an angle of 57′ at the center of the Moon, this being the horizontal parallax of the Moon. The daily rotation of the Earth therefore produces a daily libration.

In addition there is the considerably smaller physical libration, which stems from the fact that the shape of the Moon departs slightly from a sphere and so the Moon performs small oscillations in the gravitational field, mainly that produced by the Earth.

Altogether these librations result in our being able to see from the Earth 59 percent of the Moon's surface.

Having studied the motions of the Sun, Earth, and Moon, we turn our attention to the impressive spectacles of lunar and solar eclipses.

A lunar eclipse occurs when the full moon enters the shadow cast by the Earth. As in the case of shadows cast by terrestrial objects, we distinguish between the full shadow, the *umbra,* and the surrounding region of partial shadow, the *penumbra*. If the Moon is completely immersed in the region of the full shadow of the Earth, we speak of a *total lunar eclipse;* if only part of the Moon is in shadow, we have a *partial* lunar eclipse. In consequence of known geometrical relationships, a lunar eclipse can last for at most 3^h40^m and totality at most 1^h40^m. Since the light from the Sun is scattered by the Earth's atmosphere, the outer edge of the penumbra on the Moon is quite blurred, and the umbra is noticeably unsharp. Since the blue light experiences stronger absorption than the red light, the penumbra, and to a lesser extent also the umbra, appears to be pervaded by reddish coppercolored light. Exact photometric observations of lunar eclipses can give information about the high layers of the Earth's atmosphere.

If the Moon, at the time of new moon, passes in front of the Sun, then a solar eclipse takes place (Figure 4.4). This can be *partial* or *total*. If the apparent diameter of the Moon is smaller than that of the Sun, then a central coverage results only in an *annular* eclipse. In a partial eclipse of the Sun, the observer is in the penumbra of the Moon, while during totality he is in the umbra. In the case of an annular eclipse, the vertex of the umbral cone is between the observer and the Moon.

In regard to the astrophysical study of the outermost layers of the Sun and of interplanetary material in its neighbourhood, total solar eclipses are of special significance because the bright light from the solar disk is completely cut off *outside* the Earth's atmosphere. Since the zone of totality on the Earth's surface is relatively narrow, the atmosphere is nevertheless slightly illuminated by scattered light coming in from the sides, even during totality. Relative to the Sun, the Moon regresses at a rate corresponding to the duration of the synodic month, on the average $0''.51$ per second on the sky; this corresponds to a stretch of 370 km on the Sun's surface. Eclipse observations with good time resolution therefore give an angular resolution which exceeds that of available telescopes (see below).

Occultations of stars by the Moon, which like solar eclipses have to be predicted specially for any given place of observation, are very sharply defined in time, since the Moon possesses no atmosphere. They are important for checking the Moon's path, the determination of fluctuations in the Earth's rotation, and checking ephemeris time. Since the Moon regresses in the mean by $0''.55$ per second relative to the stars, by photometric observations of occultations of the stars using precise time resolution, it is even possible in favorable cases to measure the angular diameter of the infinitesi-

Figure 4.4. Eclipse of the Sun (schematic). The Moon crosses the Sun's disk from west to east. Total and partial eclipses of the Sun are observed in the regions traversed by the umbra and penumbra, respectively.

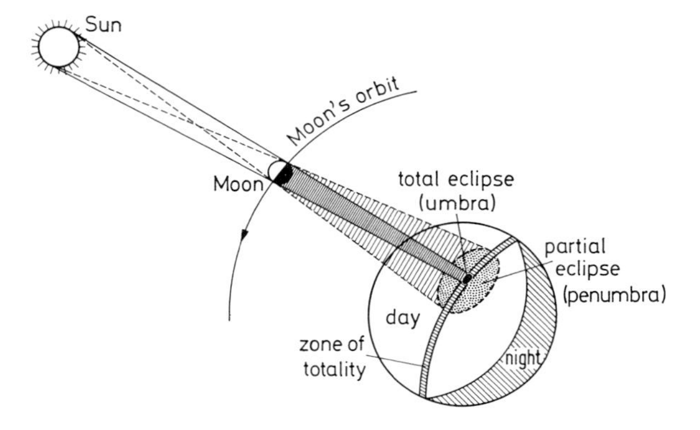

mal disks of the stars concerned. Even more important are occultations by the Moon of radio sources, whereby high angular resolution can be attained.

It was known to the cultures of the ancient East that solar and lunar eclipses (in the following, we say simply "eclipses" for short) repeat themselves with a period of $18^y\ 11\overset{d}{.}33$, the so-called Saros cycle. This cycle depends upon the fact that an eclipse can occur only if the Sun *and* Moon are fairly near to a node of the lunar orbit. The time that the Sun takes from a lunar node to return to the same node is somewhat shorter than one tropical year, on account of the regression of the lunar nodes, being equal to 346.62 days. This time is called an eclipse year. As one easily verifies, the Saros period comprises a whole number of synodic months *and* a whole number of eclipse years, in fact

$$223 \text{ synodic months} = 6585.32 \text{ days}$$

and

$$19 \text{ eclipse years} = 6585.78 \text{ days}$$

besides which we have

$$239 \text{ anomalistic months} = 6585.54 \text{ days}$$
$$(\text{anomalistic month from perigee to perigee} = 27.555 \text{ days}).$$

After $18^y\ 11\overset{d}{.}33$ the eclipse pattern in fact repeats itself with great accuracy. As one can show from the orbits of the Earth and the Moon, taking account of the diameters of these bodies, in a single year at most three lunar eclipses and five solar eclipses can occur. At any one place a lunar eclipse, which can indeed be seen over an entire hemisphere of the Earth, can be observed fairly frequently, while a total solar eclipse is an extremely rare occurrence.

5. Planetary System

The planets that have been known since ancient times (with their time-honored symbols) are Mercury ☿, Venus ♀, Mars ♂, Jupiter ♃, and Saturn ♄. Mankind has continually sought to transform the apparent capriciousness of their paths into a gradual unfolding of regularities.

In Section 1 we have briefly mentioned the concern of the ancients for the significance of planetary motions. Here we adopt the standpoint of the heliocentric world-view, as Nicholas Copernicus presented it in 1543. However, we shall drop the restriction to circular motions that Copernicus had retained as a relic of Aristotelian concepts and adopt the elliptic orbits of Johannes Kepler and his three laws of planetary motion (1609 and 1619). Thereby we arrive at the threshold of modern mathematico-physical thought that took significant shape at the hands of Galileo Galilei (1564–1642) and that in Isaac Newton's *Principia* (1687) was epitomized in the

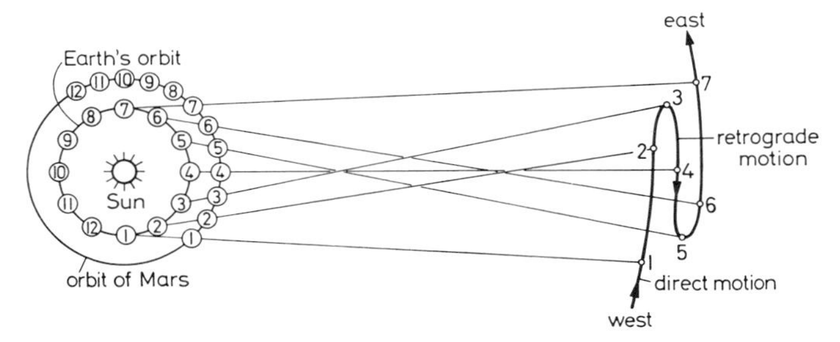

Figure 5.1. Direct (west-east) and retrograde (east-west) motion of the planet Mars. The positions of the Earth and Mars in their orbits are numbered from month to month. At 4 Mars is in opposition to the Sun; it is overtaken by the Earth at this point and is therefore in retrograde motion. This is the time when it is closest to the Earth and most favorably placed for observation. The orbit of Mars is inclined at 1°.9 to that of the Earth (that is, to the ecliptic).

statements of classical mechanics and gravitational theory. We shall concern ourselves with the physical structure of the planets in Section 7.

The occurrence of *direct* (west–east) motion with an interval of backward or *retrograde* (east–west) motion is clarified in Figure 5.1 by the example of the planet Mars.

Figure 5.2 shows the path of an *inner planet,* for example Venus, round the Sun, as seen from our standpoint on the more slowly revolving Earth. The planet is nearest to us at *inferior conjunction:* then it moves away from the Sun in the sky and reaches its *greatest westerly elongation* of 48° as a

Figure 5.2. Orbit and phases of Venus, an inner planet. The elongation of Venus on the sky cannot exceed ±48° (±28° for Mercury). The phases resemble those of the Moon. Greatest brightness occurs near maximum elongation.

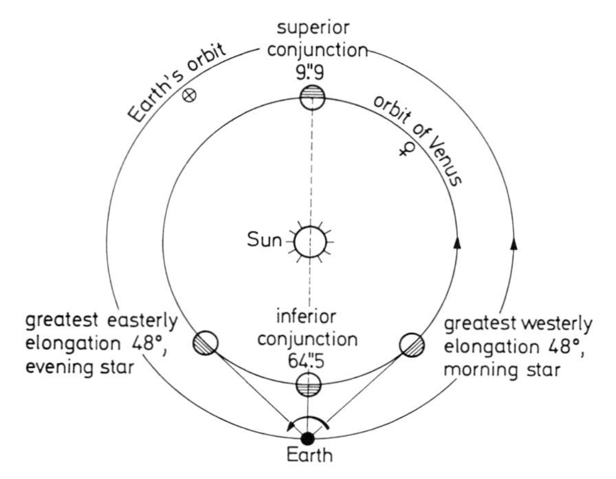

morning star. At *superior* conjunction Venus is at its furthest distance from the Earth and is again close to the Sun in the sky. It then moves further away and reaches its *greatest easterly elongation* of 48° as an *evening star*. The ratio of the orbital radii of Venus and the Earth is determined by the maximum elongation of ±48°, and in the case of Mercury ±28°. The phases of Venus and the corresponding changes in its apparent diameter (from 9″.9 to 64″.5) were discovered by Galileo with his telescope. They verify that the Sun is at the center of the (true) path of Venus. As we can see from Figure 5.2, Venus attains its greatest apparent brightness near its greatest elongations. At inferior conjunction, Venus (and Mercury) may pass in front of the Sun's disk. Such a transit of Venus was formerly of interest as a means of measuring the Sun's distance or solar parallax (see below).

An outer planet, Mars ♂ for example, is nearest to us at *opposition* (Figure 5.3); it culminates there at midnight by true local time, has its greatest apparent diameter, and is most favorably placed for observation. When in *conjunction* it is near the Sun in the sky. Unlike the cases of the Moon and the inner planets, the phases of an outer planet do not range the whole way from "full" to "new." We term *phase angle* φ the angle subtended by the Sun and the Earth at the planet. The fraction $\varphi/180°$ of the planet's hemisphere that faces the Earth is in darkness. As one easily verifies, the phase angle of an outer planet is greatest in *quadrature*, that is, when the planet and the Sun are seen in directions at 90° to each other. The greatest phase angle of Mars is 47° (Figure 5.3) and of Jupiter only 12°.

Figure 5.3. Orbit and phases of Mars, an outer planet. Greatest brightness and greatest angular diameter (25″.1) occur at opposition.

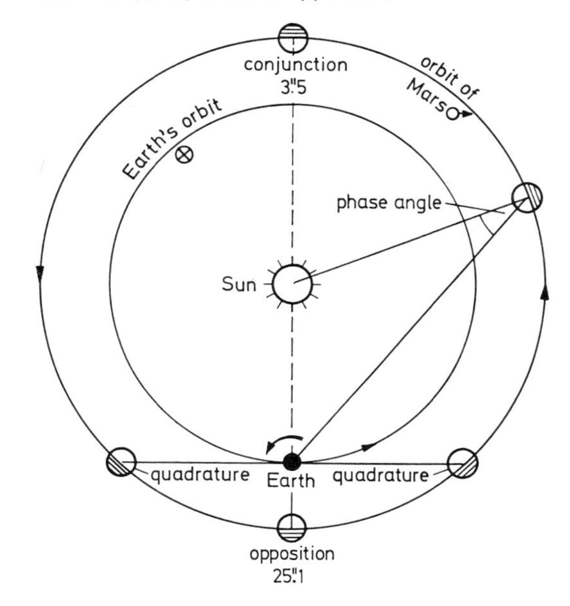

We give the name *sidereal period* to the true time of revolution of a planet round the Sun. The *synodic period* is the time of revolution relative to the Sun, that is, the time between two successive corresponding conjunctions, etc. As in the case of the Moon, we have for a planet, by subtraction of angular speeds,

$$\frac{1}{\text{synodic period}} = \left| \frac{1}{\text{sidereal period}} - \frac{1}{\text{sidereal period of the Earth}} \right|. \qquad (5.1)$$

In the case of Mars, for example, from the direct observation of the synodic period of 780 days and knowing the sidereal year to be 365 days, the sidereal period of the planet is found to be 687 days.

Kepler first determined the true shape of the orbit of Mars by combining pairs of observations of Mars that were separated in time by the sidereal period of Mars, and in which therefore Mars must be at the same position in its orbit. From such pairs of observations of Mars separated by 687 days, he could "triangulate" the positions of Mars and so determine its true path. Two fortunate circumstances helped Kepler in due course to derive his first two laws of planetary motion: the fact that Appollonius of Perga had investigated the conic sections mathematically, and the fact that of the planets known at that time Mars happened to have the greatest eccentricity $e = 0.093$. He found his third law ten years later, proceeding from his unshakable conviction that the motions of the planets must somehow manifest the "harmony" of the universe.

The complete description of the path of a planet or a comet (see below) round the Sun requires the orbital elements shown in Figure 5.4.

(1) Semimajor axis a. We express it either in terms of the semimajor axis of the Earth's orbit = 1 astronomical unit AU or else in kilometers.
(2) Eccentricity e [perihelion distance $a(1 - e)$; aphelion distance $a(1 + e)$].

Figure 5.4. Orbital elements of a planet or comet.

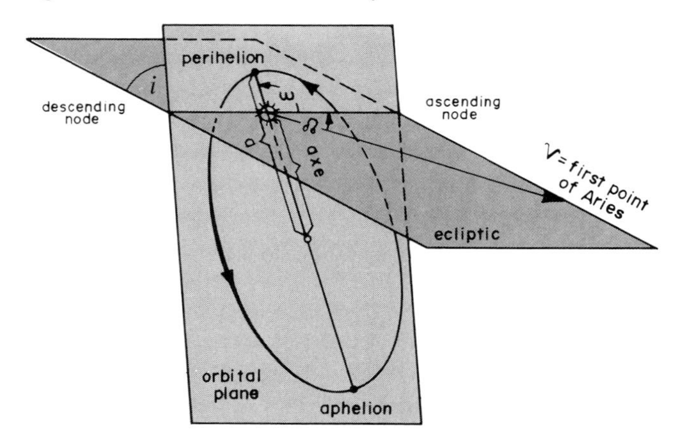

(3) Inclination of the plane of the orbit to the ecliptic i.
(4) Longitude of the ascending node ☊ (angle from the equinoctial point ♈ to the ascending node).
(5) Displacement ω of perihelion from the node (angle between the ascending node and perihelion). The sum of the angles ☊ $+ \omega$, of which the first is measured in the ecliptic and the second in the plane of the orbit, is called the *longitude of perihelion* π.
(6) Period P (sidereal period measured in tropical years) or the daily motion μ ($\mu = 360 \times 60 \times 60''/P_{\text{days}}$).
(7) Epoch E or the instant in time of perihelion passage T.

The elements a and e determine the size and shape of the orbit (Figure 3.3), i and ☊ the orbital plane, ω the orientation of the orbit in the plane. The course of the motion is determined by P and T, where, in accordance with Kepler's third law, the period P is determined by a, apart from small corrections (see below).

Table 5.1 contains the orbital elements of the planets that are of interest to us here (Figure 5.5). In addition to the planets known to the ancients, we have included the brightest of the *minor planets* (Ceres) as well as Uranus, Neptune, and Pluto. We shall later recount something of the interesting story of their discovery. To our planetary system there belong also the *comets* and *meteors* (or "shooting stars").

Nowadays we denote comets by the year of discovery and in each year we number them according to the sequence of their perihelion passages (see below); the name of the discoverer is often added. In ancient times and in the middle ages, according to the dogma of the immutability of the heavenly regions, the comets were relegated to the atmospheric surroundings of the Earth. Tycho Brahe first made exact observations of the comets of 1577 and 1585, and the parallaxes he derived showed that, for example, the comet of 1577 was at least six times as distant as the Moon. Isaac Newton recognized that the comets move round the Sun in elongated ellipses or parabolas, that is, in conic sections having eccentricity a little less than or equal to unity. His great contemporary Edmond Halley improved the techniques for computing cometary orbits and in 1705 he was able to show that the comet of 1682 (since named "Halley's comet") has a period of about 76.2 years. According to Kepler's third law, the major axis of the orbit is therefore $2 \times 76^{2/3} = 36$ astronomical units, that is, the aphelion lies somewhat beyond the orbit of Neptune. Halley's computed orbit also showed that the bright comet of 1682 was identical with those of 1531 and 1607, so he was also able to predict the return of the comet in 1758. Altogether 28 returns of Halley's comet since 240 B.C. have been witnessed.

In the main, cometary orbits fall into two groups: (1) Nearly parabolic orbits with periods greater than 100 years, perihelion passages at about 1 AU yielding the highest probability of discovery. (2) Elliptic paths of short–period comets. The aphelia are concentrated near the orbits of the major

Table 5.1. Planetary System (Epoch 1960.0)

Planet	Symbol	Discovery	Sidereal period years	Some orbital elements: Semimajor axis of orbit: Astronomical units AU	Some orbital elements: Semimajor axis of orbit: Million km 10^6 km	Some orbital elements: Eccentricity e	Some orbital elements: Inclination to ecliptic i degrees
Inner planets							
☿ Mercury			0.241	0.387	57.9	0.206	7.0
♀ Venus			0.615	0.723	108.2	0.007	3.4
⊕ Earth			1.000	1.000	149.6	0.017	0.0
Outer planets							
♂ Mars			1.880	1.524	227.9	0.093	1.8
Minor planets		Ceres: Piazzi, 1801; Gauss	4.603	2.767	413.6	0.076	10.6
♃ Jupiter			11.86	5.203	778	0.048	1.3
♄ Saturn			29.46	9.539	1427	0.056	2.5
⛢ Uranus		W. Herschel, 1781	84.01	19.18	2870	0.047	0.8
♆ Neptune		Leverrier and Galle, 1846	164.8	30.06	4496	0.009	1.8
♇ Pluto		Lowell and Tombaugh, 1930	247.7	39.44	5910	0.250	17.2

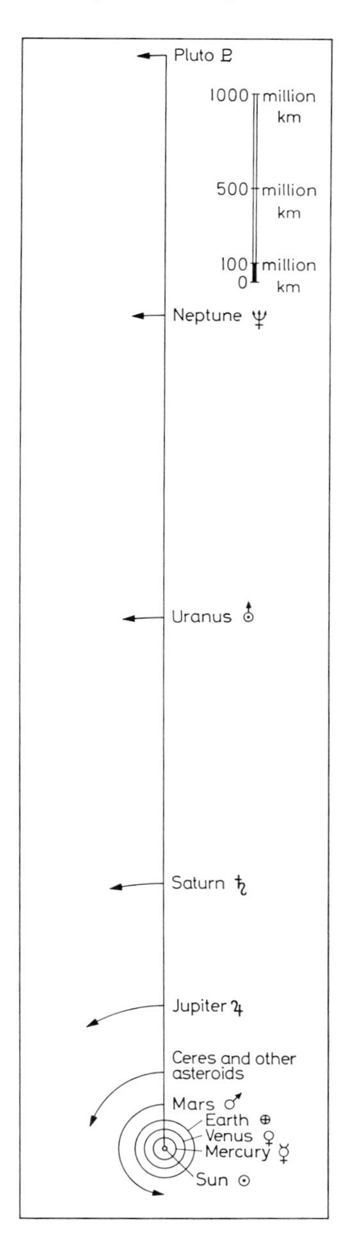

Figure 5.5. Mean orbital radii (semimajor axes) of the planets. The lengths of the arcs correspond to the mean motion in one (terrestrial) year. During this time Venus makes 1.62 revolutions and Mercury 4.15.

planets, particularly Jupiter. Such cometary families may have arisen by the "capture" of long-period comets by Jupiter or some other planet. Average values of orbital elements are about $a = 3.6$ AU, $e = 0.56$ (thus giving aphelion distance 5.6 AU almost equal to $a_{♃} = 5.2$ AU) and $i \approx 15°$.

Encke's comet has the shortest known period of 3.30 years. The comets

Schwassmann-Wachmann 1925 II ($a = 6.4$ AU, $e = 0.132$) and Oterma 1943 ($a = 3.96$ AU, $e = 0.144$) have nearly circular orbits.

The swarms of shooting stars or *meteors,* that on certain dates in the year seem, as a matter of perspective, to emanate from their so-called *radiants* in the sky, are simply fragments of comets whose orbits almost intersect that of the Earth; this is shown by the periodicities of the occurrences. In many cases, the material seems to be fairly well concentrated in the path, so that from time to time, depending on their periods, specially lively displays of shooting stars are observed. For example, there was the famous fall of Leonids (radiant RA 152°, $\delta + 22°$) that A. von Humboldt observed on 1799 November 11/12 in South America and that can be associated with the comet 1866 I of period 33 years. In addition, there are the so-called *sporadic meteors,* for which a periodicity cannot be discerned. That shooting stars are actually small heavenly bodies, which penetrate the Earth's atmosphere and so become incandescent, was first established in 1798 by two Göttingen students Brandes and Benzenberg by making corresponding observations from two sufficiently separated sites and then calculating the heights.

Even before that in 1794 E. F. F. Chladni had shown the *meteorites* are none other than (more massive) meteoric bodies that have reached the ground.

Neither amongst comets nor amongst meteors has anyone found hyperbolic orbits, that is, orbits of objects that have entered the solar system from outside.

In Sections 7 and 8 we shall consider the physical structure of planets, of their atmospheres and satellites as well as that of comets, meteors, and meteorites.

Here we must again consider the important question as to how one can measure in kilometers the distance of the Earth from the Sun, or more precisely, the semimajor axis of the Earth's orbit, which we have defined as one astronomical unit AU. Astronomers prefer to speak of the *solar parallax* $\pi_\odot$, the angle subtended by the Earth's equatorial radius $a = 6378$ km at the center of the Sun. Unlike the case of the Moon, the solar parallax is too small for direct measurement. Consequently, astronomers first determine from observations made at several observatories in both the northern and the southern hemisphere the distance of, for instance, a planet or minor planet whose orbit comes sufficiently near the Earth. Formerly, they observed Mars in opposition or a transit of Venus at inferior conjunction. More recently, they have made extensive series of observations at opposition of the minor planet Eros, which is favorable for the purpose. There has now developed a combination of these astronomical methods with radar techniques. It is now possible to make direct measurements with high precision of the distance not only of the Moon but also of Venus and of Mars by combining the time lapse of reflected radio signals with terrestrial measurements of light speed. Using the detailed measurements, one com-

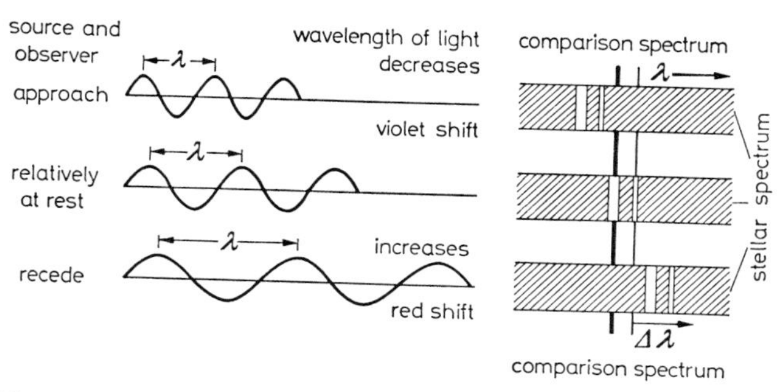

Figure 5.6. Doppler effect $\Delta\lambda/\lambda = v_{rad}/c$.

putes the radius of the Earth's orbit basically using Kepler's third law, but in practice using very difficult celestial mechanics.

Instead of the radius of the Earth's orbit, one can also measure the velocity of the Earth in its orbit with the help of the Doppler effect (Figure 5.6). If a radiation source moves with radial velocity v_{rad} (relative velocity component in the line of sight) away from an observer, then the wavelength λ or the frequency $\nu = c/\lambda$ (c = light speed) suffers a change

$$\Delta\lambda = \lambda v_{rad}/c \quad \text{or} \quad \Delta\nu = \nu v_{rad}/c. \tag{5.2}$$

A relative recession of the source from the observer (by definition, a positive radial velocity) produces an increase in the wavelength, that is, a redshift of the spectral lines, and a decrease in the frequency, and conversely.

In practice, one follows during a considerable fraction of a year either the radial velocity of a fixed star relative to the Earth, given by the Doppler effect shown by its spectral lines, or else the relative velocity of, say, Venus and the Earth, given by the frequency shift of radar signals, the shift produced by a moving reflector being double the shift from a moving source.

The historically very significant first measurement of the speed of light by O. Römer 1675 depends upon a corresponding concept. He determined the frequencies of revolution of the moons of Jupiter from their passages before and behind Jupiter's disk. Because of the finite speed c of light propagation, if the Earth was receding from Jupiter the frequencies diminished, if it was approaching they increased. Using the then available determination of the solar parallax, Römer derived a remarkably good estimate of the light speed. That he thereby anticipated the Doppler principle (5.2) some two hundred years before its first spectroscopic application is scarcely evident from the usual presentations.

The *aberration of light* discovered in 1725 by the Astronomer Royal J. Bradley in the course of his attempt to measure stellar parallaxes also

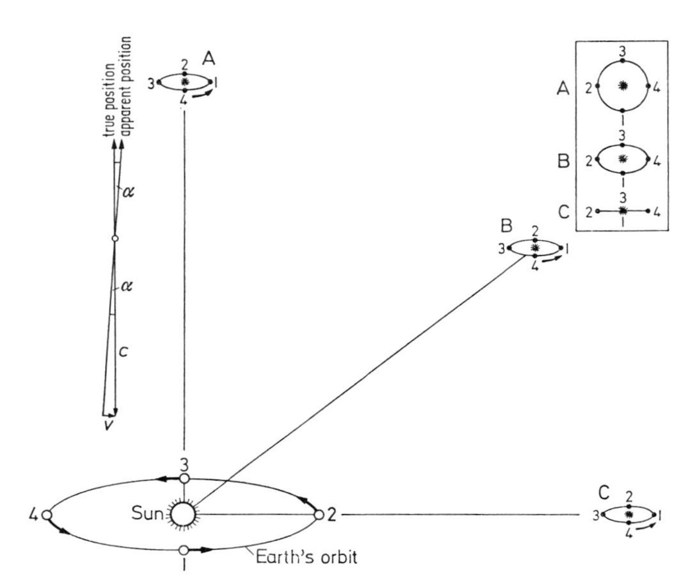

Figure 5.7. Aberration of light. The star's light appears to be deflected by an angle v/c in the direction of the velocity vector of the Earth (see left of diagram), where v is the velocity component of the Earth perpendicular to the ray of light and c is the velocity of light. A star therefore describes a circle of radius α = Earth's velocity/velocity of light = 20″.49 at the pole of the ecliptic, a straight line of maximum extension $\pm\alpha$ in the ecliptic, and in between an ellipse (shown upper right). * True position of star. An observer looking to the right sees the star at intervals of ¼ year in the positions 1–2–3–4.

depends upon the finiteness of the light speed (Figure 5.7). If from the moving Earth we observe a star that, for an observer at rest, lies always in a direction at right angles to the motion, we have slightly to incline our telescope toward the direction of the Earth's velocity v by the angle of aberration v/c in order that it should point at the star. Thus, in the course of a year the star describes a small circle about the pole of the ecliptic (Figure 5.7, upper right); a star in the ecliptic moves to and fro in a line; in between, a star describes the appropriate ellipse. Usually, one illustrates aberration by the analogy of an astronomer who hurries along carrying an umbrella through vertically falling rain.

This demonstration, as well as our elementary explanation of the Doppler effect, is open to criticism because it does not take account of the principle that the light speed is independent of the motion of the source. A consistent interpretation of all effects, and the first to agree to order $(v/c)^2$ with the experiments of Michelson and others, was first given by A. Einstein's special relativity theory (1905).

In conclusion we recapitulate the quantities we have discussed and their mutual interrelations

Equatorial Earth radius $a = 6378$ km
Solar parallax (equatorial horizontal parallax) $\pi_\odot = 8''.794$.
Astronomical unit = semi-major axis of the Earth's orbit $A = a/\pi_\odot = 149.6 \times 10^6$ km
Light speed $c = 299\,792$ km/s
Light time for 1 AU $= A/c = 498.5$ s
Mean orbital speed of the Earth $v = 29.8$ km/s
Equatorial rotational speed of the Earth 0.465 km/s
Aberration constant $v/c \approx 20''.49$.

6. Mechanics and Theory of Gravitation

After protracted and hazardous beginnings with Kepler and Galileo, Isaac Newton in his *Principia* (1687) constructed the mechanics of terrestrial and cosmical systems. Using this in combination with his law of gravitation, in the same work he deduced Kepler's laws and many other regularities in the motions of the planets. Small wonder that the further development of *celestial mechanics* remained for almost two centuries one of the chief fields of endeavor for great mathematicians and astronomers.

We state Newton's three laws of motion in modern language:

Law I. *A body continues in its state of rest or of uniform motion in a straight line, except in so far as it is acted upon by external forces* (Law of inertia).

We represent a velocity in magnitude and direction by a vector $\boldsymbol{v}$, similarly a force by $\boldsymbol{F}$. The addition and subtraction of such quantities follows the well-known parallelogram law; they can be specified by their components in a rectangular coordinate system x, y, z, that is, by their projections on the coordinate axes, for example $\boldsymbol{v} = (v_x, v_y, v_z)$. If the moving body has mass m, Newton defined its *momentum* as the vector

$$\boldsymbol{p} = m\boldsymbol{v}. \tag{6.1}$$

This important concept makes possible the statement of

Law II. *The rate of change of momentum of a body is proportional to the magnitude of the external force acting upon it and takes place in the direction of the force.*

In symbols we write for the case of a single body, denoting the time by t,

$$\frac{d\boldsymbol{p}}{dt} = \frac{d}{dt}(m\boldsymbol{v}) = \boldsymbol{F}. \tag{6.2}$$

Law I is obviously the special case of Law II when $\boldsymbol{F} = 0$. We can also express the velocity $\boldsymbol{v}$ as the rate of change of the position vector $\boldsymbol{r}$ with components x, y, z; so we write $\boldsymbol{v} = d\boldsymbol{r}/dt$ and consequently for the case of constant m

$$m\, d^2\boldsymbol{r}/dt^2 = \boldsymbol{F}. \tag{6.3}$$

The statement, force = mass × acceleration, is valid however *only* for masses that do not change with time, while equation (6.2) remains true also in special relativity theory where the mass depends upon the speed according to the formula $m = m_0(1 - v^2/c^2)^{-1/2}$ in which we call $m_0 = m_{v=0}$ the rest mass. If we have N bodies which we distinguish by the subscript $k = 1, 2, \ldots, N$, then corresponding to equation (6.2) we have the N vector equations

$$\frac{d\boldsymbol{p}_k}{dt} = \frac{d}{dt}(m_k\boldsymbol{v}_k) = \boldsymbol{F}_k \tag{6.4}$$

or the corresponding $3N$ equations for the components.

Newton's last law concerns the interaction of two bodies and asserts:

Law III. *The forces that two bodies exert upon each other are equal in magnitude and opposite in direction* (Law of action and reaction).

Then if $\boldsymbol{F}_{ik}$ is the force which body i exerts upon body k we have

$$\boldsymbol{F}_{ik} = -\boldsymbol{F}_{ki}. \tag{6.5}$$

As a simple example of Newton's law of motion we consider (Figure 6.1) a mass m that moves with uniform speed $\boldsymbol{v}$ in a (horizontal) circle at the end

Figure 6.1. Calculation of centrifugal force. (a) Circular orbit of the mass *m*. Position vector **r** at the times 1, 2, 3, . . . Velocity vector $\boldsymbol{v} = d\boldsymbol{r}/dt$ in the direction of the tangent to the orbit, and of magnitude $v = r\, d\varphi/dt$. **(b) Hodograph.** Velocity vector at the times 1, 2, 3, . . . Acceleration vector $d\boldsymbol{v}/dt$ in the direction of the tangent of the hodograph and therefore parallel to $-\boldsymbol{r}$. The magnitude of the acceleration is $v\, d\varphi/dt = v^2/r$.

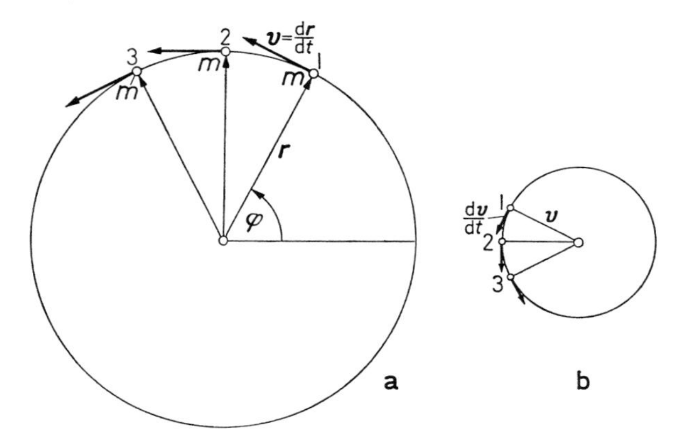

of a string of length r. (The magnitude of a vector quantity, its so-called "modulus," is usually denoted by modulus signs, thus $|\boldsymbol{v}|$, or by the use of the corresponding italic letter, thus $v = |\boldsymbol{v}|$.) The angular speed $d\varphi/dt$, where φ is in circular measure, is then equal to v/r. If we draw successive velocity vectors from a point so as to construct the so-called *hodograph*, we read off the acceleration as having magnitude $|d\boldsymbol{v}/dt| = (v/r)v$ and as being directed toward the center of the circle in Figure 6.1(a). So we derive the law of centrifugal force

$$F = mv^2/r \tag{6.6}$$

which had been discovered by Christian Huygens before Newton. Newton's third law then tells us that the string exerts the same force outward on its point of attachment as that by which it pulls the moving body inward.

If the detailed structure of a body is irrelevant for a problem in mechanics, we speak of it as a *point mass* (particle). For instance, in the theory of planetary motion we treat the Earth as a point mass; if we are interested in an atom, we treat its radiating electron as a point mass, and so on.

Starting from Newton's three laws of motion of a particle, we proceed first to the equations of motion of a *system of particles*. Thence we derive the three *conservation laws* of mechanics, to which we shall repeatedly appeal.

6.1 Linear momentum: motion of mass center

In a system of point masses m_k, denoted by subscripts i or k ($i, k = 1, 2, \ldots, N$), we distinguish between the *internal force* $\boldsymbol{F}_{ik}$, which particle i exerts upon particle k, and the *external force* $\boldsymbol{F}_k^{(e)}$, which acts upon particle k from outside the system. The equation of motion (6.4) for particle k becomes

$$\frac{d\boldsymbol{p}_k}{dt} = \boldsymbol{F}_k^{(e)} + \sum_{i=1}^{N} \boldsymbol{F}_{ik}. \tag{6.7}$$

Using the law of action and reaction (6.5) and summing over all particles of the system, we get

$$\frac{d}{dt}\Sigma\boldsymbol{p}_k = \Sigma\boldsymbol{F}_k^{(e)}$$

where the summation Σ is taken over all particles of the system from $k = 1$ to $k = N$. If we consider the N particles as a single system we may define

$$\left.\begin{array}{ll}\text{total momentum} & \boldsymbol{P} = \Sigma\boldsymbol{p}_k \\ \text{resultant external force} & \boldsymbol{F} = \Sigma\boldsymbol{F}_k^{(e)}.\end{array}\right\} \tag{6.8}$$

Then the equation of motion becomes

$$d\boldsymbol{P}/dt = \boldsymbol{F} \tag{6.9}$$

like the equation for a single particle. If there is no external force ($\boldsymbol{F} = 0$) then we obtain the law of *conservation of linear momentum*

$$\boldsymbol{P} = \Sigma \boldsymbol{p}_k = \text{constant}. \tag{6.10}$$

We can render the meaning of equations (6.9) and (6.10) a little more intuitive if we define for the system

$$\text{total mass } M = \Sigma m_k \tag{6.11}$$

and position vector $\boldsymbol{R}$ of mass center S, such that

$$M\boldsymbol{R} = \Sigma m_k \boldsymbol{r}_k. \tag{6.12}$$

Then equation (6.9) becomes the equation of motion of the mass center

$$M \, d^2\boldsymbol{R}/dt^2 = \boldsymbol{F} \tag{6.13}$$

analogous to that for a single particle. We infer that in the force-free case $\boldsymbol{F} = 0$ the mass center (corresponding to Law I) must perform uniform motion in a straight line expressed by $d\boldsymbol{R}/dt = \text{constant}$.

6.2 Moment of momentum: conservation of angular momentum

We consider first (Figure 6.2) a particle m_k that can revolve about a fixed point O at the end of a rod $\boldsymbol{r}_k$. Let a force $\boldsymbol{F}_k$ act on m_k. This "tries" to make m_k revolve about an axis through O perpendicular to the plane containing $\boldsymbol{r}_k$ and $\boldsymbol{F}_k$. For this, only the transverse component $|\boldsymbol{F}_k| \sin \alpha$ is effective, α being the angle between $\boldsymbol{r}_k$ and $\boldsymbol{F}_k$. The quantity, rod length $|\boldsymbol{r}_k|$ times the effective force component $|\boldsymbol{F}_k| \sin \alpha$ erected as a vector perpendicular to the plane of $\boldsymbol{r}_k$ and $\boldsymbol{F}_k$ is known mathematically as the *vector product* $\boldsymbol{r}_k \times \boldsymbol{F}_k$ (sometimes the "cross" product) and physically as the *moment of the force* about O, or the turning moment $\boldsymbol{M}_k = \boldsymbol{r}_k \times \boldsymbol{F}_k$. Just as we define this moment in the case of a force, so in the case of the momentum $\boldsymbol{p}_k = m_k \boldsymbol{v}_k$ we define the *moment of momentum* or *angular momentum* $\boldsymbol{N}_k$ by

$$\boldsymbol{N}_k = \boldsymbol{r}_k \times \boldsymbol{p}_k = \boldsymbol{r}_k \times m_k \boldsymbol{v}_k .$$

The order of the factors in a product like $\boldsymbol{M} = \boldsymbol{r} \times \boldsymbol{F}$ is such that a right-handed screw turning from $\boldsymbol{r}$ toward $\boldsymbol{F}$ will advance in the direction of $\boldsymbol{M}$.

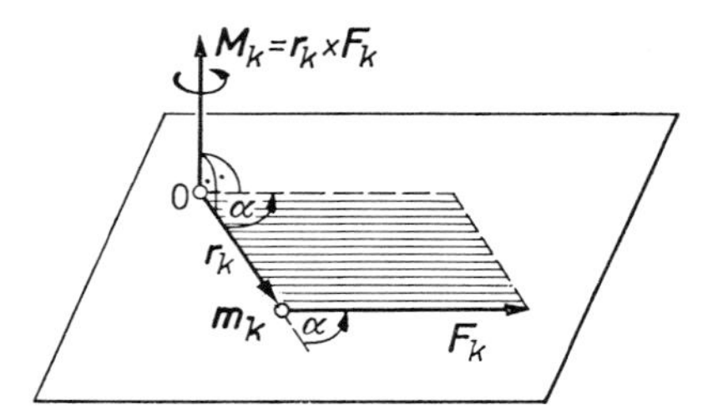

Figure 6.2
Couple $\boldsymbol{M}_k = \boldsymbol{r}_k \times \boldsymbol{F}_k$. The magnitude of $\boldsymbol{M}_k$ is $|\boldsymbol{r}_k| \, |\boldsymbol{F}_k| \sin \alpha$ and is equal to the area of the parallelogram spanned by $\boldsymbol{r}_k$ and $\boldsymbol{F}_k$.

From Newton's equation of motion (6.4) we now obtain by vector multiplication with $\boldsymbol{r}_k$ on the left

$$\boldsymbol{r}_k \times \frac{d}{dt}(m_k \boldsymbol{v}_k) = \boldsymbol{r}_k \times \boldsymbol{F}_k \quad \text{or} \quad \frac{d}{dt}(\boldsymbol{r}_k \times m_k \boldsymbol{v}_k) = \boldsymbol{r}_k \times \boldsymbol{F}_k. \tag{6.14}$$

(The second form follows because

$$\frac{d}{dt}(\boldsymbol{r} \times m\boldsymbol{v}) = \frac{d\boldsymbol{r}}{dt} \times m\boldsymbol{v} + \boldsymbol{r} \times \frac{d}{dt}(m\boldsymbol{v})$$

and, since $d\boldsymbol{r}/dt = \boldsymbol{v}$, the first term on the right is the vector product of two parallel vectors which is zero by definition.)

For a system of particles we define the resultant moment of all the internal and external forces about the fixed point O, and also the total angular momentum of the system by the equations

$$\left.\begin{aligned} &\text{resultant turning moment } \boldsymbol{M} = \Sigma \boldsymbol{r}_k \times \boldsymbol{F}_k = \Sigma \boldsymbol{r}_k \times (\boldsymbol{F}_k^{(e)} + \Sigma_i \boldsymbol{F}_{ik}) \\ &\text{total angular momentum } \boldsymbol{N} = \Sigma \boldsymbol{r}_k \times \boldsymbol{p}_k = \Sigma \boldsymbol{r}_k \times m_k \boldsymbol{v}_k \end{aligned}\right\} \tag{6.15}$$

Then the equation of motion becomes

$$d\boldsymbol{N}/dt = \boldsymbol{M} \tag{6.16}$$

or *the rate of change of angular momentum is equal to the sum of the moments of all the forces*.

If only *central forces* operate within the system, that is, forces like gravitation that act only along the lines joining the particles, the contribution to $\boldsymbol{M}$ of the internal forces vanishes and on the right-hand side of equation (6.16) there remains only the moment of all the external forces.

If now all the external forces vanish, or at least if their resultant moment is zero, we have $d\boldsymbol{N}/dt = 0$. So we get the important law of *conservation of angular momentum*

$$\boldsymbol{N} = \Sigma \boldsymbol{r}_k \times m_k \boldsymbol{v}_k = \text{constant}. \tag{6.17}$$

6.3 Energy law

If a particle of mass m_k moves under the action of a force $\boldsymbol{F}_k$ through a small displacement $d\boldsymbol{r}_k$ making angle α with $\boldsymbol{F}_k$, the work performed on the particle is

$$dW = |\boldsymbol{F}_k|\,|d\boldsymbol{r}_k| \cos\alpha = \boldsymbol{F}_k \cdot d\boldsymbol{r}_k. \tag{6.18}$$

This (scalar) quantity we call the scalar product of the two vectors, and we denote it by the dot between the factors. To evaluate the work done in a finite displacement $A \to B$ we use Newton's equation of motion (6.4) with $\boldsymbol{v}_k = d\boldsymbol{r}_k/dt$ and we obtain

$$\int_A^B \boldsymbol{F}_k \cdot d\boldsymbol{r}_k = \int_A^B \frac{d}{dt}(m_k \boldsymbol{v}_k) \cdot d\boldsymbol{r}_k = \frac{1}{2} m_k \boldsymbol{v}_k^2 \Big|_A^B. \tag{6.19}$$

We call the quantity $(1/2)m_k v_k^2$ the *kinetic energy* of the particle and we can take its sum E_{kin} over any number of particles. Further, if $\Sigma \boldsymbol{F}_k \cdot d\boldsymbol{r}_k$ is a perfect differential $-dE_{\text{pot}}$, that is, if the work done by the forces is independent of the actual paths of the particles and depends only upon their initial and final states, then the sum of the kinetic energy E_{kin} plus the potential energy E_{pot} is constant. We have the further important law of *conservation of energy*

$$E = E_{\text{kin}} + E_{\text{pot}} = \text{constant}. \tag{6.20}$$

In a system of particles moving under gravitational forces (see below) and in many other cases E_{kin} depends only on the velocities and E_{pot} only on the positions of the particles.

6.4 Law of gravitation: celestial mechanics

In order to obtain a theory of cosmical motions, Newton had to have in addition to his laws of mechanics also his law of gravitation (about 1665):

Two particles of mass m_i and m_k at distance r apart attract each other in the direction of the line joining them with a force

$$F = -G\,\frac{m_i m_k}{r^2}. \tag{6.21}$$

As a result of an integration that we shall not repeat here, Newton then showed that exactly the same law of attraction (6.21) holds good for two massive spheres (Sun, planets, . . .) of finite extent, as for the two corresponding massive particles. He next verified the law of gravitation (6.21) assuming that free fall at the Earth's surface (Galileo) and the revolution of the Moon are both governed by the force of attraction of the Earth.

The acceleration (= force/mass) in the case of free fall can be measured by experiments with falling bodies, or more accurately by use of a pendulum. Its numerical value at the equator is 978.05 cm/s², or taking account of the centrifugal acceleration of the Earth's rotation of 3.39 cm/s², it is

$$g_\oplus = 981.4 \text{ cm/s}^2. \tag{6.22}$$

Again the Moon moves on its circular path of radius r with velocity $v = 2\pi r/T$, where T = 1 sidereal month, and so experiences the acceleration (Figure 6.1)

$$g_☾ = v^2/r = 4\pi^2 r/T^2 = 0.272 \text{ cm/s}^2 \tag{6.23}$$

using r = 384 400 km = 3.844×10^{10} cm, $T = 27^{\text{d}}32 = 27.32 \times 86\,400$ s. The accelerations $g_\oplus$ and $g_☾$ are in fact inversely proportional to the squares of the radii of the Earth R and the lunar orbit r, that is,

$$g_\oplus : g_☾ = \frac{1}{R^2} : \frac{1}{r^2} = 3620. \tag{6.24}$$

The value of the universal constant of gravitation G here appears only in combination with the hitherto unknown mass of the Earth M. Similarly, in other astronomical problems G occurs only along with the mass of the attracting celestial body. One cannot in principle determine a value for G from astronomical measurements; it has to be evaluated by terrestrial measurements.

First Maskelyne in 1774 made use of the deflection of a plumb line, that is, the effect on the direction of a plumb line of the gravitational attraction of a mountain. Then in 1798 Henry Cavendish, after whom the Cavendish Laboratory in Cambridge is named, used a torsion balance, and in 1881 P. v. Jolly, the teacher of Max Planck, used a suitable form of lever balance. The result of modern measurements is

$$G = (6.673 \pm 0.003) \times 10^{-8} \text{ dyn cm}^2/\text{g}^2. \tag{6.25}$$

Since the acceleration due to gravity at the Earth's surface (for the time being neglecting rotation and flattening) is related to the mass M and radius R by the formula already used $g = GM/R^2$ we can now compute the mass M and mean density of the Earth $\bar{\rho} = M/(4/3)\pi R^3$, and we obtain

$$M = 5.98 \times 10^{24} \text{ kg} \qquad \bar{\rho} = 5.51 \text{ g/cm}^3. \tag{6.26}$$

We shall in due course mention the geophysical significance of these values.

First we return to Newton's *Principia* and derive Kepler's laws from the laws of mechanics and of gravitation in order to gain a better understanding of Kepler's laws themselves and of the constants appearing in them.

Since clearly the mass of the Sun is so much greater than that of the planets, we first treat the Sun as at rest and measure the radius vectors of the planets from the center of the Sun. Also we first leave out of account the mutual attractions of the planets, their so-called perturbations.

The motion of a planet around the Sun then proceeds under the action of the *central force* $-G\mathfrak{M}m/r^2$, where $\mathfrak{M}$ is again the mass of the Sun and m ($\ll \mathfrak{M}$) that of the planet. Consequently, the law of angular momentum (6.17) applies, that is, the angular momentum vector

$$\boldsymbol{N} = \boldsymbol{r} \times m\boldsymbol{v} \tag{6.27}$$

where $\boldsymbol{v}$ is the velocity vector of the planet, is constant in magnitude and direction. The vectors $\boldsymbol{r}$ and $\boldsymbol{v}$ remain therefore in the same plane perpendicular to $\boldsymbol{N}$, the *fixed plane* of the planet's motion. The magnitude $|\boldsymbol{r} \times \boldsymbol{v}| = rv \sin\alpha$ is twice the area traced out in unit time by the radius vector $\boldsymbol{r}$ of the planet (Figure 6.3). The law of angular momentum is then identical

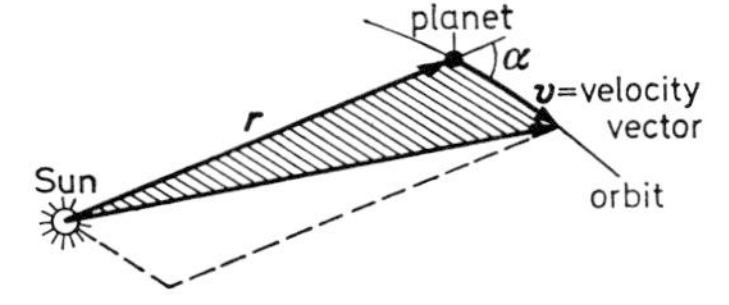

Figure 6.3
Rate of description of area by position-vector of a planet ("areal speed")$(1/2)|\mathbf{r} \times \mathbf{v}| = (1/2)rv \sin\alpha$.

with the assertion that each planet moves in a fixed plane with constant areal velocity (Kepler's second law and part of Kepler's first law).

Here we shall omit the somewhat formidable calculation which shows that the path of a particle (planet, comet, . . .) under the action of a central force proportional to r^{-2} must be a conic section, that is, circle (eccentricity $e = 0$), ellipse ($0 < e < 1$), parabola ($e = 1$), or hyperbola ($e > 1$) with the Sun at a focus (Kepler's first law).

Instead, we readily write down the *energy law* for the motions of the planets, etc. Consider a particle of mass m taken to infinity ($r \to \infty$) from rest at distance r from the Sun ($\mathfrak{M}$) under the action of the gravitational force $-G\mathfrak{M}m/r^2$. The work done is equal to its potential energy $E_{\text{pot}}(r)$, where

$$E_{\text{pot}}(r) = -G\int_r^\infty \frac{\mathfrak{M}m}{r^2}\,dr = -G\frac{\mathfrak{M}m}{r}. \tag{6.28}$$

Since no work is done in any transverse motion, we see that the expression is independent of the choice of integration path. Referred to unit mass, we call $\varphi(r) = -G\mathfrak{M}/r$ the *potential* at distance r from the Sun. This is one of the fundamental concepts of celestial mechanics as well as of theoretical physics.

The total energy of a planet $E = E_{\text{kin}} + E_{\text{pot}}$, that is,

$$E = \frac{1}{2}mv^2 - G\mathfrak{M}\,m/r \tag{6.29}$$

or reckoned per unit mass

$$E/m = \frac{1}{2}v^2 - G\mathfrak{M}/r, \tag{6.30}$$

thus remains constant in time. We again infer that the speed increases from aphelion to perihelion.

The complete calculation of planetary motion and the derivation of Kepler's third law we shall carry through only for circular orbits. Nevertheless, with a view to subsequent generalization, we no longer require the planetary mass to be small compared with the solar mass. We therefore consider the motion of two masses in the first place about their mass center S, and in the second place the relative motion referred to, say, the more massive body. Let m_1, m_2 be the masses, a_1, a_2 their distances from the mass center S and $a = a_1 + a_2$ their distance from each other. From the definition of the mass center we have then (Figure 6.4)

$$a_1 : a_2 : a = m_2 : m_1 : m_1 + m_2 \tag{6.31}$$

or

$$m_1 a_1 = m_2 a_2 = \frac{m_1 m_2}{m_1 + m_2}\,a.$$

For each mass the force of attraction Gm_1m_2/a^2 must balance the centrifugal force. Denoting the period of the system by T, the latter force for the particle m_1 is

$$m_1 v_1^2/a_1 = (2\pi/T)^2 m_1 a_1 \tag{6.32}$$

and from the law of action and reaction, or from equation (6.31), the force for the particle m_2 has the same magnitude. Again using equation (6.31) and rearranging the factors, we obtain

$$a^3/T^2 = G(m_1 + m_2)/4\pi^2. \tag{6.33}$$

If we drop the assumption of circular orbits, the masses m_1, m_2 move round their mass center S in *similar conics,* also the relative orbit is another similar conic. The *semimajor axes,* which we also denote by a_1, a_2, and a, then replace the orbital radii, but we shall not give the proof here. Thus we obtain the generalized third law of Kepler (6.33). As we shall see, in the solar system the mass of even the greatest planet Jupiter is only about one-thousandth the solar mass. To this accuracy, we can replace the mass $m_1 + m_2$ by the solar mass $\mathfrak{M}$. Then by inserting numerical values for the motion of the Earth or another planet in equation (6.33) we derive the mass of the Sun. From the apparent semidiameter 16′ in combination with the value of a we get the radius $R_\odot$. Then from $(4/3)\pi R^3 \bar{\rho} = \mathfrak{M}$ we finally evaluate the mean density. The results are for the Sun

$$\begin{aligned} &\text{mass } \mathfrak{M}_\odot &&= 1.989 \times 10^{33}\ \text{g} \\ &\text{radius } R_\odot &&= 6.960 \times 10^{10}\ \text{cm} \\ &\text{density } \bar{\rho}_\odot &&= 1.409\ \text{g/cm}^3. \end{aligned} \tag{6.34}$$

The uncertainty amounts to about ± one unit in the last place of decimals.

In corresponding manner, we compute the masses of the planets from the motions of their satellites (Table 7.1) again using Kepler's third law. If no satellite is available then, as is obviously much more difficult, we must have recourse to the mutual *perturbations* of the planets.

It is interesting to consider the Kepler problem further from the standpoint of the energy law (6.29). For a circular orbit the centrifugal force is equal to the force of attraction of the two masses and so $mv^2/r = G\mathfrak{M}m/r^2$, or

$$\frac{1}{2}mv^2 = \frac{1}{2}G\,\mathfrak{M}\,m/r \qquad \text{(circular orbit).} \tag{6.35}$$

Thus the kinetic energy E_{kin} is equal to one half the potential energy with the sign reversed

$$E_{\text{kin}} = -\frac{1}{2}E_{\text{pot}}. \tag{6.36}$$

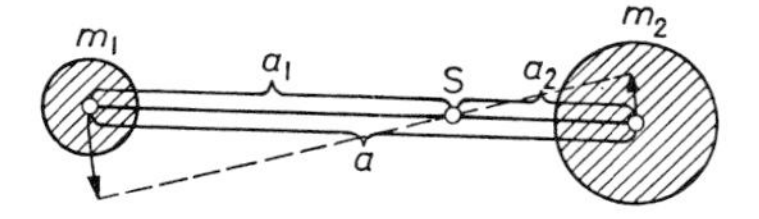

Figure 6.4
Motion of the masses m_1 and m_2 about their common center of gravity S, where $m_1a_1 = m_2a_2$.

One can show (see Section 26.8) that, averaged over time, this result holds good for any systems of point masses held together by gravitational forces between them (obeying the inverse-square law). This is the *virial theorem* which is important, for example, in the theory of stellar systems.

We now consider the case of a *parabolic* orbit, for example that of a nonperiodic comet. At infinity both the potential and kinetic energies are equal to zero. Thus $E = 0$ and from equation (6.29)

$$\frac{1}{2}mv^2 = G\mathfrak{M}\,m/r \quad \text{or} \quad E_{\text{kin}} = -E_{\text{pot}}. \tag{6.37}$$

Thus at the same distance from the Sun, the *parabolic speed* is $\sqrt{2}$ times the speed in a circular orbit. For example, the mean speed of the Earth is 30 km/s, so the speed of a comet or meteor stream that encounters us in parabolic orbit is $30\sqrt{2} = 42.4$ km/s.

Besides the Kepler problem, Newton solved many other problems of celestial mechanics.

We first discuss *precession,* at least in outline. The wandering of the Earth's axis round the pole of the ecliptic depends upon the same principle as the corresponding motion of a spinning top under the effect of terrestrial gravitation. The equatorial bulge of the Earth is attracted toward the plane of the ecliptic by the Moon and the Sun, whose masses in the mean over such long time intervals we can think of as being spread around their paths (and here we ignore the small inclination of the Moon's orbit) (Figure 6.5). The turning moment $\boldsymbol{M}$ so produced acts in accordance with equation (6.16) upon the Earth's angular momentum vector $\boldsymbol{N}$, which is practically along the axis of rotation. As we see at once from Figure 6.5, the change of $\boldsymbol{N}$ corresponding to $d\boldsymbol{N}/dt = \boldsymbol{M}$ produces the familiar motion of $\boldsymbol{N}$, or of the Earth's axis, on a cone of fixed angle. Numerical calculation yields the correct period of luni-solar precession (Figure 2.1).

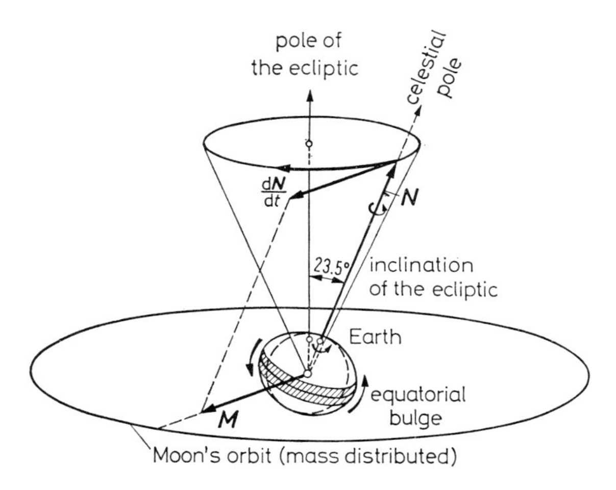

Figure 6.5
Lunar precession.

Next we must briefly mention the old problem of the tides. These being known to Mediterranean peoples only by hearsay, in his time Galileo developed in controversy a wholly incorrect theory of the 12-hour alternation of ebb and flow, which contributed to bringing about his unfortunate trial by the Inquisition. It was again Newton who presented a static theory of the tides (Figure 6.6).

In the motion of the Earth and the Moon about their common mass center, the acceleration vector (which is the difference between gravitation due to the Moon and centrifugal acceleration due to this orbital motion) of a particle of water in the ocean acts *upward* when the Moon crosses the meridian. So there the water is lifted and we have high water. In keeping with the apparent motion of the Moon (one "Moon day" = $24^{h}51^{m}$) two high-water "hills" and two low-water "valleys" continually travel round the Earth, getting later each day by 51 minutes. The tidal force of the Sun is about half that of the Moon. At new moon and full moon the tidal forces of the Moon and the Sun act together and produce "spring" tides; at first and last quarters, they oppose each other and we have "neap" tides. Actually this statical theory explains only the crude features of the phenomena. The *dynamical* theory of tides investigates the way in which forced oscillations, corresponding to the different periods of the apparent motions of the Moon and the Sun, of the various oceanic basins are excited. Following G. H. Darwin, tidal prediction consists essentially in Fourier analysis and synthesis appropriate to the stated astronomical periods. As already mentioned in Section 3, *tidal friction* in narrow seas acts as a brake on the Earth's rotation and produces a lengthening of the day. In accordance with the conservation of angular momentum (6.17), the angular momentum that the Earth loses in this way must reappear as additional orbital angular momentum of the Moon. From Kepler's third law the angular momentum per unit

Figure 6.6. Static theory of the tides. The acceleration toward the Moon of the three points shown is, corresponding to their various distances from it, $b - \Delta$ for point A (lower culmination of the Moon), b for point B (center of gravity of the Earth), and $b + \Delta$ for point C (upper culmination of the Moon). The (rigid) Earth as a whole has the acceleration b. Consequently at A and C there is an excess acceleration Δ, which produces high water at both points.

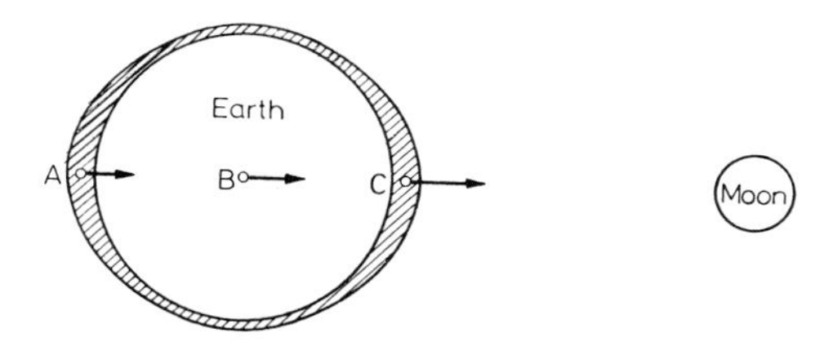

mass is proportional to the square root of the orbital radius, and so the Moon must slowly recede from the Earth.

We must not leave the theory of planetary motion without clarifying the decisive change from the Ptolemaic to the Copernican world systems by looking back from a modern viewpoint (Figure 6.7).

With the Sun as origin (heliocentric) let $\boldsymbol{r}_P$, $\boldsymbol{r}_E$ be the position vectors of a planet and of the Earth. Then the position vector of the planet referred to the Earth (geocentric) is specified by the vector difference

$$\boldsymbol{R} = \boldsymbol{r}_P - \boldsymbol{r}_E. \tag{6.38}$$

Let us look back from here for a moment to the geocentric view of Ptolemy: (1) In the case of the outer planets, for example Mars ♂, we begin by drawing the vector $\boldsymbol{r}_P$ out from the Earth and letting it rotate in the same way as it originally rotated about the Sun. To $\boldsymbol{r}_P$ in accordance with equation (6.38) we add the vector $-\boldsymbol{r}_E$, which is the position vector of the Sun seen from the Earth, and so we obtain the position vector $\boldsymbol{R}$ of the

Figure 6.7. Motion of an outer planet (Mars) and an inner planet (Venus) on the celestial sphere, represented heliocentrically and geocentrically. The broken arrow ←--- shows the position of the planet in the sky. The monthly motions of the Earth and the planet are shown by arrow points on the circles. We have for

	Outer planets	Inner planets
deferent	$\boldsymbol{r}_P$	$-\boldsymbol{r}_E$
epicycle	$-\boldsymbol{r}_E$	$\boldsymbol{r}_P$.

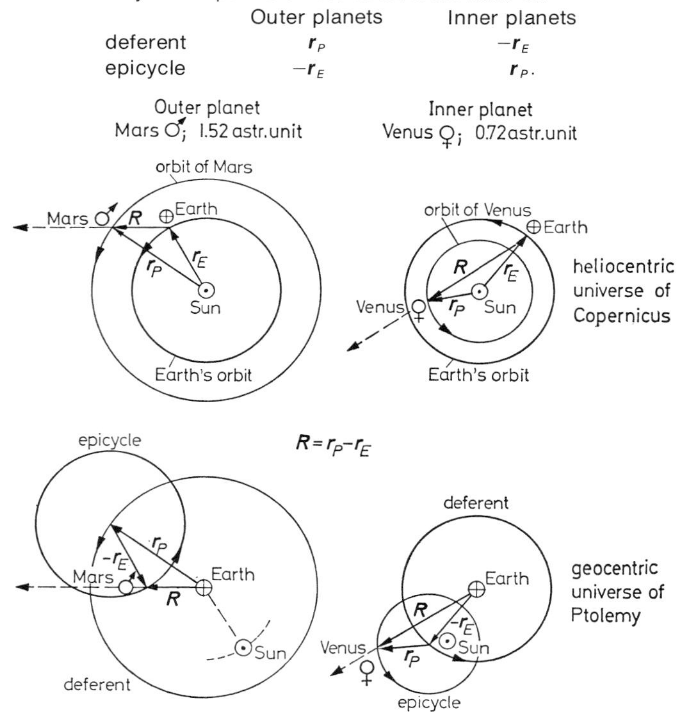

planet as seen from the Earth. The immaterial circle that $\boldsymbol{r}_P$ describes round the Earth with the sidereal period of the planet is the Ptolemaic *deferent*. The other circle that the planet at the end point of the vector $-\boldsymbol{r}_E$ describes around the point $\boldsymbol{r}_P$ with the sidereal period of the Earth is the Ptolemaic *epicycle*. (2) In the case of the inner planets it seemed better to Ptolemy first to let the greater vector revolve around the Earth as deferent with period equal to a sidereal year, and then to let the smaller vector $\boldsymbol{r}_P$ describe about the end point $-\boldsymbol{r}_E$ of the other the Ptolemaic epicycle with period equal to the sidereal period of the planet.

Thus far the geocentric construction expresses exactly the relation $\boldsymbol{R} = \boldsymbol{r}_P - \boldsymbol{r}_E$. The representation (2), applied to *all* the planets, expresses the world system of Tycho Brahe.

Actually, however, we have not completed the transformation to the Ptolemaic system. So long as mankind could measure only the position of the planets in the sky, that is, only their directions and not their distances, they were concerned only with the direction of the vector $\boldsymbol{R}$ and not its magnitude. Thus one could change the scale of $\boldsymbol{R}$ for each planet. That is, the vectors

$$\boldsymbol{R}'_P = A_P \boldsymbol{R}_P \tag{6.39}$$

with a fixed but arbitrary numerical factor A_P for each planet would give, in accordance with the Ptolemaic system, a completely satisfactory description of the motions of the planets in the sky.

We now see clearly what is lost when we retreat step by step from the Copernican to the Ptolemaic system:

(1) The change of reference system entails the renunciation of any simple mechanical explanation.
(2) While the scale factors A_P leave the position of the planets in the sky unchanged, we thereby lose the mutual relationships of the planets in space.
(3) The fact that in the Ptolemaic system the year period, corresponding to the motion of the vector $\boldsymbol{r}_E$, has to be introduced *independently* for each planet shows the artificiality of the older system.

It is important to make it clear, however, that the purely kinematic consideration of the motion of the planets in the sky did not permit a discrimination between the old and the new system. Essential progress came with Galileo's first observations with his telescope (1609): (1) One could look upon Jupiter with its freely revolving moons as a "model" of the Copernican planetary system. (2) The phases of Venus determined the relative positions of the Sun, Earth, and Venus. The smallness of the phase angle of Jupiter, say, gave at any rate qualitative support for Copernicus. The concept of a celestial *mechanics* presupposed—and this should not be forgotten—the basic uniformity of cosmical and terrestrial matter and of its physics.

6.5 Artificial satellites and space vehicles Space research

Let us now discuss briefly the orbits of artificial satellites and space vehicles. We shall consider here only the gravitational field of the Earth, restrict ourselves to circular orbits, and neglect the braking effect of the Earth's atmosphere.

A satellite which describes a circular orbit in the immediate neighborhood of the Earth (at the Earth's radius $r_0 = 6380$ km) must, from equation (6.35), possess a speed $v_0 = 7.9$ km/s and an orbital period $T_0 = 84.4$ min. For larger orbits of radius r, Kepler's third law yields a speed $v = v_0(r/r_0)^{-1/2}$ and an orbital period $T = T_0(r/r_0)^{3/2}$. In particular, the orbital period equals one sidereal day for $r = 6.6$ Earth radii. Such a synchronous satellite then remains "stationary" above the same point on the Earth's surface.

To take a space vehicle, which carries no rocket motor with it, out of the gravitational field of the Earth (alone) to infinity, one must give it at least the parabolic or escape velocity $v_0\sqrt{2} = 11.2$ km/s.

In his delightful novel "From the Earth to the Moon" (1865),[2] Jules Verne proposed solving this problem with the help of an enormous cannon. However, this does not work, as the initial velocity of a shell cannot significantly exceed the—too small—velocity of sound in the exploding gas.

Greater speeds can be achieved using *rockets*. We shall now explain the mechanics of a rocket without gravity (that is, on a horizontal testbed or in space) and without air resistance. Let the mass of the body, etc., plus rocket fuel be $m(t)$ at time t. Let the change in $m(t)$ in unit time, corresponding to the mass of burnt material ejected in unit time, be dm/dt. If the outflow speed of the burnt material relative to the rocket is v_E, then the momentum transmitted to it is $-(dm/dt)v_E$. If we now treat the acceleration of the rocket from the point of view of an observer moving with it (an astronaut), we obtain immediately the Newtonian equation of motion

$$m(t)\frac{dv}{dt} = -\frac{dm}{dt}v_E \quad \text{or} \quad \frac{dv}{v_E} = -\frac{dm}{m(t)}. \tag{6.40}$$

Using the initial conditions that the mass equals m_0 at take-off, where $t = 0$, $v = 0$, and integrating the last equation, we obtain at once the rocket equation

$$v = v_E \ln\frac{m_0}{m(t)}. \tag{6.41}$$

[2]An English translation of "De la terre à la lune" has been published by Dodd, Mead & Co., Inc. (1962). Our futurologists could turn green with envy reading Jules Verne's predictions: his launch point is only 150 km away from Cape Canaveral, where the Kennedy Space Center now is. A 200-inch reflecting telescope(!) was built to observe the projectile; one of its first tests was the complete resolution of the Crab Nebula!

If we had taken account of the (uniform) gravitational field in the neighborhood of the Earth, then, for vertical take-off, the familiar Galilean term $-gt$ would also have appeared on the right hand side.

If for a numerical estimate we take even the optimistic values v_E = 4 km/s and $m_0/m = 10$ at the end of burning we find (without air resistance) a final speed for our one-stage rocket of only $v = 9.2$ km/s. For real space travel we must therefore employ multistage rockets, whose basic princple, repeated application of the rocket equation (6.41), is clear without further explanation.

In World War II the V-2 rocket, which could reach a height of almost 200 km, was constructed on the German side. After the war, the V-2 and improved rockets were used in the United States of America for research into the highest layers of the Earth's atmosphere (ozone layer, ionosphere) and the short wavelength solar radiation ($\lambda < 2850$ Å). Nowadays research with simple rockets is of still greater significance since it opens up to us at a relatively modest cost the spectral regions completely absorbed by the Earth's atmosphere: the short wavelength ultraviolet (Lyman region), the X rays and gamma rays of cosmic origin, and at the other end of the spectrum the infrared to millimeter region, and finally the radio waves reflected by the ionosphere, with wavelengths from about 30 m to 1 km. Admittedly the duration of observation is restricted to a few minutes. The results of measurement are retrieved either by ejecting the rocket's nosecone with its measuring equipment and bringing it to Earth with a parachute or by a radio link, the so-called telemetry, made very efficient by suitable encoding.

On October 4, 1957 Soviet scientists succeeded in placing the first *artificial satellite,* Sputnik I, in an orbit with minimum height 225 km and maximum height 950 km. Since then hundreds of satellites have been launched for purposes of research and communication. Particularly important for the acquisition of long series of astronomical observations outside the Earth's atmosphere are the remarkably (to $\sim 1''$) stabilized satellites of the series OSO (orbiting solar observatory) and OAO (orbiting astronomical observatory). Figure 6.8 illustrates the unmanned flight of an American Ranger space probe to the Moon with a "hard" landing.

On April 12, 1961, J. Gagarin ventured on the first manned space flight. The problem of landing men on the Moon, as difficult as it was costly, was overcome by NASA (the National Aeronautics and Space Administration of the United States) using the following principle: the Moon was reached by means of a three stage rocket. While one of the astronauts circled the Moon many times in the third stage, the other two astronauts landed on the Moon in an auxiliary vehicle LM (lunar module). After carrying out their tasks, they took off using a rocket motor, reattached their LM to the third stage of the rocket, and set out on the return journey in that after jettisoning the now redundant auxiliary vehicle. The essential difficulties of the landing lie in the strong heating on reentry into the Earth's atmosphere, requiring a heat shield, and the temporary breakdown in radio communication due to the ionization of the air.

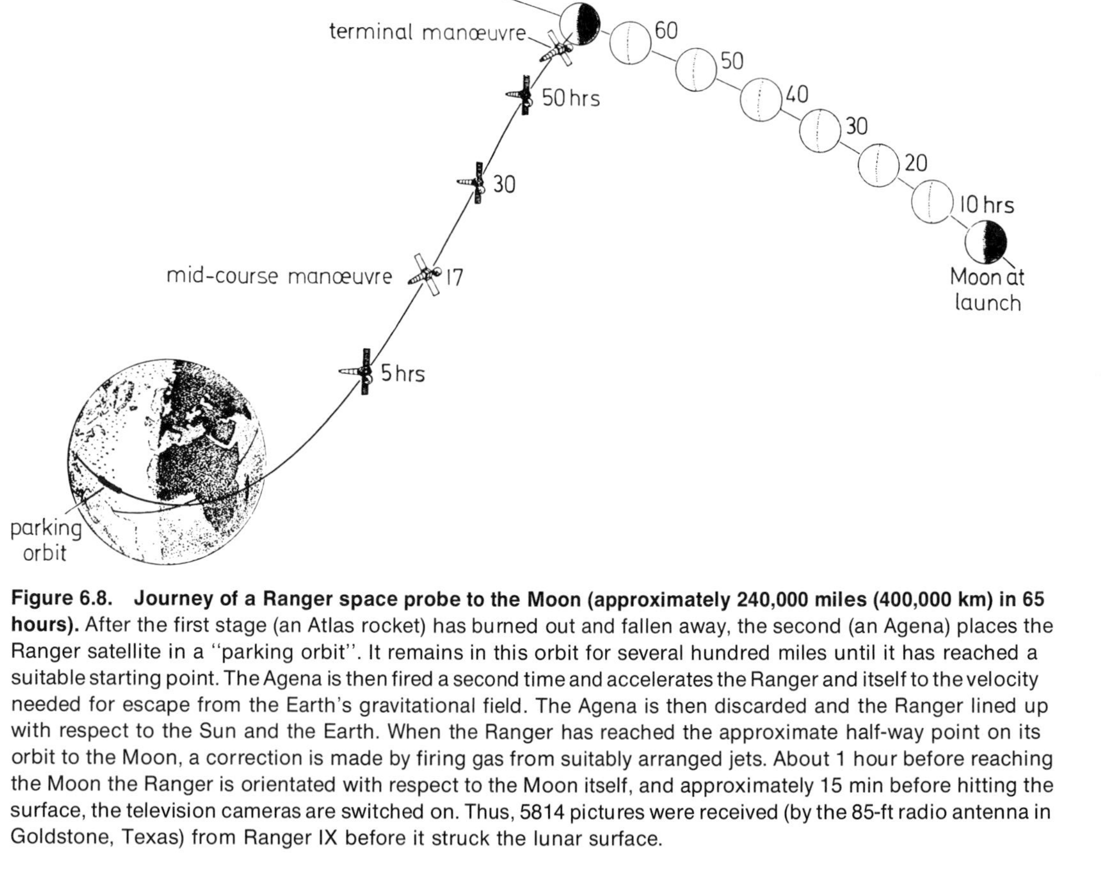

Figure 6.8. Journey of a Ranger space probe to the Moon (approximately 240,000 miles (400,000 km) in 65 hours). After the first stage (an Atlas rocket) has burned out and fallen away, the second (an Agena) places the Ranger satellite in a "parking orbit". It remains in this orbit for several hundred miles until it has reached a suitable starting point. The Agena is then fired a second time and accelerates the Ranger and itself to the velocity needed for escape from the Earth's gravitational field. The Agena is then discarded and the Ranger lined up with respect to the Sun and the Earth. When the Ranger has reached the approximate half-way point on its orbit to the Moon, a correction is made by firing gas from suitably arranged jets. About 1 hour before reaching the Moon the Ranger is orientated with respect to the Moon itself, and approximately 15 min before hitting the surface, the television cameras are switched on. Thus, 5814 pictures were received (by the 85-ft radio antenna in Goldstone, Texas) from Ranger IX before it struck the lunar surface.

Figure 6.9. The landing vehicle (Lunar Module) of Apollo 11 has landed in Mare Tranquillitatis. Astronaut E. Aldrin is setting up the seismic experiment. In the foreground many small craters are visible; a larger crater can be seen farther back on the left. On the ground lie small pieces of rock; the astronaut's shoes leave sharp impressions in the fine dust.

The astronauts N. Armstrong, M. Collins, and E. Aldrin achieved the first landing on the Moon, in Mare Tranquillitatis, on July 20, 1969 in Apollo 11 (Figure 6.9). They left behind a research station, which included, among other things, a seismograph, and they brought to Earth 22 kg of Moon rock and loose material.

In the meantime Soviet scientists have developed the technique of unmanned, automatic, or remote-controlled lunar expeditions. They have brought rock samples from the Moon to the Earth with *un*manned space vehicles; the Moon-car Lunokhod explored further regions of the Moon's surface on its remote-controlled journeys.

We shall discuss the results of the manned and unmanned flights to the Moon and of the equally interesting Mars and Venus missions in connection with the cosmic objects concerned.

7. Physical Constitution of Planets and Satellites

With the advent of space research, the investigation of planets and their satellites has in recent years developed into one of the most interesting, but also most difficult chapters of astrophysics. The interpretation of the observations demands in particular all the resources of *physical chemistry* (chemical equilibrium, phase diagrams, etc.). We begin here with a few remarks on methodology, although we must to some extent leave the discussion of the necessary instruments to Section 9. Then we review a few general theoretical topics.

After this general introduction we first give an account of the Earth and then the Moon—the most closely examined member of the planetary system—likewise the *terrestrial planets* Mercury, Venus, and Mars, and finally the asteroids. All these heavenly bodies have mean densities $\bar{\rho}$ in the range from 3.9 to 5.5 g/cm^3, corresponding approximately to terrestrial rocks or metals. Then we turn to the quite differently constructed *major planets* Jupiter, Saturn, Uranus, and Neptune. Their mean densities lie in the range 0.7 to 1.6 g/cm^3, corresponding roughly to liquefied gases in the laboratory. On Pluto, the "foundling" of our planetary system, we content ourselves with a few remarks.

The *formation and evolution* of our planetary system (cosmogony) can be usefully discussed only in the context of stellar evolution (Section 31).

7.1 Possibilities for the study of the planets and satellites

(1) We need not go further into the measurement of the apparent and true *diameters* of the planets, their *masses*, and the derived *mean densities*. Numerical values of these and other parameters are collected in Table 7.1. Figure 7.1 gives a graphic representation of the true dimensions of the planets and the Sun.

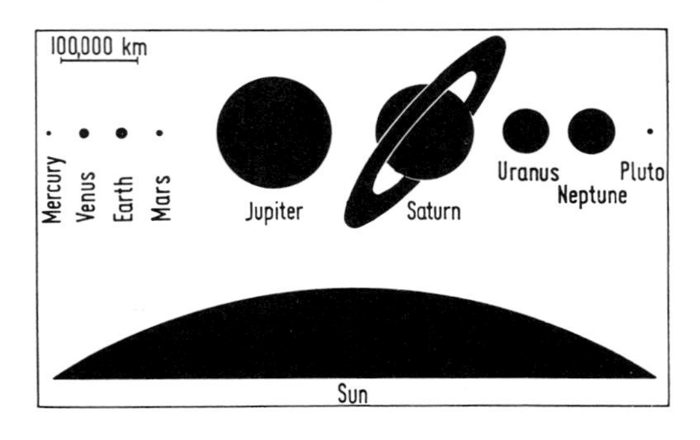

Figure 7.1
Relative sizes of the planets and Sun drawn to scale.

(2) The *rotation period* can be obtained from the telescopic observation of any sufficiently permanent surface features. Another possibility is provided by the Doppler effect in the Fraunhofer lines in reflected sunlight or in the absorption lines of the planetary atmosphere itself (for radar measurements, see point 6 in this list).

(3) The reflective power of a planet, etc. is described by the value of the *albedo,* defined as the ratio of the light reflected or scattered in all directions to the incident light. So far as it is measurable, the brightness as a function of phase angle or the surface brightness as a function of the directions of the incident and reflected light, as well as measurements of the polarization of the light, provide fairly detailed information.

(4) In the *spectrum* of a planet, the presence of at any rate some gases can be verified from their absorption bands. It is hoped that the many telluric bands of H_2O, CO_2, O_3, . . . that hamper observations made through the Earth's atmosphere may be avoided by observations from stratospheric balloons and artificial satellites.

(5) We obtain information about the *temperatures* in the atmospheres or—if these are sufficiently transparent—on the surfaces of planets and satellites by measuring the intensity of their thermal emission in the infrared (E. Pettit and S. B. Nicholson, among others) or in the millimeter to decimeter region of the radio spectrum. Which particular layers the measured radiation temperatures correspond to can be clarified however only with the help of a theoretical model of the atmosphere.

(6) Radar techniques have in the last decade achieved such precision that—as we have already pointed out—it was for the first time possible to determine the *solar parallax* and the *astronomical unit,* with undreamed-of accuracy, from measurements of the distance and velocity of Venus. Soon also annular zones round the centers of planetary disks were successfully discriminated from each other using *time-of-flight differences* of the radar waves. If the planet rotates, it is further possible to distinguish strips parallel to the projected rotation axis from the differences in the *Doppler shifts* of the reflected radar waves. Finally, the remaining ambiguity (two points symmetrically placed about the equator give equal travel times and equal Doppler shifts for the radar waves) can be completely removed by observations at different times. For such investigations, among others, American engineers have stretched a 300 m (1000 ft) diameter wire netting paraboloid antenna across a canyon at Arecibo in Puerto Rico. Then the rotation of Venus and Mercury could at last be measured unambiguously, and single mountains and craters could even be located on the surface of Venus, right through a completely opaque cloud layer.

Some results of the measuring techniques which we have described under points 1 to 6 are gathered together in Table 7.1 for the individual planets; we have also added the number of their presently-known satellites.

Table 7.1. Physical Properties of the Planets and of the Moon. Radius in Units of $R_{\oplus} = 6378.3$ km. Mass in Units of $M_{\oplus} = 5.98 \times 10^{27}$ g $= M_{\odot}/333000$. Mean Density in g/cm^3. According to W. H. McCrea (1969) the mean density of Mercury + Venus is 5.25 g/cm^3 while that of Earth + Moon + Mars is 5.27 g/cm^3.

Planet	Equatorial radius $R/R_{\oplus}$	Mass $M/M_{\oplus}$	Mean density g/cm^3	Sidereal rotation period days	Number of satellites	Atmosphere (traces)
Mercury	0.3824	0.0553	5.44	58.65	0	—
Venus	0.9495	0.815	5.24	243.0	0	$CO_2, N_2, H_2O, (HCl, HF)$
Earth	1.0000	1.000	5.51	0.9985	1	N_2, O_2, H_2O, Ar, CO_2
Moon	0.2725	0.0123	3.34	27.32	—	—
Mars	0.5334	0.107	3.90	1.026	2	$CO_2, H_2O, (CO)$
Ceres* (asteroids)	0.0604	0.00020	~5	—	0	—
Jupiter	11.116	317.89	1.33	0.4101	14	H_2, He, CH_4, NH_3†
Saturn	9.407	95.18	0.68	0.4264	10	H_2, He, CH_4, NH_3
Uranus	3.982	14.55	1.31‡	0.4507	5	H_2, He, CH_4, NH_3
Neptune	3.810	17.24	1.66‡	0.6583	2	H_2, He, CH_4, NH_3
Pluto	0.50	0.11	~4.86	6.39	0	—

*Largest asteroid. Total mass of all asteroids $\sim 4 \times 10^{-4} M_{\oplus}$.
† Also traces of ethane C_2H_6 and acetylene C_2H_2.
‡ Somewhat older data give 1.60 and 2.25 g/cm^3 respectively.

7.2 Some theoretical remarks on the atmospheres of the planets, etc.

Before we turn to the individual planets and their satellites, let us make some general remarks about temperature, structure, and the retention of their atmospheres. Finally we ask under what conditions a satellite, for example, can withstand the tidal forces of its parent body.

7.2.1. Temperature of planets

We assume that the internal energy sources of the planets are in each case very small in comparison with the incident solar radiation. The irradiation from the Sun at a distance of 1 AU is given by the solar constant (Section 12); for a planet with orbital radius r this should be multiplied by the factor $1/r^2$. It is easy to allow for the fact that the albedo is <1. For a "black" planet obeying the Stefan–Boltzmann radiation formula, the emission in the infrared would vary as: surface area $\times$ T^4. Since irradiation and emission remain in balance, then on the whole $T^4 \propto 1/r^2$ with increasing distance r of a planet from the Sun, that is, its temperature $T \propto 1/\sqrt{r}$. In particular circumstances, considerable departures from this simple relation are to be expected: for slow rotation, the balance between irradiation and emission is confined to the illuminated hemisphere; however, for faster rotation the temperatures on the day and night sides will tend to equalize. Furthermore, a strong dependence of the transparency of the atmosphere on the wavelength of the radiation can lead to the familiar "greenhouse effect" or to the local heating of particular layers, for example the ozone layer in the Earth's atmosphere.

7.2.2. Structure of planetary atmospheres

The pressure distribution in a planetary atmosphere is determined by the hydrostatic equation

$$dp = -g\rho \, dh. \tag{7.1}$$

That is: the weight of an element of a layer of thickness dh is supported against gravity g by the pressure difference between its faces. Furthermore, the pressure p is related to the density ρ and the mean molecular weight μ by the equation of state of an ideal gas

$$p = \rho \mathcal{R} T/\mu \tag{7.2}$$

where $\mathcal{R} = 8.317 \times 10^7$ erg/K mol is the universal gas constant. Substitution in equation (7.1) gives

$$\frac{dp}{p} = -\frac{g\mu}{\mathcal{R}T}\,dh \tag{7.3a}$$

$$\text{or } \frac{dp}{p} = -\frac{dh}{H} \tag{7.3b}$$

$$\text{where } H = \frac{\mathcal{R}T}{g\mu}.$$

If the so-called equivalent height or scale height H is constant, we can easily integrate and obtain for a considerable range of height the barometric formula

$$\ln p - \ln p_0 = -\frac{h}{H} \quad \text{or} \quad p = p_0 e^{-h/H}, \tag{7.4}$$

where p_0 is the pressure on the ground or in the reference layer $h = 0$. For the region between two levels of heights h_1 and h_2, with pressures p_1 and p_2, the relation

$$\ln p_2 - \ln p_1 = -\int_{h_1}^{h_2} \frac{dh}{H(h)} \tag{7.5}$$

follows quite generally from equation (7.3); according to equation (7.3b) the structure of a planetary atmosphere therefore depends on (1) g, the acceleration due to gravity on the planet, (2) the mean molecular weight μ, that is on the chemical composition and on the state of dissociation and ionization of the atmospheric gas, and (3) the temperature distribution $T(h)$. This last is in turn determined by the mechanism of energy transport.

7.2.3. Escape of gas from the atmosphere of a planet or satellite

We ask further, to what extent can a planet, satellite, etc., retain its own atmosphere? The molecules of a gas of molecular weight μ and absolute temperature T have, from the kinetic theory of gases, a most probable speed $\bar{v} = (2\mathcal{R}T/\mu)^{1/2}$ where $\mathcal{R} = 8.317 \times 10^7$ erg/K mol is again the gas constant. From equation (6.37), a molecule of speed v can escape from a body of mass M and radius R if $v^2/2 \geqslant GM/R$. Taking account of the Maxwell–Boltzmann distribution of molecular speeds for $v > \bar{v}$, one can see that Mercury, our Moon, and most satellites can have almost no atmosphere while on the other hand Titan, the largest satellite of Saturn, with a temperature calculated as in Section 7.2.1, can actually retain its atmosphere for a long time.

7.2.4. Limit for the stability of a satellite

A satellite which circles its parent body at distance r will tend to be pulled apart by tidal forces (Section 6) and for a closer approach would be disrupted or not formed in the first place. If the satellite is not too small, we may disregard its internal cohesive forces, and we can easily estimate the essential gravitational forces. Let the central body (planet) have mass M, radius R, and mean density ρ, and let the corresponding values for the satellite be M_s, R_s, and ρ_s. Now we can calculate the mutual attraction of the parts of the satellite approximately as if two masses $M_s/2$ had a mutual separation R_s; it is

$$G \frac{M_s M_s}{4R_s^2}. \tag{7.6}$$

On the other hand, the tidal force, which is pulling the two fictitious masses apart, is about equal to the difference of the attractive forces of the central body at the distances r and $r + R_s$, i.e.,

$$G \frac{MM_s}{2} \Delta \left(\frac{1}{r^2}\right) \approx G \frac{MM_s}{r^3} R_s. \tag{7.7}$$

Centrifugal (and inertial) forces are of the same order of magnitude as these forces. The condition for stability of the satellite is then

$$G \frac{M_s M_s}{4R_s^2} \geqslant cG \frac{MM_s}{r^3} R_s \tag{7.8}$$

where c is a number of order unity. Using the fact that for the satellite $M_s = (4\pi/3)\rho_s R_s^3$ and similarly for the planet $M = (4\pi/3)\rho R^3$, we obtain

$$\frac{r}{R} \geqslant (4c)^{1/3} \left(\frac{\rho}{\rho_s}\right)^{1/3} \tag{7.9}$$

A more exact calculation by E. Roche (1850) (strictly speaking, for a synchronous satellite, like our Moon) gave the stability limit:

$$\frac{r}{R} \geqslant 2.45(\rho/\rho_s)^{1/3}. \tag{7.10}$$

It follows that a satellite with the same density as its central body is not "allowed" to be nearer to it than 2.45 planetary radii.

7.3 Earth and Moon; the terrestrial planets Mercury, Venus, and Mars; and the asteroids

Within the asteroid belt, which divides the solar system into two physically different regions, all planets have less than or equal to an Earth mass and mean densities in the range 3.9 to 5.5 g/cm^3. They are obviously composed mainly of solid materials. In chemical language, their atmospheres are oxidizing; they contain O_2, CO_2, H_2O, N_2,. . . . So it seems justifiable to call them all "terrestrial planets." As an example, we first consider our Earth in somewhat more detail; obviously it is not our intention to give a compendium of geophysics. Thanks to the development of space research, the knowledge of our Moon has grown by leaps and bounds within a few years. It is also possible to observe Mercury, Venus, and Mars from close-by. For the whole inner region of the solar system, radio astronomy, especially using radar techniques, has yielded many results, some of them rather unexpected.

7.3.1. The Earth ⊕

In consequence of its rotation the Earth is, to a good approximation, an oblate spheroid with

$$\text{equatorial radius } a = 6378.2 \text{ km}$$
$$\text{polar radius } b = 6356.8 \text{ km}$$

and the flattening is

$$\frac{a-b}{a} = \frac{1}{298}.$$

The flattening and the centrifugal force result in gravity at the equator being 1/190 weaker than at the poles.

We already know the mean density of the Earth $\bar{\rho} = 5.51$ g/cm^3. The density of the Earth's crust (granite, basalt) is 2.6 to 3 g/cm^3. The moment of inertia inferred from the motion of the spinning Earth gives at any rate a summary of information about the increase of density with depth. Further knowledge comes from the propagation of earthquake waves. In an earthquake, longitudinal and transverse elastic waves are set up in the comparatively superficial epicenter. These then propagate themselves through the Earth's interior and are there refracted, reflected, and transformed into one another in accordance with the depth dependence of the elastic constants. By an exact study of seismic waves, about 1906 E. Wiechert in Göttingen discovered that there are in the Earth's interior several surfaces of discontinuity where the elastic constants and the density change abruptly. The Earth's *crust* ($\rho = 2.6$ to 3.0 g/cm^3) has a thickness of about 30–40 km beneath plain lands, which increases considerably beneath young mountain ranges to 70 km. Under the oceans it decreases again to $\sim$ 10 km. The lower boundary of the crust forms the Mohorovičič discontinuity or "Moho" (which it has been proposed to reach by boring a "Mohole" in the bed of the Pacific Ocean). Then, down to a depth of 2900 km there follows the *mantle* having $\rho = 3.3$ to 5.7 g/cm^3 which may consist mainly of silicates. In the deep interior, from 2900 km to 6370 km we have the *core* with $\rho = 9.4$ to about 17 g/cm^3. No transverse waves are propagated in the core. In this sense we may think of the outer core as fluid, but of excessively great viscosity. However, more extensive investigations suggest that the so-called *inner core* below a depth of 5000 km is again solid.

We can only indirectly infer the chemicomineralogical composition of the Earth's core. Laboratory experiments at high pressures and temperatures, along with the geophysical data, yield an acceptable phase diagram if we assume that the core, like iron meteorites, consists of 90 percent iron and 10 percent nickel, with an admixture of sulphur.

We can fairly accurately calculate the inward increase in *pressure* from the hydrostatic equation; for the center[1] we find $p \approx 3.5 \times 10^6$ atmospheres $= 3.5 \times 10^{12}$ dyn/cm^2. By comparison, Bridgman could produce about 4×10^5 atmospheres in the laboratory.

The increase of *temperature* with depth can be measured in deep bore holes and there a geothermal gradient of about 30 K/km is obtained. The temperature distribution at greater depths is determined on one hand by the generation of heat by the radioactive substances U^{238}, Th^{232}, and, to a lesser extent, K^{40}, and on the other hand by the slow outward heat transport by conduction and convection in the magma. Thus the temperature in the core must be at least some thousands of degrees, but almost certainly not as much as 10 000 K.

The best information concerning the duration of the various geological periods and the age of the Earth, defined as the time since the last thorough mixing of the Earth's material, is today given by the methods of *radioactive dating*. One uses the radioactive decay of the following isotopes, where we give only the stable end-products of each series:

	Half life T
$U^{238} \rightarrow Pb^{206} + 8\,He^4$	4.5×10^9 years
$U^{235} \rightarrow Pb^{207} + 7\,He^4$	0.71×10^9 years
$Th^{232} \rightarrow Pb^{208} + 6\,He^4$	13.9×10^9 years
$Rb^{87} \rightarrow Sr^{87} + \beta^-$	50×10^9 years
$K^{40} \nearrow A^{40} + K(\gamma)$ (K-capture); $K^{40} \searrow Ca^{40} + \beta^-$	1.3×10^9 years

For each process one determines the ratio of the end product to the initial isotope. If this ratio was zero to begin with ($t = 0$) then after time t it must be equal to $2^{t/T} - 1$.

Table 7.2 shows in brief the most important geological strata and their absolute dating. (The author is indebted for this to Professor K. Krömmelbein, Kiel.)

The oldest pre-Cambrian rocks have an age of 3.7×10^9 years.

The *age of the Earth* since the last thorough mixing or separation of its materials (by a mechanism by no means completely understood!) can be determined if one takes the initial abundance ratio of the isotopes 206 and 207 of "primaeval lead" from measurements in iron meteorites, which contain almost no uranium or thorium. With fair precision, one obtains

$$4.55 \pm 0.05 \times 10^9 \text{ years.} \tag{7.11}$$

[1]Elementary estimate: With mean gravity $\frac{1}{2}g = 5 \times 10^2$ cm/s^2 and mean density $\rho = 5.5$ g/cm^3 a column of length $R = 6.37 \times 10^8$ cm (Earth's radius) exerts a pressure $\rho = \frac{1}{2}g\rho R \approx 1.8 \times 10^{12}$ dyn/cm^2 which agrees with the exact calculation to within a factor 2.

Table 7.2. History of the Earth

Era	Times of principal mountain building	Epochs
CENOZOIC	Alpine	← Snails–mussels → ← Mammals →
MESOZOIC		← Goniatites Ammonites →
PALAEOZOIC	Hercynian (Variscan) Caledonian	← Brachiopods → ← Armored fish → ← Trilobites → ← Graptolites →
PROTEROZOIC	Assyntic Algomian	
EOZOIC	Laurentian	

Oldest rocks known (according to absolute age determinations) ~3700 million years.
Origin of Earth's crust about 4500 million years ago.

The Earth's magnetic field, the *geomagnetic field,* and its secular variations (rapid variations caused by the Sun will be considered later) may be accounted for in the following way, according to W. M. Elsässer and Sir Edward Bullard: in connection with the heat transport by convection mentioned earlier, the fluid material in the Earth's outer core forms great eddies. If in such eddies in conducting material some traces of a magnetic field are present, then these can be intensified, as in the self-exciting

Fauna	Period (age in millions of years from the present)	Flora
Earliest man →	Quaternary 1	NEOPHYTIC Age of angiosperms
→	Tertiary 70 ± 2	
Extinction of ammonites and dinosaurs	Cretaceous 135 ± 5	MESOPHYTIC Age of conifers
Earliest birds →	Jurassic 180 ± 5	
→ Earliest mammals	Triassic 225 ± 5	
→ Extinction of many groups of palaeozoic animals →	Permian 270 ± 5	PALAEOPHYTIC Age of ferns
Earliest reptiles	Carboniferous 350 ± 10	
→ Earliest amphibians	Devonian 400 ± 10	
	Silurian 440 ± 10	EOPHYTIC Age of algae
Earliest vertebrates	Ordovician 500 ± 15	
First appearance of many groups of invertebrates →	Cambrian 600 ± 20	
ALGONKIAN ARCHEAN	PRE-CAMBRIAN	

dynamo of W. v. Siemens. The details of such a self-exciting dynamo in the Earth's interior have not yet been tidied up, but in any case the dynamo theory offers the only possibility of explaining the magnetic field, its secular variations, and its reversals (see below).

Thanks to the circumstance that when certain minerals were being formed the magnetic field present at the time was, so to speak, frozen into them (P. M. S. Blackett, S. K. Runcorn, and others), it is possible to reconstruct the geomagnetic field of past geological epochs. Such *palaeomagnetic* measurements have shown that the field vectors of the past can best be explained by resorting to A. Wegener's (1912) hypothesis of *continental drift* as inferred from the map of the world. Figure 7.2 shows first how it is possible to slide together on the globe the continents round

Figure 7.2. Continental drift. After E. Bullard, J. E. Everett and A. G. Smith (1965). The continents round the Atlantic are fitted together at the steep edge of the continental shelf, at a depth of 500 fathoms = 900 m. Mercator projection, with the present-day meridian 60°W as "equator". North America has been left in its present-day position. The present-day lattice of longitude and latitude is shown in Europe, Greenland, South America and Africa.

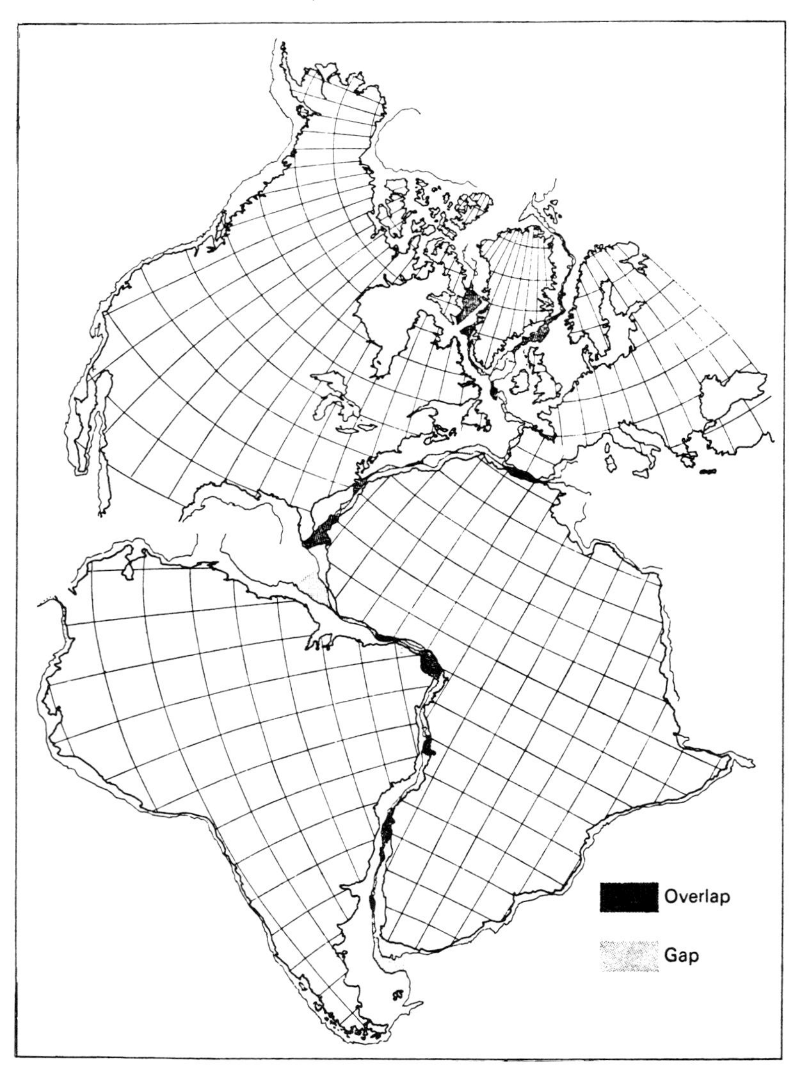

the Atlantic, including their continental shelves down to a depth of about 500 fathoms. Similarly in the east and southeast, India, Antarctica, and Australia fit together well. Making the approximate assumption that in past epochs as well the Earth possessed an essentially dipolar magnetic field, it is possible to reconstruct from palaeomagnetic measurements the relative positions of the continents in earlier geological epochs and to show, in the best possible agreement with a host of palaeontological and geological observations, how the present continents were formed bit by bit during the course of the Earth's history. For example, the Atlantic Ocean began about 120 million years ago, that is, in the Jurassic to Cretaceous period, as a small trench like the present Red Sea. Since then America and Europe have been moving apart at an average rate of a few centimeters per year. After palaeomagnetic investigations in the 1950s had made the idea of continental drift, which had been disputed for several decades, into a certainty, a direct insight into the underlying mechanism of such motions was obtained in the 1960s as follows. It was evident from palaeomagnetic measurements that strata closely following each other often had diametrically opposite directions of the Earth's field. Was this a question of a spontaneous field reversal in the rock (physicists know of such effects) or of a reversal of the polarity of the whole field of the Earth? Detailed investigations of exactly dateable sequences of strata showed that the latter is correct. This does not appear quite so surprising if one remembers that in the self-exciting dynamo of Siemens the direction of the current is determined by the weak (random) magnetization present at its onset. After it was known that the geomagnetic field reversed its polarity at irregular intervals of a few hundred thousand years, it was discovered by H. H. Hess (1962), by F. J. Vine and D. H. Matthews (1963), and by others that the floor of the Atlantic Ocean exhibited many north–south strips with alternate directions of magnetization. Magnetic dating of these strips showed that the floor of the Atlantic Ocean had been widening for about 120 million years in two directions, starting from the Mid-Atlantic Ridge which lies halfway between Europe and America, and had pushed the two continents apart. This tremendously fruitful theory of *ocean floor spreading* was quickly supplemented by the realization of H. H. Hess, T. Wilson, and others (~1965) that in all these processes gigantic lumps of the *lithosphere* (the upper part of the crust) will drift rigidly or be broken apart by fissures (which are the origin of tectonic earthquakes); one speaks therefore of *plate tectonics*.

As A. Holmes had recognized as early as 1928, it is clear that the driving mechanism for all geological events can be found in *convection currents*, particularly in the upper part of the Earth's mantle, which well up alongside the oceanic mountain ridges. Fresh Earth crust and plate material are made here and move apart on both sides. Deep sea trenches come into being at the place where an oceanic plate streams back down into the Earth's mantle, and behind them young folded mountains are produced by the compression of the continental crust. In this way, these and many other fundamental geological events find one simple explanation. The necessary energy is made available by the radioactivity of the rocks. From the geothermal gradient and the thermal conductivity of rock, it is possible to estimate that the entire Earth makes about 10^{28} erg/yr available as thermal energy. About one part in a thousand of this is used up in the generation of earthquakes. If our "thermodynamic machine" generates mechanical energy even with an efficiency of only 1 percent, that would be quite enough to drive convection in the Earth's mantle. Geological observations show further that the tectonic activity of the earth has undergone during the course of time considerable quantitative and qualitative variations, with maxima at times 0.35, 1.1, 1.8, and 2.7×10^9 years ago. S. K. Runcorn has tried to connect these times of maximum tectonic activity with transitions from one particular mode of convective flow to a more complicated one with a greater number of convection cells.

We must restrain ourselves here from going more closely into the physics of the sea, particularly as only the Earth possesses oceans.

However, we shall briefly discuss the *atmosphere* of the Earth, primarily with the intention of comparing it with other planets. Its pressure distribution is, from equation (7.5), determined primarily by the temperature distribution $T(h)$. The latter derives its origin from the mechanism of energy transport, that is, the motion of thermal energy into and out of each layer h to $h + dh$. In the lowest layer of the Earth's atmosphere, the *troposphere,* the upward transport of the heat absorbed from the Sun takes place by *convection,* which leads to a regular decrease of temperature with height. Above the so-called *tropopause* at a height of about 10 km, *radiation* takes over the transport of energy, and we find next the almost isothermal *stratosphere*. This results from a balance between, on the one hand, mainly the absorption of solar radiation and, on the other, radiation into space at long wavelengths. At a height of 25 km we find a warm layer, connected with the formation of ozone O_3. After a short decrease, the temperature climbs to more than 1000 K above about 90 km as a result of the dissociation and ionization of the atmospheric gases N_2 and O_2. As a result of the ionization, there appear the electrically conducting layers of the *ionosphere* (maximum electron density of the E layer at a height of ~115 km, of the F layer at ~300–400 km), which enables the passage round the globe of electromagnetic waves whose wavelengths are not too short.

The recombination of electrons and ions in the E layer gives rise to the emission lines and bands of the *air glow*.

As a result of observations from space vehicles, it has recently become possible to detect ionospheres and air glows in the upper atmospheres of other planets as well.

7.3.2. The Moon

The *mass* of the Moon, 7.35×10^{25} g, which is determined from Kepler's third law, the small *eccentricity* of its orbit $\epsilon = 0.055$, and its small *inclination* of about 5° to the ecliptic are all very similar to the corresponding data for the larger satellites of other planets. With regard to the mass *ratio* of satellite: planet, on the other hand, our Earth's moon, with a ratio of 1:81.30, is quite exceptional; for example, the four Galilean satellites of Jupiter all have mass ratios less than $1{:}10^4$.

Selenographic studies with larger and larger telescopes (Figure 7.3) and with space probes and finally the landings of the Apollo astronauts have made us very familiar with the formations on the Moon's surface (Figure

Figure 7.3. Moon. The crater Posidonius (lower right) with a diameter of 100 km and the smaller crater Chacornac (lower left) on the edge of Mare Serenitatis (above); in Mare Serenitatis a 180 m high well-defined mountain range; numerous small craters scattered everywhere. Lick Observatory, 1962 March 25, 120-in reflector.

Figure 7.4. Mare Nectaris with the craters Theophilus, Mädler, and Daguerre. Photographed by the Apollo 11 command module from a height of 100 km. In Mare Nectaris there are in the foreground two "ghost" craters, covered with lava, but otherwise only countless small craterlets.

7.4). There can no longer be any doubt now that all the circular structures, from the giant maria (the designation as a sea has of course only historical significance)—Mare Imbrium has a diameter of 1150 km and a depth of 20 km—through the craters right down to the microscopic dimples (Figure 7.5) of a few thousandths of a millimeter, have their origin in the impact of meteoritic bodies with planetary speeds. The depths of the meteor craters vary with diameter in the same way as for terrestrial explosion craters. It is obvious that later on the floors of the maria and of many of the larger craters were flooded by lava flows. When these had hardened, their flat surfaces were speckled anew with smaller craters.

For a long time the nature of the meandering *rilles* was one of the major puzzles of lunar research. The Apollo 15 astronauts were able to visit the so-called Hadley Rille and to observe at close quarters its steep walls with conspicuous stratification. Probably we are dealing with a lava channel which was originally partially roofed over. Similar lava channels or pipes are present in terrestrial volcanoes. Other longer rilles may be understood as fissures in cooling lava.

Pettit and Nicholson had already used infrared measurements to ascertain that during a lunar eclipse the lunar surface at the edge of the shadow changes its temperature very slowly, and in this way they measured its small heat conductivity. In fact the surface of the Moon is liberally sprinkled with fine dust and loose rock particles, which clearly owe their origin to the impacts of meteorites.

From many of the larger craters stretch out bright *ray systems,* visible even with small telescopes; they evidently consist of material thrown out when the crater was formed.

The highest elevations are found in the bright, crater-studded highlands, the *terra* regions. The greatest heights are limited by the solidity of the material, and so are of the same order of magnitude as on the Earth, as Galileo long since inferred from the length of their shadows at the edge of the dark side of the Moon, that is, at the terminator.

The *mean density* of the Moon $\bar{\rho} = 3.34\ \mathrm{g/cm^3}$ is strikingly similar to that of the upper mantle of the Earth; in no way can the Moon have the same average composition as the Earth. The *moon rocks* from the maria, which were brought back to the Earth by the astronauts, are full of hollow bubbles, like terrestrial lavas that solidified under low pressure. We are therefore dealing with *magmatic* rocks that solidified from molten material. The lunar dust mentioned previously was produced by the breaking up of such rocks. Dust and smaller fragments cake together again to form *brec-*

Figure 7.5. In the Apollo 11 moon dust were found glassy balls, which were formed out of rock melted by meteoritic impact. This electron microscope picture by E. Brüche and E. Dick (1970) shows such a tiny sphere of diameter 0.017 mm, on which the further impact of a micro meteorite has produced a "microcrater".

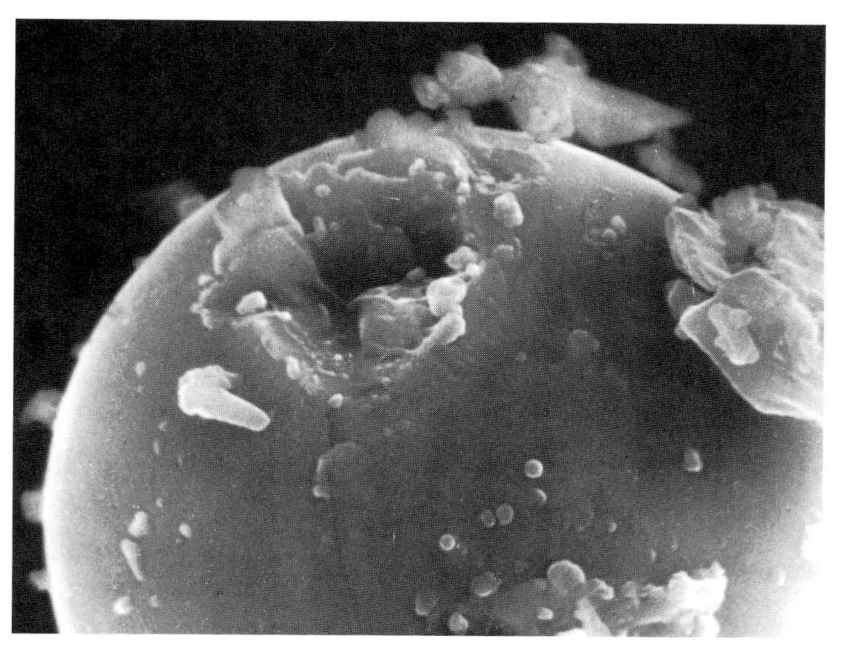

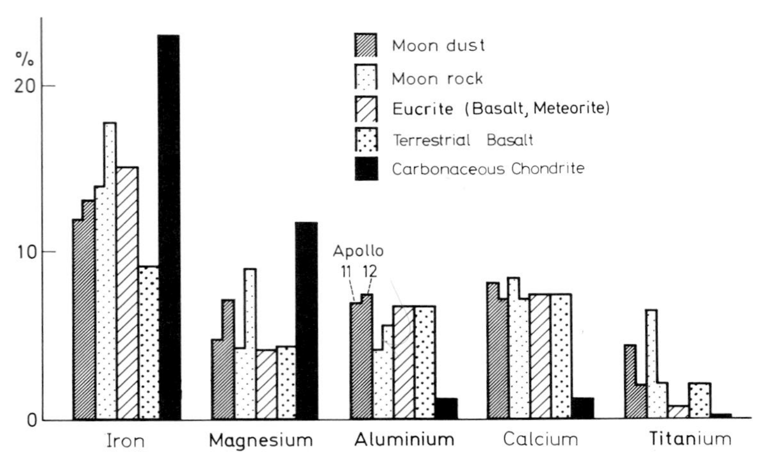

Figure 7.6. The abundance distribution (percent by weight) of the elements iron, magnesium, aluminum, calcium, and titanium in samples from the Moon probes (left half: Apollo11—Mare Tranquillitatis; right half: Apollo 12—Oceanus Procellarum), in eucrites, that is, basaltic achondritic meteorites, and in carbonaceous chondrites of Type I. In the latter, the abundance distribution of all nonvolatile elements corresponds to unaltered solar material.

cia. The lunar rocks resemble terrestrial basalts in their main mineralogical features. In Figure 7.6 we compare the abundances of a few important elements in lunar material and in terrestrial basalt with those in (see Section 8) *eucrites*, that is, basaltic achondrites, and in carbonaceous chondrites of type I, that is, those meteorites whose composition is most nearly solar. This comparison, and other studies on rare (or "trace") elements, show that in lunar rocks some elements are overabundant by about one hundredfold with respect to solar material while the siderophile elements Ni, Co, Cu, . . . are deficient by factors of about 100.

Radioactive age determinations by the methods summarized on p. 57 show that the age of the mountains is as much as 4.6×10^9 years. That is, the Moon is, to within present accuracy, the *same age* as the Earth. On the other hand the age of Mare Tranquillitatis is found to be only 3.7×10^9 years and of Mare Imbrium 3.9×10^9 years. Before 3.3 to 4×10^9 years ago, the maria were filled in stages with basalt lava which welled up from deeper regions. Their structure was not therefore completed until about a thousand million years *after* the formation of the Moon. These dates, together with the statistics of lunar craters and their "overlapping", show that the cosmic bombardment was at first extremely severe and then in the course of the first 10^9 years of the history of the Earth and Moon decreased rapidly; later the decrease was much slower.

Our Moon possesses no measurable general external magnetic field. Weak remanent magnetization of some lunar rocks, corresponding to a few percent of the Earth's field, points to an earlier lunar field. Nowadays practically all magmatic and tectonic activity on the Moon has died away. However, the great plains of the maria, "drowned" craters, and a few remarkable domes on the Moon show, as we have said, that earlier, besides the impact of meteorites, magmatic events, that is, melting of rocks and outpourings of basaltic lava, played an important role.

The signals of the seismometers placed on the Moon during the Apollo missions give us some information on the internal structure of the Moon. Surprisingly, seismic waves, which were set up on the Moon by meteorites' impacts or by the crash of the abandoned "moon ferry," persisted for a whole hour. The very small damping of moonquakes in comparison to earthquakes may depend on scattering or on dispersion in particular circumstances.

Besides the countless new observations concerning the structure and past history of the lunar surface, the Apollo flights have also introduced an entirely original innovation into the celestial mechanics of the Earth–Moon system, the "lunar cat's eye" or, to give it its official name, the "Laser Ranging Retro-Reflector." A glass or quartz prism, whose shape corresponds to the cutoff corner of a cube, has the well-known property that after reflection by all three faces of the cube a light ray will return in exactly its direction of incidence. Such reflectors were placed on the Moon. It is now possible with a large reflecting telescope to send intense laser pulses to the Moon and to determine its distance with an accuracy of about 15 cm from the return travel time. In this way the accuracy of several of the important constants of the Earth–Moon system can be improved by several powers of ten.

After our account of the Earth and Moon, let us turn to the terrestrial planets. We have already given a selection of the data in Table 7.1. We still know little of their interiors; we therefore concern ourselves mainly with their atmospheres and surfaces.

7.3.3. Mercury ☿

Mercury is very difficult to observe since it is never farther than $\pm 28°$ from the Sun. Recent radar measurements, together with previous visual observations, show that the rotation period of Mercury is *not,* as was previously believed, equal to its orbital period of 88 days but amounts to 58.65 ± 0.01 days, which corresponds to exactly ⅔ of the orbital period.

The space-probe Mariner 10, which flew past Mercury on March 29, 1974, transmitted numerous remarkable pictures of the planet. Like the Moon, its surface is densely covered by craters. The largest has a diameter of 1300 km, comparable to Mare Imbrium. Mariner 10 discovered also that Mercury possesses a magnetic field and even an extremely tenuous atmosphere.

7.3.4. Venus ♀

Recent radar measurements led to the surprising result that Venus, alone among the planets, rotates in a retrograde direction and, what is more, has a sidereal period of 243.01 days.

After W. S. Adams and T. Dunham had concluded from infrared spectra as early as 1932 that Venus possessed a dense atmosphere of carbon dioxide CO_2, further clarification was obtained from measurements of the radiation temperatures in the centimeter and decimeter regions of the spectrum, from the flybys of the Mariner satellites, and from the sending of instruments by parachute using the Soviet Venera space probes. At the foot of the Venusian atmosphere, which consists mainly of CO_2 with some N_2 and H_2O and traces of HCl and HF, the radiation temperatures at $\lambda \approx 5$ to 10 cm, themselves high, yield a temperature of about 780 K and a pressure of about 100 atm. The surface itself, as has long been known from visual observations, is completely covered by clouds, whose chemical composition is not yet known. The height of this cloud layer has been estimated to be about 50 km from the difference between the optical and radar diameters. It is known from the radar measurements that there are mountains on Venus whose heights are comparable to those on the Earth, and also craters, like those on the Moon.

Before its visit to Mercury, Mariner 10 flew past Venus on February 5, 1974 at a distance of only 6000 km. The transmitted pictures show the finest details of the cloud layer, its structure, and motions. One might almost speak of a comparative meteorology. They also gave a more detailed insight into the stratification of the atmosphere. At a height of 145 km, the ionosphere of Venus reaches a maximum density of 3×10^5 electrons/cm^3, comparable to our E layer.

7.3.5. Mars ♂

The vivid red color of the planet is due to a decrease in the reflective power at short wavelengths. This, together with polarimetric measurements by A. Dollfus, points to the presence of iron oxides like limonite $Fe_2O_3(nH_2O)$. Visual observations had already revealed a diversity of "areographic" structures. The long-popular canals of Mars however are due to a physiological-optical contrast phenomenon: our eyes have, so to speak, the tendency to join prominent points and corners together by lines (one thinks of the constellations).

Television pictures from Mariner IV (1965) showed numerous craters on the surface of Mars, whose diameters range from the limit of resolution, a few kilometers, to about 120 km. These craters correspond closely to those of our Moon; their contours are somewhat worn down by erosion. Mariner 9, which went into orbit round Mars on November 13, 1971, gave pictures with much improved resolution (Figure 7.7). These show that the surface of Mars was not formed solely by meteoritic impact but that there were

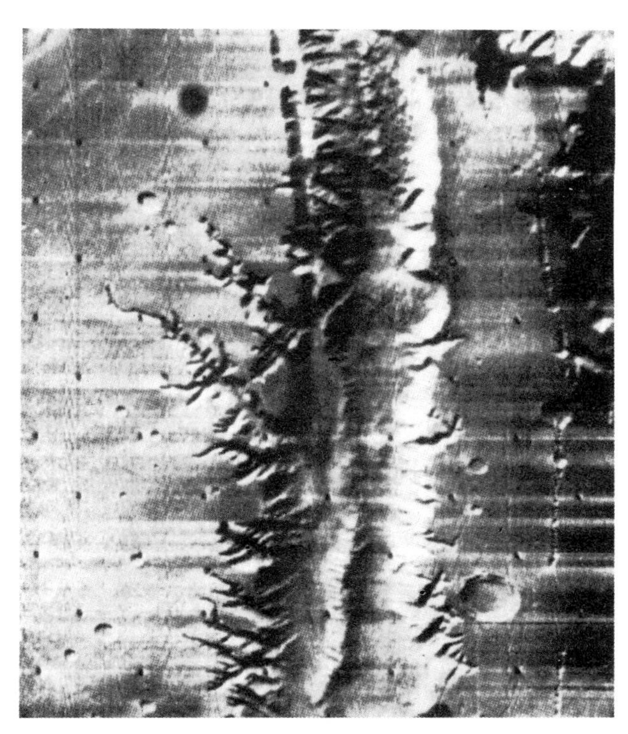

Figure 7.7. Mariner 9 photograph. This picture (January 12, 1972) of a large region (about 500 × 380 km) of the martian surface shows, besides several meteorite craters, a part of the canyon system of the Coprates region, which is altogether more than 2500 km long. This may have been formed by a complicated combination of tectonics and erosion. The chain of craterlets on the right may perhaps be thought of as volcanic lakes.

important contributions from volcanic activity (volcanic craters, shield volcanoes, caldera), tectonics, erosion (Figure 7.7), and deposition.

The Mariner IV measurements showed that Mars possesses no permanent magnetic field. Everything points to the fact that, like our Moon, it is now a tectonically dead body.

Radiation temperatures at radio frequencies show that the temperature on the surface has a mean value of 210 K with variations between about 180 and 300 K, in good agreement with theory.

Spectra of Mars show that its atmosphere consists essentially of carbon dioxide CO_2, with a slight admixture of water vapor H_2O and carbon monoxide CO. In the short wavelength "air glow" in the upper atmosphere of Mars, emission lines of hydrogen (Lyman α), atomic oxygen, and carbon can be detected, as well as CO_2^+ and CO. No trace of nitrogen can be detected either in absorption or in emission. The width of the absorption bands and the observation of an occultation of Mariner IV at the edge of the disk give the pressure at the base of the CO_2 atmosphere to be about 10 mb, that is, about 1 percent of our air pressure, with differences according to the height of the Martian terrain. From time to time violent dust storms are observed on Mars, which render the atmosphere completely opaque.

Through a telescope one can detect two white polar caps, which retreat in the Martian summer and grow again in the Martian winter. Their temperature corresponds to the sublimation temperature of solid carbon dioxide (148 K), and so they consist of "dry ice".

A. Hall discovered in 1877 the tiny moons of Mars, Phobos (Figure 7.8), and Deimos. The orbital period of Phobos—7^h39^m—is much shorter than the rotation period of the planet. Mariner 9 pictures revealed traces of violent meteorite bombardment on both satellites, whose sizes are about 12 and 20 km.

Attempts to detect any traces of life on Mars have without exception met with failure. Even if it were possible to detect absorption bands or organic

Figure 7.8. Mariner 9 picture of the martian moon Phobos. Orbital period 7^h39^m. The diameter of this irregularly shaped satellite amounts to about 20 km. It is covered with impact craters with diameters of up to 5.3 km.

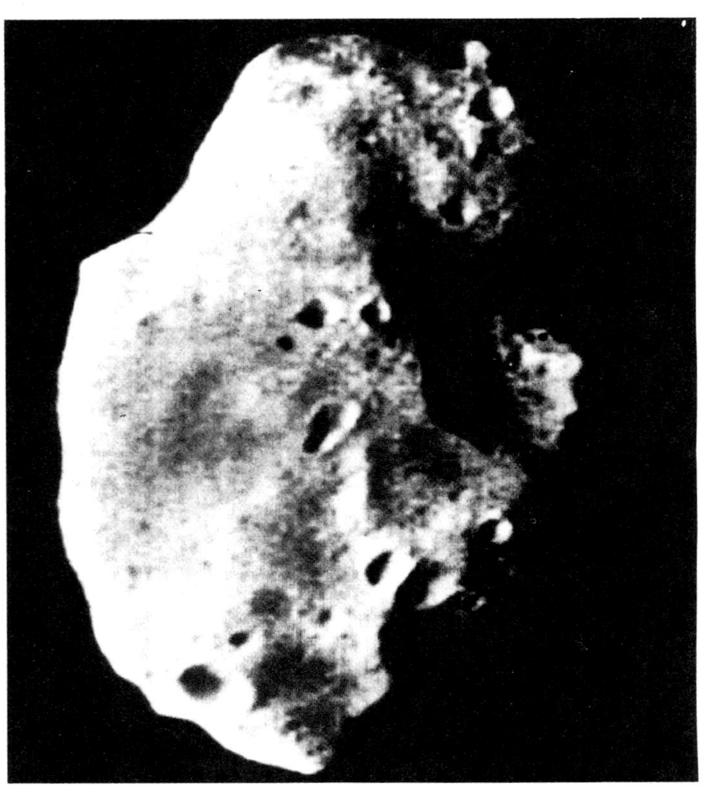

molecules, this would have no significance for our question, as investigations on certain meteorites have shown (Section 8). For the origin of organic life on Mars the most elementary prior conditions are not and were not present. An introduction of extremely primitive life forms from the Earth would perhaps not be completely excluded.

7.3.6. Asteroids or minor planets

Kepler had conjectured that there was a heavenly body in the "gap" between Mars and Jupiter (Figure 5.5). On 1801 January 1 Piazzi in Palermo discovered the first known asteroid, Ceres, but by mid-February it was lost to sight in the vicinity of the Sun. By October of the same year the 24-year old C. F. Gauss had computed its orbit and ephemerides so that Zach was able to recover it. As a sequel to this brilliant mathematical performance, the "princeps mathematicorum" in his *Theoria Motus* 1809 solved the general problem of orbit determination, that is, the evaluation of all the orbital elements of a planet or comet from three complete observations. Today many thousands of asteroids are known, most of them between Mars and Jupiter. Objects with known orbits are given a number and a name.

The distribution of their orbital eccentricities peaks at $e \approx 0.17$ and of their inclinations at $i \approx 8°$. Both values are very different from those of comets (Section 8). The unusual asteroid 433 Eros with $e = 0.233$ approached to within 0.17 AU of the Earth at its 1931 opposition and thus enabled a favorable measurement of the solar parallax to be made.

In 1973 J. Schubart was able to determine the mass of Ceres to be $5.9 \pm 0.3 \times 10^{-10}$ solar masses from perturbations in the orbit of Pallas. The total mass of all the asteroids can be estimated from photometric measurements (see below) to be about twice the mass of Ceres, or 1/2500 Earth masses.

The diameter of the largest and brightest asteroids can be measured with a micrometer. For example 1 Ceres has a diameter of 770 ± 40 km; its density of about 5 g/cm^3 corresponds roughly to that of the terrestrial planets. For the fainter asteroids—the Palomar–Leiden Survey (1970) reaches to 20.6 mag (magnitude; for definition see Section 13)—diameters as small as about 1 km can be estimated from their brightness and an assumed albedo.

Many asteroids have a periodic light variation which tells us their rotation periods. Curiously, these lie without exception between 3 and 17 hours. In particular, 433 Eros has a large amplitude light variation with a period of 5^h16^m. The light curve points to an elongated form (about 35 : 16 : 7km). The irregular shape of this and other asteroids suggests that they were formed from the breakup of larger bodies by collisions. The discovery of many families of asteroids with similar orbital elements by Hirayama and others points to the same conclusion.

With that, we conclude our very incomplete survey of the terrestrial planets and turn to the major planets (Table 7.1) which, as their masses and mean densities have already shown, are completely different. As an example we shall discuss Jupiter comparatively fully, but must condense our treatment of Saturn, Uranus, and Neptune.

7.4 The major planets

7.4.1. Jupiter ♃

Jupiter (Figure 7.9), the largest and most massive of the planets (1/1047 solar masses), has a dense atmosphere with pronounced stripes parallel to the equator, similar to the circulation system of the Earth. The so-called *red spot* has been seen for many years.

A turning point in the investigation of the major planets came in 1932 when R. Wildt identified the strong absorption bands in Jupiter's spectrum with higher harmonics of the molecules methane CH_4 and ammonia NH_3. Then in 1951 G. Herzberg succeeded in detecting in the infrared some bands of the hydrogen molecule H_2, which are weak because of their small (quadrupole) transition probability. In 1974 traces of ethane C_2H_6 and acetylene C_2H_2 were discovered. From the occultation of a star by Jupiter's disk in 1971 it was possible to gain insight into the stratification of the

Figure 7.9. Jupiter. Equatorial radius 71 350 km and mass 1/1047 solar mass. Pic du Midi, 60 cm refractor (photograph by B. Lyot and H. Camichel).

atmosphere and to conclude from the scale height [equation (7.3b)] that hydrogen and helium are its principal constituents. The detailed analysis of all observations together with attempts at the theory of the internal structure of the planet showed that Jupiter consists of practically unaltered solar material, with relative abundances (by numbers of atoms)

$$H : He : C : N = 1 : 0.1 : 4 \times 10^{-4} : 7 \times 10^{-5}. \qquad (7.12)$$

The radiation temperature of ~120 K in the microwave region $\lambda < 1$ cm may correspond to the tropopause; in the infrared it is possible to see down to layers at ~225 K. In this region visibility is hindered more and more by cirrus clouds, which may consist of NH_3 crystals. Various models of the core of the planet, still rather provisional, give densities of ~4 g/cm³ and temperatures of ~7500 K.

As we have said, the radio frequency radiation from Jupiter is thermal up to ~1 cm; in the decimeter region its intensity increases steeply and shows from its partial polarization that we are dealing with nonthermal synchrotron radiation produced in Jupiter's magnetic field. At meter wavelengths the radiation comes as intense bursts from sharply localized sources on the disk.

On December 4, 1973, the space probe Pioneer 10, after a flight of 21 months, approached Jupiter within a distance of 130 000 km (≈2 Jovian radii). Even more exciting than the fine pictures is the confirmation of a dipole-like magnetic field of ~4 G, whose axis is inclined at an angle of ~15° to the planet's rotation axis. In the Jovian magnetosphere are trapped enormous numbers of high energy electrons and protons, like the Van Allen Belts of the Earth, and also a thermal plasma. At greater heights are observed in the Lyman region the familiar emission lines of hydrogen and helium, a kind of air glow. Infrared measurements between 20 and 40 μm showed that the thermal emission of the planet exceeds the incident solar radiation by a factor of 2 to 2.5. The origin of this energy is not yet clear.

In 1610 Galileo discovered the four brightest satellites I–IV; with larger and larger telescopes, up to the present fourteen have been discovered. The innermost (Barnard, 1892) has an orbital radius only 2.54 times the equatorial radius of its planet, scarcely greater than the Roche stability limit [equation (7.10)]. The circular orbits ($e < 0.01$) of these five innermost satellites lie almost in the equatorial plane of the planet; the outer satellites by contrast have larger eccentricities (0.13 to 0.38) and inclinations. They are probably captured asteroids. The radii and masses of the four Galilean satellites seem fairly certain to be as follows:

Satellite	I	II	III	IV	
	Io	Europa	Ganymede	Callisto	
Radius	1.75	1.55	2.77	2.50	$(\pm 0.07) \times 10^3$ km
$M_{satellite}/M_{planet}$	3.8	2.5	8.2	5.1	$\times 10^{-5}$
Mean density	3.2	3.1	1.7	1.5	g/cm³

For such satellites, for example IV = Callisto, J. S. Lewis (1971) has proposed a model, which like Jupiter itself is composed of solar material: a dense core of hydrous silicates and iron hydroxides is surrounded by a mantle of an aqueous solution of NH_3 (liquid ammonia) and a thin crust of ices. The temperature in the core rises to 400 to 800 K as a result of radioactive heating. The radii of the remaining satellites are estimated from their brightnesses to be in the range 3 to 70 km.

It was discovered during the flyby of Pioneer 10 that the satellite Jupiter I = Io possesses an ionosphere of its own at a height of ~60–140 km, with a maximum electron density of $\sim 6 \times 10^4/\mathrm{cm}^3$. We must conclude from this that an atmosphere exists at the foot of which the pressure is about 10^{-5} to 10^{-7} mbar. A more precise value of 3.5 g/cm^3 was obtained for the density of this satellite.

7.4.2. Saturn ♄

Saturn (Figure 7.10) resembles Jupiter in many ways. In 1659, with a telescope which he had built himself, Christian Huygens discovered Saturn's *rings* (of which Galileo had observed some indications). Keeler's measurement (1895) of the speed of rotation of Saturn's rings from the Doppler effect in reflected sunlight showed that the various zones of the ring system revolve according to Kepler's third law and thus that they are composed of small particles. As Roche has shown (Section 7.2.4), a larger satellite in the position of the rings would be disrupted by tidal forces. The infrared spectrum shows that the rings consist at least partly of ice. The divisions (gaps) in the rings arise from resonance, that is, the orbital periods

Figure 7.10. Saturn. Equatorial radius 60 400 km and mass 1/3498 solar mass. Pic du Midi, 60 cm refractor (photograph by H. Camichel).

of particles in the gaps would be rational fractions of the periods of the inner satellites. The "gaps" in the asteroidal belt arise in analogous manner by resonance with the revolution of Jupiter. Titan, the brightest of Saturn's satellites and also discovered by Huygens, has a radius of 2425 ± 150 km and 2.41×10^{-4} Saturn masses. Hence the mean density is 2.3 g/cm^3.

The spectrum of Saturn is very similar to that of Jupiter. However it came as a surprise when G. P. Kuiper also found absorption bands of methane CH_4 in the spectrum of Titan and later L. Trafton discovered bands of hydrogen H_2. In fact, the estimates of Section 7.2.3 show that the gravitational field of the satellite is sufficient to retain an atmosphere. Altogether 10 satellites are now known, the smaller ones have radii of from 600 to 100 km. In Saturn's case too the orbital eccentricities and inclinations of the inner satellites, including Titan, are small, while those of the outer ones are considerably larger.

7.4.3. Uranus ⛢

Uranus was fortuitously discovered by W. Herschel in 1781. We know five large satellites and a ring of small ones (discovered in 1977). As with Neptune, particular features on the disk are scarcely recognizable.

7.4.4. Neptune ♆

From perturbations of the orbit of Uranus Adams and Leverrier inferred the existence of a planet with a longer period and calculated its orbit and ephemeris. In 1846 Galle found Neptune near the predicted position. Two satellites are known.

Uranus and Neptune resemble Saturn and Jupiter with regard to their spectrum and in many other features.

7.4.5. Pluto ♇

Perturbations of the orbits of Uranus and Neptune led to the conjecture of a transneptunian planet. In 1930 C. Tombaugh discovered Pluto at the Lowell Observatory as a star-like object of magnitude 14.9. The orbital elements, with eccentricity $e = 0.25$ and inclination $i = 17°$ fall completely out of the sequence of the other outer planets. Sometimes Pluto is within the orbit of Neptune. From perturbations of Neptune, R. L. Duncombe and others (1971) computed a mass of 0.11 Earth masses. If we estimate the radius (an upper limit) from its brightness as about 3200 km, we find a mean density $\bar{\rho} \approx 4.86$ g/cm^3. We are tempted to infer that Pluto was "captured" in some way.

8. Comets, Meteors and Meteorites, Interplanetary Dust; Structure and Composition

We begin with a short review of cometary orbits and some immediate inferences.

1. The orbits of long-period comets with periods of the order 10^2 to 10^6 years show randomly distributed inclinations i, direct and retrograde motions being about equally likely. The eccentricities e are a little less

Figure 8.1. Comet Mrkos, 1957 d. 1957 August 23. Mt. Wilson and Palomar Observatories 48-in Schmidt camera. Above is the elongated, highly structured Type I or plasma tail; below, the wider, almost structureless Type II or dust tail.

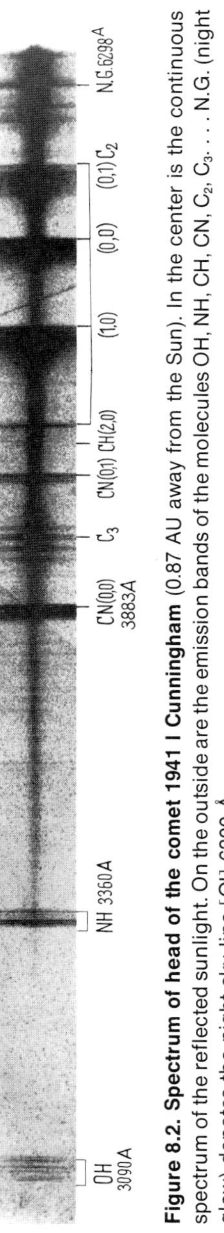

Figure 8.2. Spectrum of head of the comet 1941 I Cunningham (0.87 AU away from the Sun). In the center is the continuous spectrum of the reflected sunlight. On the outside are the emission bands of the molecules OH, NH, CH, CN, C_2, C_3. . . . N.G. (night glow) denotes the night sky line [OI] 6300 Å.

than or nearly equal to 1 so that we have to do with elongated ellipses, or parabolas as limiting cases. Hyperbolic orbits ($e > 1$) are produced only occasionally as derivative consequences of perturbations by the large planets. Since at a large distance from the Sun the velocity is very small, the comets must come from a cloud of such bodies that accompanies the Sun in its journey through the stellar system.

2. The short-period comets mostly move in direct elliptic orbits of small inclination $i \approx 15°$, whose aphelia lie near the path of one of the large planets. The cometary families of Jupiter, Saturn, . . . evidently result from the capture of long-period comets. Since in the course of time such comets are dissipated by breaking-up and evaporation of their material, the swarm of short-period comets must be continually replenished by future captures.

Photographs with suitable exposure-times (Figure 8.1) show that a comet in the first place possesses a *nucleus* (not always detectable) of only a few kilometers diameter. As a diffuse nebulous envelope around this, often in the form of parabolic shells, sometimes as rays emanating from the nucleus, there is then the *coma*. The nucleus and the coma together are called the *head* of the comet; its diameter is about 2×10^4 to 2×10^5 km. Somewhat inside the orbit of Mars, the comet develops the well known *tail,* which may have a visible length from 10^7 km up to occasionally even 1.5×10^8 km (=1 AU).

The spectrum of the *head* of a comet (Figure 8.2) shows partly sunlight whose intensity distribution indicates scattering by particles of the dimensions of the wavelength of visible light (about 0.6 μm = 6000 Å). In addition there are emission bands of the molecules, radicals and radical ions

$$CH, NH, OH, CN, C_2; NH_2, C_3; OH^+, CH^+ \qquad (8.1)$$

and, when the comet is near the Sun, the atomic spectral lines of

Na and occasionally Fe, Ni, Cr, Co, K, Ca II, [OI].

Besides the optically known OH, microwave observations in 1973 revealed the more complicated molecules of cyanogen HCN and methyl cyanide CH_3 CN. Spectra of cometary *tails,* corresponding to their lower density (making recombination difficult, see below), show chiefly the ions of radicals and molecules

$$N_2^+, CO^+, OH^+, CH^+, CO_2^+ \text{ and again CN.} \tag{8.2}$$

Satellite observations of the bright comets 1969 g and 1969 i showed that their heads were surrounded by a halo of atomic hydrogen, which fluoresces brightly in the Lyman-α line λ 1215. This hydrogen halo may under some circumstances reach a diameter of many millions of kilometers. The discovery of this hydrogen halo, together with the fact that all cometary molecules are combinations of the cosmically abundant light elements H, C, N, O suggests that comets were originally composed essentially of solar material.

The development of a comet and its spectrum can be pictured in the following way:

At a great distance from the Sun, only the nucleus is present; it contains smaller and larger fragments of stone and nickel-iron—like meteorites (see below)—mixed with compounds of the light elements mentioned above, especially hydrides, forming a sort of "ice" (F. Whipple), chemically comparable to the major planets. On approaching the Sun, substances like H_2O, NH_3, CH_4, etc., evaporate. These parent molecules, streaming away at about 1 km/s, when in the coma are further reduced by photochemical processes, and the molecules, etc., listed in equation (8.1) are excited to fluorescence by solar radiation. This is shown by the fact that gaps in the solar spectrum caused by the crowding together of Fraunhofer lines are observed to be reproduced also in the band spectra. In the cometary tail, the molecules and radicals are further ionized by the short-wavelength solar radiation; because of the low density, the rate of recombination of positive ions with electrons is small.

The characteristic forms and motions of cometary tails can be explained, as Bessel, Bredichin, and others long since realized, by the hypothesis of some force of repulsion emanating from the Sun, often much stronger than solar gravitation.

As shown by their spectra, the broad diffuse tails of so-called type II consist in the main of colloidal particles of size about the wavelength of light. Radiation pressure on such particles may in fact attain a value many times the force of gravity, as the observations require (every absorbed or scattered photon $h\nu$ imparts an impulse of order $h\nu/c$).

The narrow, elongated tails of so-called type I, on the other hand, as their spectra show, consist mainly of molecular ions like CO^+, etc. Here the computed radiation pressure no longer suffices to account for the large values of the ratio radiational acceleration: gravity. According to L. Biermann these *plasma tails* are to a greater extent blown away from the Sun by an ever-present corpuscular radiation, the so-called *solar wind* (see below). At the distance of the Earth, this consists of a stream of ionized hydrogen, that is, protons and electrons, with 1 to 10 such particles per cm^3 and with velocity about 400 km/s. Maybe in this way we may account for the often, but not always, observed influence of solar activity on comets. Investigations with the aid of space vehicles have already provided interesting information about the solar wind and in future may well make possible more definite assertions on the subject.

As has become apparent in recent times, the *meteors* or shooting stars represent only a selection of the aggregate of the small bodies in the solar system. We now distinguish rather sharply between a meteor, the short-lived luminous object in the sky—ranging from the telescopic shooting star to the fireball making things bright as day—and the parent body, the (small) meteoroid, or the (larger) meteorite. Since cosmic bodies reaching the vicinity of the Earth in almost parabolic orbits have velocity 42 km/s and since the velocity of the Earth in its orbit is 30 km/s, then according to the direction of approach they arrive with speeds between 12 and 72 km/s in the evening and morning, respectively. The bodies become heated on entering the Earth's atmosphere. In the case of larger portions of matter the heat cannot penetrate the interior sufficiently quickly, the surface becomes pitted by fusion or burns off, and these bodies reach the ground as *meteorites*. The largest known meteorite is "Hoba West" in Southwest Africa of about 50 tons. The impact of much more considerable masses must have produced meteoritic craters as on the Moon. For example, the famous crater of Canyon Diablo in Arizona has a maximum diameter of 1300 meters and at the present time a depth of 174 meters. According to geological indications, it must have come some 20 000 years ago from the impact of an iron meteorite of about 2 million tons. Its kinetic energy corresponds to about a 30-megaton hydrogen bomb! Recent investigations have tended to confirm that the Nördlinger Ries basin in southern Germany, with a diameter of about 25 km, is also a meteorite impact crater, formed 14.6 million years ago in the Upper Miocene (Tertiary). Small meteors burn away in the atmosphere, being the common shooting stars at a height about 100 km. In their flight through the atmosphere they ionize the air in a cylindrical volume. In the case of the great meteor streams they make a contribution to the ionosphere, the so-called anomalous E layer at about 100 km height. Also, such a conducting cylinder scatters electromagnetic waves as does a wire, predominantly at right angles to its own direction. The great strides made in meteor astronomy since the application of radar techniques by Hey, Lovell, and others depend upon this. On the radar screen one sees only the larger bodies themselves, but one easily obtains the direction normal to the ionized trail even for shooting stars that are below the limit of visual detection. One can also measure the velocities by radar methods. The outstanding advantage over visual observation is that these methods are independent of time of day and of cloudiness, so that the falsification of statistics from such causes is avoided. The most exact results for the brighter meteors at night are obtained from photographic observations, using wide-angle cameras of high light-gathering power, made as nearly as possible simultaneously at two stations suitably far apart. In this way one obtains the position of the path in space. Rotating sectors are used to interrupt the exposure and so to make it possible to compute the velocity in the path. The strength of the image shows the brightness and its often rapid fluctuations. Using an objective prism (see below) one can obtain spectra of the brightest meteors and so gain insight into the mechanism of their radiation and into their chemical composition.

Since the air resistance is proportional to the cross section $\sim(\text{diameter})^2$ while the weight is proportional to the mass $\sim(\text{diameter})^3$, one sees easily that for ever smaller particles the resistance so predominates that they are no longer made to glow, and they simply drift undamaged to the ground. These *micrometeorites* are smaller than a few micrometers (10^{-4} cm). Using suitable collecting devices, they are found in large quantities on the ground and also in deep-sea sediments; naturally there is difficulty in distinguishing them from terrestrial dust. In recent times, investigations with rockets (from which instruments are recovered) and satellites have yielded important new results.

As the only cosmic material immediately accessible to us, meteorites have been closely studied first by the methods of mineralogy and petrology

Figure 8.3. The Toluca iron meteorite (named after its place of discovery). The polished and etched cross section shows the *Widmannstetter figures.* These are formed by Fe-Ni plate crystals of Kamazite (7 percent Ni) and Taenite (with larger nickel content) which are juxtaposed parallel to the four surface pairs of an octahedron; such meteorites are called Octahedrites.

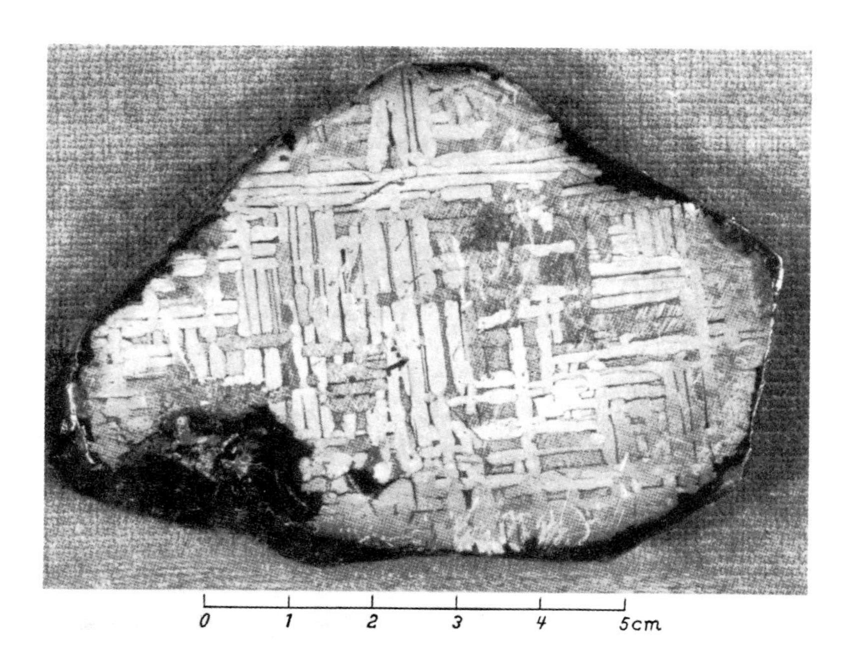

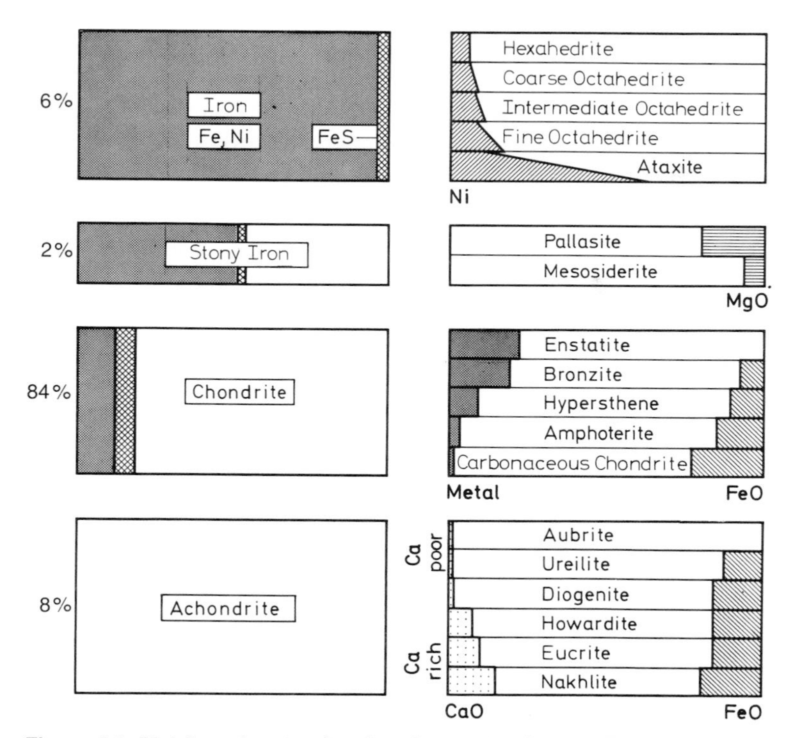

Figure 8.4. Division of meteorites into four main classes. These are divided (left) according to the ratio of metals (shaded) to silicates (white). The finer subdivision (right) is based on different chemical and structural criteria. On the extreme left is written the percentage of the falls which are of each type.

and more recently for the traces of radioactive elements and anomalous isotopes.

While it was originally believed that in the chemical analyses of many meteorites by V. M. Goldschmidt and the Noddacks we should have before us the *cosmical abundances* of the elements and their isotopes,[1] we now seek rather to use them along with the quantitative analysis of the Sun to obtain clues to the early history of the meteorites and of our planetary system.

In the first place, we distinguish between *iron meteorites* (density ≈ 7.8 g/cm^3), the nickel-iron crystals of which in their characteristic Widmannstetter etched figures (Figure 8.3) preclude confusion with terrestrial iron, and *stony meteorites* (density ≈ 3.4 g/cm^3). The latter we subdivide further into two subclasses: the common chondrites, characterized by millimeter-sized silicate spheres, the chondrules, and the rarer achondrites. Figure 8.4 shows the detailed classification. According to their chemical composition, the carbonaceous chondrites of Type I correspond essentially to unaltered

[1]A serious difficulty arises from the fact that iron meteorites predominate among the larger bodies, and stony meteorites among the smaller ones.

solar material (cf. Table 19.1); only the inert gases and other light, volatile elements are rare or absent. The matrix in which the chondrules of the carbonaceous chondrites are embedded contains many kinds of organic compounds, for example, amino acids and complicated ring systems. These are *not,* as has sometimes been supposed, of biological origin. One reason is that many amino acids are present which are not used by living organisms; furthermore, optically dextro- and laevorotatory amino acids occur equally often, while it is well known that living organisms possess only laevorotatory molecules. This matrix of the Type I carbonaceous chondrites can only have been formed at temperatures less than 350 K.

The formation of the chondrules (which are older), and even more the separation of metals (Fe, Ni, . . .) and silicates, presupposes complicated separation processes which we are barely beginning to understand. H. C. Urey (1952), and then J. W. Larimer and E. Anders have worked out the successive formation of different chemical compounds and minerals as a function of pressure and temperature. According to these calculations, the chondrites require formation temperatures of about 500 to 700 K. A high temperature fraction must have been produced at about 1300 K. The discovery of small diamonds in some meteorites caused much racking of brains. However, these may have arisen not under conditions of very high static pressure, as on the Earth, but under the action of shock waves during the collision of two meteorites in space.

The radioactive age determination yields for meteorites a maximum age that agrees with that of the Earth to within possible errors, namely 4.6×10^9 years. In addition, one can measure how long a meteorite in space has been exposed to irradiation by energetic cosmic-ray protons (assuming constant flux). By the splitting of heavy nuclei (spallation)—down to a certain depth of penetration in larger bodies—this produces all possible stable and radioactive isotopes, from the quantities of which one can calculate an *irradiation age*. The irradiation times for iron meteorites work out at some 10^8 to 10^9 years; for stony meteorites only about 10^6 to 4×10^7 years. The latter may well give the time since the body examined was broken off a larger body in a collision.

Parts of some meteorites contain trapped rare gases. Their isotope and abundance distributions suggest that, just like the rare gases in moon dust and the surfaces of lunar rocks, they stem from the solar wind. The investigations of the *xenon* contained in meteorites showed further in some cases an overabundance of those xenon isotopes which, as a more detailed study shows, can be traced back only to the relatively short-lived, and therefore extinct, radioactive isotopes of iodine and plutonium, ^{129}I and ^{244}Pu, with half lives of 1.2 and 8.2×10^7 years, respectively. The latter were probably produced by a short period of irradiation by neutrons, which has in many cases affected only parts of the meteorites, for example, the chondrules. We shall return to this mysterious event in connection with the origin of the solar system.

A great deal of puzzlement has been caused by the tektites (F. E. Suess, 1900). These are bodies made of silicate-rich glass (70–80 percent SiO_2), a few centimeters in size, whose shape, often roundish or resembling a spinning top, shows that they must have flown through the air at high speed in a molten state. Tektites are found only in certain regions, for example the moldavites in Bohemia, and so on. W. Gentner has shown that in several cases these groups can be connected with a neighboring meteorite crater of the same age; for example the moldavites with the Nördlinger Ries. The immediate objection, that in their flight from their place of origin the bodies would have evaporated long before reaching the spot where they were found, has been removed by the remark of E. David that, during the formation of an explosion crater whose diameter exceeds the height of the Earth's atmosphere, the latter would for a short time be completely "blown away" by the evaporating meteorite.

Following G. S. Hawkins and H. Fechtig, Figure 8.5 shows a conspectus of the flux densities of all "particles" present in the vicinity of the Earth,[2]

[2]We shall discuss the interplanetary plasma in connection with the physics of the Sun (Section 20).

Figure 8.5. Log *N*(*m*) as a function of *m*. *N*(*m*) is the number of meteorites, etc., with mass $>$ *m* g which in one second outside the Earth's atmosphere cross a surface of one square meter. The other abscissa scales give the diameter of the particles and their absolute brightness (reduced to a height of 100 km at the zenith) in magnitudes (Section 13).

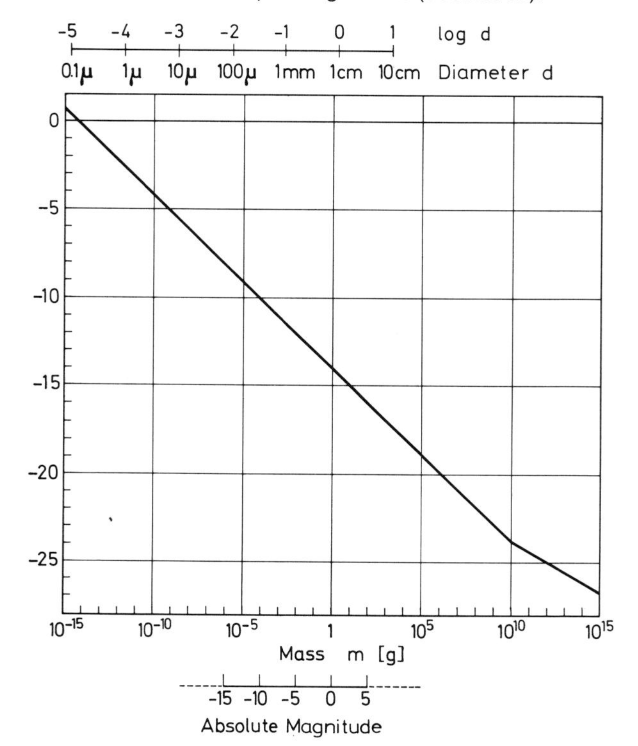

with masses from 10^{-16} g to 10^{15} g corresponding to diameters of order of magnitude from 10^{-5} cm to 10^{4} cm. In all, a mass of about 1.2×10^{9} g, predominantly in the form of micrometeorites, falls on to the Earth each day. Especially from radar measurements, we know from the orbits that a considerable proportion of meteorites is of cometary origin; another fraction, the *sporadic meteors,* move on randomly distributed ellipses having eccentricities almost equal to unity. Hyperbolic paths or velocities do *not* occur. The larger bodies more closely resemble asteroids; it should be added that we know very little about possible connections between comets and asteroids.

The *zodiacal light* arises from the reflection or scattering of sunlight by interplanetary dust. One can observe it as a cone-shaped illumination of the sky in the region of the zodiac, in the west shortly after sunset in the spring or in the east shortly before sunrise in the autumn. The weak *Gegenschein* can be observed directly opposite the Sun in the sky. During solar eclipses one observes near the Sun an extension of the zodiacal light produced by strong forward scattering (Tyndall scattering) by interplanetary dust. This is seen as an outer part of the solar corona. It is called the *F* or *Fraunhofer corona,* because its spectrum, like that of the zodiacal light, contains the dark Fraunhofer lines of the solar spectrum. In both cases the scattered light is partially polarized.

Interlude

9. Astronomical and Astrophysical Instruments

Great scientific advances are often bound up with the invention or introduction of new kinds of instruments. The telescope, the clock, the photographic plate, photometer, spectrograph, and finally the entire arsenal of modern electronics each signalizes an epoch of astronomical investigation. Equally important, however—as we must not forget—is the conception of new ideas and hypotheses for the analysis of the observations. Fruitful scientific achievements depend almost every time on the interplay of new sets of concepts and new instrumental developments, that can succeed in penetrating new domains of reality only in combination with each other. "Wonder en is gheen wonder" as we must say with Simon Stevin (1548–1620) ("A marvel is never a marvel").

The passage from classical astronomy to astrophysics—in so far as the distinction has any meaning—forms perhaps the most convenient place at which to start considering some astronomical and astrophysical instruments and techniques of measurement.

Figure 9.1 recalls the principles of Galileo's telescope (1609) and Kepler's (*Dioptrice,* 1611). In both the magnification is determined by the ratio of the focal distances of the objective and of the eyepiece. Galileo's arrangement gave an upright image and so became the prototype of opera glasses. Kepler's tube permitted the insertion of cross-wires in the common focal plane of the objective and eyepiece—so it was useful for the exact setting of angles, for example in the meridian circle. If we replace the cross wires by micrometer threads (Figure 9.1) we can measure visually the relative positions of double stars, the diameters of planetary disks, etc.

By the invention of *achromatic lenses* in 1758, J. Dollond and others overcame the troublesome color fringes (chromatic aberration) of a telescope with simple lenses. An achromatic convergent lens, for example a

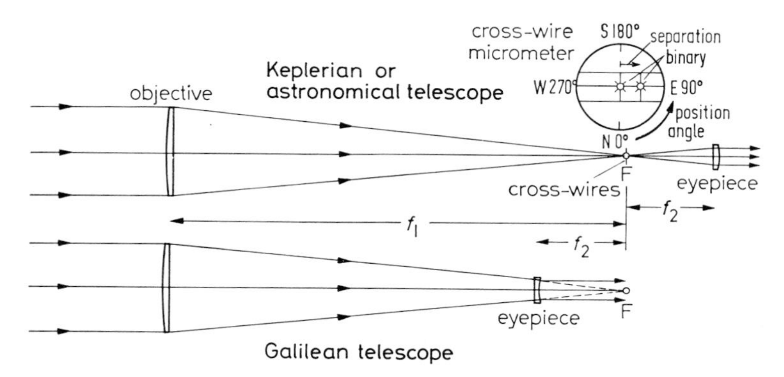

Figure 9.1. Keplerian and Galilean telescopes. The eyepiece is a convex lens in the former case and a concave lens in the latter. F denotes the common focus of objective and eyepiece. The magnification M is equal to the ratio of the focal lengths of the objective f_1 and eyepiece f_2; $M = f_1/f_2 = 5$ in the diagram. A Keplerian telescope can be used for binary measurements if a cross-wire micrometer is placed at the focus F. The curvature of the lenses is shown exaggerated.

telescopic objective, consists of a convex lens (convergent lens, positive focal length) made of crown glass, of which the dispersive power is relatively small in comparison with its refractive power, together with a concave lens (divergent lens, negative focal length) made of flint glass, of which the dispersive power is large in comparison with its refractive power. Precisely speaking, with a two-lens objective we can succeed in making the change of focal length with wavelength vanish, that is $df/d\lambda = 0$, only at a single wavelength λ_0. In the case of a visual objective we choose $\lambda_0 \approx 5290$ Å corresponding to the maximum sensitivity of the eye; in the case of a photographic objective we choose $\lambda_0 \approx 4250$ Å corresponding to the maximum sensitivity of the ordinary photographic (blue) plate.

We consider more exactly the image of a region of the sky made by a telescope on a photographic plate in the focal plane of the objective. (In the case of visual observation, the focal plane would be regarded as being enlarged by the eyepiece, as by a magnifying glass.) The operation of converting the plane wave coming from "infinity" into a convergent spherical wave is achieved by the lens by virture of the fact that in glass (refractive index $n > 1$) the light travels n times more slowly and so the light waves are n times shorter than in the vacuum. Consequently the wave surface of the light behind the objective is held back in the middle (Figure 9.2). This account, the mathematical formulation of which by means of the "eikonal" we owe to H. Bruns, W. R. Hamilton, and K. Schwarzschild, often very considerably simplifies the understanding of optical instruments compared with the direct application of Snell's law of refraction.

What the lens telescope or *refractor* achieves by the insertion of layers of different thickness with $n > 1$ in the optical path is done by the mirror telescope or *reflector* (Isaac Newton about 1670) by means of a concave mirror. This starts with the advantage that it can have no color error. As a

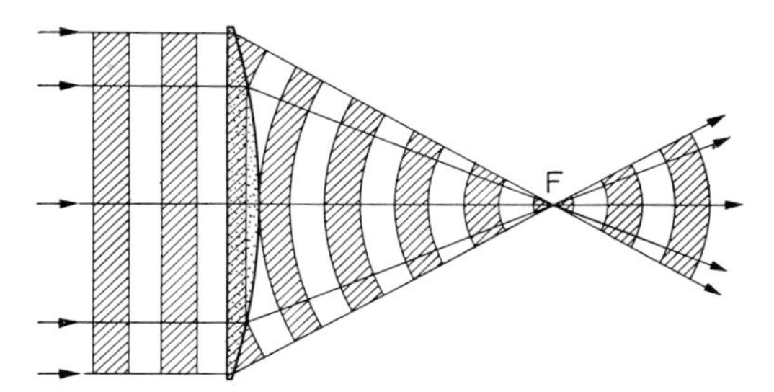

Figure 9.2. Formation of an image by a planoconvex lens. Plane wavesurfaces arrive from a star on the far left, the rays being perpendicular to these surfaces. The velocity of light is n times smaller (where n is the refractive index) in the lens, so that the wave surfaces are bent into spherical form. These spherical surfaces first converge toward and then diverge away from the focus F.

simple geometrical consideration shows, a spherical mirror [Figure 9.3(a)] focuses a parallel beam near the axis at a focal distance f equal to half the radius of curvature R. Rays further away from the axis after reflection meet the optical axis at a smaller distance from the apex of the mirror. The resulting image error is called *spherical aberration*. A *parabolic mirror* [Figure 9.3(b)] achieves the exact focusing of a beam parallel to the axis at a single focus. We see this at once, if we regard the paraboloid as the limiting case of an ellipsoid when one focus is removed to infinity. Unfortu-

Figure 9.3. (a) Spherical mirror. A bundle of rays close to the axis (see upper part of diagram) is brought to a focus F, whose distance from the axial point S of the mirror is equal to the focal length $f = R/2$, where R is the radius of curvature of the mirror. A bundle of rays further away from the axis (lower part of diagram) is brought to a focus nearer to S. This is known as spherical aberration. The plane waves incident from the right are reflected as convergent spherical waves. (**b**) **Parabolic mirror.** This brings all rays parallel to the axis to a single focus F, that is, an incident plane wave is reflected as a convergent spherical wave. This has the same curvature as the paraboloid at the axial point S of the mirror.

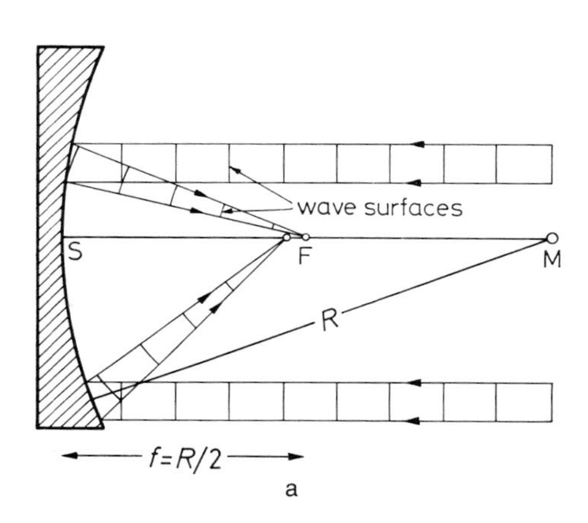

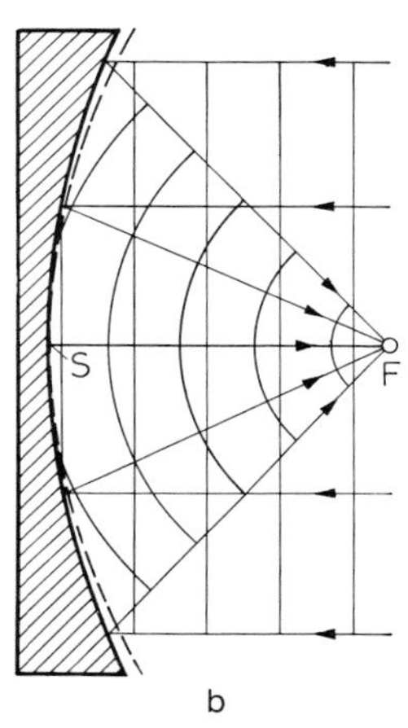

nately, however, a parabolic mirror gives a good image only in the immediate neighborhood of the optical axis. At larger aperture ratios, the usable diameter of the image field is very meager on account of the rapid increase outward of the image errors of an oblique beam. For example, the reflector of the Hamburg–Bergedorf Observatory with a mirror of diameter 1 m and focal length 3 m and so an aperture ratio $F/3$ has an image field of only 10′ to 15′ diameter.

The ingenious construction of the Schmidt camera (1930/31) satisfied the desire of astronomers for a telescope with a large image field *and* a large aperture ratio (light intensity). Bernhard Schmidt (1879 to 1935) first noticed that a spherical mirror of radius R focuses narrow parallel beams that reach the vicinity of the center of the sphere *from any direction* on to a concentric sphere of radius $\frac{1}{2}R$, corresponding to the known focal length $\frac{1}{2}R$ of the spherical mirror. With a small aperture ratio one can thus obtain a good image over a large angular region upon a curved film if one simply furnishes the spherical mirror with an entrance stop at the center of curvature, that is at twice the focal length from the mirror (Figure 9.4). If one wishes to attain high light intensity and opens the entrance stop wider, spherical aberration will make itself noticeable by the blurring of stellar images. B. Schmidt overcame this by introducing into the entrance aperture

Figure 9.4. Schmidt camera (after Bernhard Schmidt). This employs a spherical mirror with radius of curvature $R = MS$. The entrance diaphragm is located at the center of curvature M of the mirror. This ensures that all parallel bundles of rays centered by the diaphragm aperture (some of them appreciably inclined to the optical axis) are brought together under similar conditions on a spherical surface (the focal surface) centered on M and of radius $\frac{1}{2}R = MF$. The focal length is thus $f = FS = \frac{1}{2}R$. A thin aspherical correcting plate is placed in the entrance aperture in order to remove spherical aberration. (Dimensions shown for Mt. Wilson and Palomar 48-in Schmidt telescope.)

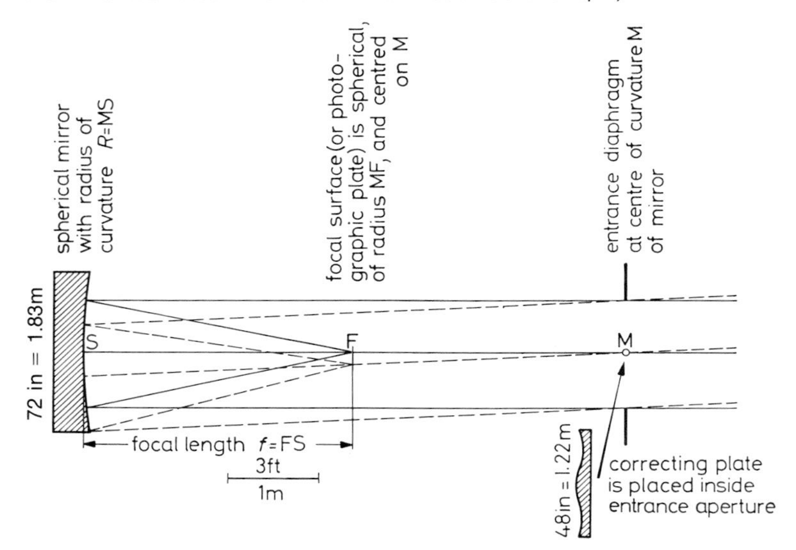

a thin aspherically figured corrector plate such that by means of appropriate glass thickness and a small displacement of the focal surface one compensates the optical path differences which correspond to the separation between the paraboloid and the spherical surface in Figure 9.3(b). On account of the smallness of these differences, this is possible for a large range of angles of entry and without chromatic errors.

The mounting of a telescope has the purpose of enabling it to follow the daily motion with the best attainable accuracy. Therefore such an equatorial or parallactic mounting has a *polar axis* parallel to the Earth's axis, which is driven by a sidereal clock, and perpendicular to this the *declination axis,* both having the appropriate graduated circles. The astronomer corrects small errors in the clockwork, refraction, etc., by guiding using electrically controlled fine adjustments.

Refractors, of which the aperture ratio is in the range $F/20$ to $F/10$, are mostly given the so-called Fraunhofer or "German" mounting as, for example, the largest instrument of this kind with a 1-m objective and 19.4-m focal length at the Yerkes Observatory of the University of Chicago (Figure 9.5).

Reflectors mostly have aperture ratio $F/5$ to $F/3$ and one employs either one of the various sorts of *fork mountings* (the declination axis passes through the center of gravity of the tube and is reduced to two pivots on the sides of the tube) or the *English* mounting in which the north and south ends of the polar axis rest on separate piers. The largest reflector at present[1]

[1] *Translator's note* (January 1977): A 6-m reflecting telescope has very recently been completed in the Soviet Union. It is located in the North Caucasus and is unique in having an alt-azimuth mounting.

Figure 9.5
Yerkes Observatory 40-in (1-m) refractor. Fraunhofer (German) mounting.

is the Hale telescope of the Mt. Wilson and Palomar Observatories with a main mirror of 200 in ≈ 5 m diameter and 16.8-m focal length (Figure 9.6). With many of the newer reflectors one can take photographs at the prime focus of the main mirror (Figure 9.7). However, one can also interpose a convex mirror and produce the image at the Cassegrain focus behind a hole drilled through the main mirror. With both arrangements one can also throw the image sideways out of the tube using a 45° plane mirror. Finally, by means of a complicated mirror system one can lead the light through a hollow polar axis to form the image of a star, for example, at the coudé focus on the slit of a large fixed spectrograph [Figure 9.7(c)].

Schmidt telescopes have, as a rule, aperture ratios *f*/3.5 to *F*/2.5 but they can go down to *F*/0.3. The Palomar 48-in Schmidt telescope (*F*/2.5) has a corrector plate of diameter 48 in = 122 cm; in order to avoid vignetting, the spherical mirror has to have a greater diameter of 183 cm.

Figure 9.6. Mount Wilson and Palomar Observatories 200-in (5-m) Hale reflector.

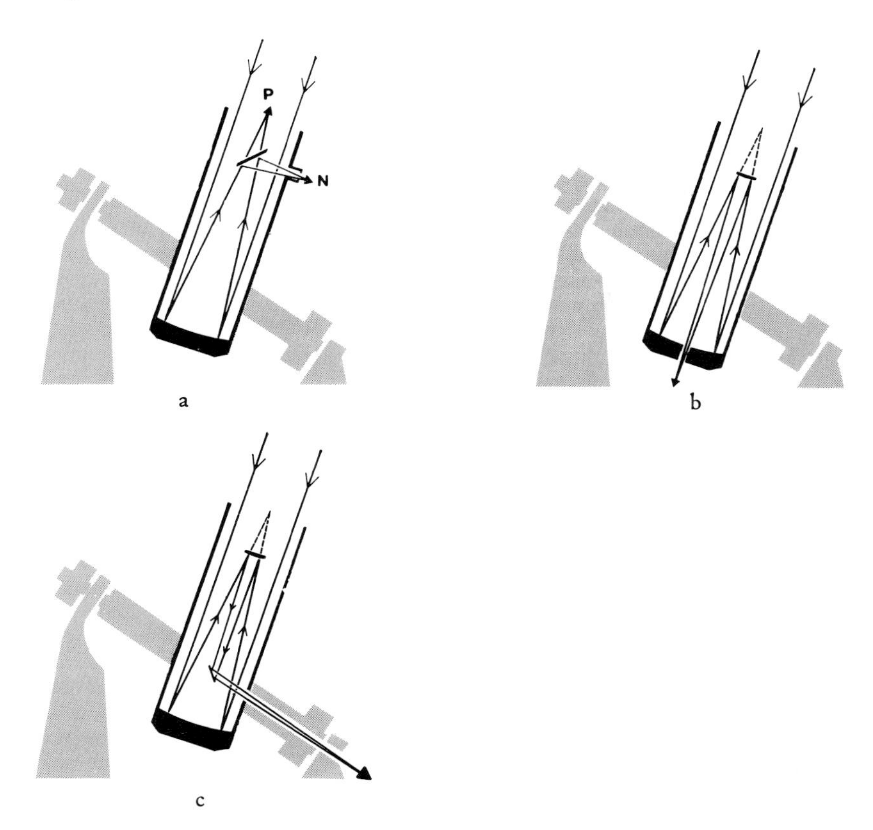

Figure 9.7 a–c. Mount Wilson and Palomar Observatories 100-in (2.5-m) Hooker reflector.

	focal length	focal ratio
(a) Newtonian or prime focus	42 ft	*F*/5
(b) Cassegrain focus	133 ft	*F*/16
(c) Coudé focus	250 ft	*F*/30

The famous *Sky Survey* was made with this instrument: some 900 fields 7° × 7° each with one plate in the blue and one in the red to limiting magnitudes 21^m and 20^m, respectively, cover the whole northern sky down to −32° declination. The Schmidt mirror of the Schwarzschild Observatory at Tautenburg in Thuringia is somewhat larger, with a corrector plate of 134 cm.

Among the special instruments of positional astronomy we must mention at least the *meridian circle* (O. Römer, 1704). The telescope can be moved in the meridian about an east–west axis. Using a pendulum clock of high precision or a quartz clock, the right ascension is determined by the time of passage of the star through the meridian at right angles to threads in the focal plane. The simultaneous determination of the culmination and so of the declination of the star is made possible by a horizontal thread in the field of view and the divided circle fixed to the axis. Modern determinations of position reach an accuracy of a few hundredths of a second of arc. On a divided circle of 1 m radius, 0.1 second of arc corresponds to half a micrometer!

We now seek to form an idea of the effectiveness of different telescopes for one task or another. The visual observer first asks about *magnification*. As we have said, this is simply the ratio of the focal lengths of the objective and eyepiece. The diffraction of light at the aperture of entry sets a limit to the perception of ever smaller objects. The smallest angular separation of two stars, of a double star for example, that one can discern is called the *resolving power*. A square aperture of side D (this being easier to discuss than a circular aperture) with parallel light as from a star forms a diffraction image that is bright in the center; on either side we first have darkness produced by interference where the illumination of both halves (Fresnel zones) cancels out. According to Figure 9.8 this corresponds to an angle in circular measure of λ/D. We approximate to a circle of radius R by a square that lies between the circumscribed square with $D = 2R$ and the inscribed square with $D = \sqrt{2}R = 1.41R$. Taking the mean, we obtain the radius in circular measure of the diffraction disk for an aperture of radius R as $\rho = 0.58\,\lambda/R$. A more exact calculation gives

$$\rho = 0.610\,\lambda/\mathrm{R}. \tag{9.1}$$

At this separation ρ the diffraction disks of two stars half overlap and are just separable. Thus equation (9.1) gives the resolving power of the telescope. Following astronomical usage, if we reckon ρ in seconds of arc and the aperture of the telescope $2R$ in inches (1 in = 2.54 cm), with $\lambda = 5290$ Å we obtain

$$\text{theoretical resolving power } \rho = 5''.2/2\mathrm{R}_{\mathrm{in}}. \tag{9.2}$$

For double-star observing Dawes found empirically $\rho = 4''.5/2R_{\mathrm{in}}$.

This limit is attained with first-class refractors in exceptionally good "seeing" conditions. With large reflectors the *thermal deformation* of the very sensitive mirror usually produces larger images. In photographs *scintillation* alone produces image diameters of order 0".5 to 3".

The theoretical resolving power of a telescope is determined by the interference of peripheral rays. A. A. Michelson achieved a somewhat greater resolving power by his *stellar interferometer* in which he placed

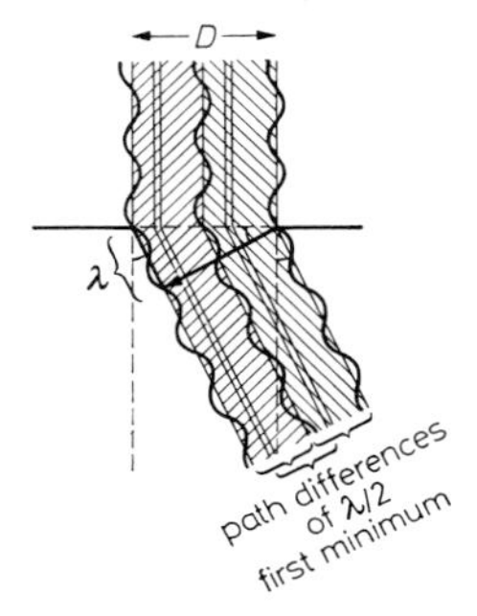

Figure 9.8
Diffraction of light at a slit or rectangular aperture of width *D*. Two equal bundles of rays a distance ½*D* apart and having a path difference of half a wavelength, that is, of ½λ will produce through *interference* a first diffraction minimum at angle λ/*D* with their original direction.

two slits at separation D in front of the object glass [Figure 9.9(a)]. Then a "point-source" star yields a system of interference fringes at angular separation

$$\rho = n\lambda/D, \qquad n = 0, 1, 2, \ldots \tag{9.3}$$

If one now observes a double star whose components have angular separation y in the direction of a line joining the two slits, the fringe systems of the two stars are superimposed. One has maximum fringe visibility if $y = n\lambda/D$. In between, the fringe visibility drops to zero if the component stars are equally bright, otherwise it passes through a minimum. Conversely, if the two slits in front of the object glass are slowly moved apart, we get

$$\text{maximum fringe visibility for } y = 0, \lambda/D, 2\lambda/D, \ldots$$

$$\text{minimum fringe visibility for } y = \frac{1}{2}\lambda/D, \frac{3}{2}\lambda/D, \ldots$$

If one observes a small disk of angular diameter y', a precise calculation shows that this is effectively equivalent to two point sources at separation y

Figure 9.9. (a) Stellar interferometer (after A. A. Michelson). A point-source star produces a system of interference fringes, whose distances $\rho = n\lambda/D$ from the optical axis are given by $n = 0, \pm 1, \pm 2, \ldots$ The fringe systems of two (equally bright) stars become superposed and produce uniform intensity, that is, *fringe visibility* zero at angular distances $y = \lambda/2D$, $3\lambda/2D \ldots$ **(b) 20-ft (6-m) interferometer (Mount Wilson and Palomar Observatories).** A steel beam over the aperture of the 100-in reflector carries two fixed inner mirrors U and V and two moveable outer mirrors A and B (all inclined at 45°). The separation AB, whose maximum value is 20 ft (6 m), corresponds to the distance D between the two slits in Figure 9.9(a). Observations are made visually at the Cassegrain focus E.

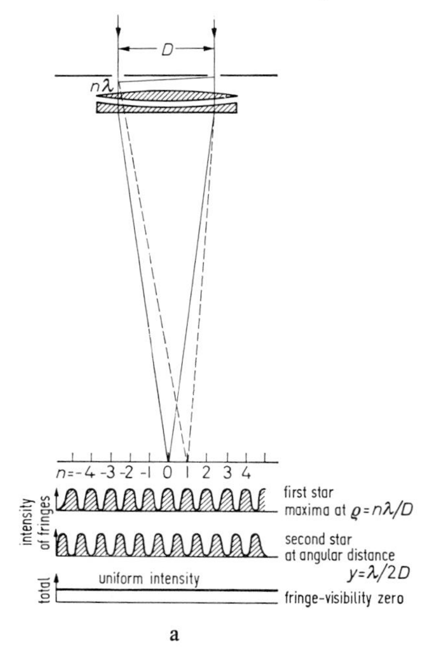

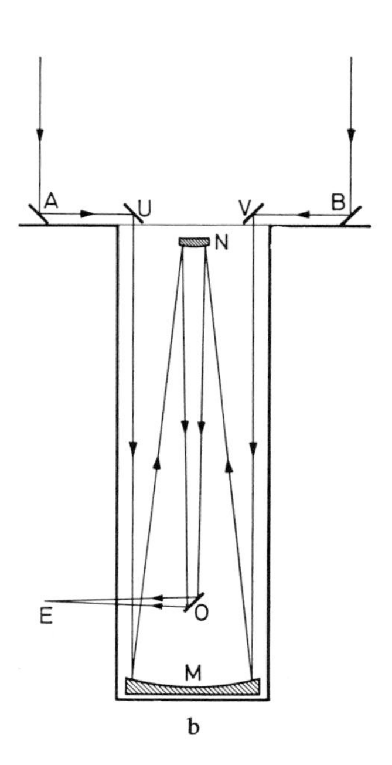

$= 0.41y'$, and one obtains the first visibility minimum at slit separation D_0 corresponding to $0.41y' = \lambda/2D_0$, or

$$y' = 1.22\lambda/D_0. \tag{9.4}$$

Since $D_0 \leqslant 2R$ in equation (9.1), it seems that little is gained as compared with the ordinary use of the telescope; actually, however, the judgment of fringe visibility is less impaired by poor seeing than is measurement with a thread micrometer. Thus Michelson and others were first able to measure the diameters of Jupiter's satellites, close double stars, etc.

Later, however, Michelson placed in front of the 100-in reflector a mirror system as in a stereotelescope so that he could make $D_0 > 2R$ [Figure 9.9(b)]. Thereby it was possible to measure directly the angular diameters of some red giant stars, the greatest being about $0''.04$.

In the Michelson interferometer both rays must be brought together correctly in phase. This difficulty, which makes it impossible to construct larger instruments, is overcome by the correlation interferometer of R. Hanbury Brown and R. Q. Twiss in the following way:—Two concave mirrors collect the light of a star on to one photomultiplier each. The correlation of the current fluctuations shown by the two photomultipliers for a given frequency interval is measured. As the theory shows, this correlation measure is connected with D and y or y' in exactly the same way as the visibility of the fringes in the Michelson interferometer. The measurements carried out in Australia by Hanbury Brown, Twiss and others make the new method appear very promising.

The effectiveness of an instrument for the photography of sources of weak surface brightness (for example, gaseous nebulae) depends, as in a camera, first upon the aperture ratio and second on absorption and reflection losses in the optics. In both respects the Schmidt mirror and its variants are unrivaled.

The question as to how faint a star can be photographed is much more complicated. The limiting brightness for a telescope is obviously determined by the requirement that the small stellar disk, affected by scintillation, diffraction, plate-grain, etc., should be in recognizable contrast to the plate background (plate haze), which is affected by night-sky illumination and other disturbances. It follows that one reaches faint stars first with larger aperture; then with given aperture one goes further with small aperture ratio, that is, long focal length. In practice one must take account also of the exposure times.

If we seek to delimit the domains of competence of telescopes using lenses or mirrors relative to each other, apart from the difference in aperture ratio, we must remember that for refractors the quality of the image is less affected by temperature fluctuations, and the theoretical resolving power is closely approached, while a mirror is highly temperature sensitive. Therefore, the refractor is primarily appropriate for visual binary stars, planetary surfaces, trigonometric parallaxes, . . . , the reflector for spectroscopy, direct photography at any rate of faint objects. . . . Photoelectric photometry and other tasks may be undertaken with both types of instrument.

The auxiliary equipment for registering and measuring the radiation of stars, nebulae, etc., is just as important as the telescope.

Today *visual observations* play a part only where we have to do with the rapid discernment of small angles or small details near to the limit permitted by scintillation, thus for the observation of visual binaries, planetary surfaces, solar granulation, etc.

The *photographic plate* is still one of the most important tools of the astronomer. Figure 9.10 shows the ranges of sensitivity of certain frequently used Kodak plates. The most highly sensitive blue plates *O* are those used most for astronomical photography. *D* and *F* are sensitized emulsions; in combination with suitable color filters they serve for photography in the visual and red parts of the spectrum. *N* and *Z* plates for infrared photography have to be hypersensitized with ammonia.

The relation between light intensity and plate darkening can only be determined individually and empirically for each plate. Photographic photometry must therefore operate so far as possible differentially. Either one measures the deflection of a microphotometer as a measure of the blackening at the center of the star image, or using an iris photometer one closes an iris diaphragm around the small stellar image (in particular that obtained with a Schmidt camera) until the deflection of the photometer takes a suitable preselected value. In both cases the interpretation of the deflections must be carried out in accordance with an otherwise determined brightness scale. This can be obtained, for example, using two exposures on the same plate, one of which is made with a measured reduction ratio by means of a neutral filter ("half-filter method"). Or else one places in front

Figure 9.10. Spectral sensitivity ranges of Kodak plates with varying sensitization. The solid black areas are the spectral regions for which the sensitization class considered is especially valuable. The striped areas give the total range of sensitivity. Wavelengths are given in nm.

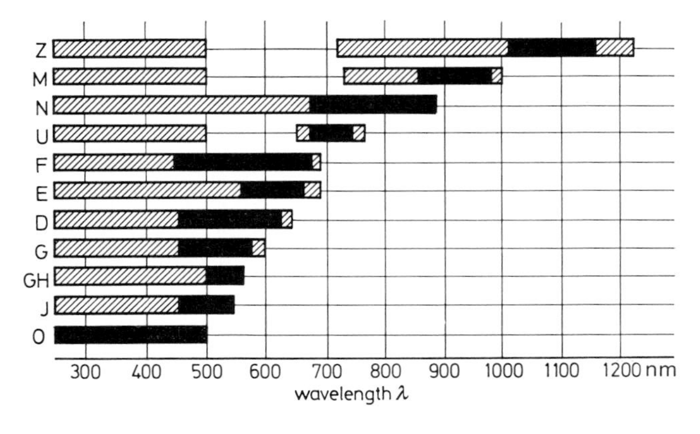

of the instrument a coarse grating (made of metal rods or the like) that produces on either side of the star diffraction images with a known reduction ratio. In recent times these methods have been almost superseded by the more exact and simpler photoelectric measurement of magnitude sequences (Section 13).

Photoelectric photometry employs as a measuring device either photocells or photomultipliers in combination with appropriate electronic amplifiers and registering galvanometers, and at the present time also with automated electronic data reduction. The spectral sensitivity can be adjusted by choice of a suitable photocathode for measurements from the ultraviolet (at the limit of transparency of our atmosphere $\lambda \sim 3000$ Å) down to the infrared ($\lambda \sim 12\ 000$ Å). At somewhat shorter wavelengths there starts the region of use of lead-sulphide cells and similar semiconducting systems. In astronomy, if one speaks of measuring the total radiation with a thermoelement or a bolometer one must remember that for terrestrial observations the atmosphere and maybe also the telescope completely absorb considerable parts of the spectrum.

It is important to be clear as to the relative advantages and disadvantages of photographic and photoelectric photometry and as to their effectual cooperation. Photoelectric measuring devices are of higher constancy and can be used easily and precisely along with laboratory methods over great ranges of brightness, while a single photographic plate covers only a very restricted range of brightness, about 1 : 20. The instrumental errors of a photoelectric measurement are about ten times smaller than those of photographic photometry. On the other hand, a single photographic exposure includes an incredible number of stars (for example, in clusters) while the observer working photoelectrically must set and measure each single star for itself. By and large, the present division of tasks is therefore:

Photoelectric photometry: brightness scales, exact light curves of individual variable stars, exact color indices (see below.)

Photographic photometry: photometry of larger star fields (clusters, Milky Way, . . .), surveys of certain kinds of stars, etc., connected with photoelectrically measured standard sequences of magnitudes.

The further analysis of cosmic light sources is taken over by *spectroscopy*. Here we treat first its instrumental side while we somewhat artificially defer consideration of its fundamental concepts and its applications.

The familiar laboratory prism—or grating—spectrograph can be fixed to a telescope, the only purpose of which is to throw the radiation of a cosmic source, such as a star, through the slit of the spectrograph. Thence it follows: (1) The aperture ratio of the collimator must be equal to that of the telescope, or at any rate not smaller, for the illumination of the prism or the grating. (2) In good seeing the slit must admit most of the star image. (3) Since an infinitely sharp spectral line would give an image of the slit on the plate (diffraction effects being generally negligible in stellar spectrographs), one arranges for the sake of economy that the image of the slit on the plate should be matched to the resolving power of the photographic emulsion (~0.05 mm). If one wishes to secure spectra of stars of a certain limiting magnitude within a given exposure time (times exceeding 5 hours being inconvenient in practice) using a telescope of given size, then the focal length of the camera and the dispersion (Å per mm) are determined. Finally it is a great advantage if the image field of the spectrograph camera is so little curved that one can work with *plates* (bent if necessary): the subsequent measuring of films brings with it many technical inconveniences.

Nowadays one uses almost only *grating spectrographs* (Figure 9.11) since it is possible so to shape the grooves in the diffraction grating that a particular angle of reflection ("blaze" angle) is strongly favored. One chooses the collimator focal length to be as large as the dimensions of the available grating permit. One constructs the camera on the principle of the Schmidt camera, the advantages of which we already know—large field of view, small absorption and reflection losses, small curvature of the image field, small chromatic errors. Figure 9.11(c) shows the large fixed coudé spectrograph (T. Dunham, Jr.) at the 100-in Hooker reflector of the Mt. Wilson Observatory. The collimator lies in the direction of the polar axis of the telescope; its aperture ratio is matched to the large focal length of the coudé system. After reflection at the grating the light can be led at choice into Schmidt cameras of focal lengths 8, 16, 32, 73, and 114 inches. For example, the most-used 32-in camera gives in the second order in the photographic region (3200–4900 Å) a dispersion of about 10 Å/mm; with this combination one can readily reach stars of magnitude 7 (see below).

Important auxiliary devices are: (1) An iron arc, that is, an electric arc between electrodes made of iron, the spectrum of which appearing above and below that of the star provides standards for the measurement of wavelengths and in particular of Doppler effects. (2) An auxiliary light entrance with a stepped slit provides continuous spectra of a filament lamp with precisely known intensity ratios. This makes it possible to assign the blackening curve of the plate for each wavelength, that is, in effect the relation between the microphotometer reading and the intensity at a given wavelength, and thus to evaluate the spectrum photometrically.

Fast spectrographs for the study of faint nebulae, stars, etc., with low dispersion are operated at the prime focus of the main mirror.

The *objective prism* is used for spectral surveys of whole starfields, that is, a prism is placed in front of the telescope in the position of minimum deviation, and so a spectrum of each star is obtained on a plate in the focal plane. In this way, for example, E. C. Pickering and A. Cannon at the Harvard Observatory produced the *Henry Draper Catalogue,* which along with position and magnitude contains the spectral type of something like a quarter of a million stars. In recent times the combination of a Schmidt mirror with a thin prism, for example at Warner and Swasey and Tonantzintla Observatories (dispersion ≈ 320 Å/mm), has proved to be very productive. The objective prism does not permit the actual measurement of

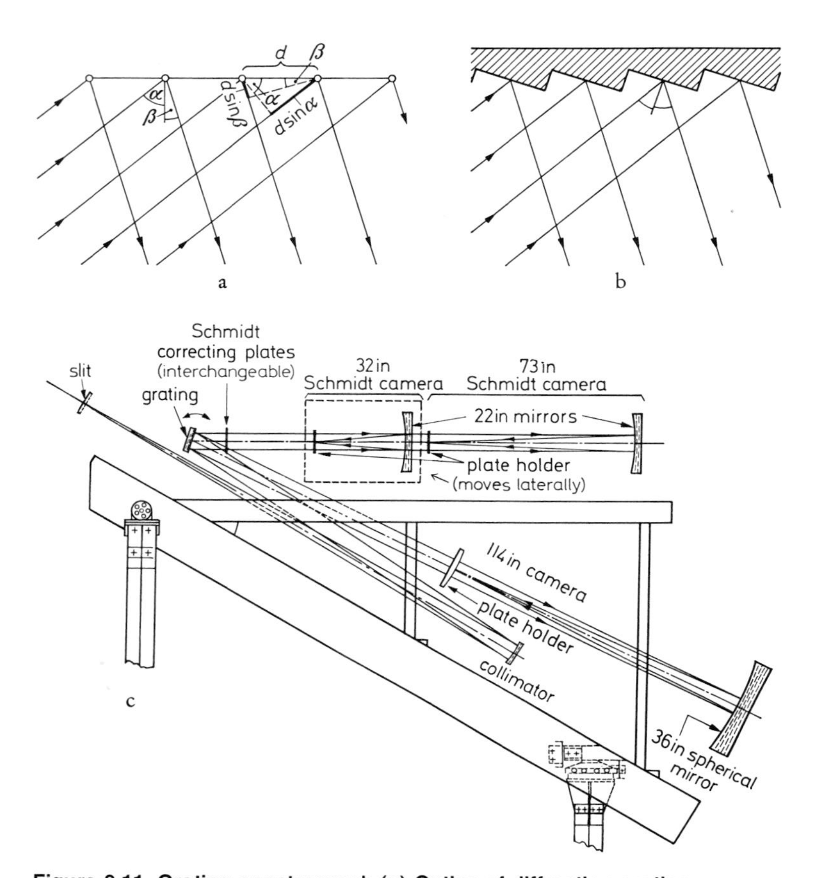

Figure 9.11. Grating spectrograph (a) Optics of diffraction grating. Interference maxima (that is, spectral lines) occur where the path difference of neighboring rays is $d(\sin\alpha - \sin\beta) = n\lambda (n = \pm 1, \pm 2, \ldots)$ where d is the grating constant. The dispersion (Å/mm) and resolving power are (other things being equal) proportional to the order n of the spectrum. **(b) Blazing.** By coating the grating with, for example, aluminium it is possible to produce specular reflection from suitably cut rulings or steps of the grating for certain angles of incidence and emergence so as to concentrate most of the light into the spectrum of one particular order. This leads to greater intensity of spectra. **(c) Coudé spectrograph at Mount Wilson.** The coudé optical system of the 100-in telescope forms the stellar image on the slit of the spectrograph. The collimator mirror forms the incident light into a parallel beam and directs this on the grating. The spectrally decomposed light is now ready to be formed into an image by one of the interchangeable Schmidt cameras (focal lengths 32, 73, 114 in; also 8- and 16-in cameras—not shown). The grating acts as the entrance stop of the camera. Correcting plates are only necessary for the short-focus cameras with greater focal ratio.

wavelengths or radial velocities. This is made possible by the *Fehrenbach prism*. It is a direct-view prism so arranged that light of a mean wavelength of the spectrum goes through as if it were a plane plate. One then makes two exposures one after the other on the same photographic plate, the prism being rotated through 180° about the optical axis between the exposures. For each star one thus obtains two spectra running in opposite senses, the mutual displacement of which makes possible the measurement of the Doppler effect.

The wish to photograph ever more distant galaxies and other faint objects, together with their spectra, with tolerable exposure times led to the development over the last 30 years by A. Lallemand and others of image converters or image tubes. The incident photons strike a thin photocathode in the focal plane and produce photoelectrons, whose energy is increased many times by an electric field. By means of electron optics the desired picture is obtained on a film with exposure times which are cut by factors of 20 to 50 by comparison with a photographic plate (which has poor quantum efficiency). Today therefore the further improvement of image tubes may be more rewarding than the questionable attempt to construct mirrors larger than 200 in. Figure 9.12 illustrates the Spectracon, which is particularly suitable for faint spectra. The SIT (Silicon Intensified Target) Vidicon, developed by J. Westphal (1973) at the Hale Observatories from the familiar television camera, also seems very promising. The data from each picture element are stored on magnetic tape and so can be processed directly by a computer.

Figure 9.12. Spectracon image tube designed by J. D. McGee. A thin photocathode lies in the focal plane of the telescope or spectrograph (on the left). The photoelectrons are accelerated in the 28-cm long evacuated cylinder by a voltage of 40 kV and are focused by a homogeneous magnetic field of about 150 G onto the Lenard window (right), a mica sheet about 5 μm thick and 1×3 cm across. They penetrate this and produce the picture on a film (right) pressed against the window by a roller.

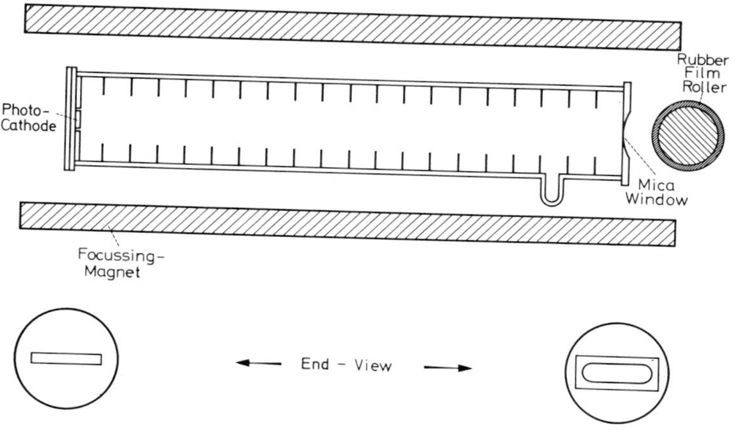

Radio astronomy opened up entirely new possibilities for observation in the region of

Wavelength λ	1 mm	10 cm	10 m	300 m
Frequency $\nu = c/\lambda$	3×10^{11}	3×10^{9}	3×10^{7}	$10^{6}\ s^{-1}$ or Hz
	300 GHz	3 GHz	30 MHz	1 MHz.

The region is bounded on the short wavelength side at about $\lambda \approx 1$ to 5 mm by absorption particularly that of atmospheric oxygen, and on the long wavelength side at about $\lambda \approx 50$m by reflection at the ionosphere.

K. G. Jansky in 1931 discovered the radio emission of the Milky Way in the meter-wavelength region. Using the subsequently improved receivers of radar devices, J. S. Hey and J. Southworth during World War II (c. 1942) discovered the radio emission of the disturbed and of the quiet Sun. After H. C. van de Hulst had predicted it, various investigators in Holland, U.S.A., and Australia discovered almost simultaneously in 1951 the 21-cm line of interstellar hydrogen. The Doppler effect of this line opened up enormous possibilities for investigating the motion of interstellar matter in our Milky Way system and in other cosmic structures. Here we shall not yet pursue the well-nigh explosive development of radio astronomy but only summarize its most important types of instrument:

(1) The *radio telescope* (Figure 9.13) with a parabolic mirror made of sheet metal or wire netting (mesh $\leqslant \lambda/5$). At the focus the radiation is received by a dipole (of which the side away from the mirror is shielded by a reflector dipole or a plate, etc.), a horn, or the like. The high-frequency energy is amplified and rectified; finally its intensity is indicated by a registering device, or it is digitized for automatic data processing. Even for the largest paraboloid (at the present time, the 100-m telescope of the Max Planck Institute for Radioastronomy, Bonn), according to equation (9.1) which is valid also here, the resolving power is small compared with that of Galileo's first telescope!

(2) In the procedure of M. Ryle (Cambridge), high resolving power is attained with a *radio interferometer* that corresponds exactly to Michelson's stellar interferometer; signals in correct phase from two radio telescopes are brought together and further amplified. Also the principles of the linear *diffraction grating* and of the two-dimensional *cross grating,* here using a fixed wavelength, have been successfully taken over into antenna techniques in order to achieve high angular resolution. In its optical counterpart, we have already become acquainted with the principle of the *correlation interferometer,* which was first applied by R. Hanbury Brown and others in radio astronomy. M. Ryle has further shown how, according to the principle of *aperture synthesis,* instead of the information obtained from one large instrument during one time interval, we can equally well use the information obtained successively from several smaller antennae in suitable prearranged positions. Then in the last few years the *very long baseline*

Figure 9.13. Australian National Radio Astronomy Observatory, 210-ft (64-m) radio telescope (Parkes, New South Wales).

interferometer was developed; here one uses two or more widely separated radio telescopes, for example Green Bank (U.S.A.) and Parkes (Australia), whose separation amounts to about 0.95 Earth diameters. These instruments independently record at exactly the same frequency the radiation (noise) from the same radio source as a function of time and also of effective baseline. The signals, recorded on tape, can then later be made to interfere if both recordings carry extremely precise time markers from two caesium atomic clocks. In this way it is possible to attain in the centimeter or decimeter range a resolution and positional accuracy of better than $0''.001$, about a thousand times better than optical measurements.

(3) Here we cannot concern ourselves with amplification techniques, although we must at least mention the enormous increase in precision of measurement using *masers* and *parametric amplifiers*. With such "low noise" amplification one can penetrate to significantly weaker radio sources and so, other things being equal, to greater distances in space than by the sole use of conventional amplifiers.

Observation from outside the atmospheric envelope of our Earth with *rockets* (the pay-load of which can return to the Earth) or *satellites* and *space vehicles* (the results of whose measurements are transmitted by telemetry) opens to astrophysics, and particularly to the study of the Sun, the whole of the spectral region that is completely absorbed by the Earth's atmosphere: the *short-wavelength ultraviolet* beyond the transparency limit of atmospheric ozone at λ 2850 Å, the adjoining *Lyman region,* where the main absorber is atmospheric oxygen O_2, then the *X rays* and finally the *γ rays*. While we can trace the solar spectrum continuously from radio wavelengths down to a few angstroms in the X-ray range, in galactic and extragalactic regions we must remember the Lyman continuum of interstellar hydrogen (Section 18), whose absorption sets in very strongly at 912 Å and allows us to "peep through" again only in the X-ray region below a few angstroms.

X-ray astronomy, with its completely new and unfamiliar instrumentation, has developed within a few years into one of the most interesting and promising branches of research.

The selection of more or less narrow wavelength regions is achieved with filters, taking advantage of the absorption edges, bounded sharply on the long wavelength side, of the elements used in the filter (cf. Sections 17 and 18). The Bragg crystal spectrometer has also been used with success.

X-ray photographs were first achieved for the Sun, using a primitive pinhole camera. Then collimators were built, behind which an extensive scintillation counter counted the X-ray photons (Figure 9.14). The precession of the rocket was used to scan successively larger regions of the sky. Only in 1964 did R. Giacconi revert to the *X-ray reflecting telescope* which had been invented in 1952 by H. Wolter in Kiel, primarily for X-ray microscopy. It is governed by the following principles. However well

Figure 9.14. Cellular collimator for observing cosmic X rays.
Resolution in the γ direction about 1°.5. A photon counter of the same area should be imagined under the collimator. The sky is scanned by the spinning of the rocket.

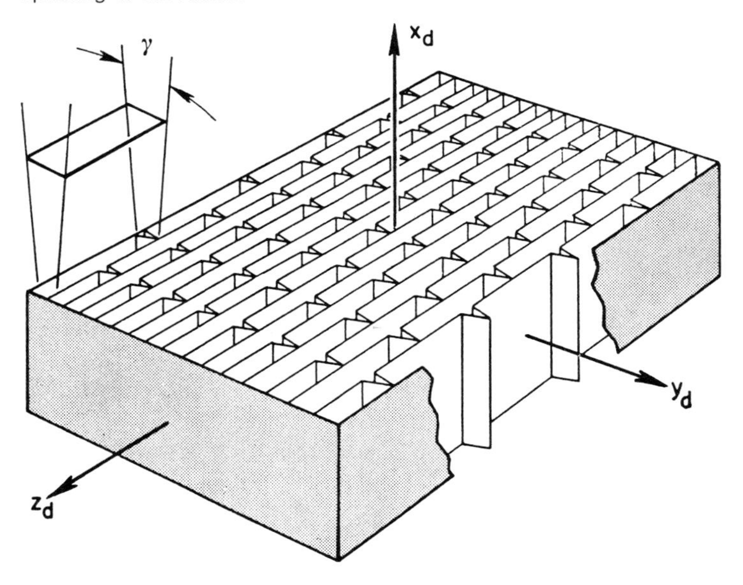

polished a metal surface may be, it will reflect X rays as if from a mirror only if they are at glancing incidence (angle between ray and surface less than a few degrees). One could therefore first try to form an image of a distant object by using a ring-shaped section of an elongated paraboloid, in which the unused part of the aperture was covered by a round metallic disk [Figure 9.15(a)]. However, such a telescope would give a very bad image because of the enormous distortion due to coma. This can be greatly reduced by following Wolter's design, in which after one reflection in the paraboloid there is a second reflection on an adjacent coaxial and confocal hyperboloid [Figure 9.15(b)].

We can here do no more than mention in passing the advancing techniques of *γ-ray astronomy,* which are still in their early stages of development.

In addition to photons (light quanta), streams of *high energy particles* are assuming ever greater significance in astrophysics. We have already

Figure 9.15. (a) The X-ray reflecting telescope of H. Wolter (1951). This consists first of a grazing incidence elongated paraboloid with a ring-shaped aperture. **(b)** The enormous distortion (coma) produced by this arrangement is made almost to vanish by a further reflection in a confocal and coaxial hyperboloid. Even better correction of the distortion can be achieved by somewhat more complicated axially symmetric mirror systems.

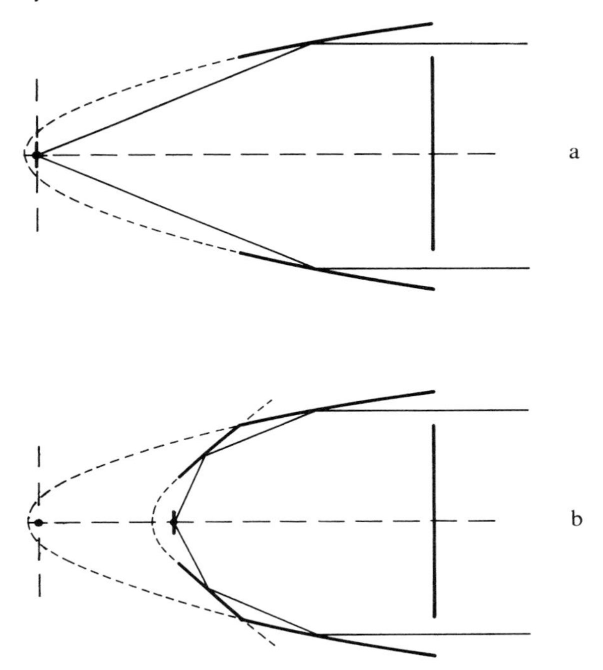

mentioned the solar wind, which shoots off into space solar material, that is, for the most part hydrogen and helium ions and their attendant electrons, with velocities of about 200 to 1000 km/s. Protons with (on average) speeds of 400 km/s have an energy of 0.8 keV. The region of higher energies, from a few times 10^8 to at least 10^{18}eV, is occupied by the *cosmic rays*. For several years, people have been trying to count *neutrinos* which arise as a byproduct, so to speak, of the energy generation in the interior of the Sun. We will postpone discussion of this whole area of high energy astronomy until Section 29. As far as instrumental aids are concerned we must restrict ourselves to a few remarks.

However, we shall point out here a direction of development which, one may well say, is opening up a new era in astronomical measurement. With the help of the programmable electronic computer it has become possible to automate the handling of any amount of data, including reduction, tabulation, and evaluation of results. For example, machines have been built which can measure the *proper motions* of stars—by comparison of two plates of sufficiently different epoch—with about the same accuracy as an "old-fashioned" astronomer but with much greater speed and persistence. The GALAXY (General Automatic Luminosity And XY) machine developed by V. C. Reddish in Edinburgh (c. 1970) can measure the positions of 900 stellar images in an hour with an accuracy of $\pm 0.5 \mu$m, photometer the images, and so on. The computer can then search for all stars in a given area whose magnitudes, color indices, etc., lie in a particular range and print a complete catalogue of them. The fantastic "work-capacity" of such machines on the one hand opens up undreamed-of possibilities for research but on the other hand presents astronomers with a great responsibility.

II. Sun and Stars
Astrophysics of Individual Stars

10. Astronomy + Physics = Astrophysics

Historical introduction

In Section 1 we sought to give an introduction to classical astronomy with some historical remarks. In the same spirit we now turn to the astrophysical investigation of the Sun and stars. The latter we consider in the first place as individuals. Part III will deal with the internal constitution and evolution of stars and with stellar systems, galaxies, etc.

At the end of Section 1 we recalled the first measurements of trigonometric parallaxes by F. W. Bessel, F. G. W. Struve and T. Henderson in 1838. They denoted a final vindication of the Copernican world system, the correctness of which nobody doubted any more. Above all one had thereby gained a sure foundation for all cosmic distance determinations. Bessel's parallax of 61 Cygni $p = 0''.293$ asserted that this star had a *distance* of $1/p = 3.4$ pc or 11.1 light years. Thence we could, for example, compare the luminosity of this star directly with that of the Sun. However, trigonometric parallaxes first became an effective aid to astrophysics only when F. Schlesinger in 1903 with the Yerkes refractor and later at the Allegheny Observatory developed their photographic measurement to an incredible degree of precision ($\sim 0''.01$).

Double stars give information about the *masses* of the stars. Sir William Herschel's observations of Castor (1803) left no doubt that here two stars are moving round each other in elliptical orbits under the influence of their mutual attraction. As long ago as 1782 J. Goodricke had observed in Algol (β Persei) the first eclipsing variable. The work of H. N. Russell and H. Shapley (1912) exploited the many-sided information that these double stars have to offer. Pickering in 1889 discovered in the case of Mizar the first spectroscopic binary (motions measured by the Doppler effect).

After its beginnings in the eighteenth century (Bouguer 1729, Lambert 1760, and others), *stellar photometry* or the measurement of the apparent luminosities of the stars gained a secure basis about a hundred years ago. For one thing, N. Pogson in 1850 introduced the definition that 1^m (one magnitude) should denote a decrease of the logarithmic brightness by 0.400, that is, a luminosity ratio $10^{0.4} = 2.512$. For another thing J. C. F. Zöllner in 1861 constructed the first *visual stellar photometer* (with two Nicol prisms to give measurable light diminution) with which *stellar colors* were measured as well.

About the same time the great star catalogs of magnitudes and positions brought about an immense enlargement of our knowledge of the stellar system: In 1852/59 there appeared the *Bonner Durchmusterung* by F. Argelander and others with about 324 000 stars down to about $9^m.5$; later for the southern sky came the *Cordoba Durchmusterung*.

Karl Schwarzschild initiated *photographic photometry* with the *Göttinger Aktinometrie* 1904/08. He recognized at once that the color index = photographic minus visual magnitude forms a measure of the color and so of the temperature of a star. Soon after the invention of the photocell by Elster and Geitel 1911, H. Rosenberg and then P. Guthnick and J. Stebbins began to develop *photoelectric photometry*, the possibilities of which were greatly enlarged by the replacement of the filar electrometer by electronic amplifiers and registering galvanometers, and then also by the invention of the photomultiplier. Thus today one can measure stellar luminosities with a precision of a few thousandths of a magnitude in usefully selected wavelength intervals from the ultraviolet down to the infrared and derive the corresponding color indices. We must mention at least the six-color photometry of J. Stebbins and A. E. Whitford (1943) and the internationally adopted system of UBV-magnitudes (*U*ltraviolet, *B*lue, *V*isual) of H. L. Johnson and W. W. Morgan (1953).

The *spectroscopy of the Sun and stars* developed in parallel with stellar photometry. In 1814 J. Fraunhofer discovered the dark lines in the solar spectrum that are named after him. With extremely modest equipment, in 1823 he succeeded in seeing similar lines in the spectra of certain stars and he noted their differences. Astrophysics proper, that is, the investigation of the stars by physical methods, began when in 1859 G. Kirchhoff and R. Bunsen in Heidelberg discovered spectral analysis as well as the meaning of the Fraunhofer lines in the solar spectrum (Figure 10.1), and in 1860 Kirchhoff formulated the foundations of radiation theory, in particular Kirchhoff's law, which lays down the relation between the emission and absorption of radiation in thermodynamic equilibrium. This law together with the Doppler principle ($\Delta\lambda/\lambda = v/c$) formed for forty years the entire conceptual equipment of astrophysics. The spectroscopy of the Sun and stars first turned its attention to the following exercises:

(1) Photographing the spectra and measuring the wavelengths for all the elements *in the laboratory*. Identification of the lines in stars and other

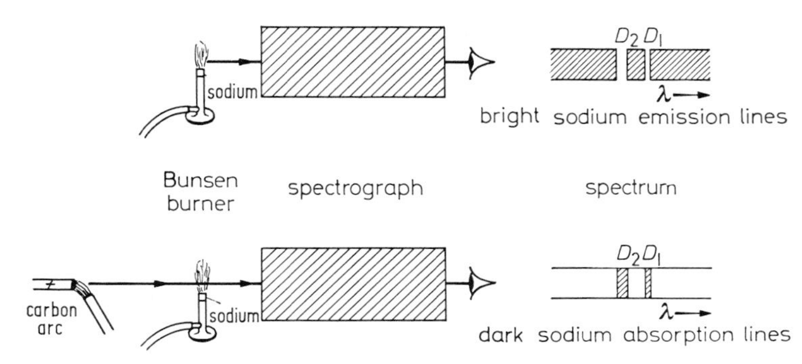

Figure 10.1. Illustrating the basic experiment in spectral analysis (after G. Kirchoff and R. Bunsen, 1859). A bunsen flame containing a trace of sodium (table salt) shows bright sodium emission lines in its spectrum. However, when light from the positive crater of a carbon arc (whose temperature is appreciably higher than that of the flame) is allowed to pass through the sodium flame, it produces a continuous spectrum crossed by dark sodium absorption lines, similar to those in the solar spectrum.

cosmic light sources (Sir William Huggins, F. E. Baxandall, N. Lockyer, H. Kayser, Charlotte E. Moore-Sitterly, and many others).

(2) Photography and ever more exact measurement of the spectra of stars (H. Draper 1872; H. C. Vogel and J. Scheiner 1890, and others) and of the Sun (H. A. Rowland, production of good diffraction gratings, photographic map of the normal solar spectrum 1888, Preliminary Table of Solar Spectrum Wavelengths 1898 with about 23 000 lines).

(3) Classification of stellar spectra, first in a one-dimensional sequence, essentially according to decreasing temperature. After preparatory work by Huggins, Secchi, Vogel, etc., E. C. Pickering with A. Cannon from 1885 onward produced the Harvard classification and the Henry Draper Catalogue. Later advances brought: The discovery of luminosity as a second parameter for the classification and thence the determination of *spectroscopic parallaxes* by A. Kohlschütter and W. S. Adams in 1914, and much later with a modern point of view the "Atlas of Stellar Spectra" (1943) by W. W. Morgan, P. C. Keenan, and E. Kellman giving the MKK classification.

(4) For a long time the measurement of *radial velocities* of stars, etc., using the Doppler principle, claimed the predominant attention of astronomers. After visual attempts by W. Huggins in 1867, H. C. Vogel in 1888 succeeded in making the first useful photographic measurements of radial velocities, the rotation of the Sun having been verified spectroscopically shortly before that. W. W. Campbell (1862–1938) at the Lick Observatory, in particular, contributed to the furtherance of the technique of radial-velocity measurement. We shall discuss the motions and dynamics of stars and stellar systems in Part III.

The discovery in 1913 of the *Hertzsprung–Russell diagram* was the culmination of this epoch in astrophysics. About 1905 E. Hertzsprung had already recognized the distinction between "giant" and "dwarf" stars. On the basis of improved measurements of trigonometric parallax, carried out partly by himself, H. N. Russell constructed the famous diagram, with spectral type as abscissa and absolute magnitude (see below) as ordinate, which showed that most stars in our neighbourhood fall in the narrow strip of the *main sequence* (Figure 15.2) while a smaller number occupy the region of the *giant stars*. Russell first associated with this diagram a theory of stellar evolution (start as red giants, compression and heating to reach the main sequence, cooling along the main sequence) which had, however, to be given up some ten years later. We shall be able to pursue such problems when we come to Part III.

What astrophysics most needed at the beginning of our century was an enlargement of its physical or conceptual basis. The theory of "cavity" radiation or, as it is called, "blackbody" radiation, that is, the radiation field in thermodynamic equilibrium, begun by G. Kirchhoff in 1860, had been brought to a conclusion by M. Planck in 1900 with the discovery of quantum theory and of the law of spectral energy distribution in blackbody radiation. Astronomers then set themselves to estimate the temperatures of the stars by the application of Planck's law to their continuous spectra. However, at the same time the gifted Karl Schwarzschild (1873–1916) erected one of the pillars of a future theory of the stars, the theory of the *stationary radiation field*. He showed in 1906 that in the Sun's photosphere (the layers that send out most of the radiation) the transport of energy outward from the interior is performed by *radiation*. He calculated the increase of temperature with (optical) depth on the assumption of radiative equilibrium and showed that one thereby derives the correct center-to-limb darkening of the solar disk. K. Schwarzschild's work on the solar eclipse of 1905 August 30 is a masterly interlocking of observation and theory. In 1914 he investigated theoretically and by spectrophotometric measurements the radiative exchange in the broad H and K lines (3933, 3968 Å) of the solar spectrum. He clearly saw that the further prosecution of his undertakings required an atomistic theory of absorption coefficients, that is, of the interaction of radiation and matter. So he turned with great enthusiasm to the quantum theory of atomic structure founded by N. Bohr in 1913. There then resulted the famous work on the quantum theory of the Stark effect and of band spectra. In 1916 Schwarzschild died all too soon at the age of 43 years.

The combination of the theory of radiative equilibrium with the new atomic physics was presented on the one hand by A. S. Eddington during 1916–1926 within the scope of his theory of the internal constitution of the stars (see Part III). On the other hand, with his theory of thermal ionization[1] and excitation M. N. Saha in 1920 created a point of departure for a

[1]This had been developed by J. Eggert in 1919 in regard to stellar interiors.

physical interpretation of the spectra of the Sun and stars. Thus stimulated, the whole of the fundamentals of the present day theory of stellar atmospheres and of solar and stellar spectra were soon developed. We must mention the work of R. H. Fowler, E. A. Milne, and C. H. Payne on ionization in stellar atmospheres (1922–25), the measurement of multiplet intensities by L. S. Ornstein, H. C. Burger, H. B. Dorgelo and their calculation by R. de L. Kronig, A. Sommerfeld, H. Hönl, H. N. Russell, followed by the important work of B. Lindblad, A. Pannekoek, M. Minnaert, and many others. By welding together the meanwhile further-developed theory of radiative energy transport and the quantum theory of line- and continuum-absorption coefficients, one could by 1927 begin in earnest to construct a rational theory of the spectra of the Sun and stars (M. Minnaert, O. Struve, A. Unsöld). This made it possible to infer from their spectra the *chemical composition* of the outer parts of the stars and so to study empirically the *evolution of the stars* in relation to the generation of energy by nuclear processes in stellar interiors.

With the discovery of the hydrogen convection zone, in 1930 A. Unsöld founded the theory of *convective streaming* in stellar atmospheres and especially in the solar atmosphere. A little later H. Siedentopf and L. Biermann put forward the connection with hydrodynamics (mixing-length theory), shortly after S. Rosseland had called attention to the astrophysical significance of turbulence.

We know that ionized gases—today we often speak of plasma—in stellar atmospheres and other cosmic structures possess high electrical conductivity. In the 1940s T. G. Cowling and H. Alfvén remarked that consequently cosmic magnetic fields could be dispersed by the ohmic dissipation of the related currents only in the course of very great time intervals. Magnetic fields and flow fields are in perpetual interaction; one must combine the fundamental equations of electrodynamics and of hydrodynamics to get those of magnetohydrodynamics (or hydromagnetics). In 1908 G. E. Hale, the founder of the Mt. Wilson Observatory, using the Zeeman effect in Fraunhofer lines, discovered magnetic fields in sunspots of up to about 4000 G (despite starting with physically incorrect hypotheses!). Using much more sensitive equipment, in 1952 H. W. Babcock was first able to measure the much weaker fields of a few gauss over the rest of the solar surface. Sunspots, the neighboring "plages faculaires" (faculae), flares (eruptions), prominences, and many other solar manifestations are statistically bound up with the 2×11.5-year cycle of *solar activity*. According to our current interpretation, all this—and it is what we least understand at present—belongs to the domain of *magnetohydrodynamics*. The study of streaming and of magnetic fields in the Sun also leads to some understanding of the heating of the corona, the outermost envelope of the Sun, to a temperature of 1 to 2 million degrees. Extremely complex processes occur here that lead to the production of the variable part of the radio emission of the Sun as well as the various kinds of *corpuscular radiation* and the *solar wind* already mentioned in connection with comets.

Again deferring consideration of the internal constitution and of the evolution of stars, we can briefly sum up the development of the physics of a single star with the help of the following key terms:

(1) Radiation theory, interaction of radiation and matter.
(2) Thermodynamics and hydrodynamics of streaming-processes.
(3) Magnetohydrodynamics and plasma physics.
(4) Cosmic corpuscular radiation of superthermal energy; astrophysics of cosmic rays, X rays and γ rays—this domain of investigation, scarcely less exciting than the rest, should rapidly gain momentum in the near future.

11. Radiation Theory

In regard both to the radiation fields within stellar atmospheres and stellar interiors and to the emitted radiation whose analysis gives us information about the structure and composition of the atmospheres, we have to do with the basic concepts of radiation theory.

In the radiation field to be studied—we make no special assumptions about it at this stage—we take a surface element $d\sigma$ with normal n and we consider the radiant energy flowing per unit time through $d\sigma$ within a small solid angle $d\omega$ about a direction defined by polar angles θ, φ, where θ is the inclination to n (Figure 11.1). In the spectral distribution of this energy we consider that within the frequency interval ν to $\nu + d\nu$ and write it as

$$dE = I_\nu(\theta,\varphi)\, d\nu \cos\theta\, d\sigma\, d\omega \tag{11.1}$$

the cross section of our radiation bundle being $\cos\theta\, d\sigma$.

Accordingly the *intensity* $I_\nu(\theta,\varphi)$ denotes the energy flow in unit time per unit frequency interval per unit solid angle about the direction θ,φ across unit area perpendicular to this direction. The usual units are: time, second; frequency, one cycle per second, Hertz (Hz); solid angle, steradian; area, square cm. Instead of frequency, we can refer the spectral distribution to wavelength (with 1 cm as unit). Since $\nu = c/\lambda$ we have $d\nu = -(c/\lambda^2)\, d\lambda$. So from the defining property $I_\nu\, d\nu = -I_\lambda\, d\lambda$ we have

$$I_\lambda = (c/\lambda^2) I_\nu \tag{11.2}$$

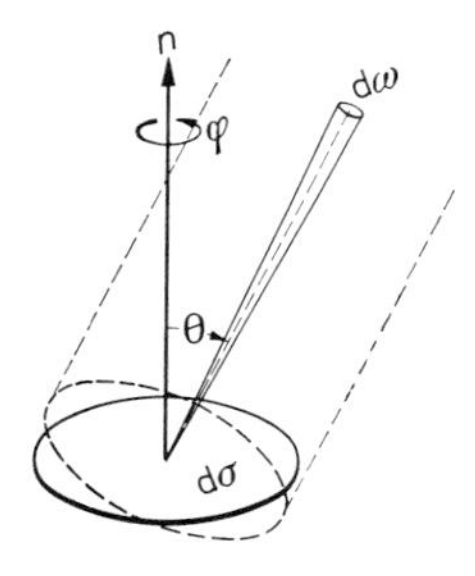

Figure 11.1
Definition of the intensity of radiation.

or in symmetric form

$$\nu I_\nu = \lambda I_\lambda.$$

The intensity of total radiation I is got by integrating over all frequencies or wavelengths

$$I = \int_0^\infty I_\nu \, d\nu = \int_0^\infty I_\lambda \, d\lambda. \tag{11.3}$$

As a simple application we calculate the energy dE that passes in unit time through the element $d\sigma$ and also through a second element $d\sigma'$ at distance r. The normals to $d\sigma$, $d\sigma'$ make angles θ, θ' (say) with the line joining $d\sigma$, $d\sigma'$ (Figure 11.2). The element $d\sigma'$ subtends at $d\sigma$ the solid angle $d\omega = \cos\theta' \, d\sigma'/r^2$. We have therefore

$$dE = I_\nu \, d\nu \cos\theta \, d\sigma \, d\omega \quad \text{or} \quad dE = I_\nu \, d\nu \frac{\cos\theta \, d\sigma \cos\theta' \, d\sigma'}{r^2}. \tag{11.4}$$

Also the element $d\sigma$ subtends at $d\sigma'$ the solid angle $d\omega' = \cos\theta \, d\sigma/r^2$. Therefore from equation (11.4) we can write

$$dE = I_\nu \, d\nu \cos\theta' \, d\sigma' \, d\omega'. \tag{11.5}$$

Thus the intensity I_ν along any ray of sunlight, for example, in accordance with the definition (11.1), has the same value in the immediate vicinity of the Sun and anywhere else outside it in space.

What in everyday language is rather vaguely called the "strength" of, say, sunlight corresponds more to the exact concept of *radiation flux*. We define the flux πF_ν in the direction n by writing the total energy of ν-radiation crossing the element $d\sigma$ in unit time as

$$\pi F_\nu \, d\sigma = \int_0^{2\pi} \int_0^{\pi} I_\nu(\theta,\varphi) \cos\theta \, d\sigma \sin\theta \, d\theta \, d\varphi \tag{11.6}$$

where from Figure 11.1 we have used $d\omega = \sin\theta \, d\theta \, d\varphi$. In an isotropic radiation field I_ν is independent of θ and φ and so $F_\nu = 0$. It is often useful to separate πF_ν into

$$\text{outward flux } (0 \leqslant \theta \leqslant \pi/2) \quad \pi F_\nu^+ = \int_0^{2\pi} \int_0^{\pi/2} I_\nu \cos\theta \sin\theta \, d\theta \, d\varphi$$

and (11.7)

$$\text{inward flux } (\pi/2 \leqslant \theta \leqslant \pi) \quad \pi F_\nu^- = -\int_0^{2\pi} \int_{\pi/2}^{\pi} I_\nu \cos\theta \sin\theta \, d\theta \, d\varphi$$

whence

$$F_\nu = F_\nu^+ - F_\nu^-.$$

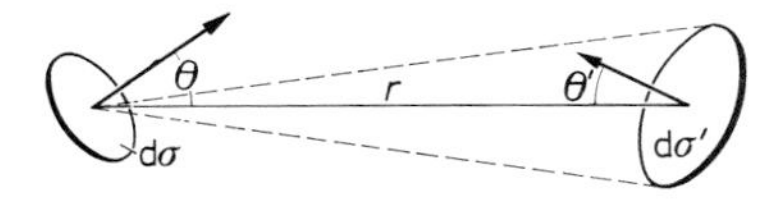

Figure 11.2
Mutual radiation of two surface elements.

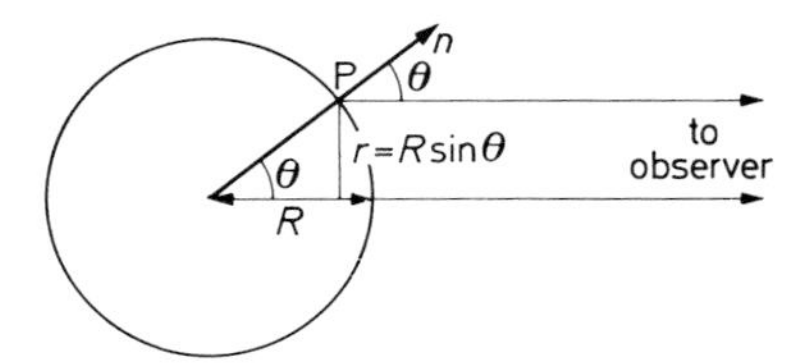

Figure 11.3
Flux of radiation πF_ν from a star. The mean intensity is $\bar{I}_v = F_v$.

Analogously to equation (11.3) we define

$$\text{total radiation flux} \quad \pi F = \int_0^\infty \pi F_\nu \, d\nu = \int_0^\infty \pi F_\lambda \, d\lambda. \tag{11.8}$$

Consider next the radiation from a spherical star (Figure 11.3). The intensity I_ν of the radiation emerging from its atmosphere will depend only on the angle of emergence θ reckoned from the normal at the point P concerned, that is, $I_\nu = I_\nu(\theta)$. As seen from Figure 11.3, the same angle θ is also the angle between the sight line and the radius through P. The distance of P from the center of the star as seen in projection on a plane perpendicular to the sight line is $\sin\theta$ in units of the radius.

The *mean intensity* $\bar{I}_\nu$ of the radiation emitted toward the observer from the visible disk, using equation (11.1), is given by

$$\pi R^2 \bar{I}_\nu = \int_0^{2\pi} \int_0^{\pi/2} I_\nu(\theta) \cos\theta \, R^2 \sin\theta \, d\theta \, d\varphi. \tag{11.9}$$

If we cancel R^2 and compare with equation (11.7) we learn that

$$\bar{I}_\nu = F_\nu^+. \tag{11.10}$$

Thus the mean intensity of radiation over the apparent disk is equal to $1/\pi$ times the radiation flux at the surface. If the star of radius R is at a great distance r from the observer, then he sees the disk as subtending solid angle $d\omega = \pi R^2/r^2$ and consequently he infers from equation (11.7)

$$\text{radiation flux } S_\nu = \bar{I}_\nu \, d\omega = F_\nu^+ \, \pi R^2/r^2. \tag{11.11}$$

Thus the theory of stellar spectra must have in view the calculation of the radiation flux. As K. Schwarzschild was the first to remark, the advantage of observing the Sun is that in this case we can measure I_ν directly as a function of θ.

In the first instance phenomenologically, we must describe the emission and absorption of radiation: A volume-element dV emits per second in solid angle $d\omega$ and in frequency interval ν to $\nu + d\nu$ the quantity of energy

$$\epsilon_\nu \, d\nu \, dV \, d\omega. \tag{11.12}$$

The *emission coefficient* ϵ_ν depends in general upon the frequency ν as well as the nature and state of the material (chemical composition, temperature and pressure) and also upon the direction. The total energy output of an isotropically radiating volume element dV per second is

$$dV\, 4\pi \int_0^\infty \epsilon_\nu\, d\nu. \tag{11.13}$$

We set the emission against the energy loss through absorption suffered by a narrow beam of radiation of intensity I_ν when it traverses a layer of matter of thickness ds. This is

$$dI_\nu/ds = -\kappa_\nu I_\nu. \tag{11.14}$$

We call κ_ν the *absorption coefficient;* again it depends in general upon the nature and state of the material, and also upon direction. In equation (11.14) we have referred the absorption to a layer of thickness 1 cm. Instead of this, we can consider a layer that contains 1 g of material per square centimeter of the surface. Then we obtain the *mass absorption coefficient* $\kappa_{\nu,M}$. If ρ is the density of the material, then $\kappa_\nu = \kappa_{\nu,M}\rho$. Finally, if we reckon the absorption coefficient per atom, we have the *atomic absorption coefficient* $\kappa_{\nu,\text{at}}$. With n atoms per cm³, we obtain $\kappa_\nu = \kappa_{\nu,\text{at}}\, n$. As regards dimensions, we have

		Dimensions
Absorption coefficient	κ_ν	[cm^{-1}] or [cm^2/cm^3]
Mass absorption coefficient	$\kappa_{\nu,M} = \kappa_\nu/\rho$	[cm^2/g]
Atomic absorption coefficient = absorption cross section of the atom	$\kappa_{\nu,\text{at}} = \kappa_\nu/n$	[cm^2]

(11.15)

If our pencil of radiation traverses a nonemitting layer of thickness s then from equation (11.14)

$$dI_\nu/I_\nu = -\kappa_\nu\, ds \tag{11.16}$$

and the intensity I_ν of the radiation after passage through the absorbing layer is related to the intensity $I_{\nu,0}$ of the incident radiation by

$$I_v/I_{v,0} = e^{-\tau_\nu} \tag{11.17}$$

where the dimensionless quantity

$$\tau_\nu = \int_0^s \kappa_\nu\, ds = \int_0^s \kappa_{\nu,M}\, \rho\, ds \tag{11.18}$$

is called the *optical thickness* of the layer traversed. For example, a layer of optical thickness $\tau_\nu = 1$ reduces a ray to $e^{-1} = 36.8$ percent of its original

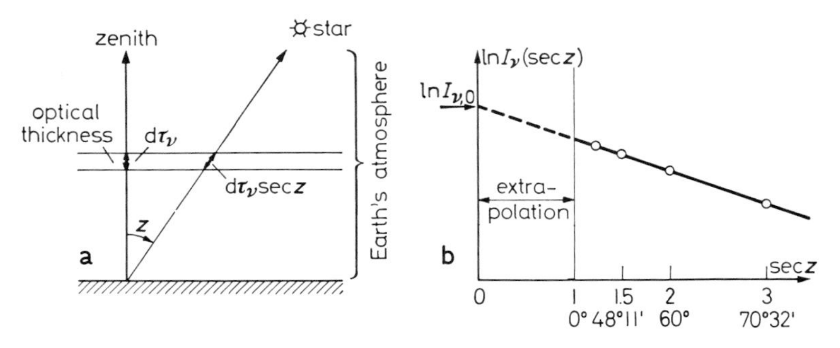

Figure 11.4. (a) Atmospheric extinction of radiation (frequency ν) of a star at zenith distance z. **(b)** Extrapolation to sec $z = 0$ gives the intensity of radiation $I_{\nu 0}$ outside the Earth's atmosphere.

intensity. The *extinction* of the ν-radiation of a star (Figure 11.4) at zenith distance z corresponds to a reduction factor

$$I_\nu/I_{\nu,0} = e^{-\tau_\nu \sec z} \tag{11.19}$$

where τ_ν is the optical thickness, measured vertically, of the Earth's atmosphere at frequency ν. Provided τ_ν is not too large, we can derive the intensity outside the atmosphere by linear extrapolation of the values of $\ln I_\nu$, measured for different values of the zenith distance, to (formally) $\sec z \to 0$.

We encounter particularly instructive relations if, following G. Kirchhoff (1860), we consider a radiation field that is in *thermodynamic equilibrium*, or temperature equilibrium, with its surroundings. We obtain such a radiation field if we immerse an otherwise arbitrary cavity in a heat bath of temperature T. Since by definition in such conditions all objects have the same temperature, it seems justifiable to speak of cavity radiation of temperature T. (Here we always assume a refractive index of unity for the cavity; it would be easy to modify this hypothesis.) In an isothermal enclosure, every body and every surface element emit and absorb equal quantities of radiant energy per unit time. Starting from this we can easily show that the intensity I_ν of cavity radiation is independent of the material contents and of the constitution of the walls of the cavity and of direction (the radiation is isotropic). We need consider only a cavity with two different chambers H_1 and H_2 (Figure 11.5). Were $I_{\nu,1} \neq I_{\nu,2}$ we could gain energy at an opening through which the chambers are put in communication by using, say, a radiometer (the familiar ''lightmill'' of opticians' shop windows) and we should have perpetual motion of the second kind. That $I_{\nu,1} = I_{\nu,2}$ must hold good for all frequencies, directions of propagation, and of polarization we can readily show by introducing into the opening between H_1 and H_2 in succession a color filter, a tube of adjustable direction, or a nicol prism. Also, in accordance with the second law of thermodynamics, we know that perpetual motion of the second kind is not possible. Thus the intensity of cavity radiation I_ν is a universal function $B_\nu(T)$ of ν and T.

We can produce and measure this quantity by "tapping" a cavity of temperature T through a sufficiently small opening. The important function $B_\nu(T)$, whose existence G. Kirchhoff recognized in 1860 and which M. Planck first explicitly calculated in 1900, we call the *Kirchhoff–Planck* function $B_\nu(T)$.

In thermodynamic equilibrium, that is, in a cavity at temperature T, the emission and absorption by an arbitrary volume element must be equal to each other. We write them down for a flat volume element of base-area $d\sigma$ and height ds for the ν-radiation in solid angle $d\omega$ about the direction normal to $d\sigma$. Using equations (11.12) and (11.14) we obtain

$$\begin{aligned} \text{emission per second} &= \epsilon_\nu \, d\nu \, d\sigma \, ds \, d\omega \\ \text{absorption per second} &= \kappa_\nu \, ds \, B_\nu(T) \, d\sigma \, d\omega \, d\nu. \end{aligned} \tag{11.20a}$$

Equating these quantities we have Kirchhoff's law

$$\epsilon_\nu = \kappa_\nu B_\nu(T). \tag{11.20}$$

This asserts that *in a state of thermodynamic equilibrium the ratio of the emission coefficient ϵ_ν to the absorption coefficient κ_ν is a universal function $B_\nu(T)$ of ν, T which at the same time expresses the intensity of cavity radiation.*

As an important application we calculate the intensity I_ν of the radiation emitted by a layer of material of constant temperature T (for example, the plasma in a gas discharge) which has thickness s in the direction of the sight line:

For a volume element that, as seen by the observer, has cross section 1 cm² and that extends from x to $x + dx$ in the direction of the sight line ($0 \leqslant x \leqslant s$) the amount of radiation emitted per second per steradian is $\kappa_\nu B_\nu(T)\, dx$. Up to its escape from the layer, this portion is reduced by the factor $\exp(-\kappa_\nu x)$. Thus we obtain

$$I_\nu(s) = \int_0^s \kappa_\nu B_\nu(T) e^{-\kappa_\nu x} \, dx$$

Figure 11.5. Cavity radiation is unpolarized and isotropic. Its intensity is a universal function of ν and T, the Kirchhoff–Planck function $B_\nu(\nu,T)$.

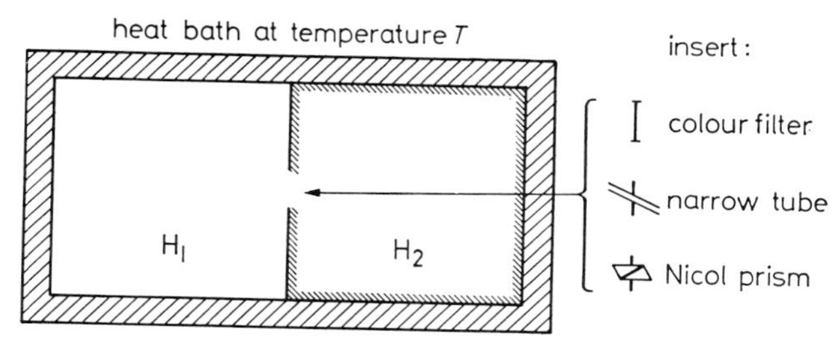

or, introducing again the *optical depth* by $d\tau_\nu = \kappa_\nu\, dx$ or $\tau_\nu = \kappa_\nu s$,

$$I_\nu(s) = B_\nu(T)(1 - e^{-\tau_\nu}) \approx \begin{cases} \tau_\nu B_\nu(T), & \tau_\nu \ll 1 \\ B_\nu(T), & \tau_\nu \gg 1. \end{cases} \tag{11.21}$$

As is easily seen these formulas hold good also when κ_ν depends upon x. Thus the emission by an isothermal optically thin layer ($\tau_\nu \ll 1$) is equal to its optical depth times the Kirchhoff–Planck function. The radiation intensity from an optically thick layer ($\tau_\nu \gg 1$), on the other hand, approximates to that of a black body and cannot exceed this. Applied to a spectral line, this gives the familiar appearance of self-absorption: While two equally broad spectral lines from an optically thin layer show an intensity ratio $\approx \kappa_1 : \kappa_2$, in the case of an optically thick layer the ratio approaches unity. (See further Section 19.)

Instead of the emission and absorption by a volume element, we can consider also those of a surface element. If we denote the emissive power by E_ν and the absorptive power by[1] A_ν (the reflective power is then $R_\nu = 1 - A_\nu$), then analogously to equation (11.20a) we have

$$\begin{aligned} \text{emission per second} &= E_\nu\, d\nu\, d\sigma \cos\theta\, d\omega \\ \text{absorption per second} &= A_\nu B_\nu(T)\, d\nu\, d\sigma \cos\theta\, d\omega \end{aligned} \tag{11.22a}$$

and the Kirchhoff law receives the somewhat different form

$$E_\nu = A_\nu B_\nu(T). \tag{11.22}$$

Thus *in thermodynamic equilibrium the ratio of emissive to absorptive powers is equal to the intensity of cavity radiation* $B_\nu(T)$.

For a body that completely absorbs all frequencies ($A_\nu = 1$), a so-called *black body,* we have $E_\nu = B_\nu(T)$. Consequently we call cavity radiation also *black-body radiation,* or *black radiation* for short. A small opening into a cavity swallows up all the radiation entering it, by multiple reflexion and absorption in the interior, whence we may again infer that cavity radiation and black radiation are the same.

As regards the actual calculation of the function $B_\nu(T)$ Max Planck set about it in 1900 in the following way: He considered the interaction between the cavity radiation field and, for instance, an electron bound elastically to an equilibrium position, a *harmonic oscillator.* Its vibrations are agitated by the electromagnetic waves of the radiation field until, in thermodynamic equilibrium, emission and absorption are equal to each other. Furthermore, however, according to the Boltzmann law of equipartition, in thermodynamic equilibrium the mean energy of the harmonic oscillator is equal to kT, where k is the Boltzmann constant ($k = 1.38 \times 10^{-16}$ erg/K). Clearly one can thence work backward to compute the intensity $B_\nu(T)$ of the cavity radiation in thermodynamic equilibrium at temperature T. Planck then recognized that he could correctly reproduce the measured intensities of cavity radiation only if he extended the basic laws of classical mechanics and electrodynamics to meet the known

[1]For the gas layer just considered, for example, we would have $A_\nu = 1 - e^{-\tau_\nu}$.

requirements of *quantum theory*. Thus there resulted the *Planck radiation formula* that we write on the ν and λ scales as

$$B_\nu(T) = \frac{2h\nu^3}{c^2}\frac{1}{e^{h\nu/kT}-1} \quad \text{and} \quad B_\lambda(T) = \frac{2hc^2}{\lambda^5}\frac{1}{e^{hc/k\lambda T}-1} \tag{11.23}$$

with the two important limiting cases

$$\frac{h\nu}{kT} \gg 1 \quad B_\nu(T) \approx \frac{2h\nu^3}{c^2}\,e^{-h\nu/kT} \quad \text{Wien's law} \tag{11.24}$$

$$\frac{h\nu}{kT} \ll 1 \quad B_\nu(T) \approx \frac{2\nu^2 kT}{c^2} \quad \text{Rayleigh–Jeans law.} \tag{11.25}$$

In the Rayleigh–Jeans law the characteristic quantity h of quantum theory has disappeared. Quite generally, quantum theory passes over into classical theory for light quanta $h\nu$ whose energy is considerably smaller than the thermal energy kT (Bohr's correspondence principle). The radiation constant that appears in the exponent in the radiation law (11.23) is written as c_2. We have

$$c_2 = hc/k = 1.4388 \text{ cm K}. \tag{11.26}$$

We obtain the total radiation of a black body by integrating equation (11.23) over all frequencies $B(T) = \int_0^\infty B_\nu(T)\,d\nu$. The total flux, that is, the emission of a black body from 1 cm² into free space, is $\pi F^+ = \pi B(T)$. If we carry out the integration of Planck's formula (using the variable $x = h\nu/kT$) we derive the Stefan–Boltzmann radiation law

$$\pi F^+ = \pi B(T) = \sigma T^4 \tag{11.27}$$

with the radiation constant

$$\sigma = \frac{2\pi^5 k^4}{15c^2h^3} = 5.670 \times 10^{-5} \text{ erg/cm}^2 \text{ s K}^4 \tag{11.28}$$

This law was found experimentally by J. Stefan in 1879 and derived theoretically by L. Boltzmann in 1884 using an ingenious calculation of the entropy of cavity radiation.

12. The Sun

We first collect together some known results for the Sun. These will then often provide the most useful units for dealing with the stars.

Using the solar parallax of 8″.794 we first derived the mean Sun–Earth distance, or the astronomical unit,

$$1 \text{ AU} = 149.6 \times 10^6 \text{ km} = 23\,456 \text{ equatorial Earth radii.} \tag{12.1}$$

The corresponding apparent radius of the solar disk is

$$15'59''.63 = 959''.63 = 0.004\ 652_4 \text{ rad.} \tag{12.2}$$

Thence we obtain

$$\text{solar radius} = 696\ 000 \text{ km.} \tag{12.3}$$

Thus on the Sun

$$1' = 43\ 500 \text{ km} \qquad 1'' = 725 \text{ km.} \tag{12.4}$$

On account of atmospheric scintillation the limit of resolution is about 500 km.

The astronomical unit also corresponds to 215 solar radii. A flattening of the Sun resulting from its rotation is not quite detectable ($<0''.1$).

Using Kepler's third law we obtain

$$\text{solar mass } \mathfrak{M}_\odot = 1.989 \times 10^{33} \text{ g.} \tag{12.5}$$

Thence we find the mean density of the Sun $\bar{\rho} = 1.409$ g/cm^3. Also we find

$$\text{gravity at the Sun's surface } g_\odot = 2.74 \times 10^4 \text{ cm/s}^2 \tag{12.6}$$

which is 27.9 times gravity at the surface of the Earth.

Disregarding for the moment radiation in far ultraviolet and in radio frequencies, the solar spectrum as observed using a tower telescope with a large grating spectrograph (Figure 12.1) is seen as a continuous spectrum crossed by many dark Fraunhofer lines (Figure 12.2). The *Rowland Atlas* and *The solar spectrum 2935 Å to 8770 Å; second revision of Rowland's preliminary table of solar spectrum wavelengths* (1966) contains wave lengths λ and identifications of these lines. The microphotometrically recorded intensity distribution in the solar spectrum, referred to the continuum as 100, is contained in the *Utrecht photometric atlas of the solar spectrum* by M. Minnaert, G. F. W. Mulders, and J. Houtgast (1940) and in more recent atlases, some at higher dispersion or with better resolution and also in other regions of the spectrum. The Utrecht measurements refer to the center of the solar disk. The solar disk shows a considerable center-limb darkening; also the lines relative to the continuum at the same place exhibit slight center-limb variations.

We denote the position on the solar disk to which an observation refers by giving the distance ρ from the center of the disk in terms of the radius as unity or, for theoretical purposes, the angle θ between the sightline and the normal to the Sun's surface (Figure 11.3). Thus

$$\rho = \sin\theta \quad \text{and} \quad \cos\theta = \sqrt{(1 - \rho^2)}. \tag{12.7}$$

In particular, the center of the disk is $\rho = \sin\theta = 0$ and the limb is $\rho = \sin\theta = 1$, $\cos\theta = 0$. The radiation intensity at the Sun's surface (optical depth $\tau_0 = 0$, see below) at distance $\rho = \sin\theta$ from the center of the disk, and referred to the wavelength scale, we denote by

$$I_\lambda(0,\theta)\ [\text{erg/cm}^2 \text{ s sr; } \Delta\lambda = 1]. \tag{12.8}$$

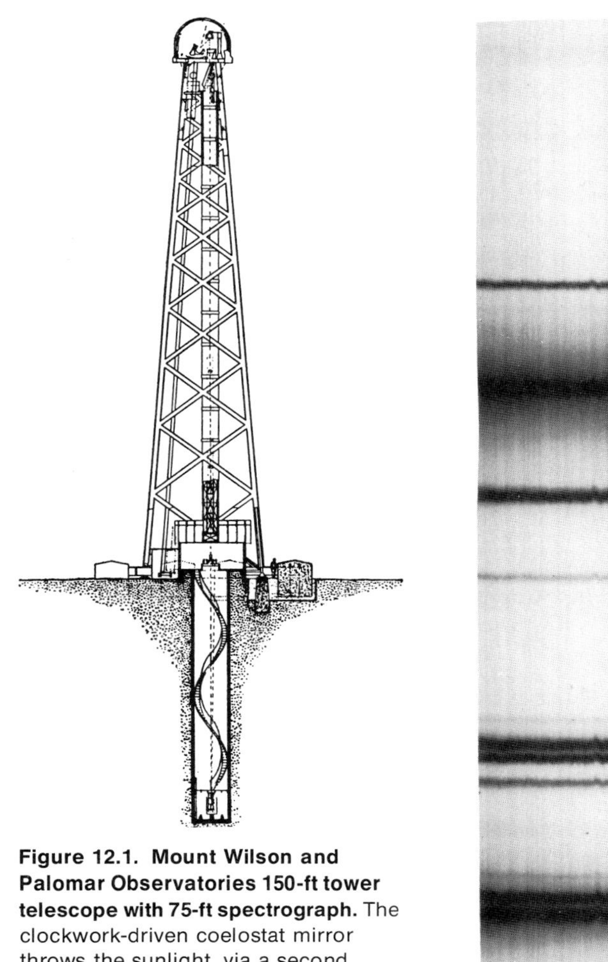

Figure 12.1. Mount Wilson and Palomar Observatories 150-ft tower telescope with 75-ft spectrograph. The clockwork-driven coelostat mirror throws the sunlight, via a second (fixed) mirror, on to the objective, whose focal length is 150 ft. This produces a solar image of 16½-in (40-cm) diameter on the slit of the spectrograph at about ground level. The grating and lens of the spectrograph are located underneath the tower in a constant-temperature well 75 ft below ground level. The spectrum can be photographed or observed visually near the slit.

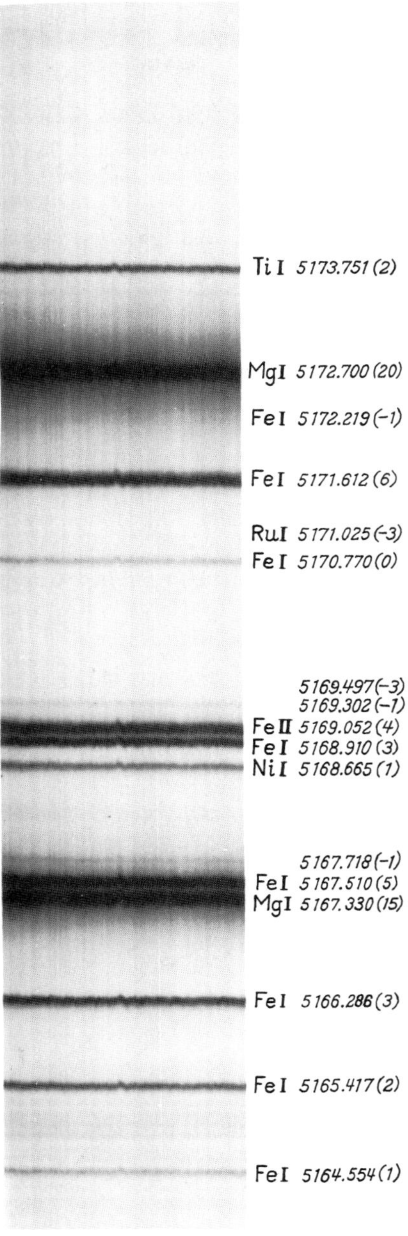

Figure 12.2. Solar spectrum (center of disk) λ 5164–5176 Å, 5th order, taken with the vacuum grating spectrograph of the McMath-Hulbert Observatory. Wavelength, identification and (estimated) Rowland intensity of the Fraunhofer lines are marked at the side.

One measures the center-limb variation of the radiation intensity

$$I_\lambda(0,\theta)/I_\lambda(0,0) \qquad (12.9)$$

by allowing the solar image to move across the slit of a spectrograph with, for instance, a photoelectric recording system. The essential difficulty in this and all other measurements of details of the solar disk is in the elimination of scattered light in the instrument and in the Earth's atmosphere. Figure 12.3 shows a summary of recent results. The ratio of the mean intensity F_λ over the solar disk to the intensity $I_\lambda(0,0)$ at the center is given by

$$\frac{F_\lambda}{I_\lambda(0,0)} = 2\int_0^{\pi/2} \frac{I_\lambda(0,\theta)}{I_\lambda(0,0)} \cos\theta \sin\theta \, d\theta = \int_0^1 \frac{I_\lambda(0,\rho)}{I_\lambda(0,0)} \, d(\rho^2). \qquad (12.10)$$

Nowadays one measures absolute values of the radiation intensity, say $I_\lambda(0,0)$ for the center of the disk, by comparison with a black body of known temperature. The highest fixed point on the temperature scale which has been precisely determined by a gas thermometer is the melting point of gold T_{Au}. The temperature scale above T_{Au} is based on this fixed point and on pyrometer measurements, using Planck's law and the radiation constant

Figure 12.3. Limb darkening of the Sun. The ratio of radiation intensities $I_\lambda(0,\theta)/I_\lambda(0,0)$ is plotted against $\cos\theta$ for several line-free wavelengths, ranging from the blue to the infrared, shown in angstrom units. The nonlinear $\sin\theta$ scale gives the distance from the center of the disk in solar radii.

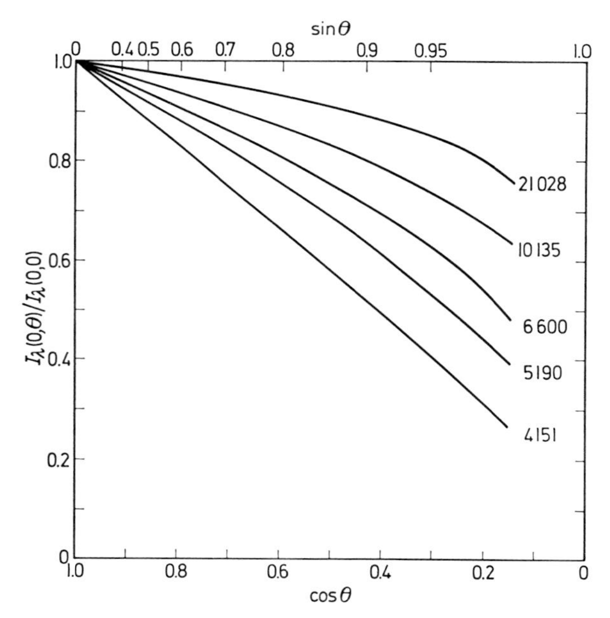

c_2 (11.23 and 11.26). The International Practical Temperature Scale (1968) is based on

$$\text{melting point of gold } T_{\text{Au}} = 1337.58 \text{ K (or } 1064.43°\text{C)}$$

and

$$\text{radiation constant } c_2 = 1.4388 \text{ cm K}.$$

The zero point of the Celsius scale is at 273.15 K. Compared to the older temperature scale of 1948 the new one is, for example, about 6°C higher at 3000°C.

Obviously the extinction in the Earth's atmosphere (Figure 11.4) must be accurately determined for numerous wavelengths and allowed for in the measurements.

Since the solar spectrum (Figure 12.2) is crossed by many Fraunhofer lines, it is best to make measurements for sharply defined wavelength intervals, for example each of width $\Delta\lambda = 100$ Å, of the mean intensity in the spectrum including the lines $I_\lambda^L(0,0)$. However, for the physics of the Sun the true continuum between the lines $I_\lambda(0,0)$ is of greater interest. Its determination in the long wavelength region $\lambda > 4600$ Å presents no difficulty. On the other hand, in the blue and violet spectral region, about $\lambda < 4600$ Å, the lines are so closely packed that the determination of the true continuum $I_\lambda(0,0)$ becomes more and more difficult until finally in the ultraviolet this is possible only in conjunction with a fully developed theory. The ratio

$$1 - \eta_{\bar{\lambda}} = I_{\bar{\lambda}}^L(0,0)/I_\lambda(0,0) \tag{12.11}$$

for a suitable restricted wavelength interval $\Delta\lambda$ of width, say, 20 to 100 Å centered on the wavelength $\bar{\lambda}$ is measured by using a planimeter to obtain the area under the microphotometer tracing of a spectrum of high dispersion such as that in the Utrecht Atlas. In the ultraviolet (λ 3000 to 4000 Å) $\eta_{\bar{\lambda}}$ is of the order 25 to 45 percent; it falls to a few percent for $\lambda > 5000$ Å. Figure 12.4 shows the results of the latest and probably the most accurate measurements by D. Labs, H. Neckel, and others; earlier values by D. Chalonge, R. Peyturaux, A. K. Pierce, and others are in good agreement with these.

By integrating over the solar disk according to equation (12.10) and integrating over all wavelengths, allowing by extrapolation for the parts of the spectrum cut off at the ends, 3.9 percent in the ultraviolet ($\lambda < 3420$ Å) and 4.8 percent in the infrared ($\lambda > 23\,000$ Å), we obtain the total flux at the surface of the Sun

$$\pi F = \int_0^\infty \pi F_\lambda^L \, d\lambda = 6.28 \times 10^{10} \text{ erg/cm}^2 \text{ s}. \tag{12.12}$$

Hence we easily derive the total emission of the Sun per second

$$\text{Luminosity } L_\odot = 4\pi R^2 \pi F = 3.82 \times 10^{33} \text{ erg/s}.$$

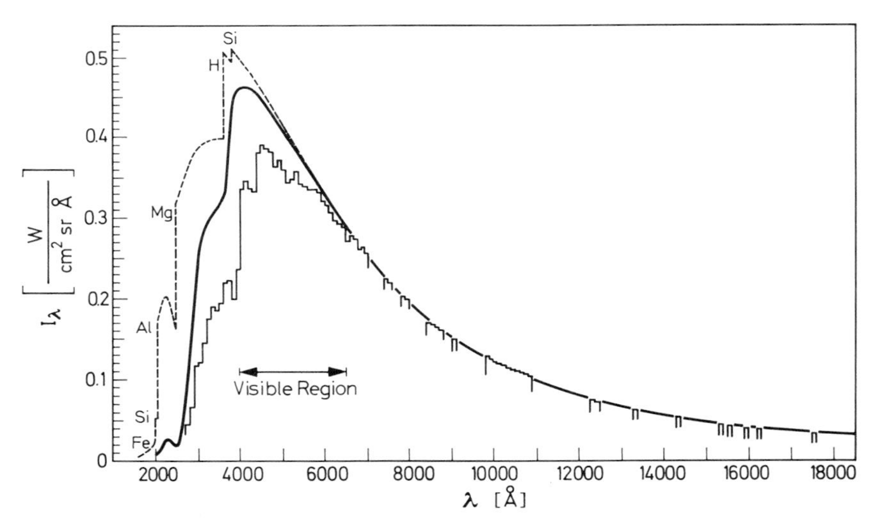

Figure 12.4. Intensity of the spectrum of the center of the solar disk in W/cm² sr Å, after D. Labs and H. Neckel (1968–70). In the ultraviolet the data of J. Houtgast (1970) are used and in the infrared those of A. K. Pierce (1954) are used. The histogram shows the measured intensity of the spectrum including the lines $I_\lambda^L(0,0)$, averaged over wavelength bins 100 Å wide. The continuous smooth curve shows the quasi-continuum, that is, it joins the points of highest measured intensity between the lines (D. Chalonge's "window"). The dashed curve depicts the continuum at the center of the disk, without lines, calculated by H. Holweger and others from models of the solar atmosphere. The difference between the calculated and unbroken curves may be explained, at least partially, by a "veil" of extremely weak Fraunhofer lines. The Balmer jump is recognizable at $\lambda \approx 3700$Å. (For this figure, the author is indebted to Mr. D. Labs (Heidelberg).)

Further we obtain the flux S at the Earth's distance ($r = 1$ astronomical unit) following equation (11.11) by multiplying the mean intensity F of the solar disk by the solid angle subtended at the Earth

$$\pi R^2/r^2 = 6.800 \times 10^{-5} \text{ sr} \tag{12.13}$$

whence

$$S = 1.36 \times 10^6 \text{ erg/cm}^2 \text{ s}. \tag{12.14}$$

After pioneering attempts by S. S. Pouillet (1837), this important quantity was first measured accurately by K. Ångström (c. 1893) and C. G. Abbot (from c. 1908) when they measured with a "black" receiver, the so-called pyrheliometer, the total solar radiation reaching the Earth's surface. However, this measurement has to be supplemented by absolute measurements of the radiation as distributed through the spectrum since only from these can one eliminate the atmospheric extinction in accordance with Figure 11.4. More recently, S has been measured, with smaller and smaller corrections for extinction, from aircraft, rockets, and spacecraft. The sum total of *all* measurements (up to 1971) again yields $S = 1.36 \times 10^6$ erg/cm² s with an uncertainty of about ±1 percent. In accordance with a long-established habit, S is often expressed in calories; we then obtain the so-called

$$\text{solar constant } S = 1.95 \text{ cal/cm}^2 \text{ min.} \qquad (12.15)$$

In view of the rapidly expanding energy requirements of mankind, it is interesting to express the energy flux supplied by the Sun in commercial units. We derive

$$S = 1.36 \text{ kW/m}^2 \text{ or } 1.52 \text{ hp/yd}^2. \qquad (12.16)$$

This supply is available outside the Earth's atmosphere, but even after allowing for the extinction, a considerable energy contribution does reach the ground, at any rate in countries with a favorable climate. Nevertheless, direct practical use of solar energy has been made hitherto only for generating electric currents in satellites and to a small extent for cooking and hot water installations in tropical countries.

We return once again to consider the total flux πF and the radiation intensity $I_\lambda(0,0)$ at the Sun's surface, equation (12.12) and Figure 12.4. As our first rather formal and provisional approach to the temperature in the solar atmosphere, we interpret the flux in the sense of the Stefan–Boltzmann radiation law and thence we define

$$\text{Sun's effective temperature: } \pi F = \sigma T_{\text{eff}}^4; \; T_{\text{eff}} = 5770 \text{ K} \qquad (12.17)$$

with a mean error of ±15K. Furthermore, if we interpret the radiation intensity $I_\lambda(0,0)$ in the sense of Planck's radiation law (11.23) we define the radiation temperature T_λ for the center of the disk as a function of wavelength.

Since the Sun does not radiate as a black body—otherwise $I_\lambda(0,\theta)$ would be independent of θ and the Sun would show no darkening toward the limb—we must not interpret T_{eff} and the T_λ quite literally. Nevertheless T_{eff} does to some extent correctly indicate the temperature of the layer of the solar atmosphere from which the total radiation comes to us, and T_λ that of the layer from which the radiation of wavelength λ comes. The effective temperature is thus an important parameter of the solar atmosphere (and in a corresponding way of stellar atmospheres) since in combination with the Stefan–Boltzmann radiation law it represents the total radiation πF, that is, the total energy flux that comes out of the interior through 1 cm² of the Sun's surface.

The layers of the solar atmosphere from which the continuous radiation originates we call the *photosphere*. Formerly one distinguished between this and the reversing layer which was supposed to lie above it and to produce the dark Fraunhofer lines as in the well-known Kirchhoff–Bunsen experiment. Nowadays we know that at any rate considerable parts of the lines arise in the same layers as the continuum, so that it is better not to use the term "reversing layer" any longer. At total eclipses of the Sun one observes at the solar limb the higher layers of the atmosphere which yield no appreciable continuum, but practically only *emission lines* corresponding to the Fraunhofer lines. This part is called the *chromosphere*. As we now know, this makes only a small contribution to the intensity of the absorption lines. Above it lies the *solar corona* which, as we shall see, merges into the *interplanetary medium*. Later on, we shall concern ourselves further with these extreme outer layers of the Sun as well as with the whole manifestation of solar activity (sunspots, prominences, eruptions, etc.), that is, with the "disturbed" Sun, as we now often call it.

13. Apparent Magnitudes and Color Indices of Stars

Hipparchus and many of the earlier astronomers long ago cataloged the brightnesses of stars. We may speak of stellar photometry in the modern sense since N. Pogson in 1850 gave a clear definition of magnitudes and J. C. F. Zöllner in 1861 constructed his visual photometer with which one could accurately compare the brightness of a star with that of an artificial starlike image by employing two nicol prisms. (Two such prisms whose planes of polarization are inclined at angle α to each other reduce the intensity of light passing through them both by a factor $\cos^2 \alpha$.)

If the ratio of the fluxes from two stars as measured with a photometer is S_1/S_2, then according to Pogson's definition, the difference of their apparent magnitudes is

$$m_1 - m_2 = -2.5 \log (S_1/S_2) \text{ magnitudes} \tag{13.1}$$

or conversely

$$S_1/S_2 = 10^{-0.4(m_1 - m_2)}. \tag{13.2}$$

The following are corresponding magnitude differences and flux ratios and the corresponding brightness ratios:

$-\Delta m$	1	2.5	5	10	15	20	magnitudes
$\Delta \log S$	0.4	1	2	4	6	8	
Brightness ratio	2.512	10	100	10^4	10^6	10^8	

The zero point of the magnitude scale was originally fixed by the international polar sequence, a set of stars in the vicinity of the north pole which had been accurately measured and the constancy of their brightness established. Greater brightness corresponds to smaller and ultimately negative magnitudes, and so it is recommended to say, for instance, "The star α Lyrae (Vega) of apparent (visual) magnitude $0^{\text{m}}.14$ is 1.19 magnitudes brighter than α Cygni (Deneb) of magnitude $1^{\text{m}}.33$, or α Cyg is 1.19 magnitudes fainter than α Lyr."

The older photometric measurements were made visually. With K. Schwarzschild's *Göttinger Aktinometrie* photographic photometry was added in the years 1904 to 1908, first using ordinary blue-sensitive plates. One soon learned to imitate the spectral sensitivity distribution of the human eye using sensitized plates made sensitive in the yellow and with a yellow filter in front of them. Thus one had, in addition to the visual magnitudes m_v, the photographic m_{pg} and then the photovisual magnitudes m_{pv}. By combining suitable plates or photocells and photomultipliers with appropriate color filters, one can nowadays adjust the sensitivity maximum of one's measuring devices, whether photographic or photoelectric, to be in any wavelength interval from the ultraviolet down to the infrared.

We now seek to clarify our ideas by asking: What precisely do our various magnitudes mean?

Suppose that a star of radius R emits at its surface the flux πF_λ or mean intensity F_λ at wavelength λ in the true continuum; suppose that a fraction η_λ of this is removed by Fraunhofer lines. The actual flux, including lines, is then

$$\pi F_\lambda^L = \pi F_\lambda(1 - \eta_\lambda). \tag{13.3}$$

Neglecting interstellar absorption for the time being, if the star is at distance r from the Earth then we receive outside the Earth's atmosphere a flux [see equation (12.13)]

$$S_\lambda = \pi R^2 F_\lambda^L / r^2. \tag{13.4}$$

Now let the contribution to the stimulus of our measuring device made by a standard spectrum with $S_\lambda = 1$ be described by a sensitivity function E_λ. Then in the case of the star the stimulus is proportional to the integral

$$\frac{1}{r^2}\int_0^\infty \pi R^2 F_\lambda^L E_\lambda \, d\lambda \tag{13.5}$$

and the apparent magnitude m becomes

$$m = -2.5 \log \frac{1}{r^2}\int_0^\infty \pi R^2 F_\lambda^L E_\lambda \, d\lambda + \text{constant} \tag{13.6}$$

to within a normalizing constant which will have to be fixed by some convention. In equation (13.6) we have taken no account of extinction by the Earth's atmosphere. Since in practice E_λ has an appreciable value only in a small range of wavelength, the extinction can be determined following Figure 11.4 as for monochromatic radiation at the center of gravity of E_λ.

The *visual magnitude* m_v is defined by (13.6) using for the sensitivity function that of the human eye multiplied by the power of transmission of the instrument, of which the center of gravity lies at the so-called isophotal wavelength of about 5400 Å in the green. In the same way, the center of gravity for the *photographic magnitude* m_{pg} is about 4200 Å.

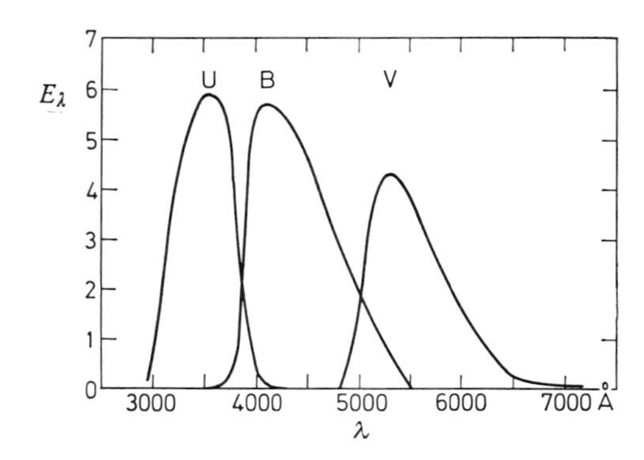

Figure 13.1
Relative response functions E_λ (referred to a light source with I_λ = constant) of the UBV photometric system, after H. L. Johnson and W. W. Morgan.

As a standard system of magnitudes and colors one nowadays mostly uses the *UBV* system developed by H. L. Johnson and W. W. Morgan (1951) (U = ultraviolet, B = blue, V = visual); the corresponding magnitudes are written for short as

$$U = m_U \quad B = m_B \quad V = m_V. \tag{13.7}$$

The corresponding sensitivity functions E_λ, which can be realized either photographically or photoelectrically, are shown in Figure 13.1. Their centers of gravity for average stellar colors are

$$\lambda_U \approx 3650 \text{ Å} \quad \lambda_B \approx 4400 \text{ Å} \quad \lambda_V \approx 5480 \text{ Å}. \tag{13.8}$$

For hot (blue) stars these are displaced toward shorter wavelengths; for cool (red) stars toward longer wavelengths.

By definition, the three magnitudes U, B, V are so related to each other, that is, the constants in equation (13.6) are so chosen that for A0V-stars (for example α Lyr = Vega, see Section 15)

$$U = B = V. \tag{13.9}$$

For practical use, including the transfer from one instrument to another with (possibly) a slightly different sensitivity function, the *UBV* system is determined by a larger number of very precisely measured *standard stars,* whose magnitudes and colors cover a wide range.

As K. Schwarzschild first recognized, color indices, such as, for example,

$$U - B \quad \text{and} \quad B - V \tag{13.10}$$

provide a measure of the energy distribution in the spectrum of a star. If as an approximation to the energy distribution we take Planck's law for a black body or, so long as $h\nu/kT \gg 1$, Wien's law

$$F_\lambda \propto e^{-c_2/\lambda T} \quad \text{and} \quad \eta_\lambda = 0 \tag{13.11}$$

and if we concentrate the integration in (13.6) to the center of gravity of the sensitivity function, we obtain

$$m_V = \frac{2.5 c_2 \log e}{\lambda_V T} + \text{constant}_V \tag{13.12}$$

or, with $c_2 = 1.4388$ cm K and $\log e = M = 0.4343$,

$$V = \frac{1.562}{\lambda_V T} + \text{constant}_V. \tag{13.13}$$

Using (13.8) and the normalization (13.9) and taking for a A0V star $T = 15\,000$K, we find

$$B - V \approx 7000 \left(\frac{1}{T} - \frac{1}{15\,000} \right). \tag{13.14}$$

The values of the *color temperatures* calculated, for example, from the color indices $B - V$ have no profound significance because of the considerable departure of the stars from black bodies. Today the significance of color indices lies in a different direction: As the theory of stellar atmospheres shows, the color indices (or the color temperature) depend upon the fundamental parameters, in particular the effective temperature T_{eff} (total flux) and gravity g or absolute magnitude of the stars (see Section 18). Since color indices can be measured photoelectrically to an accuracy of $0^{\text{m}}.01$, we expect from (13.14) that near, say, 7000 K *temperature differences* can be determined to an accuracy of about 1 percent, which cannot be attained by any other method. Naturally, the temperatures themselves are far less accurately determined.

In addition to the *UBV* system, significance has been achieved by the *UGR* system used by W. Becker with characteristic wavelengths 3660, 4630 and 6380 Å and by the six-color system of J. Stebbins, A. E. Whitford, and G. Kron which extends from the ultraviolet into the infrared ($\lambda_U = 3550$ Å . . . $\lambda_I = 10\,300$ Å). The conversion of magnitudes and color indices from one photometric system to another with not greatly different characteristic wavelengths is performed by using empirically established, mostly linear relations.

Besides the radiation in various wavelength intervals, we are interested in the total radiation of stars. Analogously to the solar constant, in the sense of equation (13.6) we therefore define apparent bolometric magnitude

$$\begin{aligned} m_{\text{bol}} &= -2.5 \log \frac{1}{r^2} \int_0^\infty \pi R^2 F_\lambda^L \, d\lambda + \text{constant} \\ &= -2.5 \log (\pi R^2 F / r^2) + \text{constant} \end{aligned} \tag{13.15}$$

where

$$\pi F = \sigma T_{\text{eff}}^4 \tag{13.16}$$

is again the total radiation flux at the surface of the star, including the effect of the spectral lines. Usually we define the constants so that, say, at about the solar temperature, the

$$\text{bolometric correction BC} = m_{\text{bol}} - m_{\text{v}} \tag{13.17}$$

is equal to zero. Since the Earth's atmosphere completely absorbs considerable parts of the spectrum, we cannot measure the bolometric magnitude directly from ground-based observations; even the term bolometric magnitude in the strict sense may be misleading. Until observations from satellites are available, we can evaluate m_{bol} or BC only with the help of a theory as a function of other measurable parameters of stellar atmospheres.

14. Distances, Absolute Magnitudes, and Radii of the Stars

As a result of the revolution of the Earth round the Sun, in the course of a year a nearer star must describe a small ellipse in the sky relative to the much more distant fainter stars (Figure 14.1: to be distinguished from aberration, Figure 5.7). Its semimajor axis, which is the angle that the radius of the Earth's orbit would subtend at the star, is called the heliocentric or annual *parallax p* of the star ($\pi\alpha\rho\acute{\alpha}\lambda\lambda\alpha\xi\iota\sigma$ = to and fro motion).

In the year 1838 F. W. Bessel in Königsberg succeeded, with the help of Fraunhofer's heliometer, in measuring directly (trigonometrically) the parallax of the star 61 Cygni $p = 0\overset{''}{.}293$. We must mention also the simultaneous work of F. G. W. Struve in Dorpat and of T. Henderson at the Cape Observatory. The star α Centauri, with its companion Proxima Centauri, observed by the latter is our nearest neighbor in space. According to recent measurements, their parallaxes are $0\overset{''}{.}75$ and $0\overset{''}{.}76$. It was a fundamental advance when in 1903 F. Schlesinger achieved the photographic measurement of parallaxes with an accuracy of about $\pm 0\overset{''}{.}01$. The *General catalog of trigonometric stellar parallaxes* and the *Catalog of bright stars* (Yale 1952 and 1964) are amongst the most important aids to astronomers.

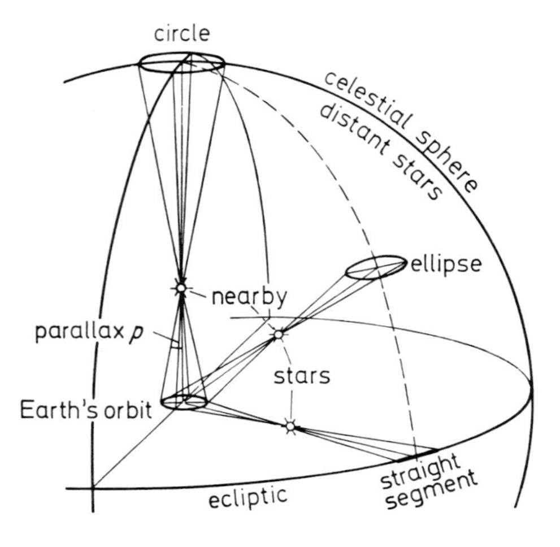

Figure 14.1
Stellar parallax *p*. Relative to the much more distant background stars, a nearby star in one year describes on the sky a circle of radius *p* if it is at the pole of the ecliptic, a straight segment $\pm p$ if it is in the ecliptic, and an ellipse if it is in between.

The first measurements of parallax signified not only the by then scarcely necessary vindication of the Copernican world system, but above all the first quantitative advance into space. We next define the appropriate units of measurement:

A parallax $p = 1''$ corresponds to a distance $360 \times 60 \times 60/2\pi = 206\,265$ astronomical units or radii of the Earth's orbit. We call this distance 1 parsec (from *par*allax and *sec*ond) contracted to pc. Thus we have

$$\begin{aligned} 1\ \text{pc} &= 3.086 \times 10^{13}\ \text{km} = 3.086 \times 10^{18}\ \text{cm or } 3.26\ \text{light years} \\ 1\ \text{light year} &= 0.946 \times 10^{18}\ \text{cm} \end{aligned} \tag{14.1}$$

Thus light traveling at 300 000 km/s takes 3.26 years to traverse 1 pc. The distance $1/p$ pc or $3.26/p$ light years corresponds to parallax p''. Corresponding to the accuracy with which trigonometric parallaxes can be measured, these take us out to distances of "only" some 15 pc or maybe 50 pc. We can reach out to about 2000 pc using "streaming" or cluster parallaxes (to be discussed later) for groups of stars that "stream" through the Milky Way system with a common velocity vector.

The observed brightness of a star (in any photometric system) depends according to equation (13.6) on its true brightness and its distance. We now define the *absolute magnitude M* of a star as the magnitude it would appear to have if it were transferred from its actual distance $r = 1/p$ pc to a standard distance of 10 pc. By the $1/r^2$ law of photometry [for example, equation (13.5)], the brightness is then changed by the factor $(r/10)^2$; in magnitudes we have therefore

$$m - M = 5 \log (r/10) = -5(1 + \log p). \tag{14.2}$$

We call $m - M$ the *distance modulus*. The following are corresponding values of the modulus and the distance:

$m - M =$	-5	0	$+5$	$+10$	$+25$ magnitudes
$r =$	1 pc	10 pc	100 pc	1000 pc = 1 kpc (1 kiloparsec)	10^6pc = 1 Mpc. (1 megaparsec)

(14.3)

In calculating in accordance with the $1/r^2$ law we have neglected interstellar absorption. We shall see that this can become significant for distances exceeding even 10 pc; then the relation (14.2) between distance modulus and distance (or parallax) is to be correspondingly modified.

Basically we can specify absolute magnitudes in every photometric system and we then attach the appropriate suffix to M; for example, visual absolute magnitude is written M_V. If no suffix is written then we always imply M_V.

We now collect the important properties of the Sun as a star; these serve also to make it a standard of comparison. We need scarcely emphasize the enormous technical difficulty of comparing the Sun photometrically with stars that are fainter by at least 10 powers of ten. The distance modulus of the Sun is determined by the definition of the parsec at the beginning of the Section directly as $m - M = -31.57$. Thus we obtain:

SUN: Apparent magnitude		color index	absolute magnitude	
Ultraviolet U	$= -26.06$	$U - B = +0.10$	$M_U = +5.51$	
Blue B	$= -26.16$	$B - V = +0.62$	$M_B = +5.41$	(14.4)
Visual V	$= -26.78$	$\mathrm{BC} = -0.07$	$M_V = +4.79$	
Bolometric m_{bol}	$= -26.85$		$M_{\text{bol}} = +4.72$	

The errors in these figures could be several units in the second decimal place. In particular, several astronomers prefer $B - V \approx 0.65$. The absolute bolometric magnitude of a star M_{bol} is a measure of its total energy emission per second by radiation. We call this its *luminosity* L, and we usually express it in terms of that of the Sun $L_\odot$ as unit. Thus

$$M_{\text{bol}} - 4.72 = -2.5 \log L/L_\odot \quad \text{with } L_\odot = 3.83 \times 10^{33} \text{ erg/s}. \tag{14.5}$$

Since the energy radiated away from the stars is generated by nuclear processes in their interiors, the luminosity is amongst the basic data for the investigation of the internal structure of the stars.

From equations (13.3) and (13.4) the flux S_λ that we receive from a star is $S_\lambda = F_\lambda(1 - \eta_\lambda)\pi R^2/r^2$. Here R is the radius and r the distance of the star. The ratio of these two quantities has a simple meaning: the very small angle α that the radius R subtends at distance r is $R/r = \alpha$ in radians or 206 265 $R/r = \alpha''$ in arc-seconds.

Then if we know the mean intensity of radiation $F_\lambda(1 - \eta_\lambda) = F_\lambda^L$ of a star, or its average value over a photometric band (for example, U, B, V), we can easily calculate the angle α from the apparent magnitude. Comparing with the Sun ($\alpha''_\odot = 959.6$) we have

$$m - m_\odot = -2.5 \log \{F_\lambda^L \alpha''^2 / F_{\lambda\odot}^L \alpha''^2_\odot\} \quad \text{or}$$
$$(\alpha''/\alpha''_\odot)^2 = (F_{\lambda\odot}^L/F_\lambda^L) 10^{-0.4(m - m_\odot)}. \tag{14.6}$$

For the radii themselves we have the corresponding result

$$M - M_\odot = -2.5 \log \{F_\lambda^L R^2 / F_{\lambda\odot}^L R_\odot^2\} \quad \text{or}$$
$$(R/R_\odot)^2 = (F_{\lambda\odot}^L/F_\lambda^L) 10^{-0.4(M - M_\odot)}. \tag{14.7}$$

As a crude estimate we may calculate the mean intensity F_λ according to the Planck radiation law, and in the Wien approximation ($c_2/\lambda T \gg 1$) we obtain

$$\frac{F_{\lambda\odot}}{F_\lambda} = \exp \frac{c_2}{\lambda}\left(\frac{1}{T} - \frac{1}{T_\odot}\right) \tag{14.8}$$

where λ is the isophotal wavelength and T, $T_\odot$ denote "the" temperatures of the star and the Sun.

If we wish to claim higher accuracy, however, we must calculate F_λ^L from the theory of stellar atmospheres and in terms of the parameters used in the theory. For certain bright red giant stars it has been possible to use the Michelson stellar interferometer to confirm the estimates got from

(14.6)–(14.8). For example, it was found for α Orionis (Betelgeuse) with $m_v = 0.9$, $p = 0\overset{''}{.}017$, and $T \approx 3200$ K

$$\alpha \approx 0\overset{''}{.}024 \text{ or } R \approx 300 R_\odot \text{ (somewhat variable).} \tag{14.9}$$

Thus the dimensions of this star correspond to about those of the orbit of Mars. Recently, using the correlation interferometer at Narrabri, Australia (Section 9), R. Hanbury Brown, R. G. Twiss, and others have measured the angular diameter $2\alpha''$ of more than a dozen stars with significantly improved accuracy.

15. Classification of Stellar Spectra: Hertzsprung–Russell Diagram and Color-Magnitude Diagram

When, following the discoveries of J. Fraunhofer, G. Kirchhoff, and R. Bunsen, the observation of stellar spectra was begun, it soon appeared that in the main these could be arranged in a one-parameter sequence. The associated change of color of the stars, or of their color indices, showed the arrangement to be in order of decreasing temperature.

Starting from the work of Huggins, Secchi, Vogel, and others, in the 1880s E. C. Pickering and A. Cannon developed the Harvard classification of stellar spectra, which formed the basis for the *Henry Draper* catalog. The sequence of spectral classes or "Harvard types"

$$\begin{array}{c} \qquad\qquad\qquad\qquad\quad \nearrow S \\ O—B—A—F—G—K—M \\ \qquad\qquad\qquad\qquad\quad \searrow R—N \\ \text{blue} \qquad \text{yellow} \qquad \text{red} \end{array} \tag{15.1}$$

where we have noted the colors of the stars, resulted after many modifications and simplifications. H. N. Russell's students in Princeton composed the well-known mnemonic *O*h *B*e *A* *F*ine *G*irl *K*iss *M*e *R*ight *N*ow.[1]

Between two letters a finer subdivision is denoted by numbers 0 to 9 written after the first letter. For example, a B5 star comes between B0 and A0 and has about the same amount in common with both these types.

The Harvard sequence is established primarily by use of photographs of the spectra of certain standard stars (Figure 15.1: the two spectra A0 I, A0 II should be disregarded for the moment). We describe it briefly in Table 15.1, associating with it the lines used as classification criteria, the chemical elements to which they belong, and the ionization levels (I = neutral atom, arc spectrum; II = singly ionized atom, for example Si^+, spark spectrum; III = doubly ionized atom, for example Si^{++} . . .). The temperatures listed

[1]Footnote for experts only: S is for Smack.

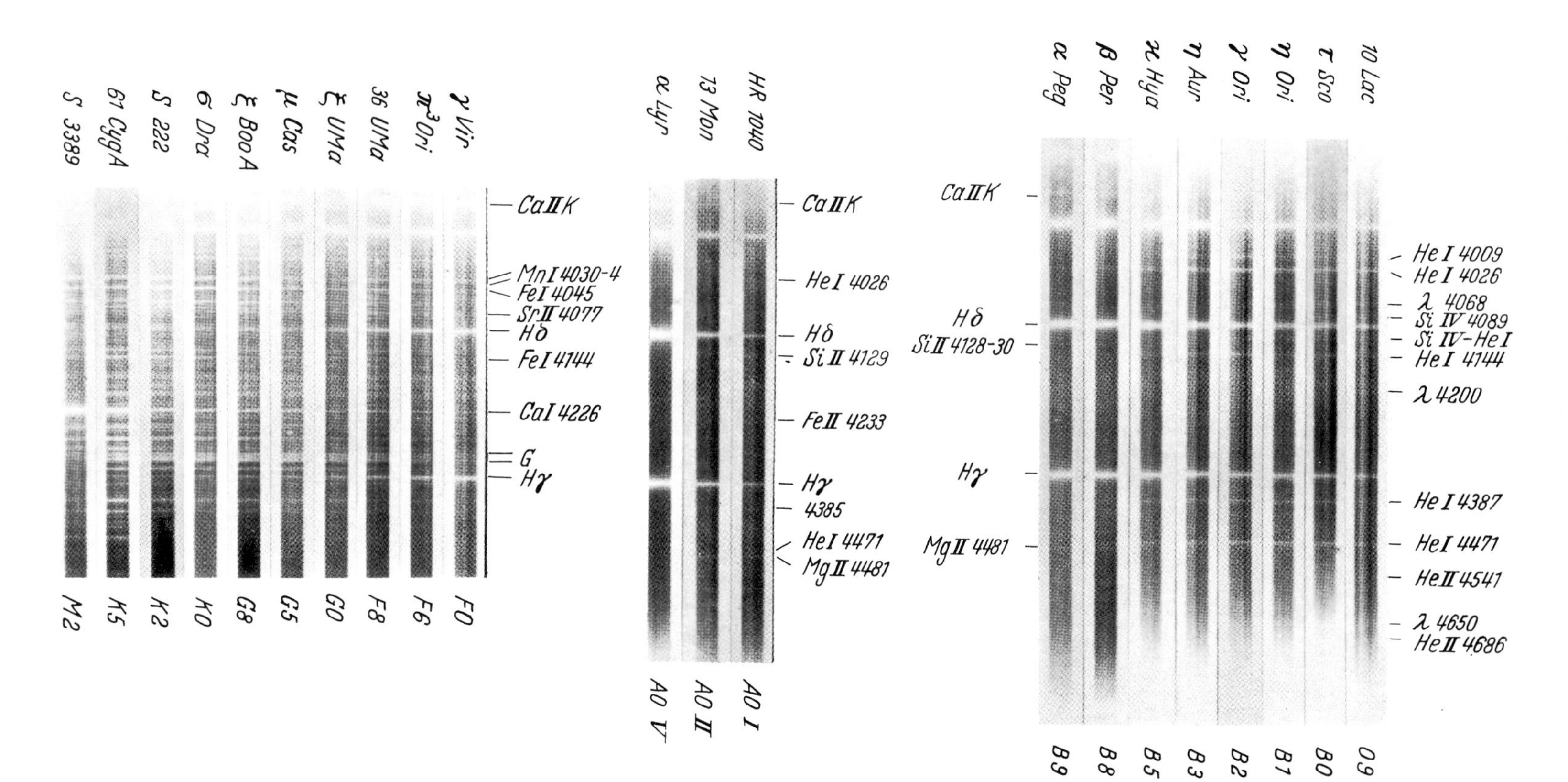

Figure 15.1. MK classification of stellar spectra. From "An Atlas of Stellar Spectra" by W. W. Morgan, P. C. Keenan, and E. Kellman (1942). The main sequence (luminosity class V) is shown from O9 to B9 (above), then A0V (center), and F0–M2 (below). The *luminosity classes* I (supergiants) and II (luminous giants) have been added at A0 in order to show the effect of absolute magnitude (spectroscopic parallax determinations are based on this effect).

Table 15.1. Classification of Stellar Spectra

Spectral type	Temperature (K)	Criteria for classification
O	50 000	Lines of highly ionized atoms: He II, Si IV, N III . . . ; hydrogen H relatively weak; occasionally emission lines.
B0	25 000	He II absent; He I strong; Si III, O II; H stronger.
A0	11 000	He I absent; H at maximum; Mg II, Si II strong; Fe II, Ti II weak; Ca II weak.
F0	7 600	H weaker; Ca II strong; the ionized metals, for example Fe II, Ti II, reach maximum about A5; the neutral metals, for example Fe I, Ca I, now reach about the same strength.
G0	6 000	Ca II very strong; neutral metals Fe I . . . strong.
K0	5 100	H relatively weak, neutral atomic lines strong; molecular bands.
M0	3 600	Neutral atomic lines, for example Ca I, very strong; TiO bands.
M5	3 000	Ca I very strong; TiO bands stronger.
R, N (C)	3 000	Strong CN, CH, C_2 bands; TiO absent; neutral metals as in K, M.
S	3 000	Strong ZrO, YO, LaO bands; neutral atoms as in K, M.

(R, N currently designated C)

correspond somewhat to the colors of the stars and are meant to serve only for provisional orientation.

A series of standard stars for the establishment of spectral types is given in Figure 15.1. Certain special features of many stellar spectra, which cannot be brought within the scope of a one-parameter classification, are denoted by the following symbols:

(1) The prefix c signifies specially sharp lines, particularly of hydrogen (Miss Maury's c stars, for example α Cyg cA2).
(2) The suffix n (nebulous) signifies a specially diffuse appearance of the lines, and s (sharp) a specially sharp appearance without the other criteria of c stars.
(3) v indicates a variable spectrum and applies to most variable stars. p (peculiar) indicates a peculiarity of any sort, for example anomalous strength of the lines of a particular element.

In the year 1913 H. N. Russell had the happy thought of studying the relationship between spectral type Sp and absolute magnitude M_v, and he

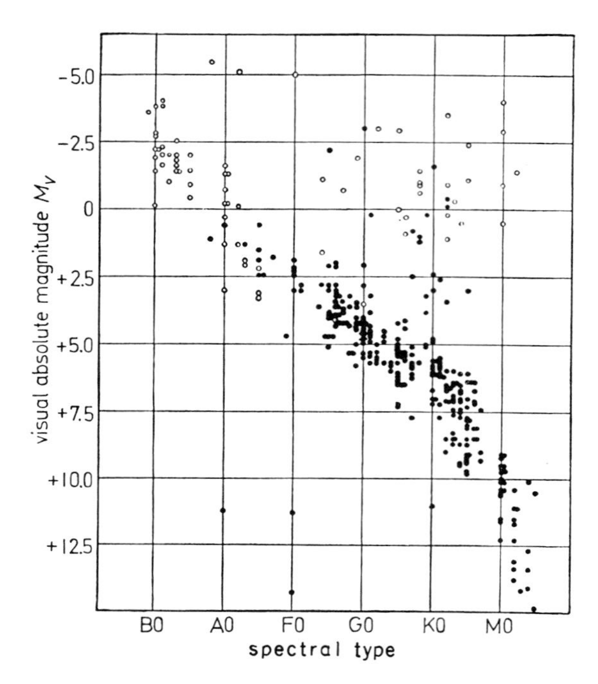

Figure 15.2. Hertzsprung–Russell diagram. Absolute visual magnitude M_v plotted against spectral type. The Sun corresponds to $M_v = 4.8$ and G2. The dots represent stars within 20 pc, which have reliable parallaxes. Open circles represent the less frequently occurring stars of greater absolute magnitude; for these the trigonometric parallaxes have been supplemented by spectroscopic and cluster parallaxes.

plotted in a diagram with Sp as abscissa and M_v as ordinate all the stars for which the parallax was sufficiently well determined. Figure 15.2 shows such a diagram which, using much improved observational material, Russell drew in 1927 for his textbook that served a whole generation as their "astronomical bible."

The majority of stars populate the narrow strip of the *main sequence* which runs diagonally from the (absolutely) bright blue-white B and A stars (for example, the stars in the belt of Orion) through the yellow stars (for example, Sun G2 and $M_v = +4.8$) to the faint red M stars (for example, Barnard's star M5 and $M_v = +13.2$).

In the right-hand upper part of the diagram there is the set of *giant stars;* by contrast, the stars that possess the same spectral type but much smaller luminosity are called *dwarf stars.* Since for about equal temperatures the difference in absolute magnitude can depend only upon a difference in stellar radius, this nomenclature seems very appropriate. The recognition and naming of giants and dwarfs goes back to the earlier work of E.

Hertzsprung (1905), for which reason we now call the Sp, M_v diagram the Hertzsprung–Russell diagram (HR diagram).

Instead of spectral type Sp one can also employ a color index, say $B-V$, and one then obtains a color-magnitude diagram which is equivalent to the HR diagram. Figure 15.3 shows the $(B-V, M_V)$ color-magnitude diagram, with now a very sharply defined main sequence, certain yellow giants (upper right) and white dwarfs (lower left), drawn for field stars and cluster stars in our vicinity in space that have well-determined parallaxes.

Since color indices can be measured with high accuracy even for faint stars, the color-magnitude diagram has become one of the most important working tools of astronomy.

The extremely bright stars along the upper boundary of the HR diagram or color-magnitude diagram are known as supergiants. For example α Cygni (Deneb; cA2) has absolute magnitude $M_v = -7.2$; therefore it exceeds the brightness of the Sun ($M_v = +4.8$) by 12 magnitudes, that is, a factor of about 63 000.

Another readily recognizable set consists of the *white dwarf* stars in the lower left of the diagram. Since in spite of relatively high temperature they

Figure 15.3. Color magnitude diagram (M_V against B-V) after H. L. Johnson and W. W. Morgan, showing main sequence derived from stars with trigonometric parallaxes $p \geqslant 0''.10$ and from stars taken from several galactic clusters with well-determined parallaxes and corrected for interstellar absorption and reddening. Also included are five white dwarfs (lower left) and several yellow giants (upper right). The stars lying above the main sequence (from Praesepe) are probably binaries.

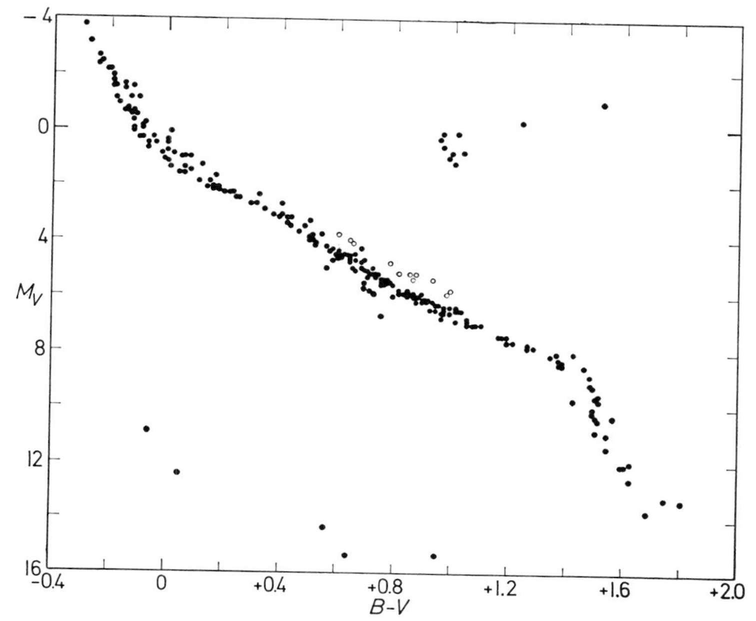

have small luminosity, they must be very small; one can easily calculate the radius, which is scarcely larger than that of the Earth. For the companion of Sirius α CMa B and certain similar objects one also knows the mass and thence one derives mean densities of the order of 10^4 to 10^5 g/cm^3. The internal constitution of such stars must be quite different from the rest. In 1926 R. H. Fowler showed that in white dwarfs the material (more precisely, the electrons) is degenerate in the sense of Fermi statistics in the same way as shortly thereafter W. Pauli and A. Sommerfeld showed to be the case for electrons in metals. That is to say, almost all the quantum states are occupied in the sense of the Pauli principle, as are the inner shells of heavy atoms.

We shall mention other, mostly smaller and more special, sets of stars in the HR diagram in other contexts.

E. Hertzsprung remarked as long ago as 1905 that Miss Maury's c stars with sharp hydrogen lines are distinguished by their great luminosity. Then in 1914 W. S. Adams and A. Kohlschütter showed that the spectra of stars of a given spectral type could be further subdivided according to luminosity on the basis of new spectroscopic criteria. In absolutely bright stars, for example, the lines of ionized atoms (spark lines) are enhanced relative to those of neutral atoms (arc lines); amongst the A stars, as has been said, the sharpness of the hydrogen lines serves as a luminosity criterion.

If we calibrate such a criterion, which of course applies only to a limited range of spectral types, with the use of stars of known absolute magnitude, then with the help of the resulting calibration curve we can determine further absolute magnitudes by spectroscopic means. If we can neglect, or correct for, interstellar absorption (which had still not been dreamt of in 1914), then by combining these absolute magnitudes with the apparent magnitudes we derive *spectroscopic parallaxes* of the stars concerned [equation (14.2)]. In Part III we shall mention the significance of these for the investigation of the Milky Way system. Here we pursue the important insight afforded by the fact that we can classify the great majority of stars according to *two parameters*.

On the basis of the Harvard classification, etc., W. W. Morgan and P. C. Keenan developed the now generally used MK classification presented in *An atlas of stellar spectra with an outline of spectral classification* (1943 with E. Kellman and 1953). Its general principles hold good for any such classification:

(1) Only empirical criteria, that is, directly observed absorption and emission phenomena are the basis of the classification.
(2) The observational material is uniform. In order on the one hand to include sufficiently refined criteria and on the other hand to reach sufficiently far into the Galactic system, a uniform dispersion of 125Å/mm at Hγ is employed, even for bright stars.[2]

[2]If we work with other spectrographs, in particular with higher dispersion, we must first photograph spectra of the standard stars (given in part in Figure 15.1) and then classify the spectra of other stars by comparison with these.

(3) The transferability of the system of classification from one instrument to another is validated by using a list of suitable standard stars, that is, by direct application, and not by using specifications, that may very well be semitheoretical.

(4) The classification is according to

(a) *Spectral type* Sp largely in agreement with the Harvard classification and with the same notation.

(b) *Luminosity class* LC. Its calibration in terms of absolute magnitude will be considered in retrospect. An LC criterion over at any rate a large range of spectral types is to depend in the first place as sensitively as possible upon the luminosity of the stars. Morgan's luminosity classes give at the same time the location of the stars in the HR diagram; they are

Ia–0 I	Supergiants
II	Bright giants
III	Giants
IV	Subgiants
V	Main sequence (dwarfs)
VI	Subdwarfs

Figure 15.4. Spectral type Sp and luminosity class LC of the MK classification; dependence on color index *B-V* and visual absolute magnitude M_V.

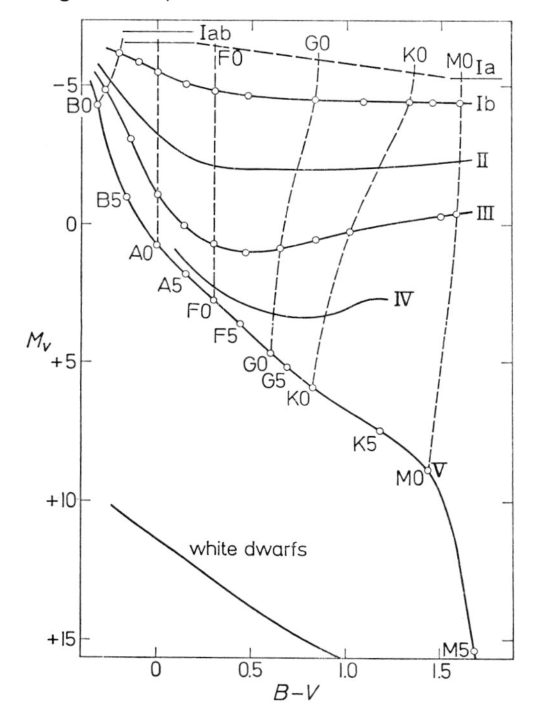

As needed, the luminosity classes I to V may be subdivided using suffixes a, ab, and b. Figure 15.1, extracted from the *Atlas of stellar spectra,* shows the important spectral types of the main sequence stars (LC = V) and, at least for spectral type A0, the division into luminosity classes I to V according to the width of the hydrogen lines.

About 90 percent of all stellar spectra can be dealt with by the MK classification; the rest are partly composite spectra of unresolved double stars and partly peculiar (p) spectra of pathological individuals.

The relationship between the parameters of the MK classification Sp and LC on the one hand, and the color index $B-V$ and absolute magnitude M_V on the other, according to the best available calibration, is shown in Figure 15.4.

In recent astronomy the two-color diagram of W. Becker (1942) plays a role second in importance only to the color-magnitude diagram. In this

Figure 15.5. Two-color diagram for the main sequence. The reddening-free color indices $(U\text{-}B)_0$ and $(B\text{-}V)_0$ are after H. L. Johnson, W. W. Morgan and others. The MK spectral types and absolute magnitudes of the stars are written along the main sequence. The almost straight line above the main sequence represents black radiators, the numbers are $T_{\text{eff}} \times 10^{-3}$ K. Interstellar reddening shifts the position of a star in the diagram parallel to the line in the upper right, which is drawn in particular for O stars.

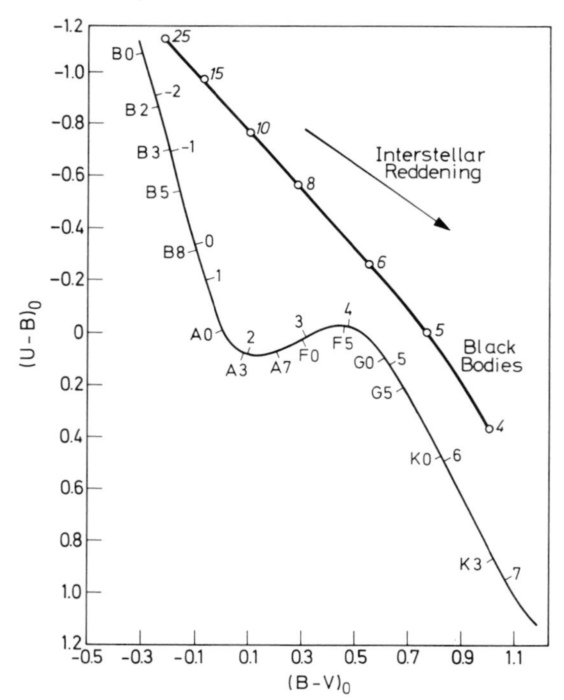

(Figure 15.5), the short-wavelength color index $U-B$ is plotted (increasing downwards) against $B-V$. As can easily be shown using equation (13.4) and the corresponding one for $U-B$, black bodies are approximately represented in this diagram by a 45° straight line. In Figure 15.5 we have used the more precise numerical calculations of I. Bues. The relation between $U-B$ and $B-V$ for the main sequence, with the corresponding spectral types and absolute magnitudes, is also shown.

The large difference between the spectral energy distributions of stars and of black bodies we shall explain in Section 18 using the theory of stellar spectra. In Part III we shall be able to consider the application of the two-color diagram to the determination of interstellar reddening as well as to the recognition of particular kinds of stars.

16. Double Stars and the Masses of the Stars

In 1803 F. W. Herschel discovered that α Gem = Castor is a visual double star whose components move round each other under the influence of their mutual attraction. Thus the observation of double stars offers the possibility of determining the masses of stars or at any rate of making quantitative assertions about them. Since the masses of stars are detectable only through their gravitational interaction, the investigation of double stars is even now of fundamental significance for the whole of astrophysics.

We first separate the optical pairs from physical pairs or true binary stars using statistical criteria, also when possible by obtaining proper motions and radial velocities. The apparent orbit of the fainter component, the "companion", around the brighter component is observed with a refractor provided with a filar micrometer (Figure 9.1) in terms of the separation (in seconds of arc) and the position angle (N 0°—E 90°—S 180°—W 270°). If one plots this apparent orbit one obtains an ellipse. Were we to look at right angles to the plane of the orbit, we should see the bright component at the focus of the orbit. In general this does not occur because the plane of the orbit forms with the plane of the sky an angle of inclination i. Conversely, one can obviously determine the inclination i so that the orbit satisfies the Kepler laws. Let

a = semimajor axis of the true (relative) orbit in seconds of arc,
p = parallax of the binary star in seconds of arc

then a/p = semimajor axis of the true orbit in astronomical units. If, further, P is the period of description of the orbit in years, then from Kepler's third law (6.33) we can easily write down the combined mass of the two stars $\mathfrak{M}_1 + \mathfrak{M}_2$ (in units of the solar mass) as

$$\mathfrak{M}_1 + \mathfrak{M}_2 = a^3/p^3P^2. \tag{16.1}$$

If by use of a meridian circle or photographically (following E. Hertzsprung) one measures the absolute motion of the two components (that is, relative to the background stars, after allowing for parallactic motion and proper motion), one derives the semimajor axes a_1 and a_2 of their true paths around the mass center and one has from equation (6.31)

$$a_1 : a_2 = \mathfrak{M}_2 : \mathfrak{M}_1 \quad \text{and} \quad a = a_1 + a_2 \tag{16.2}$$

so that one can now calculate the individual masses $\mathfrak{M}_1, \mathfrak{M}_2$.

If the fainter component of a binary star is not visible, its presence can nevertheless be inferred from the motion of the brighter component about the mass center, measured in an absolute fashion. If a_1 is the semimajor axis of this orbit, again in arc-seconds, since $a_1/a = \mathfrak{M}_2/(\mathfrak{M}_1 + \mathfrak{M}_2)$ we obtain

$$(\mathfrak{M}_1 + \mathfrak{M}_2)\left(\frac{\mathfrak{M}_2}{\mathfrak{M}_1 + \quad {}_2}\right)^3 = \frac{a_1^3}{p^3 P^2}. \tag{16.3}$$

Using meridian observations, F. W. Bessel in 1844 found in this way that Sirius must have a dark companion. A. Clark in 1862 actually discovered Sirius B, which is about $10^{\text{m}}.14$ fainter than Sirius itself. Sirius B has absolute magnitude only $M_{\text{v}} = +11.54$ although its mass is 0.96 solar mass. Since the surface temperature is quite normal, as already remarked, it must be very small. In 1923 F. Bottlinger concluded that "here we have to do with something entirely new," namely a *white dwarf* star.

In recent times K. A. Strand and P. van de Kamp have found dark companions of certain nearer stars that represent an intermediate stage between stars and planets (that is, between bodies with and without their own energy sources). In particular, our second nearest neighbor at a distance of 1.8 pc, Barnard's star, which has spectral type M5V and a mass of ~0.15 solar masses, has a kind of planetary system. Van de Kamp's original discussion suggested a companion of 0.0016 solar masses or 1.7 times the mass of Jupiter, with a period of 25 years. A more recent study (1969) with many more observations points to *two* companions with 1.1 and 0.8 times the mass of Jupiter which revolve about the primary star in the same sense in almost circular and coplanar orbits with periods of 26 and 12 years, respectively.

In 1889 E. C. Pickering observed that in the spectrum of Mizar $= \zeta$ U Ma the lines twice become double in time intervals of $P = 20^{\text{d}}.54$. Thus Mizar shows itself to be a spectroscopic binary. In this particular system two similar A2 stars revolve around each other; their angular separation is too small for telescopic resolution. In other systems only one component is recognizable in the spectrum; the other is obviously considerably fainter. If we plot the radial velocity as derived from the Doppler effect for either or both components as a function of time, we get the *velocity curve*. After removing the mean velocity or the velocity of the mass center, we can read off the sight-line component of the orbital velocity. Without going into the details here, we can calculate thence, not the semimajor axis itself, but the quantity $a_1 \sin i$ if only component 1 is detectable in the spectrum, and also $a_2 \sin i$ if component 2 is detectable as well, where i is the (unknown) inclination of the orbit.

If only one spectrum is visible, we derive from Kepler's third law and from the mass-center formula (16.2)

$$\frac{(a_1 \sin i)^3}{P^2} = (\mathfrak{M}_1 + \mathfrak{M}_2)\left(\frac{\mathfrak{M}_2}{\mathfrak{M}_1 + \mathfrak{M}_2}\right)^3 \sin^3 i = \frac{\mathfrak{M}_2^3 \sin^3 i}{(\mathfrak{M}_1 + \mathfrak{M}_2)^2}. \qquad (16.4)$$

The last quantity is called the *mass function*. For statistical purposes we can make use of the fact that the mean value of $\sin^3 i$ for all possible inclinations is 0.59, or, taking account of the probability of discovery, about 2/3. Since $\mathfrak{M}_2 < \mathfrak{M}_1$, in all cases the factor $(\mathfrak{M}_2/(\mathfrak{M}_1 + \mathfrak{M}_2))^3 < 1/8$.

If both spectra are visible then we obtain $\mathfrak{M}_1 \sin^3 i$ and $\mathfrak{M}_2 \sin^3 i$ and thence also the mass ratio $\mathfrak{M}_1 : \mathfrak{M}_2$.

If the inclination of the orbital plane of a spectroscopic binary is near to 90° then eclipses occur and we observe an eclipsing variable or *eclipsing binary*. The classic example is β Persei = Algol, with period $P = 2^d20^h49^m$ which J. Goodricke interpreted as such in 1782. From the magnitudes of an eclipsing variable measured (preferably photoelectrically) over a long enough time, we obtain first the period P and then the light curve. Hence we derive the radii of both stars in terms of the radius of the relative orbit as unit and also the inclination of the orbit i (Figure 16.1). If we can determine in addition the velocity curve for one or both components, we derive also the absolute dimensions of the system as well as the masses, and thence the mean densities, of both stars. In favorable cases we can derive even the ellipticity (flattening) and the center-limb darkening of the component stars. Thanks to the methodology for the determination of the elements of

Figure 16.1. Apparent relative orbit and light curve of the eclipsing binary IH Cassiopeiae. Corresponding points of the orbit and light curve are indicated by numbers. In this case the primary eclipse of the brighter by the fainter and smaller component is annular.

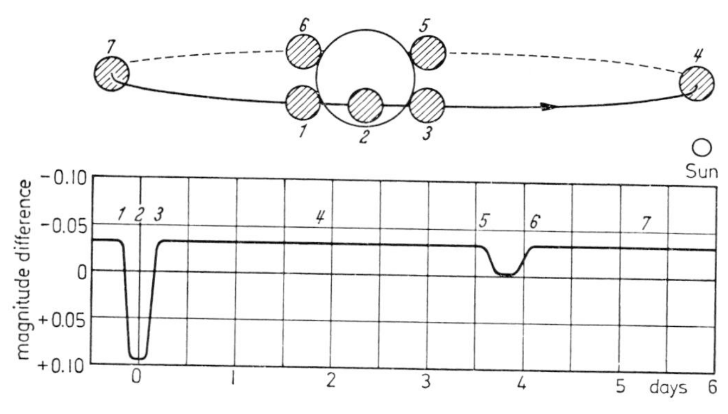

eclipsing variables developed to the greatest completeness by H. N. Russell and H. Shapley these are at the present time among the most accurately known stars. As a result of penetrating analyses of the spectra, O. Struve later showed that in close pairs the two components are in a state of material interaction with each other. Common gaseous envelopes and streams of gas flowing from one component to the other sometimes give direct insight into the *evolution* of such systems.

We now seek to give a sketch in broad terms; on account of unavoidable selection effects a detailed statistical discussion would be of doubtful significance:

The various sorts of visual, spectroscopic, and eclipsing binaries, which are distinguished only by the way in which they are observed, join together continuously and with a certain degree of overlapping. The periods range from a few hours to many thousands of years. The binaries with short periods mostly have circular orbits; long-period systems favor larger eccentricities. Besides double stars, multiple systems also frequently occur, which mostly include one or more close pairs. For instance, the "double star" α Gem = Castor first discovered by F. W. Herschel was later shown to consist of three pairs A, B, C with periods 9.22, 2.93, 0.814 days; A and B move round each other in several hundred years, Castor C moves round A and B in several thousand years. According to P. van de Kamp (1969) our neighborhood within a distance of 5.2 pc contains, including the Sun, 31 apparently single stars (6 with invisible companions), 11 double stars (61 Cyg has an invisible companion), and 2 triple systems. Thus almost half of the total of 59 stars are members of double or multiple systems.

In the spectra of double stars and eclipsing variables of short period, the components of which revolve around each other in close proximity, the Fraunhofer lines are mostly extraordinarily broad and diffuse. This depends on the fact that the two components rotate together like a rigid body, as a result of tidal friction. The periods of revolution and rotation are equal. If the component of the equatorial velocity along the sight line is $v \sin i$, then at wavelength λ there is a Doppler displacement $\Delta\lambda = \pm\lambda(v/c)\sin i$. Were the spectral line sharp for a stationary star, it would now appear as spread out into a band of width $2\Delta\lambda$, the profile of which would indicate the brightness distribution across the stellar disk, for example, in the case of no limb darkening we get a line profile of elliptic shape. For instance, if a B star of radius $5R_\odot$ rotates with period 1 day and if $i = 90°$, then its projected equatorial velocity $v \sin i = 250$ km/s, and half the resulting width of the line MgII 4481 is $\Delta\lambda = \pm 3.73$ Å.

O. Struve and his collaborators then discovered that there are also single stars in whose spectra all the lines are strongly widened in this way, and that therefore rotate with equatorial velocities of up to about 300 km/s. Like the rapidly rotating binaries, the rapidly rotating single stars belong predominantly to spectral types O, B, and A in the upper part of the main sequence.

We shall later return to consider the significance of rotation of single and binary stars and therewith to the role of angular momentum in problems of stellar evolution.

To conclude this section we take a look at the *masses* of the stars. Their values as determined from binaries of all kinds range from about $0.15\mathfrak{M}_{\odot}$ to $50\mathfrak{M}_{\odot}$. Their relation to other stellar parameters long remained obscure until A. S. Eddington in 1924 in connection with his theory of the internal constitution of the stars discovered the *mass-luminosity relation*. We can understand this in principle from a present-day standpoint as follows: The stars of the main sequence are evidently in analogous states of development (their energy requirements being supplied by the conversion of hydrogen into helium); in the main, they are therefore built to the same pattern. To a given mass, therefore, there belong energy sources of a well-determined amount, and these in turn determine the luminosity of the star. We therefore expect there to be a relation between the mass $\mathfrak{M}$ and the luminosity L or absolute bolometric magnitude M_{bol}. As the analysis of all the observational material shows (Figure 16.2), there is actually such a relation for main sequence stars. As expected, the white dwarfs do not conform to this. Also "the" mass-luminosity relation cannot without further discussion be applied to the red giant stars for which no reliable empirical masses are yet available.

Figure 16.2. Empirical mass-luminosity relation. The absolute bolometric magnitude M_b or the luminosity L plotted as a function of the mass $\mathfrak{M}$ (after G. P. Kuiper). ● visual binaries, ○ spectroscopic binaries, + Hyades binaries, □ white dwarfs.

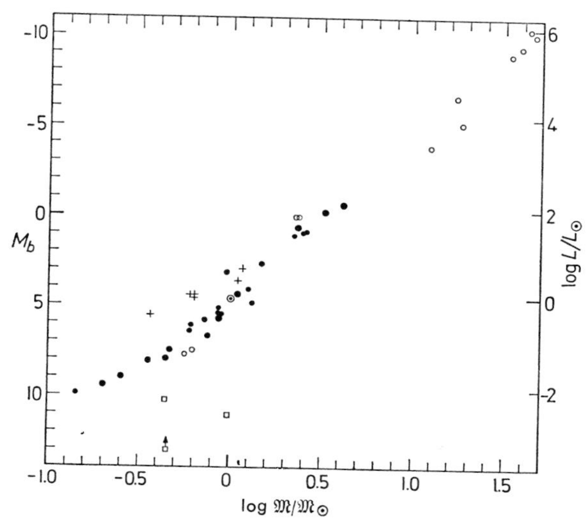

If we regard the radii R of the stars as known (calculated from the absolute magnitude and, roughly speaking, the temperature) we can evaluate the

$$\textit{surface gravity } g = G\mathfrak{M}/R^2 \text{ cm/s}^2, \tag{16.5}$$

which is important for the theory of the spectra. As we shall verify more exactly, we find that for main sequence stars of spectral types from B0V to M3V the value is constant to within about a factor 2 with

$$g \approx 20\,000 \text{ cm/s}^2 \quad \text{or} \quad \log g \approx 4.3. \tag{16.6}$$

For giants and supergiants the value is considerably less (down to $\log g \approx 0.5$), and for white dwarfs considerably greater ($\log g \approx 8 \pm 0.5$).

The important results of Sections 15 and 16 are summarized in Figure 16.3: the relations between luminosity L, bolometric magnitude M_{bol},

Figure 16.3. Relationships between luminosity L or absolute bolometric magnitude M_{bol} (vertical scales on left or right, respectively), effective temperature T_{eff} (abscissa) and stellar radius R as well as gravity g. The latter is given as g_1 for stars of 1 $\mathfrak{M}_\odot$; these numbers correspond roughly to old evolved stars. For main sequence and slightly evolved stars the mass $\mathfrak{M}$ is given approximately on the far right as a function of M_{bol}.
Units: $L_\odot = 3.82 \times 10^{33}$ erg/s, $R_\odot = 6.96 \times 10^{10}$ cm,
$\mathfrak{M}_\odot = 1.99 \times 10^{33}$ $\mathfrak{M}g$ $g_\odot = 2.74 \times 10^{4}$ cm/s².

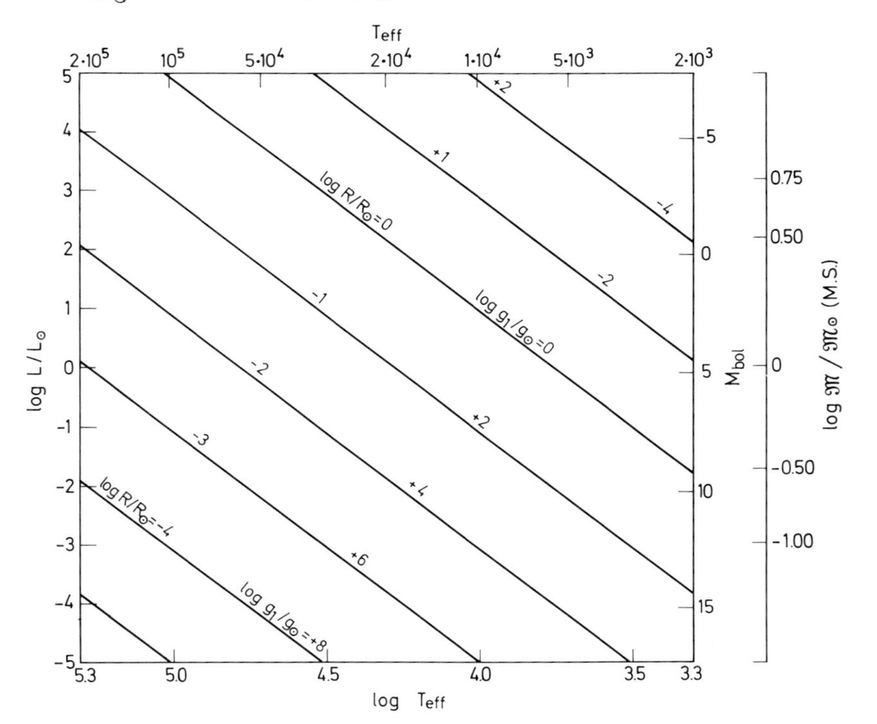

effective temperature T_{eff}, stellar radius R, and surface gravity g, where we again use the familiar equations

$$L = 4\pi R^2 \sigma T_{\text{eff}}^4 \tag{16.7}$$
$$g = G\mathfrak{M}/R^2. \tag{16.8}$$

We refer L and R to $L_\odot = 3.82 \times 10^{33}$ erg/s and $R_\odot = 6.96 \times 10^{10}$ cm as units. We calculate the gravity for stars of one solar mass, referred to $g_\odot = 2.74 \times 10^4$ cm/s^2 as unit. The scale of mass on the right is approximately correct for main sequence and slightly evolved stars (Section 26).

In Section 19 we shall see that the effective temperature and surface gravity can be obtained from the analysis of stellar spectra. Then, from equations (16.7) and (16.8), the *mass-to-light ratio* can be calculated:

$$\frac{\mathfrak{M}}{L} = \frac{1}{4\pi G\sigma} \frac{g}{T_{\text{eff}}^4}. \tag{16.9}$$

If we want to determine $\mathfrak{M}$ and L separately, we must use either the theory of the internal constitution of the stars (Section 25) or corresponding empirical data.

17. Spectra and Atoms: Thermal Excitation and Ionization

After important pioneering work by N. Lockyer, in the year 1920 M. N. Saha's *Theory of thermal excitation and ionization* led to the interpretation of stellar spectra and their classification. It rests essentially upon the quantum theory of atoms and their spectra developed by N. Bohr, A. Sommerfeld, and others from 1913 onward. Here we briefly recall the fundamentals without their full derivation:

We represent the possible energy levels of an atom graphically in an energy level or Grotrian diagram (Figure 17.1). We distinguish:

(1) *Discrete negative energy levels* $E < 0$ corresponding to the bound or elliptic orbits of the electrons in the Bohr model of the atom or the discrete eigenfunctions (standing de Broglie waves) of quantum mechanics. Each energy level is characterized by a set of several integral or half-integral quantum numbers which we represent by a *single* symbol n, m, s, or the like.

(2) *Continuous positive energy values* $E > 0$ corresponding to the free or hyperbolic orbits of electrons in the Bohr model or to "continuous" eigenfunctions (progressive de Broglie waves) in quantum mechanics. At a great distance from the atom such an electron has only kinetic energy $E_{\text{kin}} = \frac{1}{2}mv^2$, where m is its mass and v its velocity.

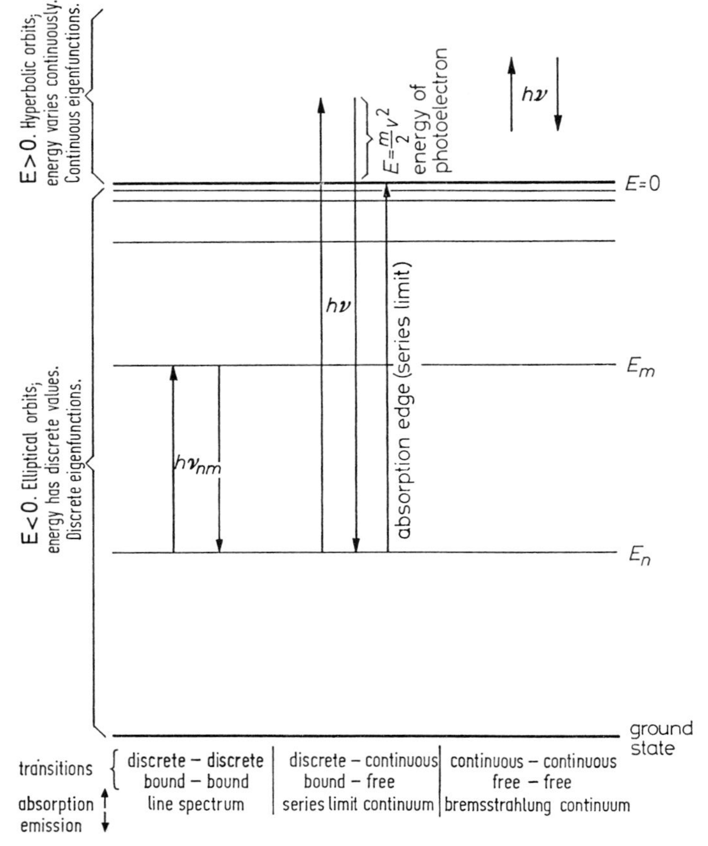

Figure 17.1. Energy level or Grotrian diagram of an atom (schematic only) showing various transitions.

In a transition between two energy levels E_m, E_n a photon of energy

$$h\nu = |E_m - E_n| \tag{17.1}$$

is absorbed (↑) or emitted (↓), where $h = 6.62 \times 10^{-27}$ erg s is Planck's quantum of action. The frequency ν s^{-1} or Hz (Hertz) corresponds to a wavenumber (number of light waves per centimeter in vacuum) $\bar{\nu} = (\nu/c)$ cm^{-1} or Kayser and a wavelength $\lambda = 1/\bar{\nu} = c/\nu$ cm. Besides the centimeter we use as unit 10^{-8} cm = 1 Å. We often reckon the energy values, not from $E = 0$ but from the ground state of the atom. As unit we use generally not 1 erg but 1 cm^{-1} or 1 Kayser and then we speak of the terms and the term scheme of the atom, or else we use 1 eV (electronvolt) or the energy that an electron gains on traversing a potential increase of 1 V. In thermal equilibrium we have always to deal with energies of the order of magnitude kT ($k = 1.38 \times 10^{-16}$ erg/K). In this context we specify the temperature in Kelvin corresponding to an energy E. There correspond thus

$$1 \text{ eV} \mathrel{\hat{=}} 1.602 \times 10^{-12} \text{ erg} \mathrel{\hat{=}} 8066 \text{ cm}^{-1} \mathrel{\hat{=}} 12\,398 \text{ Å} \mathrel{\hat{=}} 11\,605 \text{ K}. \quad (17.2)$$

We naturally divide the transitions undergone by an atom as a result of the absorption or emission of a single photon into the following classes:

(1) $E_m < 0$, $E_n < 0$; elliptic-elliptic, discrete-discrete, or bound-bound transitions under absorption or emission of a spectral line whose wavenumber $\bar{\nu}$ cm^{-1} we obtain as the difference of the term-values in C. E. Moore's *Atomic energy levels.*

(2) $E > 0$, $E_n < 0$; hyperbolic-elliptic, continuous-discrete, or free-bound transitions. At the series limit or absorption edge $h\nu_n = E_n$ the continuous-discrete absorption $\nu > \nu_n$ is accompanied by the ejection of a photoelectron having kinetic energy $\frac{1}{2}mv^2 = h\nu - |E_n|$ which is Einstein's photoelectric formula. Thereby the atom becomes ionized or goes into the next higher state of ionization. We denote the spectra of neutral, singly ionized, doubly ionized . . . atoms, for example calcium, by Ca I, Ca II, Ca III, and so on.

The inverse process is the capture of a free electron of energy $\frac{1}{2}mv^2$ with the emission of a photon

$$h\nu = \tfrac{1}{2}mv^2 + |E_n|$$

the so-called *two-body recombination.*

(3) $E' > 0$, $E'' > 0$ gives hyperbolic-hyperbolic, continuous-continuous, or free-free transitions. A photon $h\nu = |E' - E''|$ is absorbed or emitted; the free electron gains or loses the corresponding amount of kinetic energy in its encounter with the atom.

Bound-free and free-free absorption and emission were first found in the X-ray region and they were explained theoretically by N. Bohr and H. A. Kramers.

From now on[1] we shall be concerned with the *discrete terms* ($E_n < 0$) of atoms and ions.

We describe a particular energy level of an atom or ion having a single emitting or valence electron (that is, the remaining electrons undergo no transition) by *four quantum numbers:*

n the principal quantum number. In the case of hydrogen-like (coulomb-field) orbits, in the language of Bohr's theory, n^2a_0/Z is the semimajor axis of the orbit, $-e^2Z^2/2a_0n^2$ erg is the corresponding energy, and $R_\infty Z^2/n^2$ cm^{-1} the term value. Here $a_0 = 0.529$ Å is the radius of the first Bohr orbit for hydrogen, $R_\infty = 109\,737.30$ cm^{-1} is the Rydberg constant, and Z the effective nuclear charge number ($Z = 1$ for a neutral atom, arc spectrum; $Z = 2$ for a singly ionized atom, the first spark spectrum, and so on).

[1]The newcomer to the subject need not work through all the details of the following short introduction to the quantum theory and classification of atomic spectra (about the next two pages). It is important that he should understand the meaning of the energy level or term scheme in Figures 17.1, 17.3, and 17.4.

l is the orbital angular momentum of the electron measured in the quantum unit $\hbar = h/2\pi$, and l can assume integral values 0, 1, 2, . . ., $n - 1$.

$$l = 0\ 1\ 2\ 3\ 4\ 5$$

gives respectively a

s p d f g h electron.[2]

s is the spin angular momentum (Goudsmit–Uhlenbeck) in the same units. For a single electron $s = \pm 1/2$

j is the total angular momentum, again in units of $\hbar$. j is given by vector addition of l and s and can only have one of the values $l \pm 1/2$.

As an example, one electron having $n = 2$, $l = 1$, and $j = 3/2$ is denoted as a $2p_{3/2}$ electron.

In atoms or ions with several electrons, as H. N. Russell and F. A. Saunders found in 1925 in the case of the alkaline earths, the angular momentum vectors are normally coupled as follows (Russell–Saunders or *LS coupling)*:

The orbital angular momenta $\vec{l}$ are added vectorially to give the resultant orbital angular momentum $L = \Sigma\vec{l}$; similarly the spin angular momenta $\vec{s}$ give the resultant $S = \Sigma\vec{s}$. Then L and S combine, again vectorially, to give the total angular momentum J (Figure 17.2) so that

$$|L - S| \leqslant J \leqslant L + S. \tag{17.3}$$

[2]The notation referred originally to the upper term of the series: s = sharp secondary series; p = principal series; d = diffuse secondary series; f = fundamental series.

Figure 17.2. Vector model for an atom with Russell–Saunders coupling. The total orbital angular momentum vector L and the spin angular momentum vector S combine to give the total angular momentum vector J (all being measured in units of $\hbar = h/2\pi$). M_J denotes the component of J along the direction of an external field $\boldsymbol{H}$. The vectors L and S precess about J, which itself precesses about $\boldsymbol{H}$. The diagram corresponds to the particular energy level $L = 3$, $S = 2$, $J = 3$, that is, to the $^5F_3$ level.

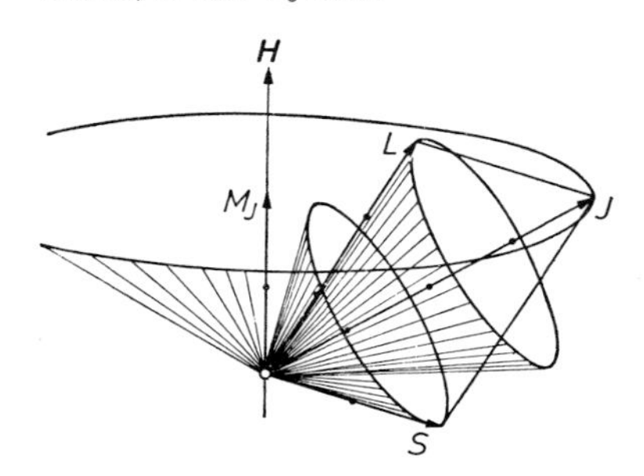

L is always integral; S and J are half integral (integral) for atoms with an odd (even) number of electrons.

A particular pair of values of S and L constitutes, as we say, a *term*. By analogy with the single-electron system, the orbital angular momentum quantum number L corresponds to a term designated as follows:

$$\begin{array}{cccccc} L = 0 & 1 & 2 & 3 & 4 & 5 \\ \mathrm{S} & \mathrm{P} & \mathrm{D} & \mathrm{F} & \mathrm{G} & \mathrm{H}. \end{array}$$

So long as $L \geqslant S$, the term splits into $r = 2S + 1$ energy levels with different J. Then, and also if $L < S$, the number r is called the multiplicity of the term, and it is written to the upper left in the term symbol; J is written to the lower right as a subscript, in order to designate the particular levels in the term. Table 17.1 gives a synopsis of possible terms of various multiplicities, their energy levels, and the usual designation.

In the presence of an external field (for example, a magnetic field) the vector of total angular momentum J sets itself so that its component M_J in the direction of the field is half integral or integral. M_J can thus assume the values $J, J - 1, \ldots, -J$; thus the directional quantization of J gives $2J + 1$ possible configurations (Figure 17.2). When the external field vanishes, these $2J + 1$ energy levels fall together; we say that the J level is $(2J + 1)$-ply *degenerate*. Furthermore we divide the terms according to their parity into two groups, the *even* and *odd* terms according as the arithmetic sum of the l values of the participating electrons is even or odd. Odd terms are distinguished by an affix ^{o} written to the upper right.

A *line* arises from a transition between two energy levels; the possible transitions between all the levels in two *terms* produce a group of neighboring lines called a *multiplet*. The possibilities for transitions (in the emission or absorption of electric-dipole radiation, analogous to that of the familiar Hertz dipole) are restricted by the following selection rules:

(1) Transitions occur only between odd and even levels.
(2) J changes only so that $\Delta J = 0, \pm 1$. The transition $0 \rightarrow 0$ is forbidden.

Table 17.1. Terms and J values of their levels for different quantum numbers L and S for Russell–Saunders coupling.

		$S = 0$	$\frac{1}{2}$	1	$\frac{3}{2}$
		$r = 2S + 1 = 1$	2	3	4
		singlet	doublet	triplet	quartet
$L = 0$	S term	$J = 0$	$J = \frac{1}{2}$	$J = 1$	$J = \frac{3}{2}$
1	P term	1	$\frac{1}{2}$ $\frac{3}{2}$	0 1 2	[$\frac{1}{2}$ $\frac{3}{2}$ $\frac{5}{2}$]
2	D term	2	$\frac{3}{2}$ $\frac{5}{2}$	1 2 3	$\frac{1}{2}$ $\frac{3}{2}$ $\frac{5}{2}$ $\frac{7}{2}$
3	F term	3	$\frac{5}{2}$ $\frac{7}{2}$	2 3 4	$\frac{3}{2}$ $\frac{5}{2}$ $\frac{7}{2}$ $\frac{9}{2}$

Example [ ] Quartet P term with energy-levels $^4P_{1/2}$, $^4P_{3/2}$, $^4P_{5/2}$. Statistical weight of the term $g(^4P) = 4 \times 3 = 2 + 4 + 6$.

For LS coupling we have further:

(3) $\Delta L = 0, \pm 1$.
(4) $\Delta S = 0$, that is, there are no intercombinations (for example, singlet triplet).

Russell–Saunders or LS coupling is recognized by the fact that multiplet splitting, arising from the magnetic interaction between orbital momenta and spin momenta, is small compared with the separation between neighboring terms or multiplets.

If the selection rules 1, 2 are *not* satisfied, forbidden transitions can occur with much smaller transition probability through electric quadrupole radiation or magnetic dipole radiation (the analog of the frame antenna).

As an illustration, Figure 17.3 shows the term scheme of neutral calcium CaI. The most important multiplets are designated by their numbers in *A multiplet table of astrophysical interest* or *An ultraviolet multiplet table* by

Figure 17.3. Term scheme or Grotrian diagram for the (arc) spectrum of neutral calcium Ca I.

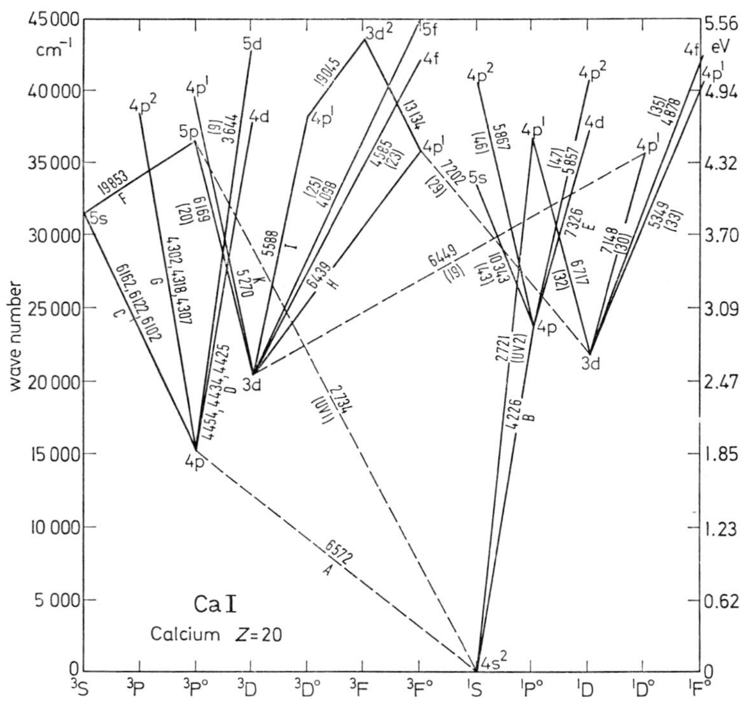

Charlotte E. Moore (Washington) and the wavelength in angstroms of the strongest lines is shown.

The theory of atomic spectra sketched here made it possible in the first place to classify the wavelengths λ or wavenumbers $\bar{\nu}$ of most of the elements in their various stages of ionization (I = arc spectrum, II = first spark spectrum, . . .) as measured in the laboratory. This means that for each spectral line we can give the lower and upper terms (usually measured in cm^{-1} from the ground state) and their classification. For example, Fraunhofer's K line, the strongest line in the solar spectrum λ 3933.664 Å, is Ca II $4\,^2S_{1/2} - 4\,^2P^o_{3/2}$.

Using the expression first in a qualitative sense and not yet in a precisely defined sense, the *intensity* of a line must depend upon what fraction of the atoms of the element concerned are in the appropriate state of ionization and what fraction of these are in the appropriate state of excitation in which they can absorb the line. Saha's theory supplies the answer to this question in so far as we may assume that the gas is in a state of thermodynamic equilibrium, that is, one that corresponds sufficiently closely to the conditions inside a cavity at temperature T (Kelvin).

We first consider an ideal gas at temperature T that is composed of neutral atoms (Figure 17.4). In unit volume ($1\ \text{cm}^3$) let

N = total number of atoms of a particular element
N_0 = number of atoms in the ground state 0
N_s = number of atoms in excited state s with excitation energy χ_s.

(We shall shortly extend this provisional notation by a suffix r inserted before suffix s to distinguish neutral, singly, doubly . . . ionized particles by $r = 0, 1, 2, \ldots$)

If all the quantum states are simple (that is, if each occupies a volume h^3, one quantum cell, in phase space) then according to the fundamental principles of statistical thermodynamics developed by L. Boltzmann

$$N_s/N_0 = e^{-\chi_s/kT} \tag{17.4}$$

where k again denotes the Boltzmann constant.

[We can think of equation (17.4) as a generalization of the barometric formula given by equations (7.4) and (7.5) according to which the density distribution in an isothermal atmosphere at temperature T is given as a function of height y by

$$N(y)/N_0 = \exp\left\{-\frac{\mu g y}{\mathcal{R} T}\right\} \equiv \exp\left\{-\frac{m g y}{kT}\right\}$$

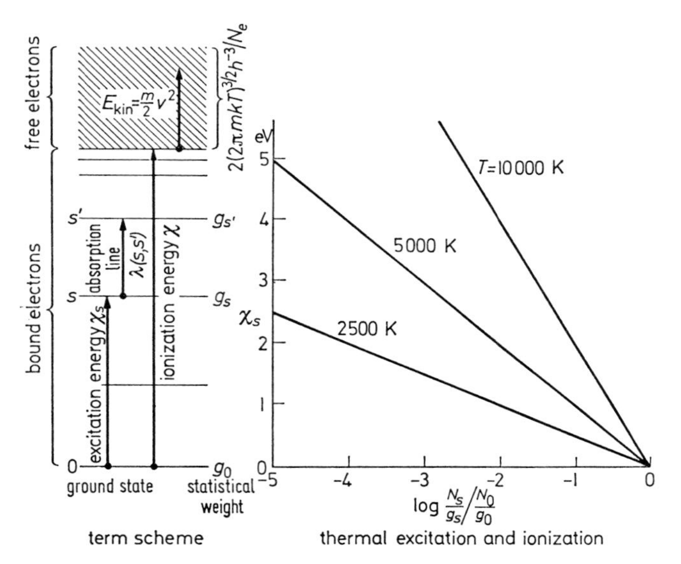

Figure 17.4. Thermal excitation and ionization of neutral atoms. The term scheme (schematic only) on the left illustrates the basic concepts. The diagram on the right shows the relative number of atoms in a single (that is, statistical weight 1) quantum state as a function of its excitation energy χ_s (χ_s in electron volts is plotted as ordinate). The ionization potential $\chi = 5.14$ eV would correspond to Na I.

Here μ means the molecular weight and m the mass of a molecule, $\mathfrak{R}$ is the usual gas constant, and k is the Boltzmann constant, g is the value of gravity. Thus mgy is the potential energy of a molecule at height y above the ground; in (17.4) the excitation energy χ_s corresponds to this. While in classical statistics the potential energy can be varied continuously, in quantum statistics there are quantized states and all simple quantum states have statistical weight 1.]

If the energy level is g_s-ply degenerate, that is, if it would split into g_s simple levels under the influence of a suitable magnetic field, or equivalently if it occupies volume $g_s h^3$ in phase space, we must assign it multiplicity or statistical weight g_s. Correspondingly, let the ground state have weight g_0. Then the general Boltzmann formula applies

$$\frac{N_s}{N_0} = \frac{g_s}{g_0} e^{-\chi_s/kT}. \tag{17.5}$$

The content of this formula is illustrated graphically on the right in Figure 17.4. If we wish to relate N_s to the total number of atoms $N = \Sigma_s N_s$, instead of the number in the ground state N_0, we have

$$\frac{N_s}{N} = \frac{g_s e^{-\chi_s/kT}}{\Sigma_s g_s e^{-\chi_s/kT}}. \tag{17.6}$$

As denominator we have the important

$$\text{sum over states} \quad u = \Sigma_s g_s e^{-\chi_s/kT} \tag{17.7}$$

or *partition function.*

We obtain the statistical weights g_s from the theory of spectra: a level with angular momentum quantum number J shows, for example, in a magnetic field $2J + 1$ different M_J (Figure 17.2) and so has weight

$$g_J = 2J + 1. \tag{17.8}$$

If we treat the levels of a multiplet term with quantum numbers S and L together, then this term has weight

$$g_{S,L} = (2S + 1)(2L + 1). \tag{17.9}$$

The summation of the corresponding g_J in Table 17.1 obviously leads to the same result.

The thermal excitation of the atoms into quantum states of higher and higher energy χ_s, described by the Boltzmann formula (17.5) or (17.6), passes over continuously into the excitation of states of positive energy $E > 0$. The atom then receives the ionization energy χ (Figure 17.4), which suffices to remove an electron from the atom, plus the kinetic energy $E = \frac{1}{2}mv^2$ with which the electron is ejected.

We now denote the number of (singly) ionized atoms with a first suffix, thus N_1, which corresponds to the number of neutral atoms which we now write as N_0 and so on for later stages, reckoning per cubic centimeter throughout:

Ionization stage	Neutral	Singly ionized	Doubly ionized	r-ply ionized
Free electrons per atom	0	1	2 . . .	r
Ionization energy	χ_0	χ_1	χ_2 . . .	χ_{r-1}
Spectra, for example iron Fe	FeI	FeII	FeIII	Fe(r + 1)
All atoms in the ionization stage	N_0	N_1	N_2	N_r
Atoms in the ground state of the ionization stage	$N_{0,0}$	$N_{1,0}$	$N_{2,0}$	$N_{r,0}$
Atoms in level $s,s',\ldots$	$N_{0,s}$	$N_{1,s'}$	$N_{2,s''}$	$N_{r,s^{(r)}}$

We employ a corresponding notation for statistical weights. How now do we calculate $N_{1,0}/N_{0,0}$, that is, the ratio of the numbers of singly ionized and neutral atoms in the relevant ground states having statistical weights $g_{1,0}$ and $g_{0,0}$? (In the first place, higher stages of ionization will be supposed to play no part.)

Clearly the problem amounts to that of calculating the statistical weight of the ionized atom in its ground state plus its free electron. The atom has statistical weight $g_{1,0}$; the electron alone, corresponding to the two possible orientations of its spin in an applied field, has statistical weight equal to 2. In addition, we must reckon the weight, that is, the number of quantum cells h^3, corresponding to the motion of the one free electron. Statistical thermodynamics shows that the electron, with mass m, takes up a volume $(2\pi mkT)^{3/2}$ in momentum space;[3] in ordinary space, if we have N_e electrons per cubic centimeter, we have $1/N_e$ cm^3 per electron. Thus we have statistical weight

$$\left.\begin{array}{l}\text{ionized atom in ground state}\\ +1\text{ free electron}\end{array}\right\} \qquad g = g_{1,0} 2 \frac{(2\pi mkT)^{3/2}}{h^3 N_e} \tag{17.10}$$

Substituting in the Boltzmann formula (17.5) we obtain, referred to the ground states of the atom and ion, the

$$\text{Saha formula} \qquad \frac{N_{1,0}}{N_{0,0}} N_e = \frac{g_{1,0}}{g_{0,0}} 2 \frac{(2\pi mkT)^{3/2}}{h^3} e^{-\chi_0/kT}. \tag{17.11}$$

For the total number of ionized or neutral atoms we obtain from (17.6) the corresponding ionization formula

$$\frac{N_1}{N_0} N_e = \frac{u_1}{u_0} 2 \frac{(2\pi mkT)^{3/2}}{h^3} e^{-\chi_0/kT} \tag{17.12}$$

In a completely analogous way for the passage from the rth to the $(r+1)^{\text{th}}$ stage of ionization, where the $(r+1)^{\text{th}}$ electron is removed with ionization energy χ_r, we have quite independently of other ionization processes

$$\frac{N_{r+1}}{N_r} N_e = \frac{u_{r+1}}{u_r} 2 \frac{(2\pi mkT)^{3/2}}{h^3} e^{-\chi_r/kT} \tag{17.13}$$

[3]The a priori probability of a state with momentum $p = mv$ (mass $\times$ velocity) or of kinetic energy $\frac{1}{2}mv^2 = p^2/2m$ is $\exp(-p^2/2mkT)$. Integrating over momentum space this gives

$$\int_0^\infty \exp(-p^2/2mkT) 4\pi p^2 dp = (2\pi mkT)^{3/2}.$$

and so on. Instead of the number of free electrons per cubic centimeter we can equally well introduce the electron pressure P_e, that is, the partial pressure of free electrons[4]

$$P_e = N_e kT. \tag{17.14}$$

If we take logarithms in the Saha formula, insert the numerical constants, reckon χ_r in eV and P_e in dyn/cm^2 ($\approx 10^{-6}$ atm), we obtain

$$\log \frac{N_{r+1}}{N_r} P_e = -\chi_r \frac{5040}{T} + \frac{5}{2} \log T - 0.48 + \log \frac{2u_{r+1}}{u_r}. \tag{17.15}$$

The important temperature parameter $5040/T$ we denote following H. N. Russell by

$$\Theta = 5040/T. \tag{17.16}$$

The quantity $\log (2u_{r+1}/u_r)$ is in general small. Originally, M. N. Saha derived his formula without this last term, using thermodynamic calculations applying Nernst's heat theorem and the chemical constant of the electron. One considers the process of ionization of an atom A and the recombination of the ion A^+ with a free electron e as a chemical reaction in which, in a state of chemical (thermodynamic) equilibrium, the reaction proceeds with equal frequency in both directions

$$A \rightleftarrows A^+ + e. \tag{17.17}$$

If we restrict ourselves to sufficiently low pressures, the number of recombination processes (per cubic centimeter per second) $\leftarrow$ will be proportional to the number of encounters between ions and electrons, that is, $\propto N_1 N_e$. The number of ionizations by radiation $\rightarrow$ will be proportional to the density of neutral atoms N_0. The factor of proportionality (here left undetermined) depends only on the temperature. Thus one understands the form of the ionization equation (17.12), as well as its generalizations, as an application of the Guldberg–Waage law of mass action

$$N_1 N_e / N_0 = \text{function of } T. \tag{17.18}$$

[At higher pressure, instead of ionization by radiation, ionization by electron collisions, and instead of two-body recombination of ion + electron,

[4]At first sight it seems surprising that, in spite of their Coulomb interaction e^2/r (r = mean separation of neighboring electrons), the free electrons can be treated as an ideal gas. However, in fact $e^2/r \ll kT$, for example for $T = 10\,000$ K this is valid for electron pressures P_e of up to 70 atm.

three-body encounters involving a further electron (in order to satisfy the energy and momentum balance) play the essential parts. Thus both processes in equation (17.17) acquire an additional factor N_e, and once again we arrive at the law of mass action (17.18).]

In Figure 17.5 we show in logarithmic measure the fraction of atoms of H, He, Mg, Ca in certain states of thermal ionization and excitation, according to a combination of the formulas (17.6) and (17.12) or (17.13), for an electron pressure $P_e = 100$ dyn/cm^2 and for temperatures from 3000 to 50 000 K. This value of P_e can be regarded as about the mean value for the atmospheres of main sequence stars. The absorption lines of these atoms and ions play a part in the Harvard and MK classification of stellar spectra.

Figure 17.5. Thermal ionization [equation (17.13)] and excitation [equation (17.6)] as a function of temperature T or Θ (where $\Theta = 5040/T$) for an electron pressure P_e = 100 dyn/cm^2 (approximate mean value for stellar atmospheres). The temperature scale covers the whole range from O stars (on left) to M stars (on right). The Sun (G2) would fit in at about $T = 5800$ K. The curves demonstrate M. N. Saha's interpretation (1920) of the Harvard sequence of spectral types (see Section 15). For example, hydrogen (H I) is predominantly neutral up to $T \approx 10\,000$ K. The visible Balmer lines are produced by absorption from the second quantum state, the excitation of which increases with T. Above 10 000 K, however, the hydrogen is rapidly removed by ionization. We therefore see that the hydrogen lines will be at their maximum intensity in the A0 stars, for which $T \approx 10\,000$ K.

Spectrum	Ionization potential χ_0 eV	Excited state and excitation potential	$\chi_{r,s}$ eV
H I	13.60	$n = 2$	10.15
He I	24.59	$2\,^3P^o$	20.87
He II	54.42	$n = 3$	48.16
Mg I	7.65	—	—
Mg II	15.03	$3\,^2D$	8.83
Ca I	6.11	$4\,^1S$	0.00
Ca II	11.87	$4\,^2S$	0.00

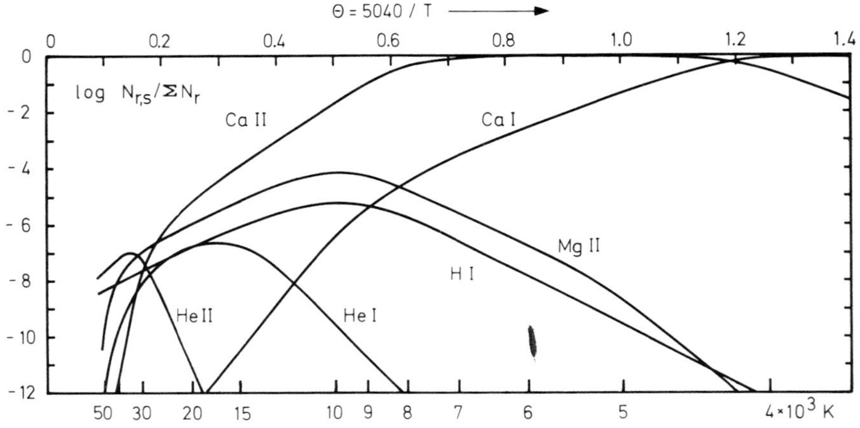

Table 17.2. Elements contributing significantly to the electron pressure $P_e = N_e kT$ in stellar atmospheres ($P_e \approx 100$ dyn/cm^2). Table shows charge number Z, atomic weight μ, solar abundance ϵ by numbers of atoms referred to hydrogen = 100 and ionization potential and statistical weight of first three ionization levels. The three sets, grouped according to the ionization potential of the neutral atom χ_0, are effective in different temperature ranges.

Z	Element		Atomic weight μ	Abundance ϵ (H = 100)	Neutral atom χ_0	g_0	Singly ionized χ_1	g_1	Doubly ionized χ_2	g_2	Temperature range of effectiveness for $P_e \approx 100$ dyn/cm^2
1	H	Hydrogen	1.008	100.	13.60	2	—	—	—	—	$T > 5700$ K
2	He	Helium	4.003	10.	24.59	1	54.42	2	—	—	
12	Mg	Magnesium	24.32	0.0040	7.65	1	15.03	2	80.14	1	$6000 \text{ K} > T > 4500$ K
14	Si	Silicon	28.06	0.0045	8.15	9	16.34	6	33.49	1	
26	Fe	Iron	55.85	0.0040	7.87	25	16.18	30	30.65	25	
11	Na	Sodium	23.00	0.00020	5.14	2	47.29	1	71.64	6	$T < 4700$ K
19	K	Potassium	39.10	0.000011	4.34	2	31.63	1	45.72	6	
20	Ca	Calcium	40.08	0.00021	6.11	1	11.87	2	50.91	1	

The maxima of the curves, which correspond approximately to the greatest strengths of the spectral lines concerned, result from the fact that at a particular stage of ionization the degree of excitation first increases with increasing T. But as T increases far enough, this stage itself gets "ionized away," so that the fraction of effective atoms decreases again. Taking the temperatures for the maxima as known (for example, the maximum of the Balmer lines of hydrogen occurs in spectral type A0V at about 9000 K) R. H. Fowler and E. A. Milne in 1923 were first able to estimate the electron pressure in stellar atmospheres.

The Saha theory was able also to account qualitatively for the increase in the intensity ratio of spark to arc lines in passing from main sequence stars to giant stars as a consequence of the increase of the degree of ionization, that is, of N_1/N_0, resulting from the lower *pressure*. On the other hand, the well-known differences between the spectra of sunspots and of the normal solar atmosphere can be accounted for by the lower *temperature* in the spots. In the laboratory, the Saha theory plays an important part in application to the King furnace, to the electric arc, to very high-temperature plasmas for nuclear-fusion experiments, and so on.

In dealing with a mixture of several elements, we can calculate the degree of ionization most simply by treating the temperature T, or else $\Theta = 5040/T$, and the electron pressure P_e or else $\log P_e$, as independent parameters, and then applying the Saha equation (17.12) to each element and to its various stages of ionization. We can afterwards calculate the gas pressure P_g as kT times the sum of all the particles, including electrons, per cubic centimeter. Then we can write down also the mean molecular weight μ. For example, fully ionized hydrogen has mean molecular weight $\mu = 0.5$ since as a result of the ionization unit mass appears in the form of one proton and one electron.

Table 17.3. Gas pressure P_g as function of electron pressure P_e and of temperature T, or of $\Theta = 5040/T$, for stellar material (Table 17.2): pressures expressed as logarithms. Below the dividing line on the right more than 50 percent of the gas pressure arises from hydrogen molecules.

$\Theta = \frac{5040}{T}$	T	$\log P_e =$ −1.0	0.0	1.0	2.0	3.0	4.0	5.0
0.10	50 400	−0.70	+0.30	1.30	2.30	3.30	4.31	5.31
0.20	25 200	−0.70	+0.30	1.30	2.30	3.30	4.31	5.31
0.30	16 800	−0.70	+0.30	1.30	2.30	3.30	4.32	5.34
0.50	10 080	−0.68	+0.32	1.33	2.36	3.61	5.35	7.32
0.70	7 200	−0.64	+0.67	2.42	4.36	6.30	8.05	
0.90	5 600	+1.37	3.26	4.82	6.13	7.68	9.92	
1.10	4 582	2.94	4.10	5.53	7.10	8.77		
1.30	3 877	3.40	4.96	6.35	7.81			
1.50	3 360	4.28	5.50	6.94	9.12			$\log P_g$

With regard to the theory of stellar atmospheres and of stellar interiors we give in Table 17.2, in percentage numbers of atoms, the composition of stellar material of which, as we shall see, the Sun and most of the stars consist. Table 17.3 gives the values of P_g for this mixture (the formation of hydrogen molecules being taken into account at the lower temperatures). For $\log P_e \approx 2$ (about the mean value for stellar atmospheres) the stellar material is almost wholly ionized for temperatures $T > 10\,000$ K and so $P_g/P_e \approx 2$ or $\log P_g \approx \log P_e + 0.3$. At the solar temperature $T \approx 5600$ K it is essentially the metals (Mg, Si, Fe) that are ionized; corresponding to their relative abundance $\approx 12 \times 10^{-5}$, we have therefore $\log P_g \approx \log P_e + 3.9$. At still lower temperatures electrons come only from the most easily ionized group, Na, K, Ca.

18. Stellar Atmospheres: Continuous Spectra of the Stars

Looking forward to the quantitative interpretation of the continuous spectra and later of the Fraunhofer lines of the Sun and stars we turn to the physics of stellar atmospheres. We can characterize the atmosphere, that is, those layers of a star that transmit radiation to us directly, by the following parameters:

(1) The effective temperature T_{eff} which is so defined that, corresponding to the Stefan–Boltzmann radiation law, the radiation flux per square centimeter of the surface of the star is

$$\pi F = \sigma T_{\text{eff}}^4. \tag{18.1}$$

For the Sun we obtained in equation (12.17) directly from the solar constant $T_{\text{eff}} = 5770$ K.

(2) The acceleration due to gravity g [cm/s²] at the surface. For the Sun we found in equation (12.6) $g_\odot = 2.74 \times 10^4$ cm/s²; according to equation (16.5) gravity for other main sequence stars is not very different.

(3) The chemical composition of the atmosphere, that is, the abundances of the elements. In Table 17.2 we have anticipated some values.

Other possible parameters like rotation or pulsation, or stellar magnetic fields, will not be considered for the moment.

We easily convince ourselves that for given T_{eff}, g, and chemical composition the structure, that is, the distribution of temperature and pressure in a static atmosphere can be completely calculated. For this purpose we need two equations:

(1) The first describes the transfer of energy, by radiation, convection, conduction, mechanical or magnetic energy, and thus determines the *temperature distribution*.

(2) The second is the hydrostatic equation, or, generally speaking, the basic equations of hydrodynamics or magnetohydrodynamics, which determine the *pressure distribution.*

The state of ionization, the equation of state, and all physical constants can all be obtained from atomic physics if we know only the chemical composition. Then the theory of stellar atmospheres makes it possible, starting from the data 1, 2, 3, to compute a model stellar atmosphere and thence to discover how certain *measurable quantities,* for example, the intensity distribution in the continuum or the color indices or the intensity of the Fraunhofer lines of a certain element in this or that state of ionization or excitation, depend upon the parameters 1, 2, 3.

If we have solved this problem of theoretical physics, then we can—and this is the decisive consideration—reverse the procedure and by a process of successive approximation determine what T_{eff}, g, and what abundance of the chemical elements the atmosphere of a particular star possesses, in the spectrum of which we have measured the energy distribution in the continuum (or color indices), and/or the intensities of various Fraunhofer lines. In this way we should have a procedure for the quantitative analysis of the spectra of the Sun and stars.

We have remarked that the temperature distribution in a stellar atmosphere is determined by the mechanism of energy transport. As K. Schwarzschild appreciated in 1905, in stellar atmospheres energy is transported predominantly by *radiation;* we speak then of radiative transfer and of radiative equilibrium.

In order to describe the radiation field, we consider (Figure 18.1) at depth t, measured from an arbitrary fixed zero level, a surface element of 1 cm² whose normal makes angle θ with the normal to the surface ($0 \leqslant \theta \leqslant \pi$). Then the energy flowing through this surface element within an element of solid angle $d\omega$ about the normal to the element in the frequency interval ν to $\nu + d\nu$ per second is $I_\nu(t,\theta)\, d\nu\, d\omega$ (see Section 11). According to equation (11.14), in a path element $ds = -dt \sec\theta$ the radiation intensity I_ν suffers by absorption a diminution of amount $-I_\nu(t,\theta)\kappa\, ds$, where $\kappa = \kappa(\nu)$ is the continuous absorption coefficient per centimeter at frequency ν. On the other hand, by the application of Kirchhoff's law to each volume element of the atmosphere assuming local thermodynamic equilibrium (LTE), the intensity I_ν is enhanced by emission by an amount which according to equation (11.20) can be written as $+\kappa B_\nu(T)\, ds$ where $B_\nu(T)$ is

Figure 18.1 Radiative equilibrium.

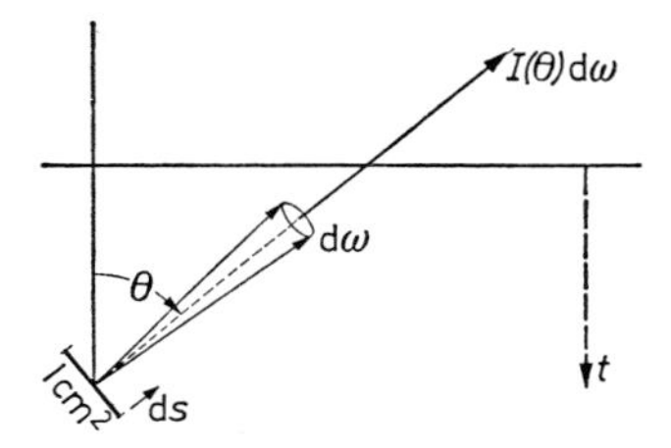

the Kirchhoff–Planck function (11.23) for the local temperature T at depth t and frequency ν. Combining these we have therefore

$$dI_\nu(t,\theta) = I_\nu(t,\theta)\kappa dt \sec\theta - B_\nu(T(t))\kappa\, dt \sec\theta. \tag{18.2}$$

In order to describe the various layers of the atmosphere, instead of geometrical depth t it is better to use the optical depth τ for ν–radiation given by (equation 11.18)[1]

$$\tau = \int_{-\infty}^{t} \kappa\, dt \quad \text{or} \quad d\tau = \kappa\, dt. \tag{18.3}$$

So we derive from (18.2) the equation of radiative transfer

$$\cos\theta\, dI_\nu(t,\theta)/d\tau = I_\nu(t,\theta) - B_\nu(T(t)). \tag{18.4}$$

If there is radiative equilibrium, that is, if all energy transfer is by radiation, then here we may bring in the energy equation which states that the total flux must be independent of the depth t. According to (11.8) it follows that

$$\pi F = \pi \int_0^\infty F_\nu(t)\, d\nu \tag{18.5}$$

$$= \int_0^\infty \int_0^\pi I_\nu(t,\theta) \cos\theta 2\pi \sin d\theta\, d\nu = \sigma T_{\text{eff}}^4.$$

The solution of the system of equations (18.4) and (18.5), with the boundary conditions that at the surface the incident radiation $I_\nu(0,\theta)$ for $0 \leqslant \theta \leqslant \pi/2$ must vanish and that at large depths the radiation field must approach black-body radiation, is relatively simple if we postulate that κ is independent of the frequency ν, or if in the transfer equation (18.4) instead of $\kappa(\nu)$ we use a harmonic mean taken over all frequencies with suitable weight factors. This is the so-called Rosseland absorption coefficient κ and we use the corresponding optical depth

$$\tau = \int_{-\infty}^{t} \bar{\kappa}\, dt.$$

For such a "gray" atmosphere, following E. A. Milne and others we obtain as a good approximate solution of equations (18.4) and (18.5) the temperature distribution

$$T^4(\bar{\tau}) = 3/4\, T_{\text{eff}}^4(\bar{\tau} + 2/3). \tag{18.6}$$

According to this the temperature is equal to the effective temperature at optical depth $\bar{\tau} = 2/3$. At the stellar surface $\bar{\tau} = 0$, T approaches a finite boundary temperature for which equation (18.6) gives the value $T_0 = 2^{-1/4} T_{\text{eff}} = 0.84 T_{\text{eff}}$.

For the actual analysis of stellar spectra the "gray" approximation (18.6) is too inexact. Before all else, therefore, in regard to the theory of stellar continuous and line spectra we must first calculate the continuous absorp-

[1]We drop the index ν on κ and τ when we are dealing with the continuum.

tion coefficient κ. Several atomic processes contribute to κ. From Section 17 and Figure 17.1 we already know:

(1) The bound-free transitions of hydrogen. From the limit of the Lyman series at 912 Å, of the Balmer series at 3647 Å, of the Paschen series at 8206 Å, and so on, there extends on the short wavelength side a series continuum for which κ falls off approximately like $1/\nu^3$. At low frequencies such continua crowd together and pass over continuously into the continuum produced by the free-free transitions of hydrogen.

(2) As R. Wildt pointed out in 1938, the bound-free and free-free transitions of the *negative hydrogen ion* H^- play an important part in stellar atmospheres. This ion can arise by the attachment of a second electron to the neutral hydrogen atom with a binding, or ionization, energy of 0.75 eV. Corresponding to this small ionization energy, the long-wave end of the bound-free continuum lies in the infrared at 16 550 Å. Free-free absorption increases beyond that point toward longer wavelengths, as in the case of the neutral atom.

The atomic coefficients for absorption from a particular energy level in the H atom and the H^- ion have been calculated with great accuracy on a

Figure 18.2. Continuous absorption coefficient $\kappa(\lambda)$ in the solar atmosphere (G2 V) at τ_0 = 0.1 (τ_0 at λ 5000 Å). Here we have T = 5040 K (Θ = 1) and P_e = 3.2 dyn/cm² (P_g = 5.8 × 10^4 dyn/cm²).

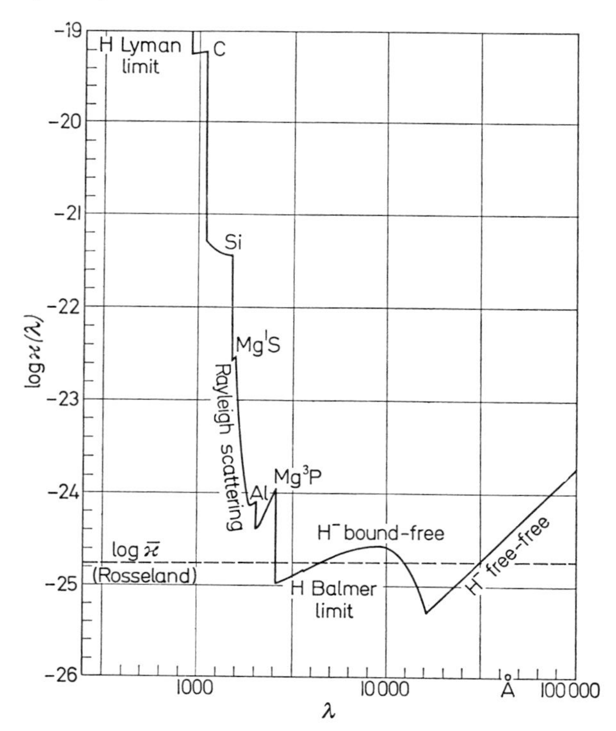

quantum-mechanical basis. If we want then to obtain the absorption coefficient for stellar material of specified composition as a function of frequency ν, temperature T, and electron pressure P_e or gas pressure P_g, we must calculate for the various energy levels the degree of ionization and excitation, using the formulas of Saha and Boltzmann. In addition we obtain the relation between gas pressure P_g, electron pressure P_e, and temperature T, as we have presented it for a particular example in Table 17.3.

We find in this way that in the hot stars (say, $T > 7000$ K) the continuous absorption by hydrogen atoms predominates, and in cooler stars that by H^- ions. Besides these two most important processes, in more exact calculations account must be taken of the following:

Bound-free and free-free absorption by He I and He II (in hot stars) and by metals (in cooler stars, see Table 17.2); furthermore, scattering of light by free electrons (Thomson scattering; in hot stars) also by neutral hydrogen (Rayleigh scattering; in cooler stars). We still know very little about the continua of molecules in cool stars. From the extensive calculations of G. Bode we reproduce in Figures 18.2 and 18.3 the absorption coefficient κ as a function of wavelength λ for mean conditions (see below) in the atmospheres of the Sun and of the B0V star τ Scorpii.

Figure 18.3. Continuous absorption coefficient $\kappa(\lambda)$ in the atmosphere of τ Scorpii (B0 V) at $\bar{\tau} \approx 0.1$. Here we have $T =$ 28 300 K ($\Theta = 0.18$) and $P_e = 3.2 \times \times 10^3$ dyn/cm² ($P_g = 6.4 \times 10^2$ dyn/cm²).

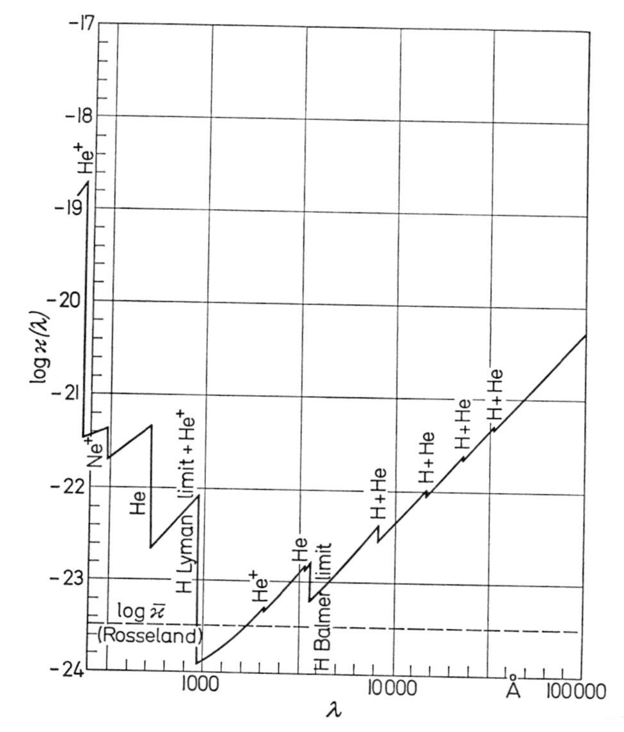

With the help of the Rosseland opacity coefficient $\bar{\kappa}$ shown in these diagrams, and the "gray" approximation (18.6) we first calculate for a given effective temperature T_{eff}, the temperature T (in the sense of local thermodynamic equilibrium) as a function of optical depth $\bar{\tau}$ corresponding to opacity $\bar{\kappa}$. The electron pressure P_e on which most features do not depend very sensitively has first to be estimated. Sometimes it is advantageous to use, instead of $\bar{\tau}$, an optical depth τ_0 that corresponds to the absorption coefficient κ_0 referring to a particular wavelength, usually $\lambda_0 = 5000$ Å. The relation between τ and τ_0 (similar also to that between optical depths for two different wavelengths) is easily obtained from

$$\left.\begin{aligned} d\bar{\tau} &= \bar{\kappa}\, dt \\ d\bar{\tau}_0 &= \bar{\kappa}_0\, dt \end{aligned}\right\} \quad \frac{d\bar{\tau}}{d\tau_0} = \frac{\bar{\kappa}}{\kappa_0} \text{ for a definite depth.} \tag{18.7}$$

For higher demands on the accuracy of model atmospheres the frequency dependence of the continuous absorption coefficient and ultimately also the very strongly frequency-dependent line absorption must be taken into account. In radiative equilibrium again the total radiation flux πF must be the same at all depths t. The theory of radiative equilibrium of such "nongray" atmospheres offers considerable mathematical difficulties, which hitherto have been overcome only by processes of successive approximation.

If we know the temperature distribution $T(\bar{\tau})$ or $T(\tau_0)$ in a stellar atmosphere, the calculation of the pressure distribution offers no further difficulty. The increase of gas pressure P_g with depth t in a static atmosphere is determined by the hydrostatic equation

$$dP_g/dt = g\rho \tag{18.8}$$

where the density ρ is related to P_g and T by the equation of state of a perfect gas. Again g means the value of gravity. If we divide both sides by κ_0, since $\kappa_0\, dt = d\tau_0$ we get

$$dP_g/d\tau_0 = g\rho/\kappa_0. \tag{18.9}$$

Since the right-hand side of this equation is known as a function of P_g and T, together with T as a function of τ_0, we can now integrate equation (18.9) numerically.

In hot stars, radiation pressure as well as gas pressure must be taken into account. If there are currents in the atmosphere whose speed v is not small compared with the local sound speed (for example, 12 km/s in atomic hydrogen at 10 000 K), then the dynamical pressure $\frac{1}{2}\rho v^2$ also contributes. In sunspots and in Ap stars with their magnetic fields of several thousand gauss, magnetic forces have to be reckoned with as well.

Important aids to the calculation of model stellar atmospheres are extensive tables that give for various mixtures of elements (for example, Table 17.2) the *gas pressure* P_g (for example, Table 17.3), the *mean molecular weight* μ (from Table 17.2 for neutral stellar material $\mu = 1.50$ and for fully ionized material $\mu = 0.70$), the *continuous absorption coeffi-*

Table 18.1. Model stellar atomosphere. The optical depth τ_{5000} is τ_0 referred to κ_0 for $\lambda_0 = 5000$ Å; $\bar{\tau}$ is calculated for the Rosseland mean $\bar{\kappa}$; temperature T is in Kelvin; gas pressure P_g and electron pressure P_e are in dyn/cm², logarithms being tabulated.

Sun (G2 V). Effective temperature $T_{\rm eff} = 5770$ K; surface gravity log $g = 4.44$.

$\bar{\tau}$	τ_{5000}	T	log P_g	log P_e
10^{-6}	2.6×10^{-6}	4030	1.84	−2.10
10^{-4}	1.4×10^{-4}	4400	3.00	−0.97
0.01	0.012	4800	4.08	+0.09
0.10	0.11	5260	4.62	0.71
0.25	0.28	5600	4.83	1.02
0.40	0.44	5890	4.92	1.27
1.0	1.10	6610	5.06	1.91
2.0	2.15	7200	5.13	2.39
4.0	4.16	7960	5.18	2.91
10.0	9.84	8540	5.23	3.26

cient κ for numerous wavelengths and its Rosseland mean $\bar{\kappa}$ all as functions of temperature T and electron pressure P_e.

The procedures for computation outlined above can be refined by successive approximations. In recent times large electronic computers have become an indispensable aid to the theory of stellar atmospheres.

In Tables 18.1 and 18.2 we summarize the models of the atmospheres of

Table 18.2. Model stellar atmosphere. The optical depth τ_{4000} is τ_0 referred to κ_0 for $\lambda_0 = 4000$ Å; $\bar{\tau}$ is calculated for the Rosseland mean $\bar{\kappa}$, temperature T is in Kelvin; gas pressure P_g and electron pressure P_e are in dyn/cm², logarithms being tabulated.

τ Scorpii (B0 V). Effective temperature $T_{eff} = 32000$ K; surface gravity log $g = 4.1$

τ_{4000}	T	log P_g	log P_e
0.001	21920	1.87	1.57
0.01	22730	2.47	2.16
0.1	25710	3.13	2.83
0.2	27220	3.32	3.02
0.4	29120	3.51	3.21
1.0	32530	3.77	3.47
2.0	35800	3.97	3.67
4.0	39840	4.16	3.87
10.0	45200	4.42	4.14
30.0	52900	4.72	4.44

the Sun and of the B0 V star τ Scorpii according to the present state of the art. Here we have anticipated the step by step adjustment of theory and observation. We shall return to this later but now we apply ourselves to the calculation of the *continuous spectrum*.

The radiation that a star sends out at its surface originates in layers at various depths in its atmosphere, in such a way that the radiation coming from the deeper layers is naturally more weakened by absorption than that coming from higher up.

The *source function*, defined as the emission coefficient ϵ divided by the absorption coefficient κ, at optical depth τ in the case of local thermodynamic equilibrium is equal to the Kirchhoff–Planck function $B_\nu(T(\tau))$ for the local temperature [equation (18.2) or (18.4)]. From (11.19), the ν–radiation that comes from this depth in a direction inclined at angle θ to the normal is weakened by an absorption factor exp $\{-\tau \sec \theta\}$ before it escapes at the stellar surface. Therefore we obtain for the radiation intensity at $\tau = 0$

$$I_\nu(0,\theta) = \int_0^\infty B_\nu(T(\tau))e^{-\tau \sec \theta}\, d\tau \sec \theta. \tag{18.10}$$

Since on the one hand according to the theory of radiative equilibrium [for example, equation (18.6)] we can calculate the temperature T as a function of optical depth τ, and on the other hand, with the help of the theory of the continuous absorption coefficient κ, we can formulate the connection between the various optical depths $\bar{\tau}$, τ, τ_0, . . . as in equation (18.7), the whole theory of the continuous spectrum of a star is contained in equation (18.10). In the case of the Sun, this includes the theory of limb-darkening ($\theta = 0$ corresponds to the center of the disk and $\theta = 90°$ to the limb).

Without going into details, with the model solar atmosphere in Table 18.1 the full calculations yield the continuous spectrum $I_\nu(0,0)$ for the center of the solar disk plotted in Figure 12.4 and the center-limb variation in Figure 12.3. With sufficient accuracy for many purposes, we can simplify the calculation of $I_\nu(0,\theta)$ if we expand the source function $B_\nu(\tau)$ in a series starting from that at a particular optical depth τ^*, which remains to be selected,

$$B_\nu(\tau) = B_\nu(\tau^*) + (\tau - \tau^*)(dB_\nu/d\tau)_{\tau^*} + \cdots \tag{18.11}$$

Substituting in equation (18.10) we obtain

$$I_\nu(0,\theta) = B_\nu(T(\tau^*)) + \{\cos \theta - \tau^*\}(dB_\nu/d\tau)_{\tau^*} + \cdots \tag{18.12}$$

If we now choose $\tau^* = \cos \theta$, the second term on the right becomes zero, and we have the so-called Eddington–Barbier approximation

$$I_\nu(0,\theta) \approx B_\nu(T(\tau = \cos \theta)). \tag{18.13}$$

This means that the intensity radiated at the surface of, say, the Sun corresponds in the sense of Planck's law to the temperature at optical depth $\tau = \cos \theta$ measured perpendicular to the surface which means $\tau \sec \theta = 1$

measured along the sight line. This is instructive; with some reservation we can say that the radiation we see originates at unit optical depth along our line of sight.

In the case of the stars we can only deal with the mean intensity over the disk, that is, apart from a factor π, the flux of radiation at the surface. From equation (11.7) we first have

$$F_\nu(0) = 2\int_0^{\pi/2} I_\nu(0,\theta)\cos\theta\sin\theta\,d\theta. \tag{18.14}$$

With the use of equation (18.12) we easily obtain the approximation

$$F_\nu(0) \approx B_\nu(T(\tau = 2/3)). \tag{18.15}$$

Thus the radiation of a star at frequency ν corresponds to the local temperature T at an optical depth $\tau = 2/3$ in this frequency, corresponding to $\cos\theta = 2/3$ giving a mean emission angle of 54°44′.

As further illustrations we show in Figure 18.4 the energy distribution in the continuum $F_\nu(0)$ calculated for the model of the B0V star τ Scorpii in Table 18.2, and immediately below the corresponding result for the A0V star α Lyrae = Vega. Here the similarity to the spectrum of a black body has disappeared. The F_ν curve looks much more like a mirror image of the

Figure 18.4. Calculated spectral energy distribution in the continuum F_ν (0) for two stars of early spectral types.

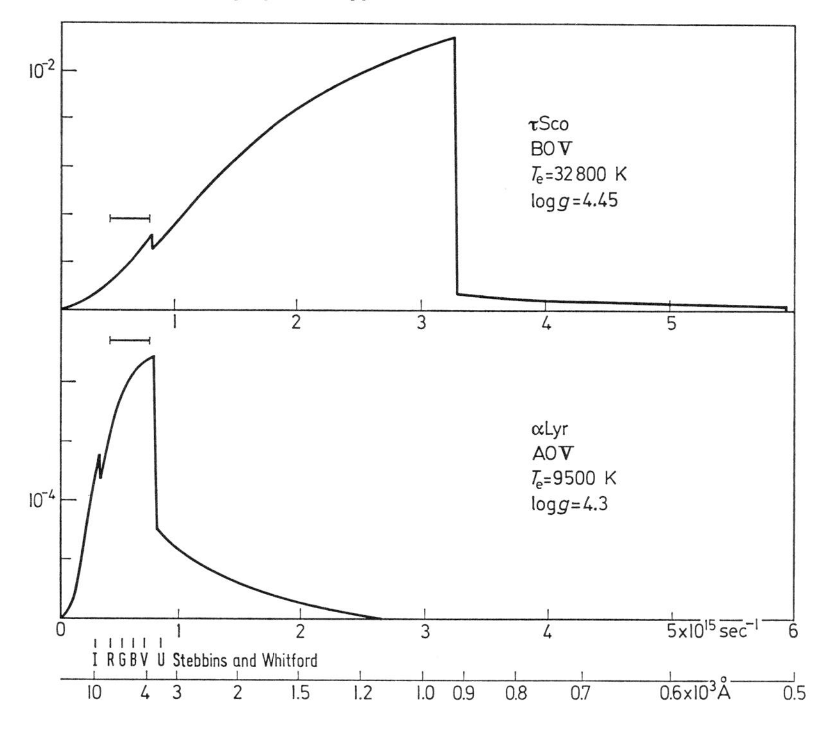

wavelength dependence of the continuous absorption coefficient κ (shown in Figure 18.3 for τ Sco). Since $d\tau/d\bar{\tau} = \kappa/\bar{\kappa}$, for large κ the radiation comes from layers near the surface having small $\bar{\tau}$; these are relatively cool and the radiation curve F_ν drops down. In this way we understand how at each absorption edge of hydrogen (λ 912 Å Lyman edge, λ 3647 Å Balmer edge . . .) F_ν drops abruptly on the short wavelength side. The hotter of these two stars τ Sco radiates mostly between the Balmer and Lyman limits, the cooler α Lyr mostly between the Paschen and Balmer limits. We indicate the small range ├─┤ of the visible spectrum and also the characteristic frequencies of the six-color photometry of Stebbins and Whitford. In order to have significant values for the comparison of theory and observation, the relative measurements (color indices) for such closely placed characteristic frequencies must obviously be made with high accuracy.

Figure 18.5. Percentages η_λ of energy absorbed by the Fraunhofer lines in stars of various spectral types.

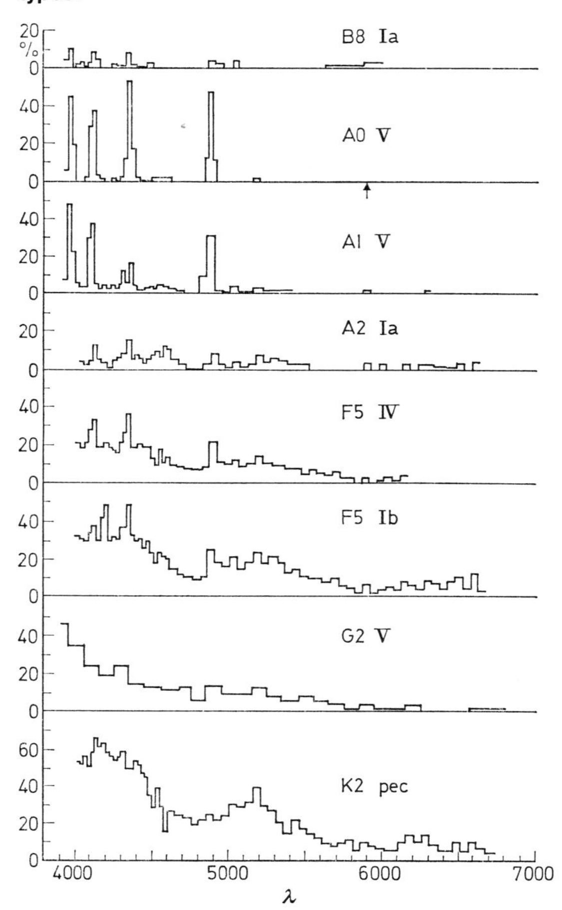

Finally, in comparing theory and observation we must remember that the theory relates to the true continuum F_ν or F_λ while stellar photometry and spectra of low dispersion measure the continuum with smeared-out lines $F_\lambda^L = (1 - \eta_\lambda)F_\lambda$, where η_λ is the fraction of the true continuum obstructed by the Fraunhofer lines. In Figure 18.5 we have assembled measured values of η_λ from coudé spectra of large dispersion for a number of stars. In hot stars the broad Balmer lines of hydrogen predominate. In cooler stars, from about F5, the numerous metal lines obstruct a considerable part of the spectrum, especially in the blue and ultraviolet. Finally, in K and M stars absorption in molecular bands also plays a part.

The integration of $(1 - \eta_\lambda)F_\lambda$ over the whole spectrum yields the total flux and thence the bolometric correction BC [see equations (13.15) (13.17)] of the star concerned. Here we cannot go into the performance of this difficult calculation. However, it need scarcely be remarked that great demands are made upon the theory when, in the case of hot stars for example, we have to infer the area under the whole curve from measurements in the tiny part of the spectrum that penetrates to the surface of the Earth (shown by $\vdash\!\dashv$ in Figure 18.4). Direct measurements of the total flux from hot stars, by far the greater part of which is in the far ultraviolet, can be made only with the help of rockets and space vehicles.

19. Theory of Fraunhofer Lines: Chemical Composition of Stellar Atmospheres

From the theory of the continuous spectra of stars we now turn to the quantitative study of the Fraunhofer lines.

Referring to observations of the solar spectrum we mentioned in Section 12 the classical work of Rowland; with modern diffraction gratings resolving powers up to $\lambda/\Delta\lambda \approx 10^6$ are attained. With the coudé spectrographs of large telescopes, the spectra of at any rate the brighter stars from about 3200 Å to 6800 Å can be obtained with a dispersion of a few angstroms per millimeter.

Besides reference to the Rowland tables and monographs on individual stars (Table 19.1), the tabulations of Charlotte E. Moore are indispensable for the identification and classification of lines (see also Section 17):

(1) *A multiplet table of astrophysical interest*—revised ed. (Nat. Bur. of Standards Washington, Techn. Note No. 36, 1959).
(2) *An ultraviolet multiplet table* (*ibid.* Circular 488, Sect. 1–5, 1950–1961).
(3) *Atomic energy levels*—several volumes (*ibid.* Circular 467, 1949–).
(4) *Selected tables of atomic spectra* (*ibid.* Nat. Standard Reference Data Series); these carry continuing revisions for individual elements.

Also, including numerous term schemes, there is Paul W. Merrill's *Lines of the chemical elements in astronomical spectra* (Carnegie Inst.

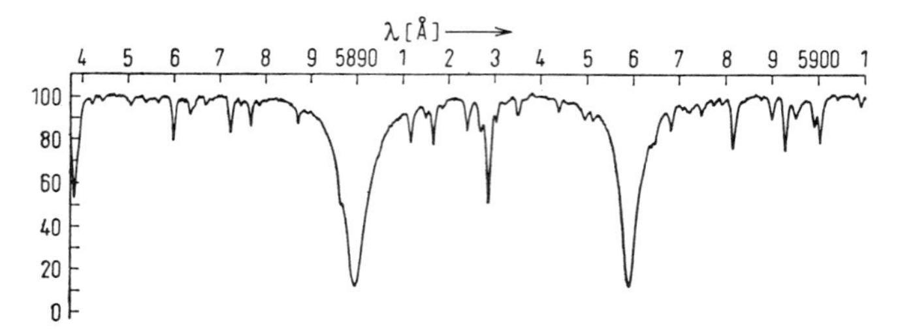

Figure 19.1. Microphotometer tracing; the Na D lines of the solar spectrum.

Washington; Publ. 610, 1956). A new edition of this was published by Charlotte E. Moore (Nat. Bur. of Standards, Washington, 23, 1968).

By automatically taking acount of the blackening curve with the help of intensity marks, the microphotometer nowadays directly yields the intensity distribution in the spectrum (Figure 19.1). Besides such photographic photometry, the direct recording of spectra by means of photocells or photomultipliers gains ever increasing significance. In practice the next step of fixing the true continuum is very difficult; its intensity is then normalized to unity, or 100 percent, from which the depression in the lines $R_\nu = 1 - I_\nu$ is measured. Aside from weak disturbing lines, so-called "blends," we then have the intensity distribution in the line given by I_ν or R_ν as a function of ν or of λ, that is, the *line profile*. The spectrograph itself would reproduce an infinitely sharp line as a line of finite width, the so-called instrumental profile. So we must note that the width and structure of weak Fraunhofer lines always include an instrumental contribution, which one can determine by measuring the profiles of sharp lines in a laboratory source (the lines of the iron arc being sufficient in reference to stellar spectra). The energy absorbed in the line is obviously independent of instrumental distortion of the line profile. As M. Minnaert has remarked, it is therefore useful to measure the equivalent width W_λ; this is the width in angstroms or milliangstroms of a rectangular strip of the spectrum having the same area as that of the line profile (Figure 19.2). The revision of the Rowland tables by Charlotte E. Moore and M. Minnaert contains equivalent widths in the

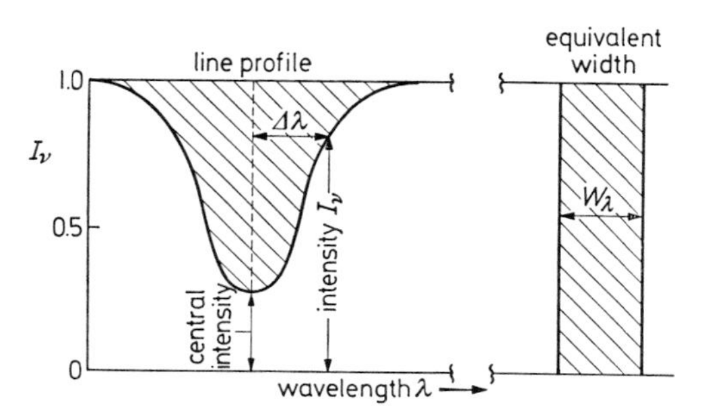

Figure 19.2 Profile and equivalent width of a Fraunhofer line.

spectrum of the center of the solar disk; some references for stellar spectra are given in Table 19.1.

From spectrophotometric measurements of lines and the continuum we must then derive the parameters that we have repeatedly mentioned as characterizing the stellar atmosphere, that is to say, the effective temperature T_{eff}, the gravitational acceleration g, and the relative abundances of the elements.

The complexity of the problem suggests the use of successive approximations. We first ask the question: For an atmosphere of given T_{eff}, g, and assigned chemical composition, what values would the theory predict for the equivalent widths W_λ of various lines? In Section 18 we have already studied the corresponding problem for the continuous spectrum. Here again we use model atmospheres. Then we investigate the way in which measurable quantities, that is, in addition to the color indices already considered, principally the W_λ for particular elements in particular states of ionization and excitation, depend upon the parameters that were chosen at the outset. Finally, we improve these parameters until we obtain the best possible agreement between the computed spectrum of the model atmosphere and the measured spectrum of the actual star under investigation.

As already indicated in Section 10, the theory of Fraunhofer lines consists of two different parts:

(1) The theory of *radiative transfer* in the lines which gives the dependence of the profiles and equivalent widths for a given model atmosphere upon the line-absorption coefficient κ_ν.
(2) The atomic theory of the *line-absorption coefficient* κ_ν itself.

We turn to the first problem:

In addition to the continuous absorption coefficient κ, which varies only slowly with wavelength and may be treated as constant in the vicinity of a line, we now consider the line-absorption coefficient κ_ν which, as a function of the interval $\Delta\lambda$ or $\Delta\nu$ from the line center (Figure 19.3), falls off more or less steeply from a high maximum to the value zero on either side. Thus we distinguish between the following absorption coefficients (which in their

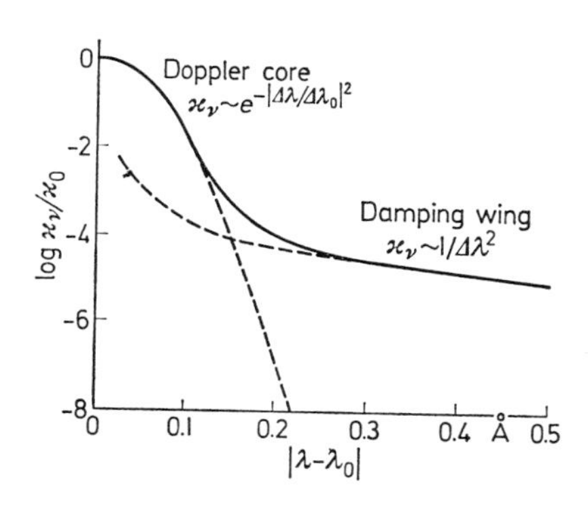

Figure 19.3
Coefficient of line absorption κ_ν (compared with κ_0 for the line center). The Doppler core and damping wings of the Na D lines have been calculated from equations (19.9) and (19.10), respectively, for $T = 5700$ K and pure radiation damping. Their superposition (full line) gives the so-called Voigt profile.

different ways depend upon the quantities T and P_e) and the associated optical depths:

$$\begin{array}{lll} \kappa_\nu = \text{line absorption coefficient} & \tau_\nu = \int \kappa_\nu \, dt & \multirow{2}{*}{$\Big\}\; x_\nu = \int (\kappa_\nu + \kappa)\, dt$} \\ \kappa = \text{continuous absorption coefficient} & \tau = \int \kappa \, dt & \\ \bar{\kappa} = \text{Rosseland opacity coefficient} & \bar{\tau} = \int \bar{\kappa} \, dt & \end{array} \tag{19.1}$$

All the integrals are to be taken from $-\infty$ to t.

If we again make the assumption of local thermodynamic equilibrium (LTE) in regard to the radiative exchange (which should be quite good in the case of atoms or ions having rather complicated term schemes),[1] then for the calculation of the line profile we may take over the whole formulation and computation in Section 18, simply replacing κ by $\kappa + \kappa_\nu$ and the optical depth τ by x_ν.

Thus the intensity of the emergent radiation that makes an angle θ with the normal to the surface of the atmosphere ($\theta = 0$ again corresponding to the center of the solar disk and $\theta = \pi/2$ to the limb) becomes

Line (frequency ν) $\quad I_\nu(0,\theta) = \int_0^\infty B_\nu(T(x_\nu)) e^{-x_\nu \sec\theta} \, dx_\nu \sec\theta$

Neighboring continuum (suffix 0)

$$I_0(0,\theta) = \int_0^\infty B_0(T(\tau)) e^{-\tau \sec\theta} \, d\tau \sec\theta, \tag{19.2}$$

where the relation between the two optical depths is expressed by

$$\frac{dx_\nu}{d\tau} = \frac{\kappa_\nu + \kappa}{\kappa} \quad \text{or} \quad x_\nu = \int \frac{\kappa_\nu + \kappa}{\kappa} \, d\tau. \tag{19.3}$$

The depression in the line is then

$$r_\nu(0,\theta) = \frac{I_0(0,\theta) - I_\nu(0,\theta)}{I_0(0,\theta)} \tag{19.4}$$

and the equivalent width defined in accordance with Figure 19.2 is

$$W_\lambda = \int r_\lambda(0,\theta) \, d\lambda. \tag{19.5}$$

We can again make use of the approximation (18.13) and so obtain

$$r_\nu(0,\theta) \approx \frac{B_\nu(T(\tau = \cos\theta)) - B_\nu(T(x_\nu = \cos\theta))}{B_\nu(T(\tau = \cos\theta))}. \tag{19.6}$$

In calculating the Kirchhoff–Planck function $B_\nu(T)$ the small frequency differences in the vicinity of a line are naturally of no significance; an essential point is that the radiation in the line comes from a higher layer $x_\nu = \cos\theta$, with a correspondingly lower temperature, than that in the

[1]We cannot here deal with the special features of the scattering of light in the continuum (Thomson or Rayleigh scattering) and in lines (resonance fluorescence at low pressures). In the outcome the departures from LTE mostly remain within moderate bounds.

neighboring continuum that comes from $\tau = \cos\theta$. If the absorption coefficient in the line center is very much greater than in the continuum ($\kappa_\nu \gg \kappa$), the central intensity of the line is simply the Kirchhoff–Planck function for the boundary temperature $T(\tau = 0)$ of the atmosphere. The line then reaches the same maximum depression $r_c(0,\theta)$ for all sufficiently strong lines in the same region of the spectrum.

For application to *stars,* in which we cannot observe the center-to-limb variation, we again need the corresponding expressions for the flux, or the mean intensity F_ν in the line and F_0 in the neighboring continuum. The depression in the line then becomes

$$R_\nu(0) = \frac{F_0(0) - F_\nu(0)}{F_0(0)} \approx \frac{B_\nu(T(\tau = 2/3)) - B_\nu(T(x_\nu = 2/3))}{B_\nu(T(\tau = 2/3))} \tag{19.7}$$

and the corresponding equivalent width in the stellar spectrum

$$W_\lambda = \int R_\lambda(0)\, d\lambda. \tag{19.8}$$

[We see from equation (19.7) that in a stellar spectrum the continuum and also the wings of the lines (where $\kappa_\nu \ll \kappa$) come mainly from those layers of the atmosphere whose optical depth is $\tau = 2/3$ in the continuum in the region of the spectrum concerned. The lines of neutral metals (Fe I, Ti I, . . .) in the spectra of the Sun and similar fairly cool stars are exceptions. On account of increasing ionization, the concentration of the atoms decreases so rapidly with increasing depth in the atmosphere that the "center of gravity" for forming the lines lies no deeper than $\tau \approx 0.05$ to 0.1. The centers of strong lines, where $\kappa_\nu \gg \kappa$, come from correspondingly higher layers where $\tau \approx (2/3)\kappa/(\kappa + \kappa_\nu)$.]

In order to be able actually to apply the theory of radiative transfer to the lines,[2] we now turn to the second point of our program, the calculation of the line absorption coefficient as a function of temperature T, electron pressure P_e, or gas pressure P_g and distance from the line center $\Delta\nu$ in frequency or $\Delta\lambda$ in wavelength. The following mechanisms affect κ_ν:

(1) *Doppler effect* of thermal velocities and maybe of turbulent motions. For instance, the thermal velocities of Fe atoms in the solar atmosphere for $T \approx 5700$ K are about 1.3 km/s which gives for, say, the line at λ 3860 Å a Doppler width $\Delta\lambda_D = 0.017$ Å. Corresponding to the Maxwell velocity distribution of the atoms, the Doppler distribution of the line absorption coefficient is

$$\kappa_\nu \propto \exp\{-(\Delta\lambda/\Delta\lambda_D)^2\}. \tag{19.9}$$

Turbulent motions in stellar atmospheres (see Section 20) often produce similar velocities and make a corresponding contribution to $\Delta\lambda_D$.

[2]In passing, we remark that in calculating the temperature distribution in a "nongray" model atmosphere account must be taken of line absorption as well as continuous absorption. Since the outward flow of radiation is held back in the lines, the radiation from the higher layers is thereby enhanced, thus causing a steeper temperature drop in the high layers.

(2) *Damping*. In classical optics a wave train of limited duration with a damping constant $\gamma(s^{-1})$, or a characteristic time γ^{-1} s, corresponds in accordance with a well-known theorem in Fourier analysis to a spectral line in which the absorption coefficient κ_ν exhibits the typical damping distribution

$$\kappa_\nu \propto \frac{\gamma}{(2\pi\,\Delta\nu)^2 + (\gamma/2)^2}. \tag{19.10}$$

Thus γ is the whole half width $2 \times 2\pi\,\Delta\nu_{1/2}$ of the absorption coefficient in units of angular frequency.

According as the temporal limitation of the radiative process is determined by the emission of the atom itself or by its collisions with other particles, following H. A. Lorentz we speak of *radiation-damping* or *collision-damping*.

(a) Radiation damping. According to quantum theory, the radiation-damping constant γ_{rad} is equal to the sum of the decay constants (reciprocal life times) of the two energy levels between which the transition takes place. Since γ is thus of the order 10^7 to 10^9 s^{-1}, we expect from this mechanism (half) half widths of the absorption coefficient, say at λ 4000 Å, of about 4×10^{-6} to 4×10^{-4} Å.

(b) Collision damping. The collision-damping constant is $\gamma_{\text{coll}} = 2 \times$ number of effective collisions per second. According to W. Lenz, V. Weisskopf, and others, we count as such those encounters of the radiating particle with perturbing particles in which the phase of the radiative oscillation is displaced by more than about one tenth of an oscillation.

In fairly cool stellar atmospheres like that of the Sun in which hydrogen is mostly neutral collision damping by neutral hydrogen atoms preponderates. The hydrogen atoms affect the radiating atom by van der Waals forces (interaction energy $\propto$ (distance)$^{-6}$).

For smaller separations of the colliding pair, repulsive forces, which decrease even faster with distance, can become important as well as van der Waals forces. We should perhaps recall that it is the interaction between attractive and repulsive forces that makes possible the existence of molecules and crystals.

So long as the radiating atom interacts with just *one* H atom at a time, the damping constant γ_{coll} is proportional to the gas pressure P_g.

In the case of spectral lines that show a large quadratic Stark effect, and in mainly ionized atmospheres, collision damping by free electrons may be predominant. Then the interaction energy varies as the square of the field strength that the electron produces at the position of the radiating particle; so it is proportional to the (distance)$^{-4}$. The damping constant is now proportional to the electron pressure P_e.

For about 10^9 effective collisions per second we expect the (half) half width of the line absorption coefficient to be of the order 10^{-3} Å. It makes no difference whether the collisions, as in the solar atmosphere, are mostly with hydrogen atoms or, as in hot stars, are mostly with free electrons.

(3) *Combination of Doppler effect and damping.* The ratio of the half damping constant $\gamma/2$ to the Doppler width $\Delta\omega_D = c\,\Delta\lambda_D/\lambda^2$ (measured in circular frequency) is

$$\alpha = \gamma/2\Delta\omega_D \tag{19.11}$$

and in stellar atmospheres almost without exception $\alpha < 0.1$. So one might suppose that collision broadening can be neglected in comparison with Doppler broadening. However, this is not correct because the Doppler distribution (19.9) falls off exponentially away from the line center, while the collision distribution (19.10) falls off only like $1/\Delta\lambda^2$. Each moving atom produces a damping distribution with a sharp central peak and broad wings, which is Doppler shifted as a whole corresponding to the velocity. Thus we obtain (Figure 19.3) a distribution of the line absorption coefficient κ_ν having a fairly sharply bounded Doppler core in accordance with equation (19.9) to which are joined almost directly the damping wings where in accordance with (19.10) κ_ν varies as $1/\Delta\lambda^2$.

The absolute value of the absorption coefficient κ_ν (cm^{-1}) is always normalized by the quantum-theoretical relation, the integral being taken over the whole line

$$\int \kappa_\nu \, d\nu = \frac{\pi e^2}{mc} Nf. \tag{19.12}$$

Here e, m again denote the charge and mass of the electron; c is light speed; N is the number per cubic centimeter of absorbing atoms in the energy level from which the absorption occurs. Within the framework of classical electron theory, which sought to ascribe the spectral lines to harmonic electron oscillators of the corresponding frequencies, the formula would be complete without f (that is, with $f = 1$). In agreement with laboratory measurements (see below), quantum theory requires that the formula be extended to include the so-called *oscillator strength* f. After some simple manipulation, we finally obtain from equations (19.9) and (19.10) in combination with (19.12) the absorption coefficient in cm^{-1} at interval $\Delta\lambda$ from the center of a line of wavelength λ_0

Doppler core $$\kappa_\nu = \sqrt{\pi}\, \frac{e^2}{mc^2} \frac{\lambda_0^2 Nf}{\Delta\lambda_D} \exp\{-(\Delta\lambda/\Delta\lambda_D)^2\} \tag{19.13}$$

Damping wings $$\kappa_\nu = \frac{1}{4\pi} \frac{e^2}{mc^2} \frac{\lambda_0^4}{c} \frac{Nf\gamma}{\Delta\lambda^2}. \tag{19.14}$$

The quantity e^2/mc^2 is the so-called classical radius of the electron $r_0 = 2.818 \times 10^{-13}$ cm.

Since the number of atoms N is proportional to the statistical weight g of the corresponding state, instead of f we usually state the value of gf. The intensity of an emission or absorption line produced by an optically thin layer is directly proportional to this quantity.

One can calculate the relative *gf* values within a multiplet (Section 17) using quantum-theoretical formulas found by A. Sommerfeld and H. Hönl, H. N. Russell, and others in connection with the Utrecht measurements made by H. C. Burger and H. B. Dorgelo (1924). The *gf* values for a doublet such as the sodium D lines $3\ ^2S_{1/2} - 3\ ^2P^0_{1/2,3/2}$ are in the ratio 1 : 2, for a triplet such as the calcium lines $4\ ^3P^0_{2,1,0} - 5\ ^3S_1$ (λ 6162, 6122, 6103 Å) they are in the ratios 5 : 3 : 1, and so on.

There exist corresponding formulas for aggregates of higher order, the so-called supermultiplets and transition arrays.

A general idea of the absolute values of the oscillator strengths is given by the f–sum rule of W. Kuhn and W. Thomas: Let absorption transitions $n \rightarrow m$ be possible with oscillator strengths f_{nm} from a particular energy level n of an atom or ion with z electrons (more precisely, we always restrict ourselves to consideration of the radiating electrons that take part in the relevant transitions) and let transitions $m \rightarrow n$ to n from lower levels occur with oscillator strengths f_{mn}. Then we have

$$\sum_m f_{nm} - \sum_m \frac{g_m}{g_n} f_{mn} = z, \tag{19.15}$$

where g_n and g_m are the statistical weights (Section 17) of the corresponding energy levels. If there is effectively a single strong transition from the ground state of an atom or ion to the next higher term, then for the multiplet as a whole we can put approximately $f \approx z$. For instance, we have about $f = 1/3 + 2/3 = 1$ for the D lines of sodium taken together.

For hydrogen and ionized helium one can calculate the f values exactly from quantum theory.

For certain hydrogen-like spectra (in particular systems with 1, 2, or 3 radiating electrons) D. R. Bates and A. Damgaard have developed a very practical quantum theoretical approximation procedure.

For the so-called complex spectra of atoms and ions with several outer electrons (such as the astrophysically very important spectra of the metals Fe I, Ti I, . . . , Fe II, Ti II, . . .) one can measure *relative* f values in emission (in the arc or King's furnace) or in absorption (King's furnace). The main difficulty lies in measuring absolute f values, even for a few selected lines of the atom or ion, since here the number of absorbing or emitting particles has somehow to be measured directly. One can, for instance, introduce into an electric furnace a sealed quartz absorption vessel in which there is established a certain vapor pressure, depending upon the temperature, of the metal under investigation. In recent times the decay constants or transition probabilities of individual atomic states, equivalent to the f values, have been successfully measured electronically. A great step forward was taken by S. Bashkin, who invented the *beam foil* method for measuring lifetimes of excited states of neutral, singly ionized and even multiply ionized atoms. Using, for example, a van de Graaff accelerator, one produces a beam of the particles to be investigated, with energies of several million electron volts. This beam is sent through a thin foil (usually carbon) and so is suddenly stimulated to radiate. Along its further path, in high vacuum, it is now possible—as in the familiar "canal ray" experiments of W. Wien—to measure the decrease with time of the radiation in whatever spectral line is of interest. This depends on the decay constant of the upper energy level and—an undesirable perturbing effect—on the subsequent arrival of electrons from higher energy levels, which mostly decay more slowly. Frequently transitions to several lower levels are possible from the upper level of interest. It is then necessary to determine their branching ratios in order to obtain finally the transition probabilities or f values of individual lines.

A bibliography of f values and atomic transition probabilities, arranged according to elements, has been published by B. M. Miles and W. L. Wiese and then by L. Hagan and W. C. Martin (Nat. Bureau of Standards Spec. Publ. 320 and 363, Washington 1970–1972). W. L. Wiese, M. W. Smith, and B. M. Glennon have further published a critical compilation in two volumes of measured and calculated f values (Nat. Standard Ref. Data Series—NBS 4 and 22, Washington 1966 and 1969).

We now consider how the profile and the equivalent width of an absorption line grows if the concentration N of the atoms producing the line (or the product of N with the oscillator strength f) increases. The absorption coefficient κ_ν in the central part of the line is determined by the Doppler effect according to equation (19.13); this part is connected to the damping wings in the outer parts, the damping constant being determined by radiation damping and collision damping.

The connection between the depression in the line R_ν and the absorption coefficient κ_ν (cm^{-1}) for an absorption tube of length H in the laboratory, without reemission, would be from equation (11.7)

$$R_\nu = 1 - e^{-\kappa_\nu H}. \tag{19.16}$$

For a stellar atmosphere, it is given by the formulas (19.2) to (19.7). With not too great concern for accuracy, we can often replace this somewhat complicated calculation with the approximate or interpolation formula

$$R_\nu = \left(\frac{1}{\kappa_\nu H} + \frac{1}{R_c}\right)^{-1}. \tag{19.17}$$

Here H denotes an effective height, or NH an effective number of absorbing atoms above 1 cm^2 of the stellar surface. For $\kappa_\nu H \ll 1$ (absorption in a thin layer) $R_\nu \approx \kappa_\nu H$, and for $\kappa_\nu H \gg 1$ (optically thick layer) R_ν tends to the limit R_c for very strong lines which has already been mentioned. One can calculate H or NH by comparing (19.17) with the formulas (19.2) to (19.7); in a not too large wavelength interval (often in practice some hundreds of angstroms) one may treat the effective thickness of an atmosphere as a constant.

We now exhibit the results of these calculations: In the lower part of

Figure 19.4 we have drawn the line profile for various values of the quantity[3] log NHf + constant. The depression in the weak line ($C \ll 1$) with $R_\nu \approx \kappa_\nu H$ reflects simply the Doppler distribution of the absorption coefficient in the core of the line. With increasing NHf or with $C \gg 1$, the line center approaches the maximum depth R_c, since here we receive only radiation from the uppermost layer at the boundary temperature $T(\tau = 0)$. At first, the line becomes only a little wider, because the absorption coefficient falls off steeply with $\Delta\lambda$. This situation changes only if with increasing NHf the effective optical depth becomes appreciable also in the damping wings (Figure 19.3). Since now $\kappa_\nu H \approx NHf\gamma/\Delta\lambda^2$, the line acquires wide damping wings and, other things being equal, its width at a given depth R_ν is proportional to $(NHf\gamma)^{1/2}$.

By integration over the line profile we easily obtain the equivalent width of the line. We show it as a multiple of $2R_c\,\Delta\lambda_D$, that is, of a strip of depth equal to the maximum depth and width equal to twice the Doppler width. Thus we obtain the *curve of growth* showing $\log(W_\lambda/2R_c\Delta\lambda_D)$ as a function of $\log C = \log NHf$ + constant, which is important for the evaluation of stellar spectra (Figure 19.4, upper part).

As one can see from our discussion, for weak lines (left) $W_\lambda \propto NHf$, and we are in the linear part of the curve of growth. Then follows the flat, or Doppler, region in which the equivalent width is about 2 to 4 times the Doppler width $\Delta\lambda_D$. For strong lines (right), corresponding to the growth in the width of the profile, $W_\lambda \propto (NHf\gamma)^{1/2}$ and we get into the damping, or square-root region of the curve of growth. Quite generally, the damping constant γ is important here. In Figure 19.4 we have given the ratio of

[3]More precisely, we are concerned with the quantity $C = \kappa_0 H/R_c$, where from equation (19.13) κ_0 is the absorption coefficient at the center of the Doppler core. $C = (1/R_c)(\sqrt{\pi}e^2/mc^2)(\lambda_0^2 NHf/\Delta\lambda_D)$ is the effective optical depth of the atmosphere for the center of the line.

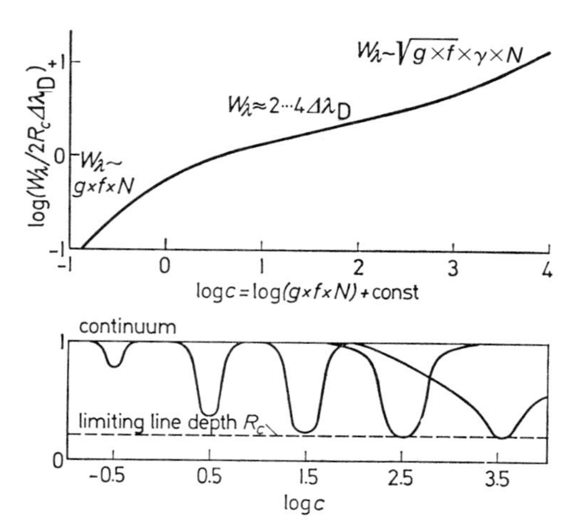

Figure 19.4
Curve of growth (upper diagram). The equivalent width W_λ is plotted as a function of the concentration of absorbing atoms, log (gfN) + constant. Here W_λ is expressed in terms of the area of a strip of dimensions twice the Doppler width $\Delta\lambda_D$ by the limiting depth R_c. The line profiles (lower diagram) illustrate how the curve of growth arises.

damping to Doppler width $\alpha = \gamma/2\ \Delta\omega_D$ the numerical value 1/30, corresponding to a mean value for metal lines in the solar spectrum. The strong D lines of Na I as well as the H and K lines of Ca II in the solar spectrum, for example, lie on the damping part of the curve of growth.

The hydrogen lines claim special consideration. Since in an electric field these show specially large linear Stark-effect splitting, their widening in a partially ionized gas results in the first place from a quasistatic Stark effect of the statistically distributed electric field produced by the slowly moving ions. Working from the consideration that in the distance interval r to $r + dr$ from a hydrogen atom a perturbing ion is present with probability proportional to $4\pi r^2\ dr$ and that it then produces a field proportional to $1/r^2$ (to which also the line splitting is proportional), we can show that in the wings of the lines the absorption coefficient $\kappa_\nu \approx 1/\Delta\lambda^{5/2}$. The theory was developed originally by Holtsmark and has recently been refined by taking account of nonadiabatic effects, of collision damping by electrons and of an improved calculation of the microfield in the plasma.

Now we turn to our main problem, *the quantitative analysis of stellar spectra*. For an initial orientation, we usually begin with an approximation procedure, the so-called coarse analysis, in which we work with constant mean values of temperature, electron pressure and effective thickness H (in some considerable part of the spectrum). Then we can apply the universal curve of growth calculated from the interpolation formula (19.17) in order to obtain the number of absorbing atoms NH above 1 cm² of the stellar surface for any particular species of atom or ion in the energy level that produces the line, using the measured equivalent width W_λ of the line and the known values of f and γ.

By comparing the NH values for different excitation energies χ_s and for different stages of ionization (for example Ca I and Ca II) for one and the same element, from the formulas of Boltzmann and of Saha (Section 17) one can now calculate the temperature T and the electron pressure P_e. Knowledge of these then leads from the numbers of atoms in particular energy levels to the total number of all particles of the element concerned (for all stages of excitation and ionization) and thus to the *abundance distribution of the elements*. If we know this and the degree of ionization of the various elements fairly completely, then we can go over from the electron pressure P_e to the gas pressure P_g and calculate the *gravitational acceleration* g from the equation of hydrostatic equilibrium. Indeed, the gas pressure is nothing other than the weight, that is the mass times the gravitational acceleration g, of all the particles above 1 cm² of the selected reference surface, or, as we usually say, the surface of the star.

Upon the basis of such a coarse analysis we can build the more exact, but also much more laborious, fine analysis of stellar spectra. As in Section 18, we construct a model of the stellar atmosphere under investigation, selecting the most plausible T_{eff}, g, and chemical composition. Invoking the whole theory of radiative transport, of continuous and line absorption coefficients (including their dependence on depth) and so on, we calculate the equivalent widths W_λ of the lines for this model. We then compare the results of these calculations for the model with the measurements for such elements as are represented by several different ionization and excitation levels. Hydrogen lines also play an important part since we know that hydrogen is the most abundant element so that its abundance does not enter here. Furthermore, we can appeal to the energy distribution in the continuum or to the color indices (Section 18). From the preceding considerations one can estimate which criteria depend more strongly on T_{eff}, or on g, or on the abundance of a particular element. Thence we can improve the starting values of these quantities using step by step approximation procedures.

The newcomer is usually dismayed by the great number of lines in a spectrum. Yet in carrying out an analysis it is often found that the available measurements scarcely suffice to determine all the parameters of interest! The situation gets worse if, for instance, we treat turbulence in the stellar atmosphere as a function of depth and then require that this function be extracted from the measurements.

Having thus sought briefly to present the methodology of a quantitative analysis of stellar spectra, we turn to the results obtained.

In Table 19.1 we collect the results of careful spectral analysis first for stars of the main sequence, in particular for the Sun, and then for two extreme supergiants. In the case of these "normal" stars we give the spectral type Sp and the luminosity class LC (cf. Figure 15.4) according to Morgan and Keenan. The table contains the logarithm of the number of atoms $\log N$, referred as usual to $\log N = 12$ for hydrogen. In the case of hot stars, helium and the lighter elements up to Si or S are represented by highly excited lines, for which f values can be calculated fairly exactly. Here the heavier elements are expected only in states of high ionization that have no lines in the accessible part of the spectrum. Below about 10 000 K the lines of ionized, and then of neutral, metals make their appearance in ever greater numbers. In the case of the Sun we have incomparably better observational material than for any other star, and also T_{eff} and g are already known. Table 19.2 contains the solar abundances of trace elements, whose weak lines are detectable only in "low-noise" spectra at very high dispersion.

Naturally, the accuracy with which the abundance of a particular element can be determined is different for stars of different temperatures. It depends upon the number of lines by which it is represented, the part of the curve of growth in which these occur and, above all, the accuracy with which the oscillator strengths f are known. It is also to be remembered that the abundances can be determined only along with T_{eff} and g; for instance, in middle spectral types an underestimate of T_{eff} mimics an underestimate of metal abundances. Generally speaking, on account of their high ionization potentials, helium and the lighter elements are best determined (relative to hydrogen) in hot stars, while the more easily ionized metals can be studied better in cooler stars. At the present time, the accuracy of a careful determination would correspond to about $\Delta \log N = \pm 0.3$.

We shall clarify the analysis of the emission spectrum of the *solar corona* in Section 20. On the left of the table we give for comparison the abundances of the elements in the *Orion nebula,* a galactic H II region (Section 24). The young, hot stars, like 10 Lac, are *born* out of such gas clouds, as we shall see in Section 26. The *planetary nebulae* (Section 24), on the other hand, represent a late stage of stellar evolution.

On the right of Tables 19.1 and 19.2 we give the abundances of the elements in *carbonaceous chondrites of type I;* apart from the very volatile elements, these meteorites are almost unaltered solar material.

Finally, in the last column of Table 19.1 is listed the distribution of abundances in *galactic cosmic rays* as given by M. M. Shapiro *et al.* (1972 personal communication). This distribution is the one corresponding to the *place of origin* of the cosmic rays and has been obtained by allowing for the influence of the interstellar medium.

For stars whose temperatures are not too different from the Sun's it is convenient to carry out quantitative analyses *relative to the Sun* and so get round the uncertainty in the absolute scale of the f values, which is still one of the main sources of error in absolute abundance determinations. For solar-type stars one can achieve in this way a relative accuracy of $\Delta \log N \approx \pm 0.1$. We have presented relative measurements of this kind in Table 19.3 in such a way that for each star, on average, the abundances of the heavy elements (C to Ba)—usually abbreviated to "metals"—are in agreement for star and Sun. Then on the first line is the *overabundance of hydrogen* in the star considered or, as we would usually say, the *metal deficiency*.

The abundances of the elements are clearly intimately connected with the story of their origin and with their transformations by nuclear processes in the course of stellar evolution. Without anticipating later explanations (Sections 25 and 29), we gather together the most important results of the quantitative analysis of the Sun and stars:

(1) The chemical composition of "normal" stars (the spiral arm population I and the disk population of the Milky Way; see Sections 22 and 27) is nearly the same for them all;[4] they agree in particular with the solar material. It therefore seems to us to be justifiable to describe this mixture of elements as *cosmic material*. Compare in particular the relative analyses of HD 219 134(K3 V), HD 136 202 (F8 IV–V), and ϵ Virginis (G8 III) in Table 19.3. By definition, we denote by "normal" such stars as can be unambiguously placed in the MK classification. Indeed, the very possibility of a two-dimensional classification presupposes that no parameter is necessary beyond T_{eff} and g, that is, that these stars have the same chemical composition.

[4]More precisely, the relative abundances of the heavy elements vary only within the limits of uncertainty $\Delta \log N = \pm 0.1$ and the abundances of this group relative to H and He vary within about ± 0.4. Occasional detailed anomalies in the elements concerned in the CNO cycle will not be considered here.

Table 19.1. Abundance distribution log N of the elements in stellar atmospheres, on a scale on which log $N = 12$ for hydrogen, for normal main sequence stars O9 to G2 and for supergiants B1 and A2. For each star is given the MK classification and the values of T_{eff} and log g determined from the spectrum (for the Sun the values are naturally the directly measured ones). For comparison, values for planetary nebulae and the Orion nebula, a galactic H II region, are given on the left (see Section 24); on the right are the element abundances in type I carbonaceous chondrites and in galactic cosmic rays (reduced to their point of origin).

	Planetary nebulae			H II-region	Hot main sequence stars				Supergiants		Middle main sequence		Meteorites	Galactic cosmic rays
Spectral type	NGC	NGC	IC	Orion-	O 9 V	B 0 V	B 3 V	A 0 V	B 1 Ib	A 2 Ia	G 2 V			
Star (HD)	7027, 2022	7662	418	nebula	10 Lac	τ Sco	ι Her	α Lyr	ζ Per	α Cyg	Sun		Carbona-ceous chondrites I	
T_{eff}		$\approx 10^5$	35 000		37 450	32 000	20 200	9 500	27 000	9 170	5 780			
log g					4.45	4.1	3.75	4.5	3.6	1.13	4.44			
											phot.	corona		
1 H*	12.0	12.0		12.0_0	12.0	12.0	12.0	12.0	12.0	12.0	12.0			10.9
2 He	11.2	11.2		11.0_4	11.2	11.0	10.8		11.3	11.6				9.6
6 C				8.7_1	8.4	8.1	8.1		8.3	8.2	8.53	8.3		8.2
7 N	8.1:	7.7		7.6_3	8.4	8.3	7.7		8.3	9.4	7.91	7.6		7.3
8 O	8.9	8.4		8.7_9	8.8	8.7	8.4	8.8	9.0	9.4	8.83	8.7		8.2
9 F	4.9										4.6		4.92	
10 Ne	7.9	7.8		7.8_6	8.7	8.6	8.6	9.3	8.6			7.6		7.5
11 Na	6.6							7.3			6.30	6.3	6.36	6.2
12 Mg					8.2	7.5	7.3	7.7	7.8	7.8	7.57	7.6	7.57	7.6
13 Al					7.1	6.2	6.1	5.7	6.8	6.6	6.4	6.5	6.48	6.4
14 Si					7.7	7.6	7.1	8.2	8.0	7.9	7.65	7.55	7.55	7.5
15 P											5.4		5.65	5.5

16 S	7.9	7.5_0	7.2	7.1		7.5		7.2	7.0	7.25	6.7
17 Cl	6.9	5.8_5						⩽5.5		4.79	
18 A	7.0			6.7					6.0		5.9
19 K	5.7							5.0		5.13	
20 Ca	6.4:			6.4	6.3		6.5	6.36	6.6	6.42	6.5
21 Sc					3.4		3.2	3.0		3.09	
22 Ti					4.8		5.1	4.8		4.91	
23 V					4.0		3.9	4.0		4.02	
24 Cr					5.6		5.7	5.6		5.63	
25 Mn					5.3		5.6	5.4		5.50	
26 Fe			7.3	7.4	(6.5)		7.6	7.6	7.6	7.50	7.6
27 Co							3.7	5.0	6.0	4.91	
28 Ni					7.0		4.8	6.25	6.3	6.24	6.2
29 Cu								4.16		4.32	
30 Zn								4.4		4.53	
31 Ga								2.9		3.26	
32 Ge								3.3		3.68	
37 Rb								2.6		2.36	
38 Sr					2.8		3.1	2.8		2.93	
39 Y					2.1			2.3		2.21	
40 Zr					2.9			2.3		3.05	

*1 D $\sim 7.3 \pm 1$.

Literature: Planetary nebulae: Aller and Czyzak (1968), Flower (1969); Osterbrock (1970). HII-regions; Orion: Peimbert and Costero (1969), Morgan (1971). 10 Lac: Traving (1957). τ Sco: Hardorp and Scholz (1970) (also λ Lep B0.5 IV). ι Her: Kodaira and Scholz (1970) (also η Hya B3 V and HD 58343 B 3 Ve). OB-stars: Scholz (1972). (Review article). α Lyr: Hunger (1960) and previous papers. ζ Per: Cayrel (1958). α Cyg: Groth (1961), Wolf (1971) (also η Leo A 0 Ib). Sun: Müller (1967), Lambert *et al.* (1968, 1969), Garz *et al.* (1969). Carbonaceous chondrites I: Urey (1967) with extensive bibliography. Galactic cosmic rays: Shapiro *et al.* (1971); Shapiro (1972) personal communication.

For a key to the references for Tables 19.1 to 19.3 see A. Unsöld: The Chemical Evolution of the Galaxies. *Proc. First Europ. Astron. Meeting,* vol. III, Ed. B. Barbanis and J. D. Hadjidemetriou. Springer-Verlag, Heidelberg 1974. Small changes have been in accordance with recent investigations.

Table 19.2. Abundances of rare elements in the solar atmosphere and in carbonaceous chondrites, on scales on which log N (H) = 12 and log N (Si) = 7.55, respectively.

Element	Sun	Carbonaceous chondrites I	Element	Sun	Carbonaceous chondrites I
3 Li	0.8	3.25	59 Pr	1.6	0.78
4 Be	1.2		60 Nd	1.8	1.44
5 B	<2.5		62 Sm	1.6	0.91
41 Nb	2.0		63 Eu	0.5	0.51
42 Mo	2.0		64 Gd	1.1	1.15
44 Ru	1.5		66 Dy	1.1	1.11
45 Rh	0.9		68 Er	0.8	0.89
46 Pd	1.3	2.17	69 Tm	0.4	0.09
47 Ag	0.2	1.53	70 Yb	0.8	0.87
48 Cd	1.9	1.93	71 Lu	0.8	0.09
49 In	1.4	0.78	81 Tl	0.9	0.80
50 Sn	1.7	1.81	82 Pb	1.85	1.75
51 Sb	1.9	1.13	83 Bi	≤0.8	0.78
56 Ba	2.1	2.22	90 Th	0.8	~0.6
57 La	1.8	1.11	92 U	≤0.6	~0.0
58 Ce	1.8	1.62			

(2) In the realm of chemical analysis, the type I *carbonaceous chondrites* most closely resemble solar material. Their condensation from the original solar material must have occurred at such a low temperature that only the very volatile elements (H, He, . . .) escaped.

(3) The *interstellar matter* of H II regions, for example the Orion nebula (Section 24), from which young, hot stars like 10 Lac or τ Scorpii are constantly being formed, has the same composition as these stars.

(4) Many, although (as we shall see) by no means all, stars in advanced stages of evolution, such as the red giant star ϵ Virginis (G8 III), have atmospheres (or envelopes) consisting of unaltered cosmic material.

(5) The "metal-poor" *high-velocity stars* belong to the so-called population II stars of the galactic halo. These are also the oldest stars of our Milky Way System (Section 27). In these stars the abundances of *all* the heavy elements, from carbon C to (at least) barium Ba are reduced by the same factor (to within the present accuracy of analysis). These metal deficiency factors cover the whole range from 1 to about 500. HD 140283 lies on the main sequence of the Hertzsprung–Russell diagram. HD 6833 and HD 122563 have reached the red giant stage and HD 161817 is at the next stage of evolution and belongs to the horizontal branch in the HR diagram (Section 26). HD 161817 is the high velocity star also known as Albitzky's star, with a radial velocity of −363.4 km/s; its space velocity relative to the solar neighborhood is only slightly larger. In this very interesting star the abundances of all heavy elements are reduced relative to the sun by the same factor, about 13.

Table 19.3. Abundances of elements (referred to hydrogen) relative to the Sun. The $\Delta \log N$ are calculated in such a way that on average the heavy element abundances are in agreement. The first line (1H) thus gives also the reduction of the mean "metal abundance" ($\log N_M/N_H$) in the star relative to the Sun. The abundance ratios of the heavy elements (C to Ba) prove to be independent of the reduction in the average metal abundance.

Spectral type	G 8 III	F 8 IV-V	K 3 V	~G 0 V	K 1 III	~K 2 III	~sd A 2	F 8 V
Star (HD)	ϵ Vir	136202	219134	140283	6833	122563	161817	β Vir
T_{eff}	4940	6030	4700	5940	4420	4600	7630	6120
$\log g$	2.7	3.9	4.5	4.6	1.3	1.2	3.0	4.3
1 H	0.00	0.00	0.00	+2.32	+0.96	+2.75	+1.2	−0.30
6 C	−0.12	+0.06		+0.5		−0.39	+0.1	−0.15
11 Na	+0.30	+0.03	+0.04	+0.30	−0.19	+0.33	−0.1	+0.06
12 Mg	+0.04	−0.01	+0.15	−0.01	+0.11	+0.03	+0.4	−0.04
13 Al	+0.14	+0.09	+0.07	+0.26	+0.37	−0.12	−0.2	+0.05
14 Si	+0.13	+0.02	+0.39	−0.07	+0.10	+0.29	0.0	−0.08
16 S	+0.09							−0.04
20 Ca	+0.10	+0.03	−0.11	+0.03	+0.28	+0.19	−0.1	+0.02
21 Sc	−0.07	−0.07	+0.22	−0.61	−0.15	+0.03	−0.1	+0.10
22 Ti	−0.07	−0.02	−0.10	−0.01	+0.02	+0.08	+0.4	+0.03
23 V	−0.08	−0.14	−0.09		+0.03	−0.06	−0.7	+0.03
24 Cr	+0.01	+0.02	−0.20	+0.09	+0.15	+0.06	+0.1	−0.01
25 Mn	+0.07	−0.08	−0.31	−0.35	−0.17	−0.19	−0.5	−0.06
26 Fe	+0.01	0.00	+0.10	+0.16	+0.11	+0.03	−0.4	−0.04
27 Co	−0.03	−0.27	+0.17	−0.02	−0.10	+0.04	0.0	−0.07
28 Ni	+0.03	+0.02	+0.10	+0.31	−0.06	+0.16	−0.5	−0.01
29 Cu	+0.06	−0.11						+0.05
30 Zn	+0.05	−0.21	−0.13		+0.14	+0.28		−0.17
38 Sr	+0.02	+0.10	−0.01	0.00		−0.53	0.0	+0.01
39 Y	−0.16	−0.14	+0.48		−0.46	−0.04	+0.6	+0.03
40 Zr	−0.15	−0.39	−0.30		−0.02	−0.08	−0.5	−0.01
56 Ba	−0.09	−0.05				−0.55	+0.5	0.00
57 La	−0.08				−0.25			+0.11
58 Ce	−0.08							
59 Pr	+0.37							
60 Nd	+0.06							
62 Sm	+0.01							
63 Eu						−0.27		
72 Hf	+0.18							

Literature: ϵ Vir: Cayrel and Cayrel (1963). HD 136 202: Zielke (1970) (also HD 105 590, 155 646, 208 776, 208 906). HD 219 134 and 6 833: Cayrel de Strobel (1966) (further K stars: HD 190 404, 48 781, 94 264, 6 497, 35 620, ϵ Vir). HD 140 283: Baschek (1962); Aller and Greenstein (1960). HD 122 563: Wolffram (1972). HD 161 817: Kodaira (1964). β Vir: Baschek *et al.* (1967).

At the other extreme are the "metal-rich" stars (for example, β Virginis). However, the metal enrichment factors are not more than about 3 and so are still barely outside the region of possible systematic errors. In particular, as a more exact treatment shows, the determination of the effects of *turbulence* on the spectrum is of critical importance.

20. Motions and Magnetic Fields in the Solar Atmosphere and the Solar Cycle

When looked at carefully, the solar surface is seen to be mottled in appearance. The *granulation* is composed of bright granules the temperature of which exceeds that of the darker interstices by 100 to 200 K. The diameters of the granulation elements range from about 5″ down to the limit of telescopic resolving power at about 1″ corresponding to 725 km on the Sun. From sequences of photographs, the lifetime of a granule is found to be about 8 min.

Then we can easily observe the dark *sunspots* that were discovered by Galileo and his contemporaries. They appear predominantly in two zones of equal northern and southern heliographic latitude. A typical sunspot has the following structure and dimensions (to order of magnitude):

	Diameter (km)	Area (in millionths of solar hemisphere)	
umbra (dark center)	18 000	80	(20.1)
penumbra (somewhat brighter fringe)	37 000	350	

The reduced brightness in the spot results from a lowering of temperature. In the largest spots the effective temperature drops from 5780 K for the normal solar surface to about 3700 K. Correspondingly, the spectrum of a large spot is much like that of a K star; we anticipated the explanation by means of Saha's theory of ionization in Section 17.

Sunspots make their appearance on the solar surface mostly in *groups*. Such a group (Figure 20.1) is surrounded by bright *faculae*. There exist also the so-called *polar faculae,* independently of sunspots. The brightness of the faculae exceeds that of the normal solar surface by a few percent only near the solar limb. Using equation (18.13), we infer that in the faculae only the outermost layers of the Sun (about $\tau \leq 0.2$) are a few hundred degrees hotter than elsewhere.

Making use of spots and—in higher heliographic latitudes (see below)—of faculae, etc., we can study the *rotation* of the Sun. We find that the Sun does not rotate as a rigid body but that the regions of high latitude lag behind the equatorial region:

heliographic latitude	0° (equator)	20°	40°	70°	
mean sidereal rotation	14°.5	14°.2	13°.5	~11°.7	per day
sidereal period of rotation	24.8	25.4	26.7	~31	days.

(20.2)

Within their limits of accuracy, measurements of the Doppler effect at the solar limb (equatorial speed 2 km/s, approximately) confirm this picture. The synodic period of rotation—seen from the Earth—is correspondingly

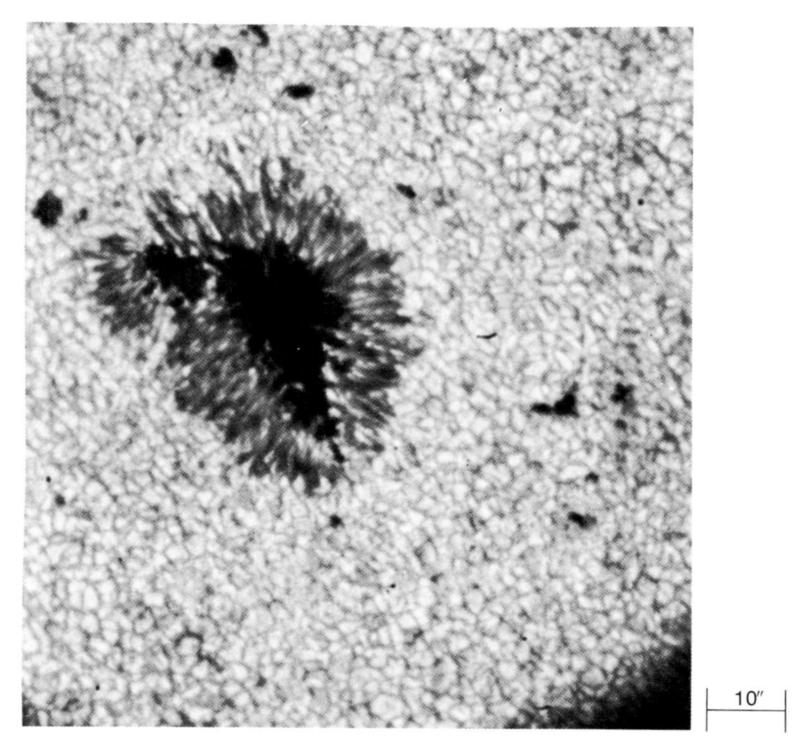

Figure 20.1. Granulation and sunspot (Mount Wilson sunspot number 14 357). Several dark pores, with diameters of a few seconds of arc, can be seen in the vicinity of the spot. Photograph: 12-in (30-cm) stratospheric telescope (M. Schwarzschild) at a height of 15 mi (24 km), 1959 August 17 $16^{h}\ 13^{m}$ UT. Exposure time 0.0015 s and passband 5470 ± 370 Å.

longer; for the sunspot zones we find a value in round figures of 27 days, which agrees with the quasi-periodical recurrence of a number of geophysical phenomena.

As the apothecary Schwabe was able to show about 1843, the spottedness of the Sun varies with a period on the average of 11.2 years. The other manifestations of solar activity, which we shall consider again later on, also follow the sunspot cycle so that nowadays we often speak of the 11.2 year *solar cycle* or cycle of solar activity. As a measure of solar activity we use the quantity R introduced by R. Wolf in Zürich:

$$\text{relative sunspot number } R = k\,(10 \times \text{number of visible spot groups} + \text{number of all spots}) \qquad (20.3)$$

where k denotes a constant depending upon the telescope employed. Or else we use the areas of umbrae, of entire spots and of the faculae as measured photographically at Greenwich since the time of Carrington (1) directly in projection, in units of one-millionth of the solar disk, (2) corrected for foreshortening, in millionths of the solar hemisphere.

A more exact insight into the higher layers of the solar atmosphere (which are obviously strongly affected by the solar cycle) is afforded us by observation in the light of their spectral lines. According to equation (19.6), the latter comes from an optical depth for continuous absorption plus line absorption given by

$$x_\nu = \int (\kappa + \kappa_\nu)\, dt \approx \cos \theta. \tag{20.4}$$

In this way we can separately observe layers having an optical depth in the continuum of only $\tau \approx 10^{-3}$ or less, that do not appear at all in ordinary photographs of the Sun. The instruments we use are as follows:

(1) The *spectroheliograph* (G. E. Hale and H. Deslandres 1891), a large grating monochromator with which the Sun is photographed piece by piece in a sharply bounded wavelength interval of about 0.03 Å to 0.1 Å inside a Fraunhofer line. We obtain the most interesting pictures using the very strong K line of Ca II (λ3933 Å) or the hydrogen line Hα (λ6563 Å). The calcium or hydrogen spectroheliograms, which relate to fairly high layers of the solar atmosphere (the chromosphere), form one of the most important aids to the study of solar activity.

(2) The *Lyot polarization filter* (B. Lyot 1933/38) which makes it possible in a single short exposure to photograph the entire solar image, for example in the red hydrogen line Hα 6563 Å in a wavelength band, to some extent adjustable, of only about 0.5 to 2 Å. Consequently the Hα "filtergram" attains somewhat better definition than the corresponding spectroheliogram. Lyot constructed similar filters also for the coronal lines (see below).

The observation of the highest layers at the solar limb is strongly encroached upon by scattered light. This comes partly from contamination

Figure 20.2 Coronograph (after B. Lyot, about 1930). Light passing through the objective *A*, an optically pure simple planoconvex lens, forms a solar image on the disk *B*. This disk is introduced in order to cut out the light coming from the solar disk, and projects 10 to 20 s of arc beyond the solar limb. Light which was originally diffracted at the edge of the objective must next be removed. This is effected by inserting an annular stop at *D*, where an image of the aperture stop has been formed by the field lens *C*. The final image is formed by the objective *E* upon either a photographic plate or a spectrograph slit, where the corona, prominences, and other features may be studied.

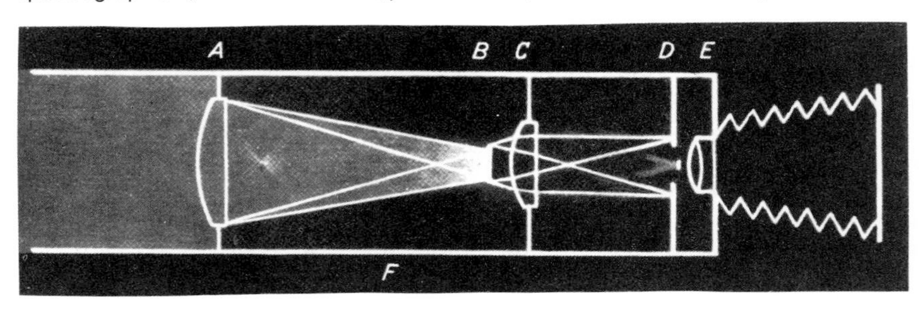

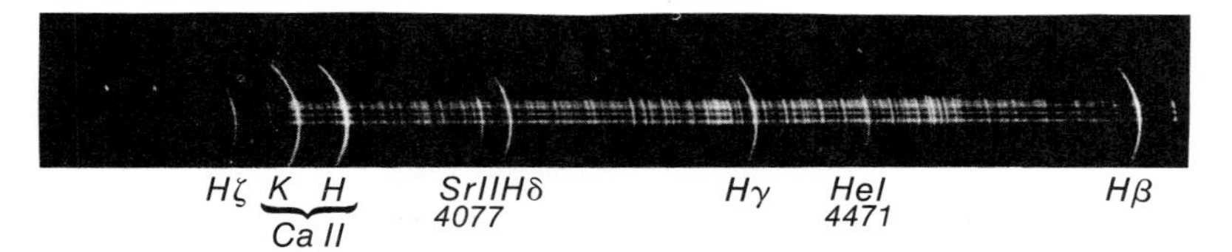

Figure 20.3. Flash spectrum (i.e. emission spectrum of the solar chromosphere). Photograph taken at total solar eclipse, Khartoum, 1952 February 25 by J. Houtgast. Objective-prism camera 9 foot (270 cm) focal length and 1½ inches (4 cm) aperture. Exposure time 0.2 to 0.9 s after second contact.

of the Earth's atmosphere and of the optics of the instrument, but also to a considerable degree from the diffraction of light at the entrance aperture. The *Lyot coronograph* to a great extent eliminates instrumental scattered light; its principle is shown in Figure 20.2. Furthermore, *total solar eclipses* (Section 4) still form an indispensable aid to the investigation of the outermost layers of the Sun. In future, observation from space vehicles may offer serious competition.

If we observe the solar limb in the continuum, its brightness falls away very rapidly toward the outside as soon as the optical thickness along the sight line becomes less than unity. The corresponding optical thickness in a strong Fraunhofer line remains greater than unity for hundreds, or even thousands, of kilometers further out. This means that the Fraunhofer spectrum with its absorption lines goes over into an emission spectrum of the highest layers of the solar atmosphere called the *chromosphere*. This was first noticed by J. Janssen and N. Lockyer at the total eclipse of 1868. When the Moon covered the Sun except for its extreme limb, the emission spectrum of the chromosphere flashed out for a few seconds and consequently it became known as the *flash spectrum* (Figure 20.3). Recent spectrographic observations using a movie camera (with time resolution about 1/20 s), in combination with the motion of the Moon relative to the Sun calculated from the ephemerides, give remarkable information about the stratification of the solar chromosphere: its scale height (corresponding to a decrease in density by a factor $e = 2.72$) is greater than that expected for an isothermal atmosphere at the boundary temperature $T_0 \approx 4000$ K, and it even increases outward.

During totality in a solar eclipse, or in the clearest possible air on a high mountain such as the Pic du Midi (2870 m) using a Lyot coronograph, the solar *corona* is observed extending out to several solar radii (Figure 20.4). Its form (flattening, ray structure, etc.) and brightness, especially for the inner parts, depend on the phase in the solar cycle. Spectroscopic analysis distinguishes the following phenomena which we attempt now to interpret.

(1) The *inner corona* ($r \approx 1$ to 3 solar radii) exhibits a completely *continuous spectrum,* the energy distribution in which otherwise corresponds to that of normal sunlight. Following W. Grotrian, we call this the K

Figure 20.4. Solar corona, near sunspot minimum. Photograph taken at eclipse, Khartoum, 1952 February 25, 9^h 10^m UT by G. van Biesbroeck. Camera 20-ft (6.1-m) focal length and 6-in (15-cm) aperture. Kodak 103a-E emulsion and yellow filter. Exposure time 1.5 min. At sunspot minimum the corona displays extended streamers in the neighborhood of the sunspot zones, and the finer "polar tufts" over the polar caps. At sunspot maximum the corona has a more rounded form.

corona. Its light is partly linearly polarized. In agreement with K. Schwarzschild and others, we ascribe it to Thomson scattering of photospheric light by the free electrons of the obviously completely ionized gas (plasma) of the corona. Thanks to the Doppler effect resulting from the high velocities of the electrons, the Fraunhofer lines are blurred out.

We can calculate the mean (that is, disregarding fluctuations) electron density N_e in the K corona, as a function of distance r from the center of the Sun, from its brightness distribution.[1] We find, for instance, at about coronal maximum:

[1]We know the distribution of light intensity and the Thomson scattering coefficient per free electron $\sigma_{el} = (8\pi/3)(e^2/mc^2)^2 = 0.665 \times 10^{-24}$ cm^2, which to within the factor 8/3 is equal to the classical cross section of the electron.

r = 1.03	1.1	1.5	2.0	2.5	3.0	solar radii (7×10^{10} cm)
N_e = 3.2×10^8	1.6×10^8	1.6×10^7	2.8×10^6	8.3×10^5	3.2×10^5	electrons/cm^3

(2) In the inner corona we observe in emission the so-called *coronal lines* whose identification remained a great puzzle in astrophysics until in 1941 B. Edlén succeeded in interpreting them as forbidden transitions of highly ionized atoms. The strongest and most important are those in Table 20.1.

Table 20.1. Coronal lines

	Red coronal line 6374.51 Å	Green coronal line 5302.86 Å	Yellow coronal line 5694.42
Spectrum	[Fe X]	[Fe XIV]	[Ca XV]
Preceding ionization potential	235	355	820 eV
Transition	$3s^2\,3p^5\,{}^2P^\circ_{3/2} - {}^2P^\circ_{1/2}$	$3s^2\,3p\,{}^2P^\circ_{1/2} - {}^2P^\circ_{3/2}$	$2s^2\,2p^2\,{}^3P_0 - {}^3P_1$
Electron temperature T_e (ionization)	1.2×10^6	1.9×10^6	2.5×10^6 K

It was possible with confidence to identify further corresponding lines of the elements A, K, Ca, V, Cr, Mn, Fe, Co. The ionization potentials of several hundred electronvolts signified emphatically an electron temperature of some millions of degrees. And then it was realized that in fact a whole series of other phenomena indicate a coronal temperature of 1 to 3×10^6 K:

(a) The *density distribution* $N_e(r)$ in combination with the hydrostatic equation, or the barometric formula [equations (7.3) and (7.4)], leads, for the equatorial region of the corona at solar minimum, for example, to a temperature of $1.4\pm0.1\times10^6$ K, in agreement with the optical and radio observations to be mentioned in a moment.

(b) Since a radiation field corresponding to 10^6 degrees is certainly not present, we cannot calculate the *ionization* in the corona from Saha's formula (LTE). The latest calculation of individual processes of ionization and recombination yields temperatures of 1.2 to 2.5×10^6 K for the maxima of the lines shown in Table 20.1. Clearly the temperature in the corona varies both spatially and temporally within about that range.

(c) If we ascribe the *widths of the coronal lines* measured by B. Lyot, D. E. Billings, and others to the thermal Doppler effect [cf. equation (19.9)], we obtain for various lines 1.7 to 3.7×10^6 K. The fact that these temperatures are higher than those deduced from the ionization conditions may be attributed to more or less turbulent motions with velocities of about ±20 km/s.

(d) Solar spectra in the *X-ray and Lyman region* (a few angstroms to ~ 1500 Å), taken using rockets and satellites (particularly the OSO series: Orbiting Solar Observatory), show primarily the emission spectrum of the inner corona with numerous allowed and forbidden lines from high ionization states of the more abundant elements. The presence of these lines and their great intensity confirm the high temperature of the corona.

In the X-ray region we find as well as the lines the free-free and free-bound continua (cf. Figure 17.1) of the coronal plasma at 1 to 3×10^6 K. X-ray photographs of the Sun, using the Wolter telescope and filters to limit the wavelength range used (for example, 8–12 Å), show at the solar limb the K corona with its characteristic streamers and on the disk the active regions, above which (as shown also by optical observations) the hotter, denser coronal condensations are found. These regions coincide almost exactly with those of the faculae or "plages faculaires" (Deslandres) seen in calcium spectroheliograms. In the X-ray region the photospheric continuum at ~ 6000 K, dominant in the visible, is quite undetectable by comparison with the coronal emission at 1 to 3×10^6 K.

(e) We can account for the *thermal radio emission* of the Sun as free-free radiation from the solar corona. We observe it in millimeter, centimeter, and decimeter wavelengths while nonthermal components mostly predominate at longer wavelengths. The changes in the thermal radiation intensity in the course of days and months admit its separation into the always-present radiation of the quiet Sun and the slowly varying radiation originating in coronal condensations. At the solar limb and in condensations in the corona an optical thickness greater than unity is attained in places, so that we can measure the black-body radiation directly. Its temperature, which again corresponds to the electron temperature, also has values from 1 to 3×10^6 K.

(3) In the *outer corona,* at some distance outside the solar limb and increasing steeply (relative to the K corona) outward from the Sun, we observe scattered photospheric radiation showing the *Fraunhofer lines unchanged.* Following W. Grotrian, who designated this component the F corona (Fraunhofer corona), C. W. Allen and H. C. van de Hulst (1946–1947) showed that it arises from "Tyndall scattering," that is, predominantly forward scattering by small particles somewhat larger than the wavelength of the radiation. These particles are far enough from the Sun not to be heated to the point of vaporization. Measurements of the distribution of brightness and of polarization in the outer corona, especially Blackwell's observations from an aircraft in the stratosphere, have shown that the F corona is simply the innermost part of the *zodiacal light.* Grotrian had already conjectured this. Thus the F or dust corona, or the zodiacal light, does not belong to the Sun and is relatively little influenced by it.

Here we shall not concern ourselves further with interplanetary dust, but pursue our study of the K corona, the true *solar corona*. Since the temperature must somehow increase from the 4000 K of the lower chromosphere to the 1 to 3 million K of the corona, it is no longer astonishing that lines of high ionization and excitation potential make their appearance with great intensity even in the upper chromosphere, including amongst others

	Excitation potential eV	Ionization potential eV
Hydrogen Balmer lines	10.15	13.54
He I lines, e.g., D_3 λ5876 Å	20.87	24.48
He II lines, e.g., 4–3; λ4686 Å	48.16	50.80

Incidentally, P. J. C. Janssen first observed the D_3 line of the "solar element" in the Sun in 1868; it was not until 1895 that W. Ramsay succeeded in isolating helium from terrestrial minerals.

The spectrum of the upper chromosphere (about 2000 to 10000 km above the solar limb) is generally like that of *prominences*. These "clouds in the corona" show in their spectra not only weak lines of neutral metals, such as Na D, Ca I 4227, etc., but also very strong lines such as the calcium H and K lines λ3933/68, Hα λ6563 and the other hydrogen lines as well as the helium D_3 line λ5876 and even (weakly) He II λ4686. The lines of H I, He I, and He II which are emitted from the *same* volume element as the familiar lines of the Fraunhofer spectrum are clearly excited nonthermally by the intense radiation in the extreme ultraviolet from the transition layer and the corona. Quiescent prominences (Figure 20.5) retain their form with little change (flow at about 10 km/s) often for weeks. From time to time, however, without warning, they may accelerate more or less jerkily to velocities of 100 km/s up to, on occasions, 600 km/s. Such eruptive prominences (Figure 20.6) can then escape into interplanetary space. All prominences have a characteristic filamentary structure. Also the upper part of the chromosphere under good observing conditions, for example using a Lyot filter, has the appearance of a "burning prairie" of small prominences, the so-called *spicules* (Figure 20.7) that move outward or inward with velocities of about 10 km/s.

In Figure 20.8 we show concisely the association of sunspots, prominences, corona, etc., with reference to the 11.2-year solar cycle. Then we must ask ourselves above all else about its meaning. Obviously the picture of a static atmosphere in radiative equilibrium is no longer adequate; rather is everything in motion and in a state of flux.

We first enquire as to what thermodynamic engine, in which heat energy flows from higher to lower temperatures in accordance with the second law of thermodynamics, generates the mechanical work needed to drive these motions. The so-called hydrogen convection zone (A. Unsöld 1931) performs this office. From the deeper photospheric layers, where $\log P_g \approx 5.2$ and $T \approx 6500$ K, downward to about $\log P_g \approx 12$ and $T \approx 10^6$ K the solar atmosphere is convectively unstable. This zone has thickness about one

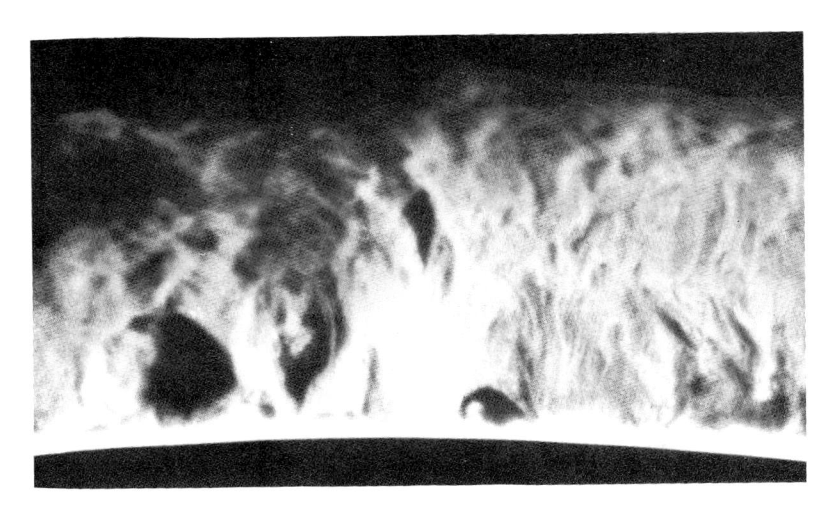

Figure 20.5. Quiescent prominences (filaments) usually have the form of a thin sheet, standing almost perpendicular to the solar surface on several "feet". The dimensions of this sheet are: thickness $\sim$ 6600 km (4000–15 000 km), height $\sim$ 42 000 km (15 000–120 000 km) and length $\sim$ 200 000 km (up to 1.1×10^6 km). Detailed photographs taken in Hα light at Sacramento Peak show filamentary structures, in which the material streams upward or downward with speeds of the order of 10 to 20 km/s. The solar magnetic fields clearly have an important influence on the structure of all prominences.

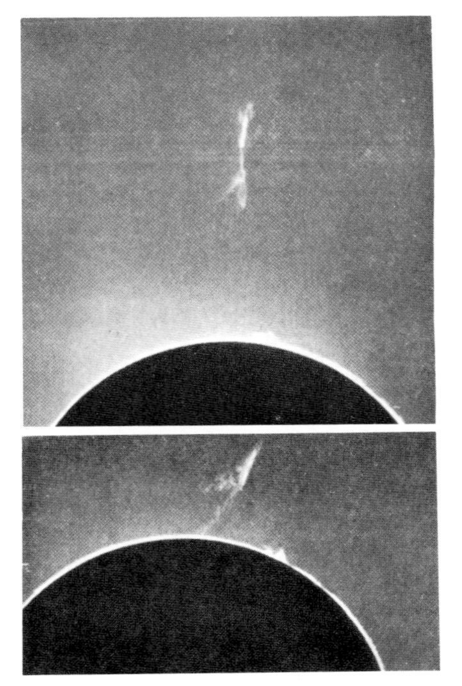

Figure 20.6
Eruptive or ascending prominence 1928 November 19. The photographs are separated by an interval of $1^h\ 11^m$. The greatest height observed is 900 000 km above the solar limb and the maximum velocity 229 km/s.

Figure 20.7. Spicules at the solar limb in Hα (the disk has been covered). These structures, which were first described by A. Secchi, have a mean height above the solar limb of ~ 8000 km and a thickness of approximately 500–1000 km. Moving with velocities of approximately 20 km/s upward or (more rarely) downward, they have a lifetime of approximately 2 to 5 min. The direction taken by the spicules is that of the local magnetic fields.

tenth the solar radius. Above it, the most abundant element, hydrogen, is practically neutral, within it the hydrogen is partially ionized, below it the ionization is complete. If a volume element of the partially ionized gas rises, the hydrogen begins to recombine and in each recombination process 13.6 eV (corresponding to $16kT$ at 10 000 K) is added to the thermal energy of $\frac{3}{2}kT$ per particle. The adiabatic cooling is thereby so very much reduced that the effective ratio of the specific heats c_p/c_v approaches unity and in radiative equilibrium the rising volume element would become hotter than its new surroundings. Thus the volume element would rise higher still.

Figure 20.8. The 11.2-year cycle of solar activity (after W. J. S. Lockyer). Area and distribution in latitude of sunspots and prominences, and variation in form of the corona.

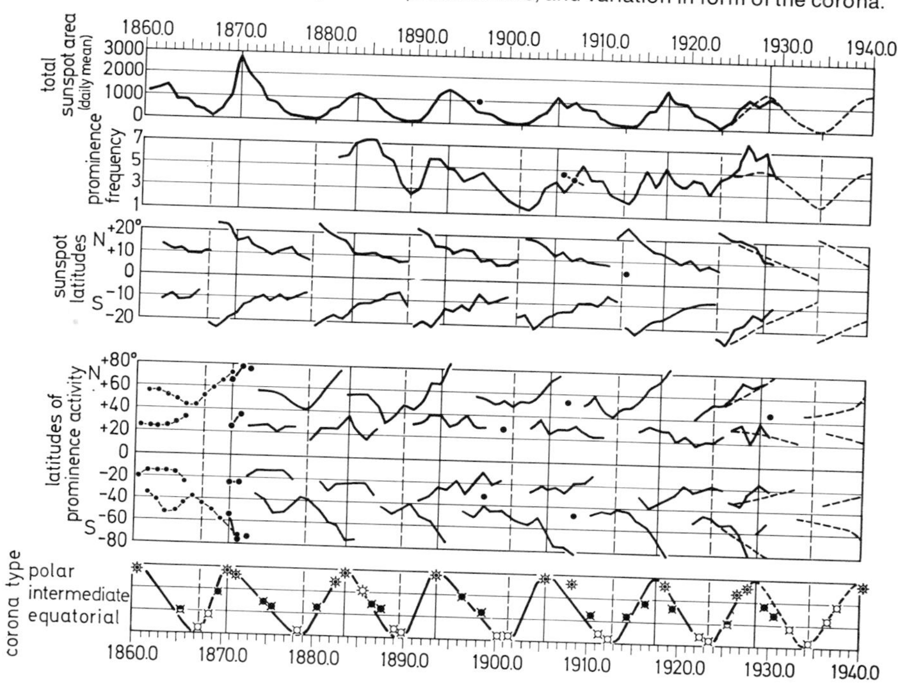

Exactly the reverse happens to a sinking volume element. This effect is further enhanced by the opposed influence of ionization upon the radiative temperature gradient. Thus we obtain a zone of convection currents. For $\log P_g \geqslant 5.3$, convection takes over practically the whole of the energy transport; in the deeper layers of the convective zone radiative energy transport becomes insignificant.

While the thermodynamics of the hydrogen convective zone is fairly simple and clear, its *hydrodynamics* includes some of the most difficult problems of the subject. Only models using the rather crude mixing-length theory of W. Schmidt and L. Prandtl have been worked out: a body of gas of dimensions l is supposed to describe a sort of free path of the same order l and then all at once to give up its temperature excess, its momentum, etc., by mixing with its surroundings—obviously a very crude schematization of highly complex convection currents.

Solar granulation can now be ascribed to the specially marked instability of a surface layer, a few hundred kilometers thick, of the hydrogen convection zone. The granules are of the order of magnitude of the thickness of this layer, but also not much greater than the scale height of the atmosphere there.

Besides the granulation, there is also a second coarser network of convection cells of which the mesh has diameters between 15 000 and 40 000 km. We see this particularly clearly in the Ca II spectroheliograms (the calcium flocculi) which apply to a fairly high layer of the solar atmosphere. The flow travels upward in the middle of a cell, with radial speed of about 0.4 km/s outward and downward round the edge. The lifetime of a cell (about equal to the time for it to turn over) is about 20 hours. Since its diameter is about equal to the thickness of the hydrogen convection zone, it is tempting to associate the cells with flow through the *whole* convection zone.

The currents so far considered of 0.5 to 2.5 km/s influence the solar spectrum by giving the Fraunhofer lines a serrated structure and increasing their width on the average. So long as the moving volume elements are optically thin, their Doppler effects are simply superimposed on those of the thermal motion. We have then the so-called microturbulence which increases the purely thermal width $\Delta\lambda_D$ by factors between about 1.2 and 2. The Doppler effects of the somewhat faster flow in the spicules of the upper chromosphere make themselves correspondingly evident in the cores of strong Fraunhofer lines.

We must not get lost in details but we must proceed at once to ask the disturbing question as to how in the transition region, the thickness of which is only some 15 000 km, between the chromosphere and the corona the temperature increases outward from about 4000 K to between about 1 and 3 million degrees. According to the second law of thermodynamics, this heating-up of the outermost atmospheric layers of the Sun against the "natural," entropy-generating, temperature drop from the inside toward the outside can occur only through the use of mechanical energy or other "ordered" forms of energy of greater negative entropy.

Following M. Schwarzschild and L. Biermann we can in fact show that in the upper layers of a partially convective atmosphere mechanical energy transport more and more supersedes radiative transport. As Proudman and Lighthill have shown more exactly, sound waves first arise in the turbulent motions in the photosphere. Magnetohydrodynamic waves, in which a variable magnetic field is coupled to the vibrations of the material, are also formed in the plasma of the solar atmosphere by its interaction with its associated magnetic field (see below). All these waves propagate into the higher, less dense layers of the solar atmosphere and as a result steepen into *shock waves*.[2] Their energy is relatively rapidly dissipated and turned back into heat energy, and since at low densities and high temperatures the radiative power of solar matter becomes progressively worse, the temperature rises until the atmosphere has found a new type of energy transport. As H. Alfvén noticed, this arises from the circumstance that, at sufficiently high temperature and steep *inward* temperature gradient, the heat conduction by the free electrons of the plasma (like the high heat conductivity of metals) suffices to convey the energy back to the inside where it is ultimately radiated away.

The transition region mentioned above, where within ~ 15,000 km the transition from a few thousand to millions of degrees is accomplished, has been further revealed by radio astronomy and space research. In the radio region, the free-free absorption coefficient of the plasma at frequency ν is proportional to $N_e^2/T^{3/2}\nu^2$. We can therefore see further and further down into the solar atmosphere at higher frequencies. The radio spectrum of the quiet Sun in the millimeter to decimeter band thus gives us very simply a picture of the temperature and pressure distribution in the transition region. In the Lyman and X-ray regions, which have become accessible from rockets and satellites, the emission lines of many ions with a wide range of ionization and excitation potentials make their appearance. From the intensities of these lines, compared with the electron-scattering continuum of the K corona, S. R. Pottasch and others were able to obtain the abundances of the elements concerned, relative to hydrogen (Table 19.1); these are in good agreement with the values obtained from the photospheric absorption lines.

By combining the observations in the radio, the visible (eclipses, coronagraph), and the extreme ultraviolet (XUV; satellites) D. Reimers (1971) has constructed a model, for the equator at sunspot minimum, of the transition region and inner corona (Table 20.2). The steep rise in temperature to a maximum of 1.4×10^6 K demands a flux of *mechanical* energy, drawn ultimately from the hydrogen convection zone, of 5.8×10^5 erg/cm^2 s, only about one part in 10^5 of the total solar energy flux.

The remarkable filamentary structure of prominences, the polar plumes of the corona at sunspot minimum, and also the enormous coronal stream-

[2]The energy density of the sound waves is proportional to the density ρ times the square of the velocity amplitude v. If the waves advance with sound speed c into a less dense medium, the v^2 increases correspondingly. If v reaches the order of magnitude of the sound speed so that the Mach number $M = v/c$ tends to unity, then a shock wave is produced.

Table 20.2. Transition region and inner corona of the Sun, from the model of D. Reimers (1971) for the equator and sunspot minimum. The height scale could perhaps be displaced by about ±1000 km. The temperature maximum, from which the temperature decreases very slowly outward, is marked in bold type.

	Height (km) above the solar limb	T (K)	Electron density N_e (cm^{-3})	
Transition layer	2970	25000	1.3×10^{10}	radio spectrum
	3000	63000	5.2×10^{9}	
	3030	160000	2.0×10^{9}	XUV-spectra
	3200	400000	8.0×10^{8}	
	8000	1.0×10^{6}	2.8×10^{8}	
Corona	**20000**	$\mathbf{1.4 \times 10^{6}}$	$\mathbf{1.7 \times 10^{8}}$	K corona (eclipse)
	100000	1.4×10^{6}	6.2×10^{7}	

ers over large centers of solar activity, and not least the flow structure that is observed in Hα spectroheliograms in the vicinity of spots (G. E. Hale's so-called hydrogen vortices) long ago prompted the thought that in solar physics hydrodynamics would not have the last word but that magnetic fields could also play an essential rôle. Thus in 1908 the ingenious G. E. Hale looked for the Zeeman effect in the spectrum of a sunspot. This effect is the magnetic splitting of, for example, the line Fe I 6173 seen with the help of a polarization device that cuts out alternately the right- and left-hand circularly polarized outer Zeeman components, the observation being made along the magnetic field lines. He discovered magnetic fields that reached about 4000 G in the largest spots. Furthermore, he showed that the two spots of a so-called bipolar spot group show, like a horseshoe magnet, always one north and one south pole. Finally, a long series of observations by Hale and Nicholson showed that in such a group the leading spot (according to the Sun's rotation) has opposite polarity in the northern and southern hemispheres and that this alternates from one solar cycle to the next. Thus the true duration of a solar cycle amounts to 2×11.2 years.

Since all purely thermodynamic attempts to explain the low temperature of sunspots have failed, we must surely suppose that in the depth of the spot the convective energy flow is strongly impeded by the magnetic field.

In 1952 H. W. Babcock and H. D. Babcock using much refined equipment succeeded in detecting Zeeman effects in the Sun corresponding to only 1 or 2 G, for which the splitting corresponds to only a tiny fraction of the line width. It appears that the facular regions, or more precisely the *plages faculaires,* of the Ca II spectroheliograms in the surroundings of groups of sunspots ordinarily exhibit fields of 10 to 100 G, and they are apparently caused by these fields. Since the more intense regions of the corona, the so-called coronal condensations, occur above the "plages," we are driven to the concept that moderately strong magnetic fields provide favorable conditions for the mechanical energy transport from the photosphere into the higher layers of the chromosphere and corona.

This may be bound up with the still baffling phenomenon of solar eruptions or flares (not to be confused with eruptive prominences) (Figure 20.9). If we observe a spot group (center of activity) in the Hα line or the calcium K line, the facular "plages," between the spots and in their surroundings, exhibit structures of the order of magnitude of a few thousand kilometers with small irregular brightness fluctuations and motions. *Suddenly* many fairly bright structures coalesce; then, in the case of a 3^+ eruption a region of 2 to 3×10^{-3} of the solar hemisphere, corresponding to an entire spot group, blazes up in the lines Hα, Ca II K, ... (Eruptions are classified as of "importance" 1, 2, 3 or 3^+.) The Hα emission line reaches a central intensity up to about three times that of the normal continuum and a width of many angstroms. The duration of the flare ranges from the order of a second for "microflares" of about 1$''$ diameter up to several hours for giant 3^+ flares. One often, somewhat vaguely, relates the occurrence of flares to some sort of instability in the magnetic field of the spot group. The author prefers the view that the flow of mechanical energy, which among other things heats the faculae and corona, is so suddenly switched on that the energy cannot escape and an explosion occurs. An ordinary facula compared with a flare is somewhat like the burning of dynamite compared with its detonation.

Figure 20.9. A large eruption (3^+ flare, 1961 July 18, see right of picture) which was accompanied by a pronounced emission of cosmic rays. Just above the center of the disk a long filament (that is, prominence, compare Figure 20.6) can be seen in absorption. Cape Observatory, Hα patrol photograph (intensity calibration marks and the time recording appear on the left).

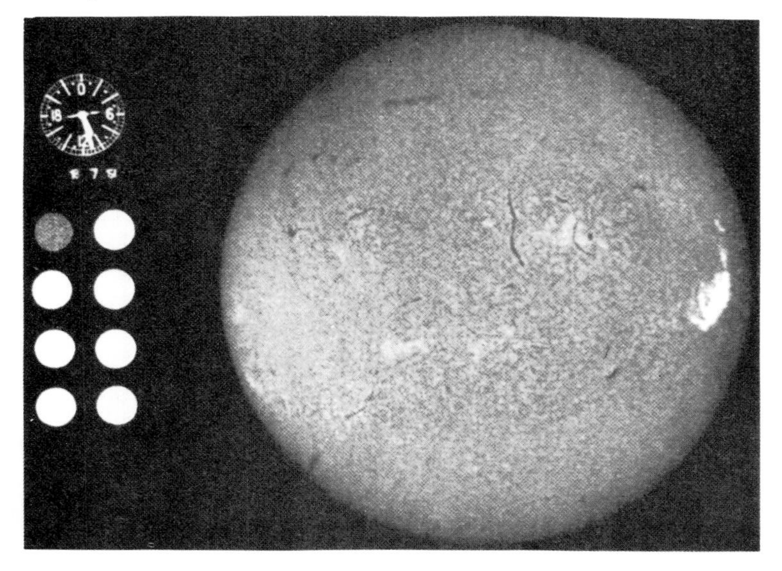

Bound up with the flare, we have the emission of corpuscular radiation in a large range of energies. Plasma clouds, which arrive at the Earth about one day later and so have a velocity about 2000 km/s, produce there magnetic storms ("sudden commencement") and polar lights. As Forbush and Ehmert found, great flares also make a solar contribution to cosmic rays with particle energies of 10^8 to 10^{10} eV. Their chemical composition largely matches that of normal solar material.

Connected with the flares, and also with less spectacular manifestations of solar activity, we have the nonthermal component of the *radio emission* from the Sun.

Its analysis with the aid of the radio-frequency spectrometer, which plots the intensity as a function of time over a large frequency range, led J. P. Wild and his collaborators in Australia to distinguish several types of *bursts* of radio emission (Figure 20.10). In order to understand this we must recall that in a plasma of electron density N_e electromagnetic waves of frequency less than the so-called critical or plasma frequency ν_0 cannot be propagated, where

$$\nu_0 = \left(\frac{e^2}{\pi m} N_e\right)^{1/2} \hat{=} 9 \times 10^{-3} N_e^{1/2} \text{ MHz} \tag{20.5}$$

because then the refractive index n would be imaginary ($n^2 < 0$). Thus radio emission of a given frequency can originate in the corona only above a particular layer.

Now the Type II and Type III bursts show, respectively (Figure 20.10), a slow or fast displacement of their frequency band toward lower frequencies. Thence we can infer that the stimulating agency sweeps through the corona with speeds of the order of 1000 km/s in the case of Type II and up

Figure 20.10. Dynamical (that is, time-dependent) radio spectrum in the meter wavelength region, after J. P. Wild. The diagram shows (schematic only) the development of "bursts" of various types following a large flare. The lower the frequency (that is, uppermost part of diagram), the higher the coronal layer in which it arises. The upward speed of the exciting agency can be read off from the dynamical spectrum; thus, this is slow for the Type II bursts and fast for the Type III bursts. Type II and Type III bursts frequently show a harmonic (2 : 1) overtone.

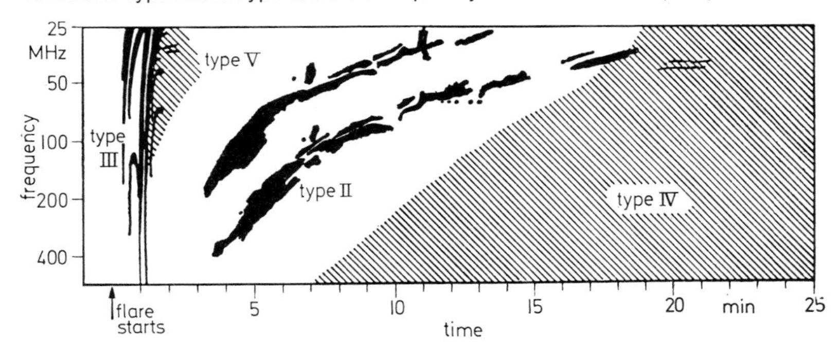

to about 40 percent of the speed of light in the case of Type III. Type IV occurrences emit a continuum in a wide frequency band over a longer time; here we very likely have to do with synchrotron emission by fast electrons.

Since 1967 J. P. Wild and his colleagues at the Culgoora observatory in Australia have been able to observe, at 80 MHz (λ3.75 m), the motions of the different kinds of burst sources on and beyond the solar disk directly on the oscilloscope screen of the radioheliograph. Using 96 suitably connected parabolic antennas, each of 13-m diameter, it is possible to examine the Sun and the corona with a field of view of 2° and a resolution of 2 to 3′. Every second a picture is obtained in two polarizations (right and left circular or two directions of linear polarization). The observations confirm the evidence of the radio spectra. Unfortunately we cannot here discuss the difficult explanation of the different types of burst in terms of plasma physics.

As the foregoing survey shows, one of the most important concerns of present-day astrophysics (as well as of the technique of nuclear fusion) is the theory of the flow of material of high conductivity in the presence of magnetic fields, that is, *magnetohydrodynamics* or *hydromagnetics*. From the fundamental work of H. Alfvén, T. G. Cowling, and others we may infer the following:

(1) Almost all cosmic plasmas have very high electrical conductivity σ.

(2) If we have a conductor at rest in a magnetic field H then the variation of the field generates induced currents which decay only slowly (as in a familiar experiment with eddy currents). In alternating-current practice the decay time τ is given by the ratio of the self-induction L to the resistance R. The latter is proportional to $1/\sigma$. Since in electrostatic units σ has dimension $[\mathrm{s}^{-1}]$, a factor of dimension $[\mathrm{s}^2]$ must enter in forming τ. This can be only x^2/c^2, where x is a linear measure of the conductor and c is the speed of light. Actually we obtain from Maxwell's equations

$$\tau \approx \frac{\pi\sigma}{c^2}\, x^2. \tag{20.6}$$

The dependence of the sort $\tau \propto x^2$ shows that the propagation and decay of a magnetic field in a conductor *at rest* has the character of a diffusion process.

(3) If we allow *motions* in the conducting medium we have to consider whether the magnetic field propagates more quickly by diffusion independently of the matter, or remains "frozen" in the material. If Alfvén's condition for the latter

$$\sigma H x / c^2 \sqrt{\rho} > 1, \tag{20.7}$$

is satisfied (ρ = density)—and that is frequently the case in cosmic plasmas—the material can move essentially only *along* the field lines, like beads on a string.

(4) Since the magnetic pressure (Maxwell stress) is of order of magnitude $H^2/8\pi$ this quantity is usually of the same order of magnitude as the dynamical pressure $\sim \frac{1}{2}\rho v^2$ (in not too special magnetohydrodynamic flows).

Magnetohydrodynamics (MHD) forms the basis for our understanding of many astrophysical phenomena. To begin with we shall stay in the realm of solar physics. Naturally we must leave on one side the complicated mathematical structure. However, in the first instance a clear understanding of the physical principles is more important.

20.1 Sunspots and the 2 × 11-year cycle of solar activity

As early as 1946, T. G. Cowling pointed out that, because of the very high electrical conductivity σ of the plasma, the magnetic field of a sunspot in a *stationary* solar atmosphere should be destroyed by "diffusion" only on a timescale of about 1000 years [cf. equation (20.6)]. However, sunspots actually have lifetimes ranging from days to a few months. It was not until ~ 1969 that M. Steenbeck and F. Krause noticed that the *turbulence* in the hydrogen convection zone makes an important contribution to the twisting of the magnetic field and (through the fundamental equations of MHD) of the attendant fields of the electric current $\boldsymbol{i}$ and velocity $\boldsymbol{v}$, so that the *effective* conductivity in the solar atmosphere is decreased by a factor of about 10^4 and equation (20.6) now gives the correct order of magnitude for the lifetime of sunspots.

Following V. Bjerknes (1926), we may imagine the origin of a *bipolar spot group* (others are closely related to these) as due to the presence in the Sun of permanent toroidal (that is, running parallel to circles of latitude) "tubes" of magnetic field lines. Since the pressure in these is partly of magnetic origin, the gas pressure and density are smaller than in the surrounding material. They rise toward the surface and are there "cut open." The two open ends of such a field tube form a bipolar spot group.

The discovery by H. W. Babcock that the weak general magnetic field in high heliographic latitudes also reverses with an 11-year rhythm is essential for the understanding of the twice-eleven-year cycle and of Spörer's law of the motion of spot zones from high latitudes (±30° to 40°) at maximum to lower latitudes (±5°) at minimum (Figure 20.8) and Hale's laws of the magnetic polarity of spot groups. The entire cycle of solar activity (Figure 20.8) is due to the fact that in the interior of the Sun (mainly in the hydrogen convection zone) the whole structure of motions and magnetic field varies with a 22-year period and (on average) maintains itself. The theory of this *solar dynamo,* whose driving force is the hydrogen convection zone, has been clarified bit by bit through the work of H. W. Babcock (1961), R. B. Leighton (1969), M. Steenbeck, and F. Krause (1969) and W. Deinzer (1971). From the *toroidal* magnetic field, whose initial existence is postulated, a *meridional* magnetic field must obviously be formed as well, in the appropriate part of the cycle, and vice versa. According to Steenbeck and Krause, turbulence makes this possible; by a mechanism which we cannot here explain in detail, the turbulence gives rise to a current parallel to the field lines and this in turn produces a meridional component of the magnetic field, at right angles to the original field lines. In other words, induction is important for our dynamo, caused not only by the differential rotation of the Sun but also by the statistically varying turbulent motions. The (very complicated) working out of these ideas seems, in fact, capable of reproducing the basic phenomena of solar activity to the right order of magnitude.

20.2 Ray structure of the corona, faculae (plages faculaires) and coronal condensations

The polar plumes of the corona at solar minimum (Figure 20.4), as Babcock's magnetograph measurements have shown, clearly arise from the fact that there are different densities of coronal plasma in different "field tubes" of the magnetic field of the Sun, which is fairly regular in the polar caps and of strength 1 to 5 G.

If we assume, rather schematically, that in the transition layer the chromosphere at $\sim$ 6000 K goes over suddenly, at a particular pressure, into the corona (also almost isothermal) at $\sim 1.4 \times 10^6$ K, then the pressure distribution is represented by the solid line in Figure 20.11. If now the "wave heating," which causes the temperature rise between the chromosphere and the corona, is stronger in a particular region of the solar surface, this transition will occur at a somewhat deeper level and—as the dashed line in Figure 20.11 shows—the pressure and density of the corona will be increased by the same factor at *all* heights. The magnetic field simply makes sure that this "too dense" material cannot move sideways but remains together in one of the "hairs" of the polar plumes. Coronal condensations, in which the density of the inner corona is increased by factors of about 5, arise in exactly the same way in the stronger fields ($\approx$ 100 G) in the region of the faculae (plages) which surround spot groups. To

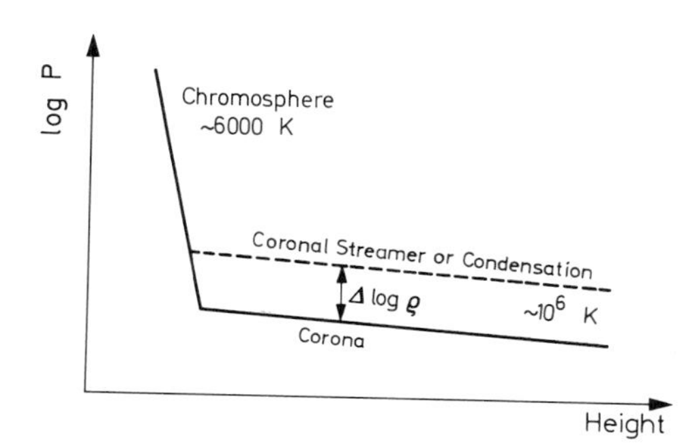

Figure 20.11
Formation of a coronal streamer or a coronal condensation as a result of the strengthening of the mechanical energy flux and confinement of the denser coronal material in a magnetic flux tube (schematic).

achieve this effect (and a raising of the coronal temperature by a factor of about 2), it is sufficient, as a more exact analysis shows, to increase the mechanical (wave) energy flux by about a factor 10. Even then it carries only 1 part in 4000 of the total energy flux of the Sun! The great coronal streamers (Figure 20.4) arise in the same way in field lines stretching high above their underlying spot groups. All the complicated phenomena of solar activity (including flares, etc.) obviously arise from the strengthening of the mechanical energy flux by moderately strong magnetic fields. Equally clearly, the actual physics of this "valve effect" requires further investigation.

20.3 Solar wind

As we mentioned in Section 8, L. Biermann (1951) put forward the hypothesis that the plasma tails of comets are not driven out by radiation pressure but by a continuous stream of particles from the Sun. Measurements using satellites and space vehicles then revealed that this plasma (of about solar composition) is streaming out from the Sun at about 400 km/s in the neighborhood of the Earth's orbit, with variations between about 250 and 800 km/s. Its density corresponds to about 5 protons (and electrons) per cubic centimeter, again with large variations. The more or less statistical distribution of particle velocities corresponds to a temperature of $\sim 2 \times 10^5$ K. Linked with the plasma is a magnetic field of order 5×10^{-5} G (also spoken of as 5γ). In 1959 E. N. Parker described this phenomenon as the *solar wind* and proposed an explanation on the basis of hydrodynamics and MHD. If one first calculates the pressure distribution $p(r)$ of a corona taken schematically to be isothermal, one finds that at a large distance r from the Sun the pressure reaches a finite boundary value which for coronal temperatures $<500\,000$ K is *less* than that of the interstellar medium (Section 24). In that case the latter would stream into the Sun, and we would have the phenomenon of *accretion,* to which long ago F. Hoyle, H. Bondi, and others ascribed great astronomical significance. However if $T >$ 500 000 K, as in the real corona, a continuous outflow must occur, which is the solar wind. Its velocity v is determined as a function of distance r from the solar center by means of *Bernoulli's theorem* (that is, the energy equation of hydrodynamics) together with the continuity equation and the equation of state of the gas $p = \rho \mathcal{R} T/\mu$, where again ρ signifies the density, $\mathcal{R}$ the gas constant, and μ the mean molecular weight (for a fully ionized plasma, with 90 percent H + 10 percent He, $\mu = 0.61$).

Parker takes the continuity equation for spherically symmetric outflow, that is, the same quantity of material flows through each spherical shell $4\pi r^2$, and so

$$4\pi r^2 \rho(r) v(r) = \text{constant}. \tag{20.8}$$

This statement has to be considerably modified if the material is guided by a magnetic field $\boldsymbol{H}$. In that case a particular stream of matter must flow along a particular tube of magnetic force, so long as the material pressure at the "walls" does not exceed the magnetic pressure. The solution of the problem this raises, one of the most difficult problems in magnetohydrodynamics, is at the present time still in its initial stages.

We therefore prefer to ask first what conclusions we can draw from *Bernoulli's equation*. This states that along a *streamline*. that is, a line of the velocity field $\boldsymbol{v}$, the sum of the kinetic energy $v^2/2$, the potential energy (or potential) $\mathrm{U}(r)$, and the pressure energy $\int dp/\rho$ (all calculated per unit mass) is a constant for stationary flow ($\partial/\partial t = 0$). The gravitational potential of the Sun (mass $\mathfrak{M}$; gravitational constant G) is from (6.28) $\mathrm{U} = -G\mathfrak{M}/r$; we therefore obtain

$$\frac{v^2}{2} + \int \frac{dp}{\rho} - \frac{G\mathfrak{M}}{r} = \text{constant}, \tag{20.9}$$

where the upper limit of the (pressure) integral is the point being considered. For example, to calculate the velocity $v_\oplus$ of the solar wind near the Earth's orbit $r \approx r_\oplus$, we use the fact that it starts in the corona at $r \approx r_\odot$ with $v \approx 0$. The potential difference $G\mathfrak{M}\left(\frac{1}{r_\odot} - \frac{1}{r_\oplus}\right)$, when converted to kinetic energy $v^2/2$, corresponds to the *escape velocity* (cf. Section 7) of the Sun, $v_E = 620$ km/s. If we then apply Bernoulli's equation (20.9) to the acceleration of the solar wind between $r_\odot$ and $r_\oplus$, we obtain

$$v_\oplus^2 = -2\int \frac{dp}{\rho} - v_E^2. \tag{20.10}$$

The numerical value of the integral depends crucially on the run of temperature $T(r)$ between Sun and Earth. If we start from the *observation* that T decreases relatively little from the corona, with $T \approx 1.4 \times 10^6$ K, to the Earth's orbit $T \approx 2 \times 10^5$ K we can make an approximate calculation with a constant mean temperature $\bar{T} \approx 10^6$ K and then it is easy to do the integration ($dp = d\rho\,\mathcal{R}\,\bar{T}/\mu$):

$$v_\oplus^2 = \frac{2\mathcal{R}\bar{T}}{\mu}\ln(\rho_\odot/\rho_\oplus) - v_E^2. \tag{20.11}$$

$\sqrt{(2\mathcal{R}\bar{T}/\mu)}$ corresponds to the thermal velocity of the particles (165 km/s) or, to within 10 percent, the velocity of sound in the corona. With the above numerical values we obtain

$$v_\oplus = 400 \text{ km/s}. \tag{20.12}$$

That our calculation leads to a result in agreement with the measurements happens—and it is important to be clear about this—because our numerical value for $\bar{T}$ has implicitly taken into account the fact that the outer corona or solar wind beyond the transition region is heated by *thermal conduction* (that plasma has an enormously high thermal conductivity) and has only a very small radiation loss. For example, if we had computed a coronal model with *adiabatic* outflow, we would have obtained as a maximum value for the integral in equation (20.10) the *enthalpy* per gram of coronal material $c_pT_\odot$ where c_p signifies the specific heat at constant pressure. It is easy to verify that adiabatically outflowing material would quickly cool and come to a standstill close to the Sun. Our hydrodynamic theory of the solar wind[3] does indeed reproduce the observations to the correct order of magnitude, but it is still unsatisfactory to this extent, that—as is easily verified—the *mean free paths* are of the same order of magnitude as the characteristic scales of the model. A computation using the kinetic theory of gases, in which it is essential to take the magnetic field into account ("collision-free plasma"), is therefore very desirable.

As it streams out into interplanetary space, the solar wind takes magnetic field lines with it. Let us use a polar coordinate system whose plane corresponds roughly to the solar equator or to the ecliptic (we use these here interchangeably and restrict ourselves to their neighborhood). Then in time t a particle with speed v reaches a distance

$$r = vt \tag{20.13}$$

from the Sun. During this time, the latter has rotated through an angle

$$\phi = \omega t \tag{20.14}$$

where the angular velocity $\omega = 2\pi$ per sidereal rotation period. Particles which start one after another from a particular point on the Sun (for example, an active region) therefore lie at a particular moment on an *Archimedean spiral*

$$\phi = \frac{\omega r}{v} \tag{20.15}$$

which cuts a radius vector at an angle α, which is determined by

$$\tan \alpha = \frac{r\, d\phi}{dr} = \frac{\omega r}{v}. \tag{20.16}$$

If $v = 430$ km/s one obtains $\alpha \approx 45°$ in good agreement with observations.

Since magnetic field lines must always close on themselves (div $\boldsymbol{B} = 0$), they stretch out in loops, whose beginning and end are situated on the Sun in regions of opposite magnetic polarity. At times such regions fill a considerable fraction of the solar surface. Giant field loops are therefore

[3]Mathematically, we can view our treatment—much simplified by comparison with Parker's—as a successive approximation procedure in which a model of a static corona with thermal conductivity (S. Chapman, 1957) serves as an initial approximation.

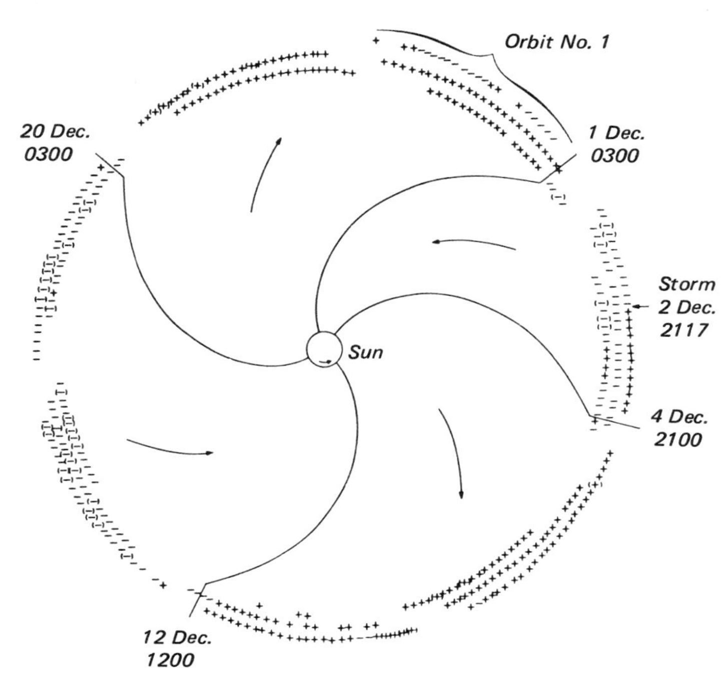

Figure 20.12. Sector structure of the interplanetary magnetic field. Observations from IMP-1, after J. M. Wilcox and N. F. Ness (1965). Plus and minus signs denote field lines which are directed outward and inward, respectively. The part of the field near the Sun is completed schematically in accordance with the theory.

continuously being driven out by the solar wind into interplanetary space. (Further out the field lines may also form closed rings.) This *sector structure* of the interplanetary magnetic field (Figure 20.12) was first determined in the neighborhood of the Earth's orbit by J. M. Wilcox and N. F. Ness (1965) by means of the satellite IMP-1. The number and distribution of the sectors varies according to the arrangement of active regions on the Sun.

20.4 Dynamics of solar prominences

According to their spectra, which still show quite a number of lines of neutral atoms, prominences have an (electron) temperature of only 4000 to 6000 K. The *quiescent prominences* often float for a whole week in the surrounding corona at about 1.4×10^6 K. How is this possible? We saw earlier that only cool material can radiate effectively. That is, cool material stays cool, and hot stays hot, even in the presence of a given energy transport (which is, moreover, seriously impeded by magnetic fields). Pressure balance between prominence and corona in a horizontal direction requires that $p \propto \rho T$ is about the same in both. That is, the density in prominences must be about 300 times greater than in the corona (H. Zanstra). That the quiescent prominences do not fall down was shown by R. Kippenhahn and A. Schlüter (1957) to be due to the fact that they lie on a cushion of magnetic field lines (which the material certainly cannot get through), just like rain water in a hollow in the tarpaulin cover of a haystack. We must however explain as well why individual prominence knots do not fall down the lines of force of their guiding magnetic field approximately according to Galilean laws but much slower, and why they often drift down for long periods at constant speed. This behavior, which is very reminiscent of clouds in our Earth's atmosphere, clearly arises from the fact that the corona has a enormous *viscosity* as a result of the large velocities and so also long mean free paths of the electrons. That the prominences do not rapidly fall down to the photosphere when they get the chance, but float like real clouds, arises (in both cases) from the dominance of viscous forces over pressure and inertial forces; we have "creeping flow" with small Reynolds number.

How does it happen that pieces of a prominence or of the chromosphere often—apparently without discernible cause—are suddenly accelerated to speeds of $\sim$ 100 km/s, and sometimes as much as 600 km/s? If for example an arched magnetic tube of force (as guidance) joins two places (see again Figure 20.11) with different pressures p_1 and p_2, but about the same temperature T and the same potential (same height), then Bernoulli's equation (20.11) predicts a stream with velocity given by

$$v^2 = \frac{2\mathfrak{R}\,T}{\mu} \ln (p_2/p_1). \tag{20.17}$$

Here the variables refer to the *corona,* which, according to our description, carries the prominence along with it. Taking $p_2/p_1 \approx 10$ (which seems to be reasonable, from observations of the corona) and $T \approx 2 \times 10^6$ K, one obtains $v \approx 300$ km/s, which in fact is of the order of magnitude of the observed prominence speeds. We now understand as well why the flow speeds in prominences and in the solar wind are of the same order of magnitude.

[The further development of hydrodynamical, and particularly magneto-hydrodynamical, investigations—especially in astrophysics—presents the fundamental difficulty, from the mathematical standpoint, that the equations involved are *nonlinear*. We can make progress with even the simplest problems only by means of special numerical computations from which it is hard to judge to what extent the result of the calculation is restricted to a particular model or to what extent it is possible to generalize it.]

21. Variable Stars: Motions and Magnetic Fields in Stars

The first observations of variable stars about the turn of the sixteenth to the seventeenth century formed at the time a powerful argument against the Aristotelian dogma of the immutability of the heavens. The observations by Tycho Brahe and Kepler of the supernovae of 1572 and 1604 have even in our own time contributed significantly to our knowledge of these puzzling objects, and they have made possible the identification of their remnants by radio astronomy. We must also recall the discovery of Mira Ceti by Fabricius.

We have already spoken of the development of the technique of photometric measurements from Argelander's method of step estimates, which is now used only by amateurs, through photographic photometry, which serves mainly in surveys of stellar clusters, Milky Way fields, galaxies, etc., for variable stars, to photoelectric photometry, that yields the light curve of an individual variable with an accuracy of a few thousandths of a magnitude. We need scarcely remind ourselves that today the measurement of color indices and the analysis of spectra play an essential part in addition to the measurement of magnitudes.

We denote variable stars by letters R, S, T, . . . , Z with the genitive of the name of the constellation, and we can use also RR, RS, . . . , ZZ; for particular stellar clusters, stellar fields, etc., we nowadays for the most part use simply the catalog numbers of the stars.

It is clear from the outset that the investigation of *variable* stars promises deeper knowledge of the structure and evolution of the stars than the study of unchanging static stars. On the other hand, however, the observation and theory of variables present vastly greater difficulties. A warning against easy *ad hoc* hypotheses may not be out of place here.

Our concern is not to describe in outline the innumerable classes of variable stars, mostly named after a prototype. We leave aside the eclipsing variables, which we have already discussed, and select certain interesting and important types of intrinsically variable stars which we now survey from the standpoint of their physical significance. We must also consider variable stars as particular stages in stellar evolution. However we shall not take account of this important point of view until Part III.

21.1 Pulsating stars

Pulsating stars include, amongst others, the following groups of variable stars; they are all giant stars (although there do exist in addition "dwarf Cepheids" more or less on the main sequence).

RR Lyrae stars or *cluster variables*. Stars with regular luminosity change having periods of about 0.3 to 0.9 days, luminosity amplitudes of about 1^m, spectral types about A–F. They belong to the halo and nucleus of the Milky Way and they are important in globular clusters.

Classical cepheids (δ Cephei stars) are likewise entirely regular, with periods of about 2 to 40 days. They have about the same luminosity amplitudes as cluster variables, but their spectral types are F–G. They occur in the spiral arms of the Milky Way system.

W Virginis variables, also regular, have similar periods, but are about two magnitudes fainter. They occur in the halo and nucleus of the Milky Way.

Then there are several groups of semiregular variables with longer periods and later spectral types, in particular:

RV Tauri variables with periods of 60 to 100 days

Long period variables of several sorts, all of late spectral types, with quasi-periods of about 100 to 500 days; Mira Ceti is an example.

The radial-velocity curves of these sketchily described groups of variable stars give the first indication of their physical nature. These are closely related to the light curves. At first people tried to refer all the regular velocity changes of, say, a classical cepheid to double-star motion. Integration over the velocity, without further hypotheses, gave the dimensions of the "orbit," since if x is the coordinate in the direction of the sight line

$$\int_{t_1}^{t_2} \frac{dx}{dt}\, dt = x_2 - x_1. \tag{21.1}$$

It was found that the star in this orbit would have no room for a companion. Therefore H. Shapley in 1914 fell back upon the possibility of a radial pulsation of the star which A. Ritter had discussed in the 1880s as a purely theoretical problem. The pulsation theory of cepheids (and related variables) was then further developed by A. S. Eddington from 1917 onward. This in turn provided the stimulus for Eddington's pioneering work on the *internal constitution of the stars* (Section 25).

A simple estimate embodies an important clue to the theory of pulsating stars, although the details are difficult: We think of the pulsation as a standing sound wave in the star. Its velocity is $c_s = (\gamma p/\rho)^{1/2}$, where $\gamma = c_p/c_v$ is the ratio of the specific heats, p is the pressure, and ρ the density. The mean pressure $\bar{p}$ inside a star is of the order of magnitude of the force of gravity upon a column of material of 1 cm^2 cross section, that extends inward from the surface of the star of mass M. Thus

$$\bar{p} \approx \underset{\text{mean density}}{\bar{\rho}} \times \underset{\text{radius}}{R} \times \underset{\text{acceleration}}{GM/R^2} \tag{21.2}$$

The period of oscillation P is then, again as regards order of magnitude,

$$P \approx R/c_s \approx R(\gamma GM/R)^{-1/2}. \tag{21.3}$$

Since $M = (4\pi/3)R^3\bar{\rho}$ we obtain the important relation between period P and mean density $\bar{\rho}$ of the star

$$P \propto (\bar{\rho})^{-1/2} \tag{21.4}$$

which has been well confirmed by extensive observational material.

W. Baade suggested a second test of the pulsation theory. The brightness of a star, other things being equal, is proportional to the area of its disk πR^2 times the radiation flux πF_λ at its surface. But we can obtain the time variation of the stellar radius R directly by integration of the radial-velocity curve according to equation (21.1). Its value is usually about 5 percent of the stellar radius. On the other hand, we can evaluate the radiative flux πF_λ using the theory of stellar atmospheres from the color indices or other spectroscopic criteria. Some results for δ Cephei are collected in Figure 21.1 following W. Becker. The required proportionality of the measured brightness with R^2F_λ is in fact well verified.

In 1912 Miss H. Leavitt at the Harvard Observatory discovered for the many hundreds of cepheids in the Magellanic Clouds a relation between the period P and, in the first instance, the apparent magnitude m_V. Since all these stars have the same distance modulus, this was the discovery of a *period-luminosity relation*. Using still rather meager observational material on proper motions (Section 23), H. Shapley determined the zero point of the scale of absolute magnitudes. Thence it was now possible (problems of

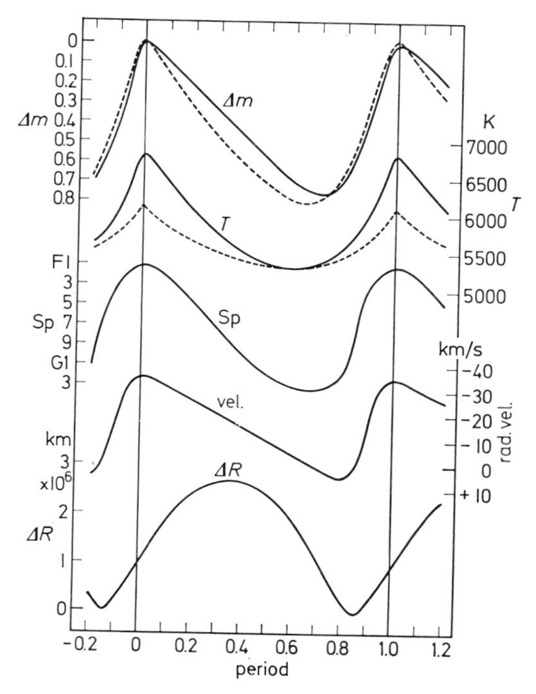

Figure 21.1
Delta Cephei (after W. Becker). Periodic variations of, reading from the top
1) Brightness, showing light curve before (continuous line) and after (dashed line) correction for the varying radius of the star
2) Temperature: color temperature (continuous line), radiation temperature (dashed line).
3) Spectral type. 4) Radial velocity.
5) $\Delta R = R - R_{min}$.

interstellar absorption being then still unknown) to discover the distance of any cosmic system in which cepheids could be detected. With the help of S. Bailey's observations (1895) of many cluster variables (RR Lyrae stars) in globular clusters, H. Shapley was able in 1918 for the first time to evaluate their distances and thus to define the outlines of the galactic system in the modern sense. Using the same method, but applying it to classical cepheids (with longer periods) E. Hubble then in 1924 determined the distances of some of our neighboring spiral nebulae, and he definitely showed that these are galaxies similar to our Milky Way system. We shall discuss this advance into cosmic space in Sections 27 and 30. Here, however, we must recount an important correction to the basis of the cepheid method, which W. Baade discovered in 1950. In fact, he was able to show that the zero point of the period-luminosity relation is differently situated for different types of cepheids. In particular, for equal periods the classical cepheids of population I (see Section 22 with reference to stellar population) are 1 to 2 magnitudes brighter than the W Virginis stars of population II. The curve for cluster variables or RR Lyrae stars lies about on the extension of the W Virginis curve.

The more exact study of light curves permits the recognition of several subclasses of the main groups here presented. Corresponding to these, there are also possibly small differences in the period-luminosity relation.

The cooler pulsating variables with longer periods of luminosity change, like the *RV Tauri variables* and the *long-period variables,* have more and more irregular luminosity variations. The theory of the internal constitution of the stars (Section 25) shows that in cooler stars the hydrogen convection zone becomes ever more extensive. Therefore it is natural to see a coupling of the pulsation with the turbulent convection currents as the cause of the observed semiregular luminosity change.

The amplitude of the variations, measured for example in visual magnitude m_{v}, increases systematically toward cooler stars. This depends essentially upon the known form of Planck's law of radiation. If we write the visual magnitude m_{v} in Wien's approximation, as in (13.12), we have

$$m_{\mathrm{v}} = \frac{1.562}{\lambda_{\mathrm{v}} T} + \mathrm{constant}_{\mathrm{v}}. \tag{21.5}$$

A given temperature fluctuation ΔT thus corresponds to a luminosity amplitude

$$\Delta m_{\mathrm{v}} = -\frac{1.562}{\lambda_{\mathrm{v}} T^2} \Delta T, \tag{21.6}$$

which increases like $1/T^2$ for cooler stars.

The maintenance of the pulsation presents an interesting theoretical problem: What "valve" ensures that the oscillation of the star—like the piston of a heat engine—always receives its push in the correct phase? The generation of thermonuclear energy near the center of the star is practically unaffected by the pulsation. We have to do much more with the temperature and pressure dependence of the *opacity,* which regulates the flow of radiative energy and so determines the temperature to which any particular layer adjusts itself at any instant. In combination with the change in the adiabatic temperature gradient, this effect shows itself to be specially important in the region of the second ionization of helium. The relevant calculations—very difficult ones—also show and make theoretically comprehensible the combination of parameters of the star, that is, the region of the color magnitude diagram, for which pulsation can occur.

21.2 R Coronae Borealis stars

In the region of the red giant stars there is another entirely different kind of slowly varying star, named after its prototype R CrB. From time to time its luminosity, normally constant, drops by several magnitudes and then slowly recovers. Spectral analysis of these relatively cool stars shows that their atmospheres are poor in hydrogen but rich in carbon (and probably helium). We may imagine that the R CrB stars occasionally eject clouds of colloidal carbon (a kind of soot), which eclipse the star.

21.3 Spectrum variables (Ap stars) and metallic-line stars

Near the main sequence various kinds of stars are found which do not fit into the two-dimensional MK classification. At the present time the physical significance of the third parameter which is needed to describe them is quite unclear. All are distinguished by peculiarities in their spectrum. The hotter *Ap stars* are variable, the cooler *metallic-line stars* are not. Whether these two neighboring groups have anything to do with each other is an open question.

The *spectrum variables* or *Ap stars* show anomalous intensities and a periodic variation in the intensity of certain spectral lines; different lines behave differently. The prototype is α^2 Canum Venaticorum with a period of 5.5 days, in which the lines of Eu II and Cr II vary in antiphase while, for example, Si II and Mg II remain almost constant. In most cases the spectral variations are accompanied by brightness fluctuations of about $0^{\mathrm{m}}.1$. H. W. Babcock was able to show by measuring the Zeeman effect that these stars have magnetic fields of several thousand gauss, of which the strength, and often the sign as well, show periodic changes. According to A. Deutsch, a considerable fraction of the observations, at least, can be explained on the supposition that these stars possess gigantic magnetic *spots,* in which—according to their polarity—one group or another of spectral lines is enhanced. The variations are attributed to the *rotation* of the star. In fact the measured rotational broadening of the lines matches the periods of the stars well, that is, rapidly varying stars mostly have broad lines and vice versa.

In the color-magnitude diagram the spectrum variables or Ap stars (peculiar A stars; however the phenomenon is also found in neighboring spectral types) lie on or near the main sequence, and each variety is associated with a definite range of color index B − V:

Helium-weak stars | Si stars | Mn stars | Eu–Cr–Sr stars

It has been observed recently in some Ap stars that lines of otherwise very rare heavy elements appear in unexpected strength; for example, in Eu–Cr–Sr stars E. Brandi and M. Jaschek (1970) have found Os I and II, Pt II, and perhaps U II. We can regard this as evidence that the anomalous abundances are caused by neutron bombardment. The connection of different element abundances with different magnetic fields may then be due to the bombarded material being immediately frozen, with its own magnetic field, into a "plasmoid" (in the magnetohydrodynamic sense).

Next to the Ap stars, along the main sequence toward cooler temperatures, are the *non*variable *metallic-line* stars. From their hydrogen lines they are classified as about A0 to F0. If one goes by their *hydrogen type,* the stars lie on the main sequence. However the lines of calcium (especially H and K) and/or scandium are too weak for that, while the metallic lines of the iron group and the heavy elements are too strong. Comparison of lines from different excitation and ionization states shows that the abundances of the elements are anomalous and that it is not a question of some anomalous excitation ratios, non-LTE or something like that. In contrast to the Ap stars, material from a nuclear event would here be thoroughly mixed. Van den Heuvel (1963) tried to connect the two classes by considering mass exchange during the evolution of close binary systems. On the other hand, F. Praderie, E. Schatzman, and G. Michaud (1967–1970) put forward explanations based on diffusion processes. At any rate, the appearance of Ap and Am stars on the main sequence of comparatively young star clusters shows that these stars must originate in the immediate neighborhood of the main sequence. All further attempts at explanation are clearly still very speculative.

21.4 Stellar activity, T Tauri and other irregularly varying cool stars and flare stars

The T Tauri or RW Aurigae stars, so-called after their prototypes, are cool stars (spectral types K0–M5e), which lie on or above the main sequence in the Hertzsprung–Russell diagram. Their brightness "flickers" irregularly, of the order of 1^{m} in a few days. V. A. Ambarzumian, G. Haro, and others showed that these variables appear in the sky preferentially in the neighborhood of dark clouds and young star clusters: in Orion, where a star cluster is forming in the region of the well-known nebula, in the Pleiades in Taurus, and so on. Recent investigations confirm Ambarzumian's hypothesis that we are dealing with stars which were formed out of interstellar material a relatively short time ago ($\lesssim 4 \times 10^8$ years) (Section 26).

Spectroscopically, the T Tauri stars are distinguished by bright emission lines, especially Ca II H + K, Hα, and other Balmer lines of hydrogen.

Similarly varying M stars, for which UV Ceti may be taken as prototype, are found in the solar neighborhood. Since their absolute magnitude is usually fainter than 10^m, they are no longer accessible in more distant parts of the Milky Way.

Many variables of the above types show *flares* at irregular intervals, which differ from those on the Sun only in their somewhat greater brightness. During the flares the photographic magnitude of the whole star increases within 3 to 100 s (!) by 6^m to 7^m; it then fades considerably more slowly. Smaller increases in brightness merge into almost continuous fluctuations in light. During an outburst, the UV spectrum of *flare stars*—as they are also called—shows (A. H. Joy *et al.*) a superimposed *continuum*, so that the magnitude in the ultraviolet can increase by up to 10^m, as well as strong emission lines of Ca II, H, He I and even He II, as in solar eruptions. However, while in the Sun the continuous spectrum of the flare reaches at most a few percent when compared to the brightness of the photosphere, in cool stars ($\sim$ 3 to 4000 K) this ratio is inverted since the radiation flux from their photospheres is several magnitudes weaker. B. Lovell and others (1963) were able to observe the related *radio flares* at meter wavelengths at the same time as the optical flares.

On the Sun, active regions or their faculae (plages) are distinguished by Ca II $H_2 + K_2$ emission lines, on which are sometimes superimposed narrower $H_3 + K_3$ absorption features. Exactly the same is observed in many giant and main sequence stars of the spectral types G to M.[1] O. C. Wilson has recently made a precise study of this stellar activity, which was discovered by K. Schwarzschild and G. Eberhard (1913) and clearly arises, just as on the Sun, from the fact that the hydrogen convection zone generates a "mechanical" energy flux. In many stars the $H_2 + K_2$ emission components show variations in time which point to rotation or a cycle of activity on the star. In some cool dwarf stars G. E. Kron has observed small periodic variations in brightness which can be attributed to "star spots," which are analogous to sun spots; for example, Ross 248 shows fluctuations of amplitude $0^m\!.06$ with a period of $\sim$ 120 days, which may be interpreted as the rotation period of the star. All signs of stellar activity, such as H and K emission lines or flares, are mostly strongly pronounced in young stars, decrease with increasing *age* of the star (Section 26) and disappear after about 4×10^8 years.

In 1957 O. C. Wilson and M. K. V. Bappu observed a totally unexpected relationship: the *width* of the Ca II emission lines is, independent of their

[1]According to O. Struve, Be stars, that is, B stars with hydrogen emission lines, are totally different. Mainly observed in connection with fast rotation, they have extended gaseous envelopes or rings, which are excited to fluorescence by the ultraviolet radiation of the star.

intensity, a function of the *absolute magnitude* of the star. This gives us an excellent method for determining *spectroscopic parallaxes* for cooler stars. Why the turbulent velocity ξ_t in the chromospheres of cool main sequence and giant stars is connected in this way with the other parameters of their atmospheres belongs to the unsolved problems to which we alluded at the end of the previous section.

21.5 Novae and supernovae

The next group including *novae,* and *nova-like variables* which undergo similar but feebler outbursts in a more or less regular sequence, P Cygni stars, U Geminorum = SS Cygni stars, Z Camelopardalis stars, etc., present quite different features. As we must state in advance, their theoretical interpretation is still uncertain.

A *nova outburst* proceeds about as follows. The initial stage, the *pre-nova,* is a hot star with absolute magnitude $M_V \approx +5$ lying in the HR diagram between the main sequence and the white dwarfs. Within at most 2 to 3 days the luminosity increases to a maximum of about $M_V \approx -6$ to -8.5, and so by a factor of 4 to 6 powers of ten. The spectrum then resembles that of a supergiant like α Cygni (cA 2). The luminosity increase thus does not depend upon a rise in temperature but, as confirmed by the radial velocities, upon an enormous expansion of the star. After passing through the luminosity maximum (and during the subsequent decrease the luminosity often shows cepheid-like fluctuations) broad emission lines make their appearance. Their Doppler effects show that the nova is now ejecting shells with velocities of the order of magnitude of 2000 km/s. In several cases, for example Nova Aquilae 1918, such an envelope and its expansion could be followed in direct photographs for more than a decade. The measured radial velocity of 1700 km/s corresponded to the expansion of the envelope by 1″ per year; the comparison of these numbers gave the distance and absolute luminosity of the nova as well as the dimensions of the envelope.

In the course of several years the nova returns to its initial luminosity and presumably to its initial state. We see this most clearly in the case of recurrent novae, like T Pyxidis which has undergone several nova-like outbursts ($\Delta m \approx 7^m$) at intervals of about 10 years and which shows almost constant luminosity in between. According to recent observations, many novae, if not all, appear to belong to binary systems. In our Milky Way Galaxy and in similar galaxies some 30 to 50 outbursts take place per year.

U Geminorum or SS Cygni stars, again so called after their prototypes, show much less violent outbreaks at irregular intervals of the order of a month to a year.

On the other hand, P Cygni has remained quiet since a lively performance at the beginning of the seventeenth century. The emission lines, especially those of the Balmer series, with their absorption components shifted to the violet, show that nevertheless an envelope is being continually ejected. More than a dozen so-called P Cygni stars behave similarly.

Even though its estimation is rather uncertain because of the need to take account of the bolometric correction, the *energy* given up during a nova outburst is of the order of magnitude 10^{45} erg. This corresponds to the thermal energy content of a thin layer at for example 5×10^6 K and only 1/1000 solar mass. Everything therefore points to the nova outburst being a "skin disease" of the star.

Supernovae represent cosmic explosions of quite a different magnitude, their special status having been recognized by W. Baade and F. Zwicky in 1934. At its maximum, a supernova can outshine the entire galaxy to which it belongs. We usually distinguish two types of supernovae:

(1) Type I supernovae reach at maximum a mean photographic absolute magnitude (corrected for interstellar absorption) of $M_{pg} = -19.0$. Their light curves (Figure 21.2) are very standard. In the first 20 to 30 days following maximum the luminosity drops by 2 or 3 magnitudes; from then on the brightness falls off approximately exponentially with time.
(2) Type II supernovae reach "only" $M_{pg} = 17$. All the same, this corresponds to about 10 000 times the brightness of an ordinary nova! The fall in brightness after the maximum at first proceeds more steeply, and later, on the other hand, more slowly than in Type I. Furthermore, the light curves show larger individual differences. The spectrum of a

Figure 21.2. Photographic light curves of Type I supernovae in different galaxies.

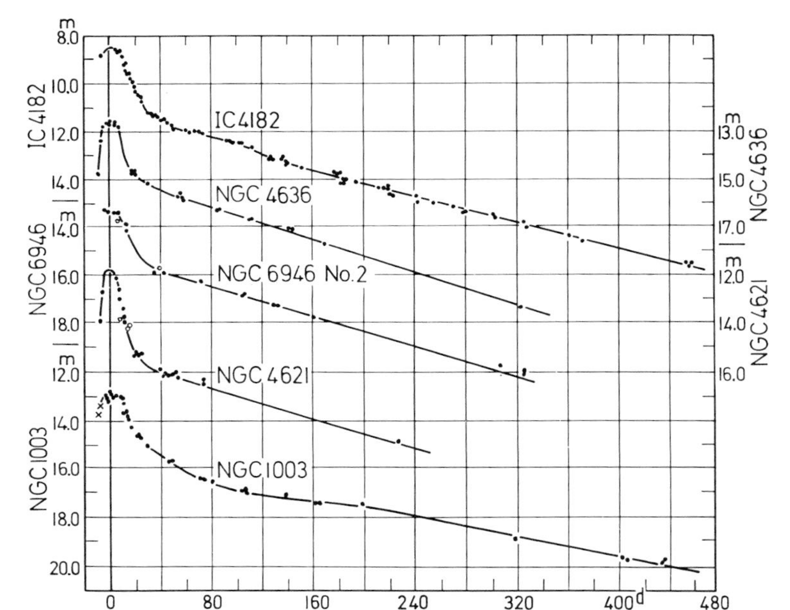

Type II supernova, and its development with time, resembles to an astonishing and far-reaching degree that of ordinary novae.

Measurements of the intensity distribution in the *continuous* spectrum suggest that the temperature of the emitting layers decreases from $\sim 10^4$ K at maximum light to $\sim$ 7000 (SN I) or 5000 K (SN II). Overlying the continuous spectrum there is a *line* spectrum. As in the old nova P Cygni, the emission lines are accompanied by blue-shifted absorption components. The Doppler shifts suggest ejection velocities of up to $\sim$ 20 000 km/s. All supernovae have lines of the metals Ca II, Na I, Mg I,. . . . In Type II these are accompanied by strong Balmer lines, which are absent or very weak in Type I supernovae. The former may have approximately "normal" composition, while in Type I supernovae the hydrogen has been consumed.

In our Milky Way system the following could be identified as supernovae: The nova of 1054 AD described in old Chinese and Japanese records, from which the Crab nebula[2] has arisen, and Bolton identified this with the radio source Taurus A. By comparing the expansion speed of $0''.21$ per year with the spectroscopically measured value of 1300 km/s we obtain a distance of about 1300 pc. Incidentally, as regards the history of ideas, it is remarkable that in contemporary Europe there was no syllable of a mention of this supernova, which at maximum was brighter than Venus. At the beginning of the section we recalled Tycho Brahe's supernova of 1572 in Cassiopeia and Kepler's supernova of 1604.

Up to 1964, mostly as a result of F. Zwicky's Palomar survey, some 140 supernovae had been discovered in distant galaxies. G. A. Tammann (1970) estimated from extensive statistics that in a galaxy like our own one supernova appears on average every 26 $\pm$ 10 years. The frequencies of Type I and Type II supernovae are in all galaxies in the ratio $\sim 1:2$.

We may attempt to estimate the total energy output E of a Type I supernova by integrating under the light curve and assuming that the ratio of total radiation to photographic (or—what makes scarcely any difference—to visual) radiation is the same as for the Sun. Under this assumption (BC ≈ 0) we obtain $E \approx 3.6 \times 10^{49}$ erg. The true value could be greater, say about 10^{51} erg as an estimate. For comparison we easily calculate that, for example, the amount of energy liberated by the conversion of one solar mass of hydrogen into helium is

$$0.0072 \times Mc^2 = 1.3 \times 10^{52} \text{ erg.}$$

We cannot discuss until Part III the related questions of the origin and further fate of supernovae.

[2]We shall return to this unique object in Section 26. See particularly Figure 26.11.

III. Stellar Systems

Milky Way and Galaxies; Cosmogony and Cosmology

22. Advance into the Universe

Historical introduction to astronomy[1] in the twentieth century

In the second and third decades of our century, there began a development of astronomy that in its significance does not rank behind the almost contemporaneous discoveries of relativity theory and quantum theory. *The universe as a whole, the cosmos in its spatial structure and in its temporal evolution, became an object for exact scientific investigation.* We pick up the thread of Section 10, and seek to follow the unfolding of the new ideas first in their historical association. At the same time this brief survey should facilitate the understanding of the detailed presentation, which we must arrange from the standpoint of the topics concerned.

About the turn of the century—following upon the "star gauging" of W. Herschel and J. Herschel—H. v. Seeliger (1849–1924), J. Kapteyn (1851–1922), and others tried to investigate the structure of the Milky Way system by the methods of stellar statistics. Even if the goal was not attained, the incredible labor of these undertakings has nevertheless proved to be very valuable in other connections.

H. Shapley's (1918) method of photometric distance determinations using cepheids (cluster variables) brought the decisive advance. The period-luminosity relation, that is, the relation between the period P of the luminosity variation and the absolute magnitude M_V, made it possible to measure the distance of every cosmic system in which one could detect any sort of cepheids.

More exact investigation of the assumptions of this procedure, (1) the absence of interstellar absorption and (2) the applicability of the same

[1]In keeping with modern usage, we regard astrophysics as part of astronomy, that is, of the comprehensive study of the stars.

period-luminosity relation for all types of cepheids, subsequently made considerable corrections necessary. In 1930 R. J. Trümpler discovered general interstellar absorption and reddening, and about 1952 W. Baade recognized that the period-luminosity relations of classical cepheids and the W Virginis stars, that is, the pulsating variables in stellar populations I and II (see below), differ from each other by some 1 to 2 magnitudes. In the following, in numerical statements we shall from the outset take full account of these new corrections, and so to that extent we depart from a purely historical standpoint.

The distances of the globular clusters determined by H. Shapley made it known that these clusters form a slightly flattened system whose center lies at a distance of about 10 kpc, or about 30 000 light years, in the direction of Sagittarius.

The present picture of our Milky Way system developed rapidly from these beginnings. The main body of stars form a flat *disk* of about 30 kpc diameter, the spiral arms being contained in this disk. As regards the *nucleus* of the system, we catch sight of its outer parts in the form of bright star clouds in Scorpio and Sagittarius. The *galactic center* itself is hidden from us by clouds of dark interstellar matter. Radio astronomy has made its direct observation possible for the first time. We ourselves are situated far out in the disk, at some 10 kpc from its center. The disk is surrounded by the much less flattened *halo,* to which the *globular clusters* and certain classes of field stars belong.

So long ago as 1926–1927 B. Lindblad and J. Oort were able to a great extent to elucidate the *kinematics and dynamics of the Milky Way system.* The stars of the disk revolve round the galactic center under the gravitational attraction of the masses that are fairly strongly concentrated there. In particular, the Sun describes its circular path of radius about 10 kpc with a speed of about 250 km/s in about 250 million years. However what we first notice is only the *differential rotation.* Stars further out from the center (as in the case of planets going round the Sun) move somewhat more slowly, stars closer in move faster, than we do. Thence we easily estimate the mass. After numerous adjustments the mass of the whole system is found to be about 2×10^{11} solar masses.

While the stars of the galactic disk describe circular orbits, the globular clusters and the halo stars move around the center in elongated elliptical orbits: as a result, their velocities relative to the Sun are of the order of 100 to 300 km/s. This is J. H. Oort's explanation of the so-called "high velocity stars."

Although it only makes up a few per cent of the mass of the Galaxy, the interstellar matter plays an important role as well as the stars. Precise distance measurements only became possible after R. J. Trümpler (1930) had grasped the nature of the interstellar absorption and reddening of starlight by cosmic dust. Somewhat earlier, A. S. Eddington (1926) had already made clear the physics of the interstellar gas and of the interstellar absorption lines, and H. Zanstra and I. S. Bowen (1927–1928) had explained the emission of the galactic and planetary nebulae. Finally, the completely unexpected discovery of the *polarization* of starlight by W. A. Hiltner and J. G. Hall (1949) led to the recognition that there is in the disk a galactic magnetic field of $\sim 10^{-5}$ G.

After considerable refinement of photographic technique used with the 100-in Hooker telescope of the Mount Wilson Observatory, in 1924 E. Hubble succeeded, to a great extent, in resolving the outer parts of the Andromeda nebula (and of others of our neighboring galaxies) into individual stars. He was able to discern (classical) cepheids, novae, bright blue O and B stars, etc. These made it possible by photometric means to determine the distance. This came to about 700 kpc or about 2 million light years (and here we have taken account of more recent corrections). After wearisome controversies, it was now made clear that the Andromeda nebula and our Milky Way system are to a great extent similar cosmic structures. W. Baade's subsequent investigations confirmed this up to fine details. Today we can therefore largely combine the studies of the Andromeda nebula (M 31 = NGC 224) and the Milky Way into a single picture; some observations are better "from outside" and others "from inside." Since the work of Hubble it has become customary to designate the "relatives" of our Milky Way as *galaxies,* and so far as possible to reserve the term nebula for bodies of gas or dust *within* the galaxies.

In 1929 E. Hubble made a second discovery of the greatest importance: the spectra of galaxies showed a *redshift* proportional to their distance. We interpret this as a uniform *expansion of the universe;* we easily realize that inhabitants of other galaxies would observe exactly the same as what we do. If we extrapolate, somewhat formally, the flight of the galaxies backward, then at a time $T_0 \approx 10^{10}$ years ago the whole cosmos would have been tightly packed together. We call T_0 the *Hubble time;* it gives a first clue to the age of the universe. What might lie further back is outside the scope of our enquiry, and in any case at time T_0 years ago the universe must have been entirely different from today. We may mention only briefly the forerunners of Hubble's discovery, V. M. Slipher, C. Wirtz, and others, as well as M. Humason's collaboration with Hubble on the 100-in telescope. Here for the first time the *universe as a whole* had become the object of exact observational scientific investigation. Within the framework of the theory of general relativity, which for the first time sought to treat gravitation and inertial forces together, from 1916 A. Einstein, and after him W. de Sitter, A. Friedman, G. Lemaître and others, had developed the principles of theoretical cosmology. From another point of view, it is of fundamental significance for the study of distant galaxies that we can obtain from the redshift of their spectral lines their distance and hence their absolute magnitude, true size, etc.

The recognition of an age of the universe of about 10^{10} years, not very much greater than the age of the Earth inferred from radioactive data as 4.6×10^9 years, gave a powerful stimulus to the study of the evolution of stars and stellar systems.

In connection with the age of the Earth and of the Sun, J. Perrin and A. S. Eddington had as long ago as 1919–1920 conjectured that the energy radiated by the Sun is generated by the conversion of hydrogen into helium. On the basis of the nuclear physics meanwhile developed, by 1938 H. Bethe and C. F. v. Weizsäcker were then able to show which thermonuclear reactions are capable of slowly "burning up" hydrogen to produce helium in, for instance, the interior of a main-sequence star at temperatures[2] of about 10^7 K. Bethe and then A. Unsöld in 1944 thence showed that nuclear energy sources would suffice for a time of the order of the age of the universe only for the cooler main-sequence stars of types G, K, M. For hot stars shorter lifetimes resulted, which in the case of O and B stars of great luminosity went as low as about 10^6 years. In such a short time, the stars could not travel far from their place of origin. So they must have been formed close to where we now see them. Apart from any speculative hypotheses, W. Baade in particular repeatedly remarked that the close association in space between blue OB stars[3] and dark clouds, for example in the Andromeda galaxy, pointed to the formation of these stars from interstellar matter.

Again to a great extent traceable to the stimulus of W. Baade, investigations by A. R. Sandage, H. C. Arp, H. L. Johnson, and others on the color-magnitude diagrams of globular clusters and galactic clusters led us further. Along with the theory of the internal constitution of the stars, they produced something like the following picture. A star, which has been formed from interstellar material, first runs through a relatively short contraction phase. Hydrogen burning starts on the main sequence; the star stays there until about 10 percent of its hydrogen has been used up. It then moves off to the right in the color-magnitude diagram (M. Schönberg and S. Chandrasekhar, 1942) and becomes a red giant. The place in the diagram where the curve for a cluster turns off to the right from the main sequence—the so-called "knee" (Figure 26.1)—shows which stars have consumed about 10 percent of their hydrogen since the cluster was formed. The time needed for this is called the *evolution time* and gives simultaneously the *age* of the cluster. Clusters possessing bright OB stars, like h and χ Persei, are therefore very young, while in old stellar clusters the main sequence is still present only below G0. Theoretical researches on stellar evolution, beginning with F. Hoyle and M. Schwarzschild in 1955, denoted also a fundamental extension of A. S. Eddington's theory of the internal constitution of the stars (Section 25) because the earlier assumption of a continual mixing of the material inside a star had to be given up in face of the observations. The new concept is that in the center of the star a burnt-out helium zone is

[2]H. N. Russell had already called attention to the circumstance that stars all along the main sequence, in spite of their very different effective (or surface) temperatures, have nearly the same central temperatures of about 10^7 K.

[3]With spectra of low dispersion, in the case of very distant stars it is often no longer possible to distinguish between the rather similar spectra of types O and B. Then we speak simply of OB stars.

formed. The resulting outward advance of the nuclei-burning zone causes the inflation of the star into a *red giant*. The study of numerous color-magnitude diagrams has now led to the fundamental result that all globular clusters have the same age of about 9 to 12×10^9 years. On the other hand among the galactic clusters there are both young and old objects. While the youngest is scarcely a million years old, the oldest galactic cluster NGC 188 has, according to A. R. Sandage and O. J. Eggen (1969), an age of 8 to 10×10^9 years, which is thus hardly different from that of the globular clusters.

The most recent research is now giving us insight into the final stages of stellar evolution. Stars which are not too massive burn out their nuclear energy sources and end as degenerate matter, that is, as *white dwarfs* (see pp. 135 and 140). A star of 0.5 solar masses, for example, has then shrunk to about the size of the Earth. Its store of thermal energy is still sufficient to maintain its low output of radiation for thousands of millions of years. In massive stars the material can be still further compressed, in which case the protons and electrons fuse together into neutrons. The *pulsars* discovered in 1967 by A. Hewish from their radio radiation are such neutron stars, with densities of $\sim 10^{14}$ g/cm^3. According to the general theory of relativity, further compression leads finally to the so-called *black holes*, whose enormous gravitational field can even prevent photons from escaping (Section 30).

After this excursion into the latest developments in astrophysics, let us pick up the thread of our historical exposition. In 1944, W. Baade had already pointed out that various regions of our Milky Way differ from one another not only in dynamical respects but also in their color-magnitude diagrams. The concept of *stellar populations* arose in this way. It soon appeared that these differ from each other essentially in their ages and in the abundances of the heavy[4] elements relative to hydrogen (we usually say for short: metal abundances). Apart from subdivisions and intermediate types, we have:

(1) *Halo population II:* Globular clusters, metal-poor subdwarfs, etc., have the same color-magnitude diagram. They describe elongated galactic orbits and form a system which is slightly flattened but strongly concentrated toward the center. Their metal abundance corresponds to about 1/500 to 1/5 of "normal." The halo population II contains almost no interstellar matter, which has clearly been used up in star formation. Its age, about 10^{10} years, is nearly equal to the age of the Milky Way.

(2) *Disk population:* Most stars in our neighborhood belong to this; they form a strongly flattened system with strong concentration toward the galactic center. The stars describe almost circular orbits; they have "normal metal abundance," similar to the Sun. The oldest star clusters of the disk population and of the halo population II are, as we have said, of similar age (within about 10^9 years). However, the disk population also merges continuously in age with the spiral population I.

[4]In this connection, we understand by this all elements apart from hydrogen and helium.

(3) *Spiral arm population I:* This is characterized by young blue stars of high luminosity. Within the disk, as we shall see, the interstellar matter is denser in the spiral arms. Associations and clusters of young stars have their origin here.

The classification of galaxies initiated by E. Hubble in 1926 (we shall get better acquainted with it later on) was first made according to their shapes but showed itself later to be basically a classification according to the preponderance of population II or population I features. The *elliptical galaxies* have only a little interstellar matter left, so they comprise mainly old stars, similar to the halo population or the disk population of the Milky Way. At the outer extreme, the very richly structured Sc and Irr I galaxies contain much gas and dust, well-defined spiral arms or other structures as well as bright blue O and B stars, all features of stellar population I.

More recently it has been discovered that the luminosities and masses of galaxies of similar appearance may differ by up to about five powers of ten. We therefore speak of *giant* and *dwarf* galaxies.

A completely new era in the investigation of our own Milky Way system and of distant galaxies was opened by *radio astronomy,* beginning with K. G. Jansky's discovery in 1931 of the meter wavelength radiation of the Milky Way.

Thermal, free-free radiation from plasma at $\sim 10^4$ K was discovered in the interstellar gas, in H II regions, planetary nebulae etc. In 1951 were seen the first successful observations of the 21-cm line of atomic hydrogen, predicted in 1944 by H. C. van de Hulst, whose Doppler shifts gave entirely new insights into the structure and dynamics of interstellar hydrogen and so of the whole Galaxy. Further lines in the centimeter and decimeter regions arise from transitions between very high quantum numbers in hydrogen and helium atoms as well as from diatomic and polyatomic molecules, some of which are surprisingly complicated.

The *radio continuum* first observed by Jansky is nonthermal, the so-called *synchrotron radiation.* As H. Alfvén and N. Herlofson conjectured in 1950 and as the Russian astronomers I. S. Shklovsky, V. L. Ginzburg, and others were soon thereafter able to render more certain, it occurs if high-energy electrons move in spiral paths around lines of force of a cosmic magnetic field, just as in a synchrotron electron accelerator.

In addition to the understanding of the mechanisms which could lead to the emission of radio-frequency radiation, of even greater significance was the achievement of ever greater *angular resolution* and more precise *radio positions.* We have already discussed in Section 9 the building of steadily improved and enlarged radio telescopes and interferometers up to the intercontinental very-long-base-line interferometers, and we now give some historical details concerning the investigation of cosmic radio sources. In 1946 J. S. Hey deduced the existence of the first radio source Cygnus A (initially radio sources were designated by their constellation and a capital letter) from time variations of the radio signal, later discovered to be scintillation of ionospheric origin. In 1949 J. G. Bolton, G. J. Stanley, and O. B. Slee identified the radio source Taurus A with the Crab Nebula M 1. By 1952 radio positions were considerably more precise and W. Baade and R. Minkowski were able to show that the radio source Cassiopeia A, like Taurus A, was the remnant of an earlier supernova. Cygnus A on the other hand was associated with a peculiar galaxy with emission lines of high excitation: the first *radio galaxy*. With this, the way was open for a development in extra-galactic radio astronomy, whose like is certainly not to be found in the whole history of astrophysics. In 1962–1963 M. Schmidt at Mount Palomar Observatory was able to identify the *quasi-stellar radio sources* or *quasars*—whose optical appearance scarcely differs from that of stars—as galaxies at great distances with very large optical and radio luminosities. Their *redshifts,* which greatly exceeded any previously known, opened to cosmology and cosmogony undreamed-of perspectives. Furthermore, it became apparent that in the nuclei of quasars and of radio galaxies and, to a lesser extent, even in normal galaxies like our own, cosmic explosions of unprecedented scale occurred from time to time, whose physical nature is still far from clear.

Since 1963 there has developed an important counterpart to the radio astronomy of synchrotron radiation, which indicates the presence of electrons in the energy range of about 10^8 to 10^{11} eV, namely, *X-ray astronomy* (H. Friedman, R. Giacconi, B. Rossi, and others). In the neighboring region of the spectrum, *γ-ray astronomy* is still in its early stages. In both wavelength regions we are dependent upon observations from rockets and space vehicles, and perhaps also from stratospheric balloons.

The heavy particles which correspond to the synchrotron electrons, that is, the protons, α particles, atomic nuclei, and their secondary particles in the cosmic rays (V. F. Hess, 1912), now also form an important research area in astrophysics. The origin and nature of the acceleration of the electrons and of the positively charged particles has at present been only very slightly clarified; we understand their guidance and storage in cosmic magnetic fields at least a little better. We have already mentioned the very interesting field of *neutrino astronomy,* still restricted to the Sun, whose origins go back to R. Davis Jr. (1964). The whole of *high-energy astronomy,* and the astrophysics of exploding and other peculiar galaxies, will without doubt play a decisive role in future theories of the evolution of galaxies, of the origin of the chemical elements and their abundance ratios and, last, but not least, of the whole of cosmology.

G. Lemaître and G. Gamow (1939) made the first daring advance into these regions with the hypothesis that the expansion of the Universe began with a "Big Bang," in whose initial stages there was prompt formation of the "cosmic" (that is, roughly solar) abundance ratios of the elements. The discovery of metal-poor stars, and above all the difficulty of continuing to build atomic nuclei beyond mass number A = 5, later brought this theory into discredit. Only after the discovery by A. A. Penzias and R. W. Wilson (1965) of the *cosmic 3 K radiation* in the microwave region, which was interpreted as a remnant of the Big Bang, did the theory return in a more modest form: the initial "Big Fireball" should produce essentially only hydrogen and helium atoms with a number ratio of about 10 : 1.

With regard to the origin of the Milky Way system and the abundance ratios of the elements (nucleosynthesis), a proposal developed in 1957 by E. M. and G. R. Burbidge, W. Fowler, and F. Hoyle (often quoted as B^2FH) became widely known. W. Fowler had already fundamentally advanced the knowledge of the important nuclear processes for energy generation in stars by his remarkable measurements of nuclear cross sections at low energy. The B^2FH theory then began with the proposition of an almost *spherical primeval galaxy* consisting of hydrogen, plus 10 percent helium and possibly already traces of heavier elements. Then the first *halo stars* were formed, produced heavy elements, were dispersed (by supernova explosions, etc.), and so enriched the interstellar medium with heavy elements. From this arose a new generation of stars, richer in metals, and so on. Before the abundance of heavy elements had reached that of present-day "normal" stars, the *galactic disk* and spiral arms were formed by the collapse of the remaining halo material which had not yet condensed into stars; dynamical considerations require that this occurred on a time scale of only about 10^8 years. We shall postpone for the moment a critical discussion of this hypothesis.

On the other hand let us notice at this point that V. A. Ambartsumian has been developing since 1958 entirely different proposals concerning the origin and evolution of galaxies, in which the *activity of their nuclei* and the ejection of gigantic amounts of matter and energy play a decisive part.

23. Constitution and Dynamics of the Galactic System

W. Herschel (1783–1822) was the first to seek to penetrate the structure of the galactic system, in his "star gauging," by counting how many stars down to a given limiting magnitude he could see in various directions.

What should we expect, were space fully transparent and uniformly populated with stars? For stars of given absolute magnitude the apparent luminosity decreases with distance like $1/r^2$ [see equations (14.2) and (14.3)]. Stars brighter than m thus occupy a sphere having $\log r = 0.2m +$ constant. Their number $N(m)$ is proportional to r^3, so we must have

$$\log N(m) = 0.6m + \text{constant}. \tag{23.1}$$

In Figure 23.1 we compare the numbers of stars per square degree according to F. H. Seares (1928) for the galactic plane and for the direction of the galactic pole ($b = 0°$ and $b = 90°$) with the numbers to be expected

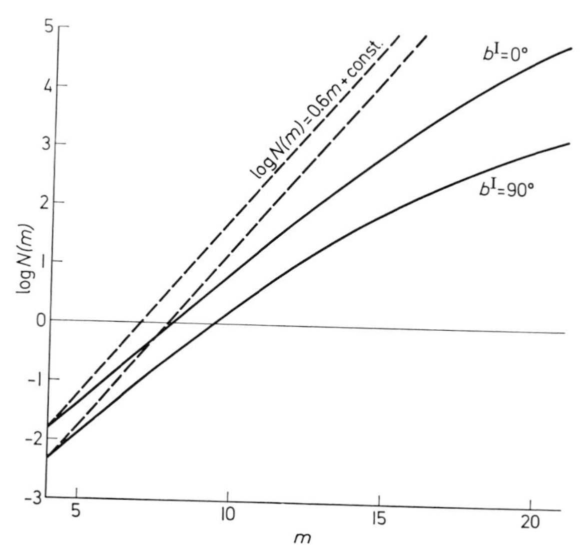

Figure 23.1. Number of stars $N(m)$ brighter than a certain magnitude m per square degree. Here m is on the international photographic scale. The solid curves are from star counts at the galactic equator ($b^{\mathrm{I}} = 0°$) and the galactic pole ($b^{\mathrm{I}} = 90°$) by F. H. Seares (1928). The broken-line curves denote the calculated functions log $N(m) = 0.6m +$ constant for constant star density, in the absence of galactic absorption. The constant was obtained by matching the calculated functions to the observations at $m = 4$.

according to equation (23.1). The much slower increase of $N(m)$ for faint magnitudes in actuality can have only two causes: (1) decrease of star density at large distances, (2) interstellar absorption, or both. H. v. Seeliger (1849–1924) and J. Kapteyn (1851–1922) took account of the dispersion of absolute magnitudes of the stars. They showed how we may represent the star numbers $N(m)$ basically by superposition (mathematically speaking, convolution) of (1) the density function $D(r)$ = number of stars per cubic parsec at distance r in a given direction with (2) the luminosity function $\Phi(M)$ = number of stars per cubic parsec in the interval of absolute magnitude from $M - \frac{1}{2}$ to $M + \frac{1}{2}$, taking account of (3) an interstellar absorption of $\gamma(r)$ magnitudes per parsec. Even if we assume the luminosity function $\Phi(M)$ to be everywhere the same, and if we determine it with the help of stars of known parallaxes in a region of 5 or 10 pc, we can nevertheless not separate the functions $D(r)$ and $\gamma(r)$. We can therefore pass over the results of older stellar statistics. The concepts introduced in this way, however, do remain important, as does the great sampling survey of the whole sky (magnitudes, color indices, spectral types) in the Kapteyn fields.

The study of the motions of the stars then leads further. We have already mentioned the spectroscopic measurement of *radial velocities* V (km/s) using the *Doppler effect* (H. C. Vogel, 1888; W. W. Campbell, and others)[1].

The *proper motions* μ of the stars on the celestial sphere, discovered much earlier by E. Halley (1718), have then to be considered. They are usually expressed in seconds of arc per year. We measure relative proper motions (referred to faint stars of small proper motion) by comparing ("blinking") two plates taken at an interval of 10 to 50 years apart, if possible with the same instrument. Reduction to absolute proper motions is achieved by using meridian instruments to measure absolute positions of certain stars at different epochs. Following C. D. Shane, remote galaxies and recently even quasars are used to furnish an extra-galactic system of reference.

The transverse component T (km/s) of stellar velocity is related to the proper motion μ as follows:

If p is the parallax of the star in seconds of arc, then μ/p is equal to T in astronomical units per year. This unit of speed is equal to $\pi/2$ times the orbital speed of the Earth and so is 4.74 km/s. Therefore

$$T = 4.74\mu/p \qquad \text{(km/s)} \tag{23.2}$$

and the space velocity of the star is

$$v = (V^2 + T^2)^{1/2}. \tag{23.3}$$

The angle θ between the space velocity and the sight line is determined by the relations

$$V = v \cos\theta \qquad T = v \sin\theta. \tag{23.4}$$

Since, for example, a star of the sixth magnitude has on the average a parallax of $0\overset{''}{.}012$ but a proper motion of $0\overset{''}{.}06$ per year, by using proper motions over, say, 20 years we can penetrate about 100 times further into space than by means of parallax measurements.

If we first make the simplifying assumption that the stars are at rest and that only the Sun is in motion relative to them with velocity $v_\odot$ in the direction of the apex, then we expect the distribution shown in Figure 23.2 of radial velocities V and transverse velocities T (or proper motions) with the angular distance λ of the star from the apex.

If the motions of the stars in space are randomly distributed, we could still apply the same considerations if we average over many stars. Thus we obtain from the proper motions and radial velocities of the stars in our neighborhood

$$\begin{array}{ll} \text{solar motion} & v_\odot = 20 \text{ km/s} \\ \text{toward apex} & \text{RA } 18^{\text{h}}00^{\text{m}},\ \delta = +30^\circ. \end{array} \tag{23.5}$$

[1]The sign is defined so that a positive radial velocity means a red shift, that is, recession from the Sun; a negative radial velocity means a blue shift, that is, approach toward the Sun. We reduce the values to the Sun as point of reference by removing the varying part that depends upon the orbital and rotational motion of the Earth.

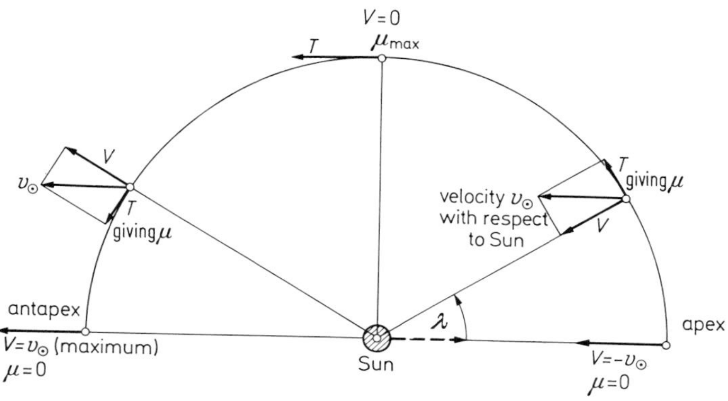

Figure 23.2. The solar motion, relative to the surrounding stars, with the velocity $v_\odot$ in the direction toward the apex. We observe the parallactic motion of the stars as a reflection of the solar motion. The diagram shows how the radial velocity V and the tangential velocity T (or the proper motion μ) of the stars depend on their angular distance λ from the apex.

Using a few proper motions, W. Herschel carried out the first determination of the solar apex as long ago as 1783. Later it was found that, to be precise, the solar motion depends upon which stars we use for its determination; this was the first indication of a systematic constituent of stellar motions. Equation (23.5) gives the so-called standard solar motion which we generally use in order to refer stellar motions to a local standard of rest.

We can now use the knowledge of the solar motion (23.5) in order to isolate the statistical part of the proper motions (the so-called *peculiar motions*), in the measured proper motions of a significant selected group of stars, from the reflection of the solar motion or the parallactic motion. The latter obviously depends upon the mean parallax $\bar{p}$ of the stellar group. Corresponding to (23.4), the part of the transverse velocity caused by the solar motion is $T = v_\odot \sin\lambda$, where λ again denotes the angular distance of the stellar group[2] from the apex. Accordingly we can now use equation (23.2) if, in averaging over the proper motions, we restrict ourselves to their components in the direction of the apex, the $\bar{\upsilon}$ components. The mean, or secular, parallax of the group of stars is thus

$$\bar{p} = -\frac{4.74\bar{\upsilon}}{v_\odot \sin\lambda}. \tag{23.6}$$

The hypothesis of statistically distributed peculiar motions is to be used with much caution.

In 1908 L. Boss discovered that, for instance, in the case of a comprehensive group of stars in Taurus, which arrange themselves around the Hyades galactic cluster, the proper-motion vectors as drawn on the sphere

[2]Here we assume that the stellar group occupies a relatively small region of the sky, so that a single mean value of λ suffices.

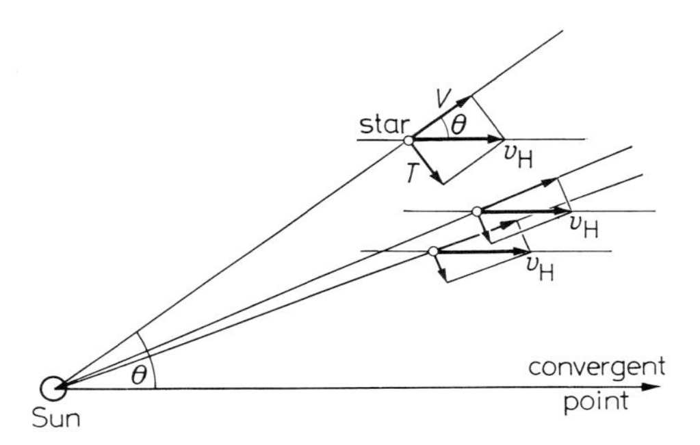

Figure 23.3. Moving cluster of star stream.

or in a star chart point toward a convergent point at $\alpha = 93°$, $\delta = +7°$. The members of this star stream, or moving cluster, thus describe parallel motions in space, like a shoal of fish, whose direction points toward the convergent point. Let the velocity of the cluster relative to the Sun be v_H. Now if we know (Figure 23.3) the proper motion μ and the radial velocity V relative to the Sun for a star of the cluster, and if θ is the angle on the sky from the star to the convergent point, then we can take over the notions in equations (23.2) and (23.4). We find

$$V = v_H \cos\theta \qquad T = v_H \sin\theta = 4.74\mu/p$$

whence we immediately obtain the parallax of the star

$$p = \frac{4.74\mu}{V \tan\theta}. \tag{23.7}$$

The Taurus cluster, for example, is about 42 pc away, and its velocity amounts to 31 km/s; most of its stars lie within a region of some 10 pc diameter. This method of stream parallaxes surpasses the method of trigonometric parallaxes in scope and often also in accuracy.

The common small proper motions of stars in galactic clusters (Pleiades, Praesepe, etc.) permit to some extent the determination of their parallaxes. However, they are important chiefly in order to test individual stars for membership of the cluster.

According to our current ideas, the diameter of the Galaxy is of the order 30 000 pc; the other galaxies are comparable structures at distances ranging from hundreds of thousands to thousands of millions of parsecs. The whole of the knowledge we have of the size and structure of the "new cosmos" depends essentially upon the methodology of photometric measurement of distance. According to the well known inverse-square law of photometry, we see a star of absolute magnitude M and parallax p, or of distance $1/p$ pc, to be of apparent magnitude m, where from equation (14.2) its distance modulus is given by

$$m - M = -5(1 + \log p). \tag{23.8}$$

Here we have not taken account of interstellar absorption. We shall return to this in the succeeding Section 24, but for the time being we shall quote values of distances, etc., in which the required correction has been made.

Ultimately we must always refer back to absolute magnitudes of particular objects, which have been derived using trigonometric, secular, moving-cluster . . . parallaxes. It was on account of their basic significance that we first explained the methods of geometrical distance determination so fully.

In order to gain insight into the structure of the Galaxy, it is well to start from aggregates of stars whose structure is immediately recognizable to the naked eye or in suitable photographs: *globular clusters* in which the stars seem to be crowded together like a swarm of bees, the less condensed *galactic* or *open clusters* of stars, stellar *associations,* and *galactic star clouds.* The classical position catalogs, for all "nebulae" which nowadays we distinguish on the one hand as galaxies and on the other hand as galactic nebulae and as planetary nebulae, are Messier's catalog (M) of 1784, and Dreyer's *New General Catalogue* (NGC) of 1890 with its continuation in the *Index Catalogue* (IC) 1895 and 1910.

(1) *Globular clusters.* The two brightest globular clusters, ω Centauri and 47 Tucanae, are situated in the southern sky. In the northern sky M 13 = NGC 6205 in Hercules is still a naked-eye object; its brightest stars are about $13^{m}_{\cdot}5$. Photographs with large reflecting telescopes (Figure 23.4) show in the brighter globular clusters more than 50 000 stars; in the center the individual stars cannot be distinguished from each other.

Figure 23.4. Globular cluster M 13 (NGC 6205) in Hercules. Distance 8 kpc.

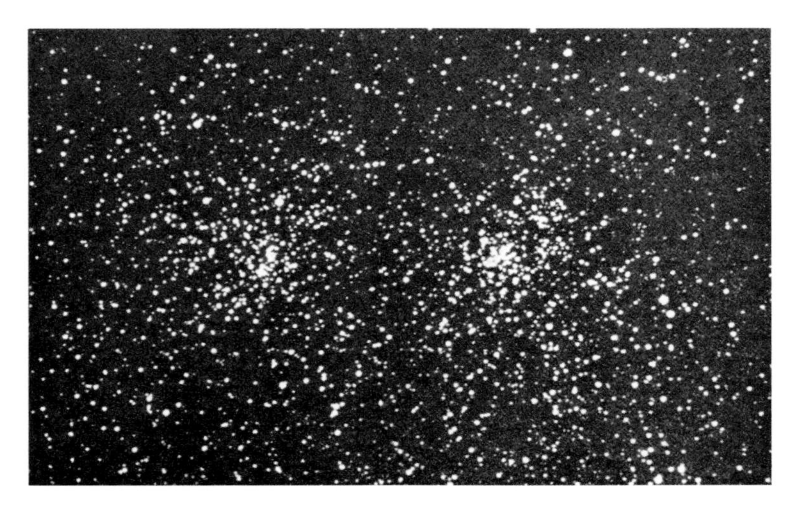

Figure 23.5. The galactic double cluster h and χ Persei

The globular clusters show a strong concentration in the sky in the Scorpio–Sagittarius direction.

(2) *Galactic or open stellar clusters* (Figure 23.5) occur in the sky right along the bright band of the Milky Way. Many are relatively rich in stars and contain hundreds of members, although still many fewer than the globular clusters; others are poor in stars with only a few dozen members. Also the concentration of the stars toward the center, the compactness of the cluster, varies a great deal. The best known are the Pleiades and the Hyades in Taurus, the double cluster h and χ Persei, and so on.

(3) The *OB associations* are relatively loose groups of bright O and B stars, which often surround a galactic cluster as, for example, the ζ Persei association surrounds the h + χ Persei cluster. On the other hand, the *T associations* are corresponding groups of T Tauri or RW Aurigae variables and other stars near the lower part of the main sequence. In 1947 V. A. Ambartsumian called attention to the cosmogonical significance of OB and T associations as very young systems.

(4) The *star clouds* of the Galaxy, for instance in Cygnus, Scutum, Sagittarius . . . are much more extended systems which we may tentatively regard as analogs of the bright "knots" in the arms of distant spiral galaxies.

About 125 globular clusters and 1000 galactic clusters are known. As well as H. Shapley's classical work "Star Clusters" (1930) we should mention especially the modern "Catalogue of Clusters and Associations" by G. Alter, J. Ruprecht, and V. Vanýsek (Budapest 1970).

We here conclude the first survey and again turn our attention to the cardinal problem of distance determination.

In 1918 H. Shapley produced a turning point in modern astronomy when he determined the distances of numerous *globular clusters* by means of the cluster variables (RR Lyrae stars) contained in them. As we have said, the essential difficulty here lay in deriving the absolute magnitudes, or in calibrating the period-luminosity relation. In the case of globular clusters that contained no variables, Shapley used as a secondary criterion the *brightest stars* of the cluster. In applying this, one purposely leaves aside the five brightest stars, as possible foreground stars; one takes as well defined the absolute magnitudes of the next brightest down to say the thirtieth. With due precautions, one may next employ as a criterion the *total brightness* of the cluster, or its angular diameter. From his observational material, which at that time concerned 69 globular clusters, Shapley was able to draw the conclusion that these form a slightly flattened system (Figure 23.6) whose center 10 kpc from us lies in Sagittarius (the distance

Figure 23.6. The Milky Way system. The diagram shows the spatial distribution of globular clusters projected on to a plane perpendicular to the galactic plane and passing through the Sun and the galactic center. Lines of equal mass density are shown, the numbers giving the density with respect to that in the solar neighborhood. The thin layer of interstellar matter lying in the galactic plane and associated with the extreme (spiral arm) population I is shown dotted. (After J. H. Oort.)

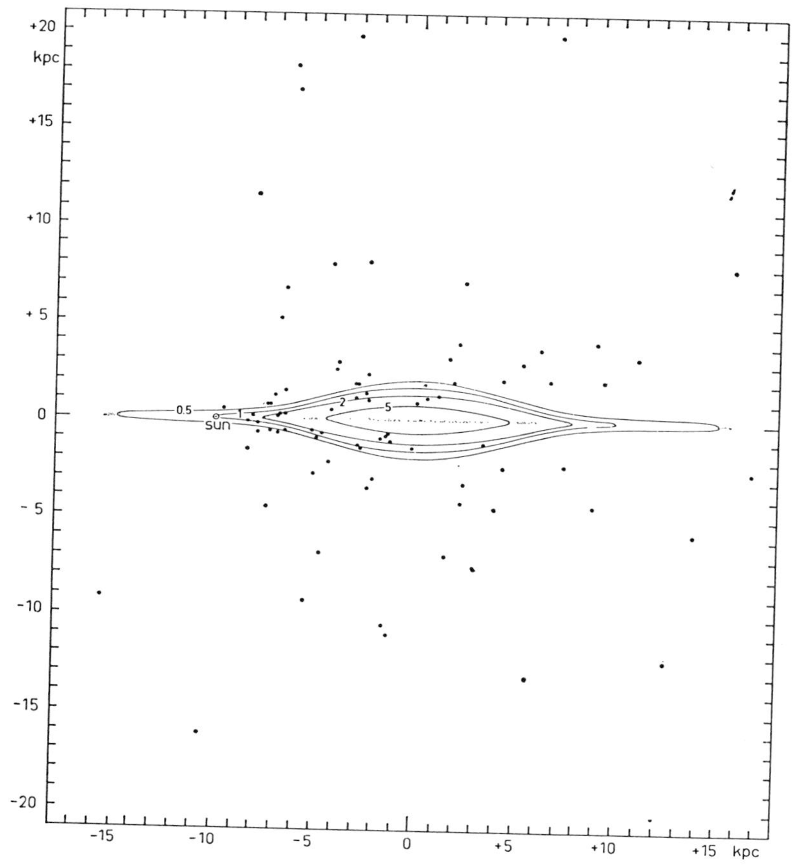

being at that time evaluated as 13 kpc). Thus the stage was set for all subsequent investigations concerning the Galaxy.

The galactic clusters that have been studied are much nearer to us than the majority of globular clusters. We can plot their color-magnitude diagrams, for example with $B-V$ color indices as abscissa and apparent magnitude m_V as ordinate. If we may assume that the main sequence is the same for all systems and for the so-called field stars of our neighborhood, which do not belong to any recognizable system, then the vertical separation between the $(B-V, m_V)$ diagram of a cluster and the $(B-V, M_V)$ diagram for our neighborhood yields directly the distance modulus $m_V - M_V$ and hence the distance of the cluster. (We shall return to consideration of finer differences between the color-magnitude diagrams and to the effect of interstellar absorption; the following results have been corrected for these effects.) We can only mention the fundamental work of R. Trümpler. In Figure 23.7 we show the distribution plotted by W. Becker (1964) of galactic clusters, that contain O to B2 stars as their earliest spectral types, together with H II regions, that is, bodies of hydrogen gas which have been

Figure 23.7. Distribution of young galactic clusters • (that is, those which still contain O-B2 stars), of H II regions ○ and of clusters with H II regions ⊙ in the plane of the Milky Way (after W. Becker). The galactic longitude $l^{\text{II}} = 0$ (below) defines the direction toward the galactic center, the Sun being situated at the center of the system of coordinates. All the objects included in the diagram arrange themselves along the *spiral arms;* they belong to extreme population I.

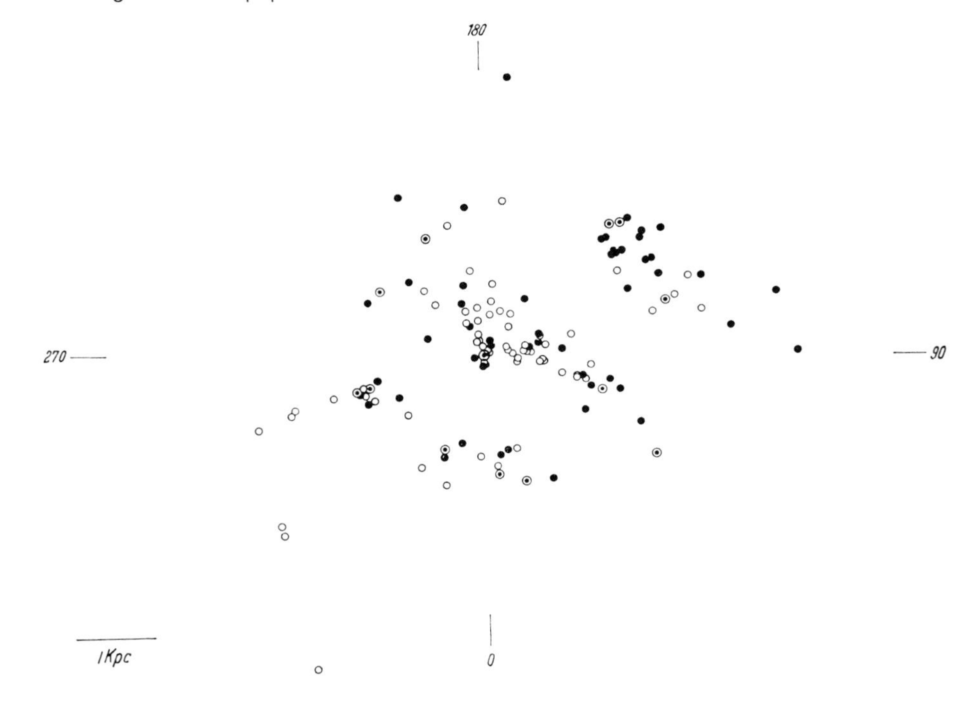

ionized and excited to give Hα emission by the radiation of the O or B stars in clusters or in OB associations or by single O or B stars (Section 24). We notice at a glance the arrangement into elongated regions in which we may perceive the neighboring portions of spiral arms.

At this point we may introduce a system of *galactic coordinates* that are useful in the study of the Milky Way, namely, the galactic longitude l in the plane of the Galaxy and galactic latitude b perpendicular to the plane, positive to the north and negative to the south. The older system l^{I}, b^{I} used the galactic north pole RA $12^{\mathrm{h}}\,40^{\mathrm{m}}$, $\delta + 28°$ (1900.0) and measured l^{I} from the intersection (ascending node) of the galactic plane with the celestial equator in 1900. An improved system of galactic coordinates was introduced in 1958 taking particular account of radio-astronomical observations:

$$l^{\mathrm{II}},\ b^{\mathrm{II}} \text{ with the galactic north pole RA } 12^{\mathrm{h}}\,49^{\mathrm{m}},\ \delta + 27\overset{\circ}{.}4 \ (1950.0) \tag{23.9}$$

where l^{II} is now measured from the

$$\text{galactic center RA } 17^{\mathrm{h}}\,43^{\mathrm{m}},\ \delta - 28\overset{\circ}{.}9 \ (1950.0).$$

In recent years the index II, which has gradually become superfluous, is often being omitted again. Old and new galactic longitudes (both of which are still used in similar work) in the vicinity of the galactic plane are therefore connected by the relation

$$l^{\mathrm{II}} \approx l^{\mathrm{I}} + 32\overset{\circ}{.}3 \quad \text{or} \quad l^{\mathrm{I}} - 327\overset{\circ}{.}7. \tag{23.10}$$

The "old" position of the galactic center $l^{\mathrm{II}} = 0$, $b^{\mathrm{II}} = 0$ is

$$l^{\mathrm{I}} = 327\overset{\circ}{.}7 \qquad b^{\mathrm{I}} = -1\overset{\circ}{.}4. \tag{23.11}$$

The Lund Observatory has published (in 1961) tables for the mutual conversion of equatorial coordinates and old, as well as new, galactic coordinates. We reproduce in Figure 23.8 the charts drawn by G. Westerhout for the transformation of the new galactic coordinates l, b or l^{II}, b^{II} into right ascension and declination α, δ for the epoch 1950.

We define the galactic components of the space velocity of stars (with positive senses)

U radially outward from the galactic center
V in the direction of galactic rotation ($l^{\mathrm{II}} = 90°$)
W perpendicular to the galactic plane toward the galactic north pole. (23.12)

It has to be stated whether the solar motion has been subtracted.

After these rather formal preliminaries, we turn to the kinematics and dynamics of the galactic system, as B. Lindblad and J. H. Oort developed them in their theory of the differential rotation of the Galaxy 1926–1927.

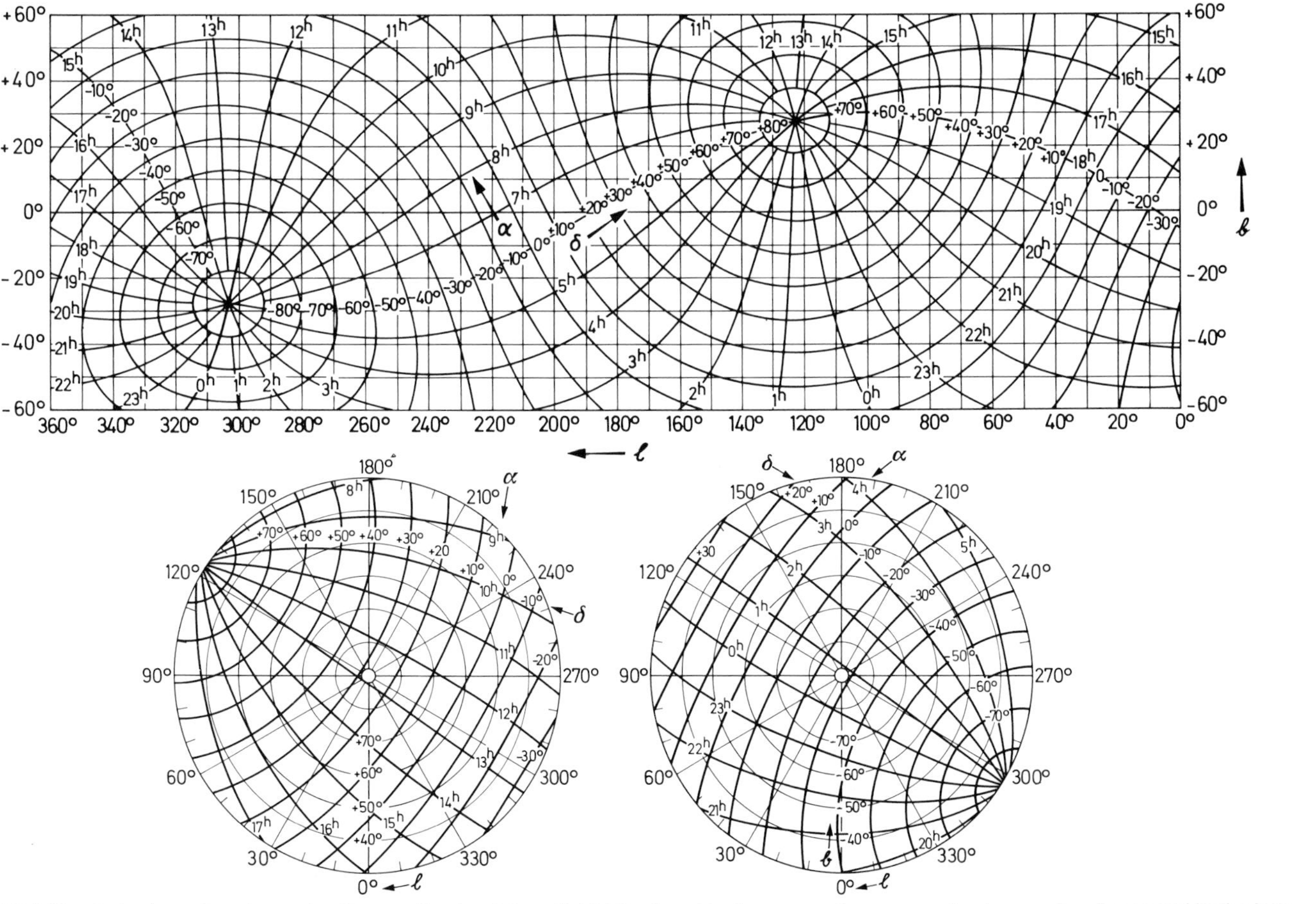

Figure 23.8 Charts for transforming galactic coordinates *l, b or* l^{II}, $\mathbf{b}^{\mathrm{II}}$ (displayed in the upper diagram as abscissa and ordinate and in the lower ones on the circumference and radius of the circle) into right ascension α and declination δ for epoch 1950, and vice versa. Upper diagram zone: centered on the galactic equator. Lower left: galactic north pole; lower right: galactic south pole. (After G. Westerhout.)

We first assume that all motions take place along circular orbits around the galactic center in the galactic plane (Figure 23.9). Let the angular velocity ω of a star P as a function of the distance R from the center be $\omega(R)$. For the Sun, let $R = R_0$ and $\omega(R_0) = \omega_0$, so that $\omega_0 R_0 = \upsilon_0$ is the velocity of the Sun in its galactic orbit. In what follows, we shall concern ourselves with the solar neighborhood, and we shall remove the solar motion (23.5) from all observations.

The velocity of the star P relative to the Sun is

$$R_0\{\omega(R) - \omega_0\} \tag{23.13}$$

as we see immediately from Figure 23.9 if we first think of P as held fixed and consider the motion of the Sun relative to P.

The corresponding radial velocity V_r of the star P (positive sign implies recession) is then

$$V_r = R_0\{\omega(R) - \omega_0\} \sin l^{\mathrm{II}}, \tag{23.13a}$$

where l^{II} is the galactic longitude. If the distance of the star from the Sun $\odot P = r \ll R_0$, we can write to a first approximation (the subscript 0 referring always to $R = R_0$)

$$V_r = -rR_0 \left(\frac{d\omega}{dR}\right)_0 \sin l^{\mathrm{II}} \cos l^{\mathrm{II}} = -\frac{1}{2} rR_0 \left(\frac{d\omega}{dR}\right)_0 \sin 2l^{\mathrm{II}}. \tag{23.14}$$

The corresponding transverse component T of the velocity of P relative to the Sun is

$$T = R_0\{\omega(R) - \omega_0\} \cos l^{\mathrm{II}}. \tag{23.15}$$

Thence we obtain the proper motion of P (positive in the sense of l^{II}) if we subtract from T/r the galactic rotation of the Sun ω_0. If we introduce the series expansion already used, we find in circular measure

$$-R_0(d\omega/dR)_0 \cos^2 l^{\mathrm{II}} - \omega_0 \tag{23.16}$$

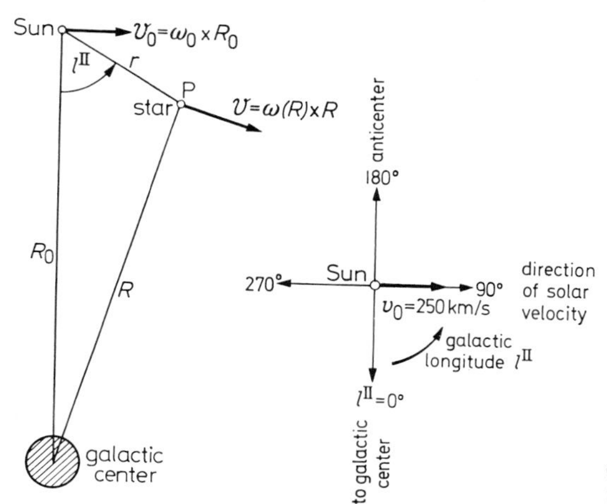

Figure 23.9

Galactic rotation and galactic longitude l^{II}.

or using $2\cos^2 l^{\mathrm{II}} = 1 + \cos 2l^{\mathrm{II}}$ and converting to seconds of arc,

$$\text{proper motion} = \frac{1}{4.74}\left\{-\frac{1}{2}R_0\left(\frac{d\omega}{dR}\right)_0 \cos 2l^{\mathrm{II}} - \frac{1}{2}R_0\left(\frac{d\omega}{dR}\right)_0 - \omega_0\right\}. \tag{23.17}$$

We call the coefficients in equations (23.14) and (23.17) the *Oort constants of differential galactic rotation*. We also express them in terms of the orbital velocity $\upsilon(R)$ where

$$\omega = \frac{\upsilon(R)}{R} \text{ so that } \frac{d\omega}{dR} = \frac{1}{R}\left(\frac{d\upsilon}{dR} - \frac{\upsilon}{R}\right) \tag{23.18}$$

and obtain

$$\left.\begin{aligned} A &= -\frac{1}{2}R_0\left(\frac{d\omega}{dR}\right)_0 = \frac{1}{2}\left\{\frac{\upsilon_0}{R_0} - \left(\frac{d\upsilon}{dR}\right)_0\right\} \\ B &= -\frac{1}{2}R_0\left(\frac{d\omega}{dR}\right)_0 - \omega_0 = -\frac{1}{2}\left\{\frac{\upsilon_0}{R_0} + \left(\frac{d\upsilon}{dR}\right)_0\right\} \end{aligned}\right\} \tag{23.19}$$

or

$$A + B = -(d\upsilon/dR)_0 \quad \text{and} \quad A - B = \upsilon_0/R_0. \tag{23.20}$$

Finally, the radial velocity V_r and the proper motion PM for our neighborhood, as functions of galactic longitude l^{II}, take the simple forms (see Figure 23.10)

$$V_r = Ar \sin 2l^{\mathrm{II}} \tag{23.21}$$

$$\mathrm{PM} = \frac{1}{4.74}\{A \cos 2l^{\mathrm{II}} + B\}. \tag{23.22}$$

After disposing of peculiar motions, observation confirms well the "double wave" ($\sin 2l^{\mathrm{II}}$) of the radial velocity and of the proper motion. While the amplitude of V_r is proportional to r, the proper motion is independent of r. After discussion of very extensive observational material, which we cannot review here, in 1964 the International Astronomical Union proposed the following values

$$A = 15 \text{ km/s/kpc} \quad B = -10 \text{ km/s/kpc}. \tag{23.23}$$

The distance of the Sun from the galactic center corresponds in the first place to that to the center of the system of globular clusters. However we can also apply the period-luminosity relation directly to RR Lyrae stars which lie in regions of the galactic center not too strongly obscured by dark clouds. With an uncertainty of about 15 percent we have

$$R_0 = 10 \text{ kpc}. \tag{23.24}$$

Then we obtain, using equation (23.20), the orbital velocity υ_0 and period P_0 of the Sun, or rather of the solar neighborhood, as

$$\upsilon_0 = 250 \text{ km/s} \qquad P_0 = 250 \text{ million years}. \tag{23.25}$$

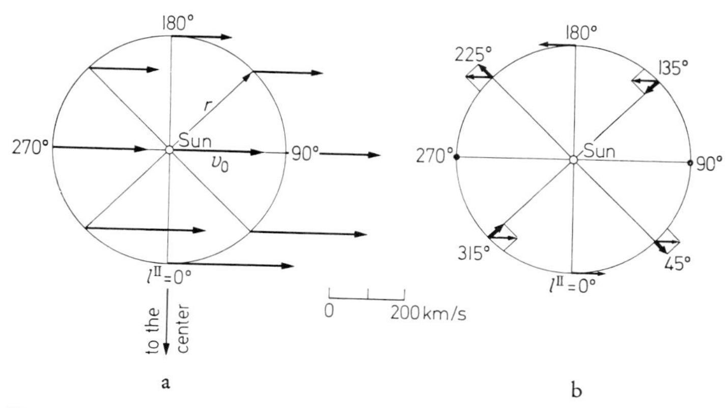

Figure 23.10 Differential galactic rotation. (a) Absolute velocities of stars at a distance r from the Sun. In the diagram $r = 3$ kpc; the lengths of the velocity vectors give the motions of the stars in 10 million years. (b) Velocities of the same stars with respect to the Sun and their radial components (thick arrows), the *double wave of radial velocities* given by equation (23.21).

Thus we have traveled once around the galactic system since the end of the carboniferous age.

Were the entire mass M, under whose influence the Sun describes its circular orbit, concentrated at the galactic center, then, as in the case of planetary motion, we must have from equation (6.35)

$$v^2 = GM/R. \tag{23.26}$$

In this way we estimate the mass of the Galaxy as

$$M = 2.9 \times 10^{44} \text{ g} = 1.5 \times 10^{11} \text{ solar masses.} \tag{23.27}$$

More exact calculations also lead to a total mass of the same order of magnitude. By logarithmic differentiation of equation (23.26) we get $dv/v = -\frac{1}{2}dR/R$ and thus from equation (23.20) $[(A - B)/(A + B)]_{\text{calc}} = 2$ while the observed rotation constants (23.23) lead to $[(A - B)/(A + B)]_{\text{obs}} = 5$. Thus the assumption of a $1/R$ potential can be only poorly satisfied[3].

What is then the situation in regard to our assumption up to now of circular galactic orbits? As soon as we admit appreciable eccentricities e of the stellar orbits, radial velocities of the order of 100 km/s and more make their appearance in the immediate vicinity of the Sun. As J. H. Oort remarked in 1928, we can in this way understand the phenomenon of high-velocity stars.

[3]The values of the constants $A = 19.5$, $B = -6.9$ km/s/kpc with $R_0 = 8.2$ kpc, which were preferred until a few years ago, giving $(A - B)/(A + B) = 2.1$ would have agreed much better with our certainly much too crude theoretical result!

If we limit ourselves first to orbits of stars in the galactic plane then their galactic velocity components U, V or the corresponding components relative to the solar neighborhood

$$U' = U \quad \text{and} \quad V' = V - 250 \text{ km/s}, \tag{23.28}$$

determine their galactic orbits. With sufficient accuracy, for the stars that are accessible to exact observation, we can set the positional coordinates equal to those of the Sun. In a diagram with coordinates U', V' we can then, for an assumed galactic force or potential field, plot curves of constant eccentricity e, curves of constant apogalactic distance R_1, and so on. In 1932 F. Bottlinger was the first to perform such calculations for a $1/R^2$ field of force. Figure 23.11 shows such a Bottlinger diagram for a field of force more appropriate to the actual Galaxy. The velocity vectors of the

Figure 23.11 Bottlinger diagram. The coordinates are the galactic velocity components U' (toward the anticenter) and V' (in the direction of rotation) with respect to the solar neighborhood. The absolute velocity components U and V may be read by reference to the coordinate axes. The orbital eccentricity e and orbital apogalactic distance R_1 in kiloparsecs may be read off from the two intersecting sets of curves. For later reference (Section 27) we note that the filled circles represent stars with an ultraviolet excess $\delta(U-B) > +0^m.16$, that is, metal-poor stars of the halo population II. These are high-velocity stars. Open circles denote stars with $\delta(U-B) < 0^m.16$. These stars represent the transition from the halo population II to the disk population, that is, the transition toward stars with more circular orbits. (After O. J. Eggen.)

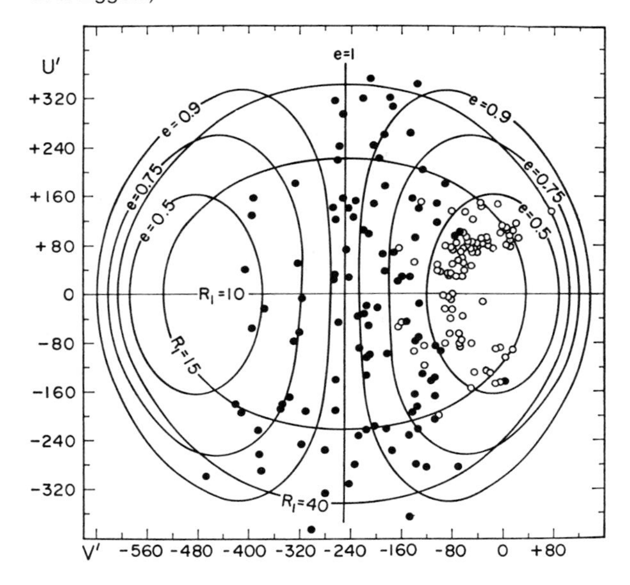

high-velocity stars demonstrate that these stars move in orbits of large eccentricity, partly with direct motion and partly with retrograde motion around the galactic center. On the other hand, the "normal" stars of our neighborhood have small U' and V', that is, like the Sun they all move with direct motion in nearly circular orbits.

As regards the motions of the stars of our neighborhood perpendicular to the galactic plane (the W components in equation (23.12), which generally used to be called Z components), according to J. H. Oort (1932 and 1960) we can largely explain them if we regard the distribution of mass density ρ in our region of the galactic disk as a plane stratified distribution. Thus we here take account of a dependence of mass density, etc., only upon the distance z from the galactic plane. Then we may first treat the W components of the stellar velocity vectors on their own account, independently of the motion parallel to the galactic plane (U and V). The stars perform perpendicular to the galactic plane oscillations about it with periods of about 10^8 years. The density distribution of stars at right angles to the galactic plane is related to the gravitational field of the galactic disk on the one hand, and to the velocity distribution of the W components on the other hand in a way analogous to that in which the density distribution of molecules in an atmosphere is related to the gravitational field and the Maxwell velocity-distribution, that is, the temperature. However, here in addition the gravitational field depends upon the mass density ρ through Newton's law of attraction (or through Poisson's equation). Therefore, Oort was able to derive the total density of matter in the galactic plane near the Sun; he obtained (1960)

$$\rho = 10.0 \times 10^{-24}\ \mathrm{g/cm^3} = 0.15\mathfrak{M}_\odot\ /\mathrm{pc}^3. \qquad (23.29)$$

A more recent discussion by R. Woolley and J. M. Stewart (1967) gave 0.11 $\mathfrak{M}_\odot$ /pc^3. According to W. Gliese (1956) the total density of observed *stars* within a region of 20 pc radius is 5.9×10^{-24} g/cm^3 to (taking somewhat stronger interstellar absorption) about 6.7×10^{-24} g/cm^3. The density of interstellar *gas* may correspond to about one hydrogen atom per cubic centimeter, that is, 1.7×10^{-24} g/cm^3. Thus dark material in the form of large fragments, whose presence we could be aware of only from Oort's method, need not play any very important part. The analysis of the W-velocity distribution again allows the two sorts of stars already mentioned to be distinguished: the disk stars with $|\bar{W}| \approx 12$ km/s and the high-velocity stars with $|\bar{W}| \approx 24$ km/s (but with the occurrence of considerably higher individual values). Thus while in fact disk stars describe almost circular orbits almost in the galactic plane, high-velocity stars mostly move in orbits of high eccentricity and of high inclination to the galactic plane.

In stellar dynamics also we again come across the fundamental concept of stellar populations introduced by W. Baade (1944). However it is better to postpone the extension of the survey given in Section 22 until we can immediately apply it in Section 27 to the stellar populations of different kinds of galaxies.

24. Interstellar Matter

Diffuse material between the stars of the Galaxy first entered the astronomers' awareness in the form of *dark clouds* whose absorption weakens and reddens the light from stars seen through them. However, only in 1930 was R. J. Trümpler able to show that, in addition to detectable dark clouds, interstellar absorption and reddening are by no means negligible anywhere in the Galaxy in regard to the photometric measurement of distances beyond a few hundred parsecs. In 1922 E. Hubble had already realized that (diffuse) galactic *reflection nebulae,* like those around the Pleiades, are produced by the scattering of the light of relatively cool stars by cosmic dust clouds, while in (diffuse) galactic *emission nebulae* interstellar gas is excited by the radiation of hot stars so as to emit a line spectrum. The study of the interstellar gas followed quickly in 1926–1927. Indeed, as long ago as 1904 J. Hartmann had discovered the stationary lines of Ca II which do not show any orbital displacement in the spectra of double stars. However, in 1926 A. S. Eddington from theory, and O. Struve, J. S. Plaskett, and others from the observations, first developed the concept that the interstellar lines of Ca II, Na I, etc., arise in a layer of gas which is partly ionized by stellar radiation, which fills the whole disk of the Galaxy, and which shares in the differential rotation. On another aspect, in 1927 I. S. Bowen succeeded in finding the long-sought identification of the "nebulium lines" in the spectra of the gaseous nebulae, as forbidden transitions in the spectra of [O II], [O III], [N II], etc., and H. Zanstra worked out the theory of nebular radiation. It was some ten years later that astronomers recognized *hydrogen* as being by far the most abundant element in the interstellar gas, as it is in stellar atmospheres. O. Struve and his collaborators first remarked that the assumption of large hydrogen abundance considerably diminished the quantitative difficulties in regard to the ionization of the interstellar gas. With the help of nebular spectrographs of great light-gathering power, they then discovered that many O and B stars, or groups of these, are surrounded by a fairly well-defined region which radiates in the red Balmer line $H\alpha$. So here interstellar hydrogen must be ionized. B. Strömgren then in 1938 developed the theory of these *H II regions*.

Neutral hydrogen—we speak of H I regions—at first seemed not to be directly observable until in 1944 H. C. van de Hulst calculated that the transition between the two levels in the hyperfine structure of the ground state of hydrogen must yield an emission line of interstellar hydrogen with a radio frequency and with measurable intensity. The transition is from $F = 1$ (nuclear spin parallel to the electron spin) to $F = 0$ (nuclear spin antiparallel to the electron spin) giving

$$\lambda_0 = 21.1 \text{ cm} \quad \text{or} \quad \nu_0 = 1420.4 \text{ MHz}.$$

Its observation for the first time in 1951—almost simultaneously at the Harvard Observatory, in Leiden and in Sydney—gave such an impetus to the study of interstellar matter and to galactic and extragalactic research, that, as against the historical development, we may begin here with an introduction to 21-cm radio astronomy.

We can calculate the absorption coefficient κ_ν of the 21-cm line from quantum mechanics. Its frequency dependence is entirely determined by the Doppler effect of the motion of the interstellar hydrogen. If there are $N(V)dV$ hydrogen atoms per cubic centimeter with radial velocity between V and $V + dV$ cm/s, then we find

$$\kappa(V) = \frac{N(V)}{1.835 \times 10^{13} T} \tag{24.1}$$

where T is the temperature that determines the velocity distribution of the hydrogen atoms and so also (as a result of collisions) the distribution of the atoms between the two hyperfine-structure levels. Then the intensity of the radiation emitted by an optically thin layer, in accordance with Kirchhoff's law, is

$$I(V) = \int \kappa(V) B_\nu(T)\, dl \tag{24.2}$$

where, since $h\nu/kT \ll 1$, the Kirchhoff–Planck function $B_\nu(T)$ is given sufficiently accurately by the Rayleigh–Jeans approximation (11.25)

$$B_\nu(T) = 2\nu^2 kT/c^2. \tag{24.3}$$

The integration is to be taken over the whole sight line.

The case of an optically thick layer arises in only a few places in the Galaxy. We recognize these by the fact that over a large range of frequency, or of velocity, the intensity reaches a constant maximum value

$$I(V) = B_\nu(T) = 2\nu^2 kT/c^2 \tag{24.4}$$

from which we can at once derive the temperature of the interstellar hydrogen

$$T \approx 125 \text{ K}. \tag{24.5}$$

Fortunately, a more exact value is of no great importance, since T cancels out of our formula (24.2) for an optically thin layer.

From equations (24.1) and (24.2), the measurement of the line profile $I(V)$ by means of a high-frequency spectrometer with a radio telescope gives the number of hydrogen atoms in a column of unit cross section along the sight line, the radial velocity V or frequency ν of which lie in the range

$$V \text{ to } V + dV \quad \text{or} \quad \nu_0 - \frac{V}{c}\nu_0 \quad \text{to} \quad \nu_0 - \frac{V + dV}{c}\nu_0. \tag{24.6}$$

The velocity distribution of interstellar hydrogen is made up of two contributions: First we have statistically distributed velocities, a kind of turbulence, whose distribution function is similar to Maxwell's. From the interstellar lines of Ca II, we already know the mean value to be about

6 km/s. The contribution of the differential rotation of the Galaxy is more important. After taking account of the solar motion, in our neighborhood, according to equation (23.21), the radial velocity V increases linearly with distance; its dependence on direction shows the characteristic double wave proportional to $\sin 2l^{\mathrm{II}}$. For greater distances from the Sun we must resort to the exact equation (23.13). From Figure 24.1 we see that the radial velocity V attains its maximum value V_m for a sight line in the direction of galactic longitude l^{II} ($|l^{\mathrm{II}}| < 90°$), such that the ray is tangent at D to a circle about the galactic center having radius $R_m = R_0 \sin l^{\mathrm{II}}$. Thus the line profile falls away steeply at V_m toward greater radial velocities. By combining the values of V_m for different galactic longitudes l^{II}, we can then derive the *rotational velocity* $\upsilon(R)$ as a function of distance from the galactic center. From this, conversely, we can calculate the distribution of radial velocity

Figure 24.1. Differential galactic rotation of interstellar hydrogen. The radial velocity $V(r)$ of the interstellar hydrogen with respect to the solar neighborhood, observed along a line of sight of galactic longitude l^{II}, is plotted as a function of the distance r. This velocity reaches a maximum V_m at D, where the line of sight just touches a circular orbit. The above illustration should be compared with Figure 23.10, which corresponds to the approximation $r \ll R_0$.

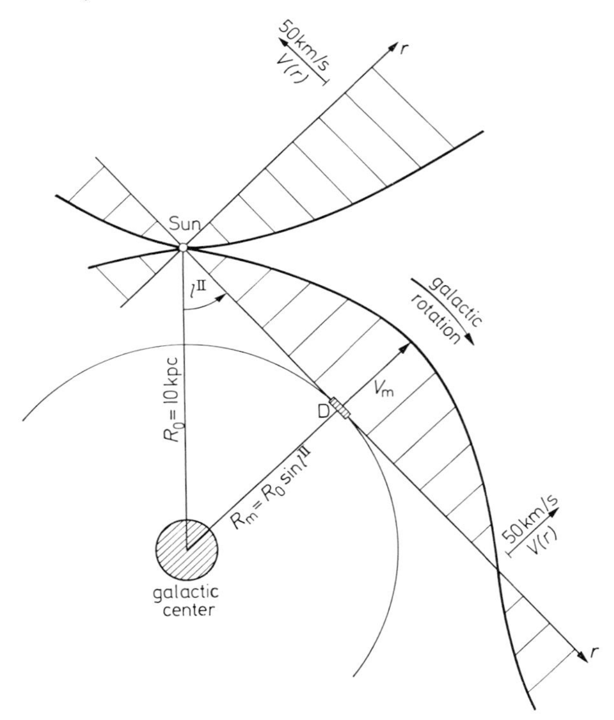

along each sight line and, using the resulting curves together with the measured line profiles, we can determine the density distribution of the hydrogen. For $|l^{\mathrm{II}}| < 90°$ the ambiguity as to whether a particular value of v belongs to the corresponding point in front of D or behind D can often be resolved by the consideration that the more remote object will in general show the smaller extension at right angles to the galactic plane, that is, in b^{II}. For the rest, in assessing the 21-cm measurements we must all the time have regard to the internal consistency of the resulting picture.

Figure 24.2 shows the distribution of neutral hydrogen in the galactic plane derived in this way from the observations of the Netherlands and Australian radio astronomers. We see clearly its concentration into spiral arms. More recent observations have shown that within a distance of 3 kpc from the galactic center there is a spiral arm that, surprisingly, expands away from the center at 50 km/s. Still further in, at ~ 0.7 kpc, there is the flat so-called *nuclear disk,* at the edge of which the rotation speed reaches a maximum of ~ 250 km/s. In this disk is located the actual nucleus of the

Figure 24.2. Distribution of neutral hydrogen in the galactic plane. Each point has been assigned the maximum density which would be seen in projection against the plane. Contours have then been drawn in accordance with the density scale (lower left) which gives the number of atoms per cubic centimeter. The galactic longitude l^{II} is indicated at the outer edge.

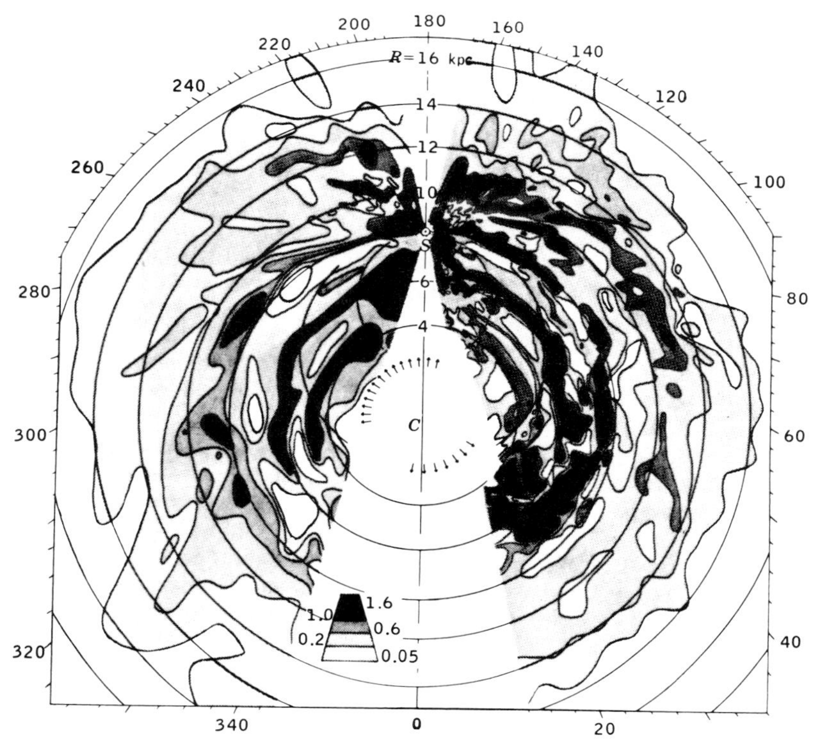

Galaxy, with a diameter of only a few parsecs. We shall come back later, in connection with the even greater activity in the nuclei of other galaxies, to the signs discovered by J. H. Oort and P. C. van der Kruit (1971) of earlier explosions in the core of our own Galaxy which about 10^7 years ago caused the ejection of about $10^7 \mathfrak{M}_\odot$ of material and the expansion of the 3-kpc arm.

The distribution of neutral hydrogen perpendicular to the galactic plane has also been studied. In the mean, it forms a flat disk. The distance between the surfaces, where the density has fallen to half its average value in the galactic plane, in the range $3 < R < 10$ kpc is about 220 pc; in the innermost 3 kpc the disk is even somewhat thinner.

In the surroundings of bright O and B stars, the interstellar gas, and particularly the hydrogen, is ionized and rendered luminous; we then see a diffuse nebula or an H II region. As H. Zanstra realized in 1927, this comes about as follows: If a neutral atom absorbs radiation from the star in the Lyman continuum $\lambda < 912$ Å it becomes ionized (see Figure 17.1). The resulting photoelectron is later recaptured by a positive ion (proton). Only in rare cases does recombination occur directly into the ground state; in most cases, cascade transitions occur through several energy levels with the emission of less energetic photons $h\nu$. As an estimate, we can say that, for every absorbed Lyman photon $h\nu > 13.6$ eV or $\lambda < 912$ Å, about one Hα photon is emitted. If we may now assume that the nebula absorbs practically all the Lyman radiation from the star, then following Zanstra we may work back from the observed Hα emission of the nebula to infer the Lyman emission from the star. If we compare this with its visual radiation, then from Planck's formula we can estimate the temperature of the star (or we may use the more exact theory mentioned in Section 18). We obtain values for the O and B stars that lie about in the range of the spectroscopically determined temperatures.

We may look at the same process from another standpoint: The number of recombinations per cubic centimeter is proportional to the number of electrons per cubic centimeter (N_e) times the number of ions per cubic centimeter making captures. However, since each hydrogen atom on ionization yields one electron, the latter number is also about equal to N_e. The Hα brightness at any particular place in the nebula must therefore be proportional to the so-called

$$\text{emission measure EM} = \int N_e^2 \, dr \tag{24.7}$$

defined by B. Strömgren. The integration is along the sightline, generally with r measured in parsecs. For diffuse nebulae, EM is of the order of some thousands or more; for weaker H II regions it is a few hundred. If we estimate the electron density N_e, taking the depth and width of the nebula to be about equal, we realize that in the H II regions the electron density is of the same order as the density of neutral atoms in H I regions (Figure 24.2), that is, $N_e \approx 10$ electrons/cm^3. In the large diffuse nebulae, like the Orion nebula, on the other hand, values $N_e \approx 10^4$ electrons/cm^3 are attained; so these are to be regarded as condensations in the interstellar gas.

In interstellar space, if an electron in its encounter with a proton is not captured but only deflected, then a free-free continuum is produced (Figure 17.1). This is too weak to be observed in the optical range but, in the range of centimeter and decimeter radio wavelengths, we observe it as thermal free-free radiation from diffuse nebulae. As we see at once, its intensity is again proportional to the emission measure EM. However, while optical observations in Hα are strongly obstructed by interstellar absorption (see below), especially near the galactic plane, radio waves pass through dark interstellar clouds without being absorbed. Taking this into account, the agreement between the emissivity as measured by optical astronomy and by radio astronomy is everywhere satisfactory.

Besides the free-free radiation of the H II regions, radio astronomers have discovered at centimeter and decimeter wavelengths transitions between neighboring energy levels in hydrogen and helium atoms with quantum numbers n in the range 100 to 200. These excited atoms must therefore be "inflated" to a radius $\sim a_0 n^2$, that is, about 1/1000 mm! By comparing the intensities of the corresponding transitions, it is possible to determine the abundance ratio H:He. It turns out that H:He $\approx$ 10 quite universally, in satisfactory agreement with recent analyses of B stars, which—as we shall see—emerged from the interstellar gas only 10^6 to 10^8 years ago.

More recently, thanks to progress in techniques of amplification in the millimeter to decimeter range, it has been possible to identify many kinds of *interstellar molecules* from their radio lines. They are mainly compounds of the abundant elements H, C, N, O, Si, S.

Table 24.1. Interstellar molecules with lines in the millimeter to decimeter range (optical identifications in italics).

H_2 *CH* OH H_2O NH_3 H_2S CN CO CS SO SiO	
as well as	
H_2CO (formaldehyde)	CH_3OH (methyl alcohol)
HCN (hydrogen cyanide)	HCOOH (formic acid)
HC_3N (cyanoacetylene)	NH_2CHO (formamide)
and many others.	

In some cases it is even possible to obtain from isotopic molecules the abundance ratios $C^{12}:C^{13}$ and $O^{16}:O^{18}$. Quantitatively, the understanding of the line intensities (in absorption or emission) is made more difficult by the fact that, by overpopulating the higher quantum states, the familiar *maser-amplification effect* can produce unexpectedly high intensities.

The occurrence of polyatomic organic molecules in interstellar space and of similar ones in carbonaceous chondrites of Type I has certainly nothing to do with some mysterious living creatures. It is probable that the concentration of gas and dust, as catalysts, in the neighborhood of nascent stars plays an important role in the creation of interstellar molecules. The formation of complicated organic molecules is favored by their large statistical weights or sums over states. Possibly organic molecules originate in the interstellar medium, just like those in the carbonaceous chondrites, by a kind of Fischer–Tropsch synthesis. In Section 31 we shall return to this problem, which is also very important in connection with the origin of organic life.

We also observe the free-free radiation of ionized hydrogen at centimeter and decimeter wavelengths in the whole extent of the galactic plane. We recognize it from the fact that, according to the theory, for emission from an optically thin layer the intensity I_ν (per unit frequency interval) is independent of the frequency. If we analyse the dependence of the measured intensity $I_\nu(l^{\mathrm{II}})$ upon the galactic longitude l^{II}, we can calculate the density of electrons, or of ionized hydrogen, N_e (taking account of its inhomogeneity) as a function of distance R from the galactic center. In Figure 24.3 we take the results obtained by G. Westerhout and compare them with the mean density distribution of neutral hydrogen (from Figure 24.2). According to this, the degree of ionization of the hydrogen has a pronounced maximum at $R \approx 3.5$ kpc, immediately outside the expanding spiral arm at $R = 3$ kpc.

Figure 24.3. Density distribution of ionized and neutral hydrogen in the galactic system. For the neutral hydrogen the mean number of atoms per cubic centimeter is plotted directly, whereas for the ionized hydrogen it is only possible to determine the emission measure [equation (24.7)] and hence the product nN, where n is the average space density and N is the average particle density within an H II region; N is about 5 to 10 particles per cubic centimeter.

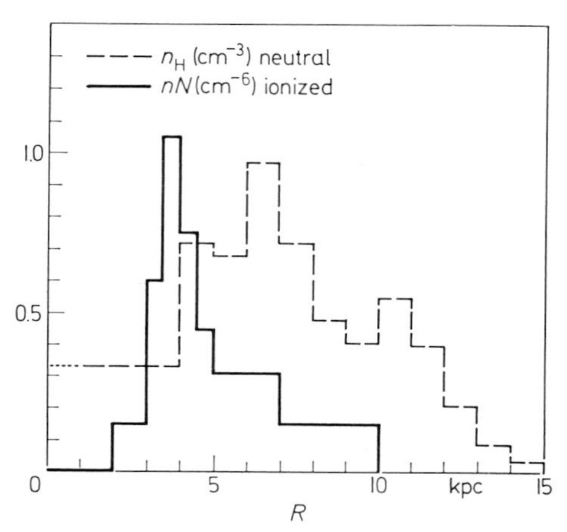

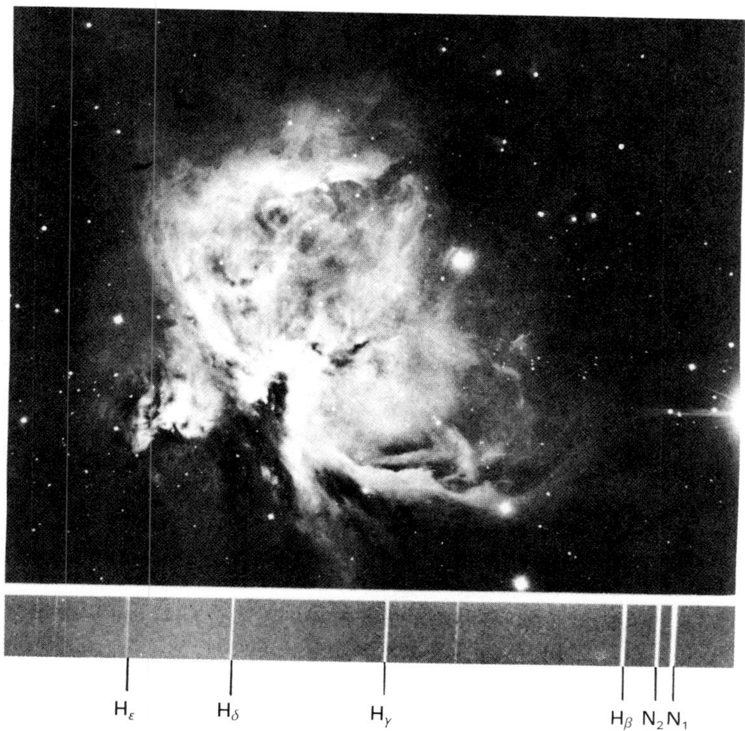

Figure 24.4. The Orion Nebula, a diffuse or galactic nebula, and its emission spectrum.

In the Galaxy there are altogether about 6×10^7 solar masses of ionized hydrogen and 1.4×10^9 solar masses of neutral hydrogen out of a total mass of about 10^{11} solar masses. The mean density of the hydrogen is about 0.6 atom/cm³ or 1.0×10^{-24} g/cm³. In our system, the 2 percent or so of interstellar matter plays no part having dynamical significance. However, we must beware of too hastily taking over this conclusion for other galaxies.

The optical spectra of gaseous nebulae (Figure 24.4) arise in conditions so far removed from thermodynamic equilibrium that a theoretical treatment of individual elementary processes becomes necessary. If r is the radius of the nebula and R that of the exciting star, the radiation from the star comes into operation only with a dilution factor $W = R^2/4r^2$. For example, with orders of magnitude $R \approx 1$ solar radius and $r \approx 1$ pc we should have $W \approx 10^{-16}$. Consequently in the atoms and ions only the ground states and long-lived metastable states (having extremely small probabilities of transition to lower energies) are appreciably occupied. According to I. S. Bowen (1928) the following processes therefore contribute to the excitation of the luminosity of the nebula:

(1) As we have seen, *hydrogen* and *helium*, the two most abundant elements, are ionized by the dilute stellar radiation in the Lyman

continuum. Recombination of ions and electrons then takes place into all possible quantum states. From these, for the most part in cascade transitions, the electrons finally fall back into the ground state. Thus we obtain the whole spectrum of H, He I, and possibly He II, as recombination radiation.

(2) *Permitted transitions* of many ions, such as O III, N III, . . . are excited by fluorescence. For example, in the recombination of He^+ (often as the final stage of a cascade process) the resonance line $1\,^2S - 2\,^2P$ λ 303.78 Å is emitted. As it happens, this can be absorbed by the O III ion in its ground state so that the 3d $^3P_2$ term is excited. From this, a whole set of O III lines are emitted, that are observed in the ultraviolet. However, from the term scheme we know further that the last of such a cascade of transitions is often given by the resonance line O III λ 374.44 Å. Again thanks to a fortuitous coincidence of term differences, this can excite a particular term of the N III spectrum, which then emits directly observable lines in the photographic region, and so on.

(3) *Forbidden transitions*. A great stir was caused in 1927 by I. S. Bowen's discovery that the lines λ 4958.91 and λ 5006.84 Å, which are strong lines in all nebular spectra and which astronomers had for a long time attributed to a mysterious element "nebulium," could be interpreted as "forbidden transitions" in the O III spectrum. These originate in transitions from a deep-lying metastable (long-lived) term in the ground term of the ion. While the usual permitted lines (dipole radiation) have transition probabilities of the order 10^8 s^{-1}, these are, for example, only 0.0071 or 0.021 s^{-1} for the [O III] nebular lines just mentioned.[1] In this case, we have to deal with magnetic dipole radiation (the analog to that of a frame antenna), in other cases with electric quadrupole radiation of the ion. The excitation of the metastable level results from collisions with electrons that have been produced by the photoionization of H and He. In such inelastic collisions the electrons obviously give up energy. Consequently the electron temperature is lower than we should at first have expected from the temperature of the star; it is usually about 8000 to 10 000 K.

In regard to the physics of their luminosity, but *not* to their cosmic status, the *planetary nebulae* (so-called because of their appearance) are related to the diffuse gaseous nebulae (Figure 24.5). The Zanstra method already explained gives for the central stars temperatures of 30 000 up to 150 000 K. The luminous envelopes. whose apparent sizes in the case of the nearest such objects are some minutes of arc, have radii of the order $\sim 10^4$ AU, electron density (for example, from the emission measure) $\sim 10^3$ to $\sim 10^4$ cm^{-3} and an electron temperature $\sim 10^4$ K. The observed splitting of the lines indicates an expansion of the envelope—the front part approaches

[1] Forbidden transitions are denoted by enclosing the symbol of the spectrum in square brackets.

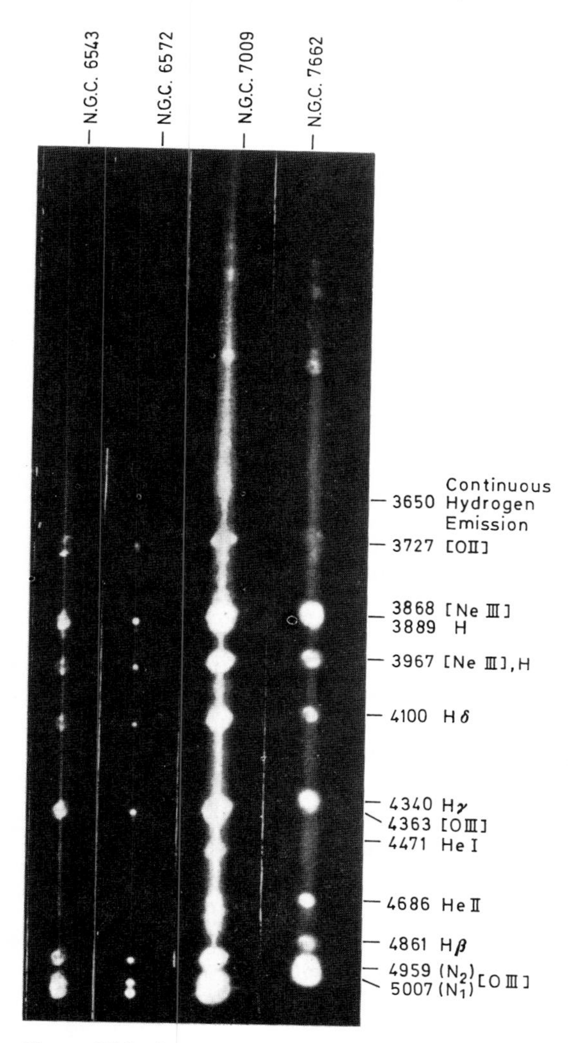

Figure 24.5. Spectra of planetary nebulae. These were taken with a slitless spectrograph, with the result that the distribution over the envelope of the light from various lines can be readily seen. In the case of NGC 6543 the continuous spectrum of the central star is also clearly visible.

us and the back part recedes—at about 20 km/s. R. Minkowski and others have found a strong concentration of planetary nebulae toward the galactic center; thus they belong to the (older) disk population of the Galaxy. Their absolute magnitudes (there being not much doubt as to their distances) show that in the HR diagram they represent the transition from hot bright stars to white dwarfs.

After this digression we return to further possible ways of studying interstellar matter. The discovery of the interstellar calcium lines Ca II λ 3933/3968 Å by J. Hartmann in 1904 gave the very first indication of the existence of interstellar atoms or ions. In fact, it was seen that in the double star δ Orionis the H and K lines did not share in the effects of orbital motion; so one spoke of "stationary lines." In the course of time interstellar lines have been discovered in the optical region that come from the following atoms, ions, molecules, and molecular ions:

$$\text{Na I, K I, Ca I, Ca II, Ti II, Fe I; CH, CN, CH}^+. \tag{24.8a}$$

By far the most intense are the H + K lines of Ca II and the D lines of Na I.

Using the satellite OAO-3 = Copernicus, a research group in Princeton discovered in 1972–1973 in the spectral region ~ 950–3000 Å interstellar lines of numerous atoms, mostly multiply ionized:

$$\begin{aligned}&\text{H I; C I, II; N I, II; O I; Mg I, II; Si II, III, IV;}\\&\text{P II; S I-IV; Cl II; Ar I; Mn II; Fe II.}\end{aligned} \tag{24.8b}$$

It is also possible to observe lines of the important molecules

$$H_2;\ \text{HD};\ \text{CO}. \tag{24.8c}$$

For all interstellar lines we are dealing with transitions from the ground term, in confirmation of our earlier considerations. In 1926, A. S. Eddington solved in its essentials the problem of the ionization of interstellar matter at great departures from thermal equilibrium. At such low pressures, ionization occurs obviously only through photoionization from the ground state, for example, in the case of neutral calcium having $\chi_0 = 6.09$ eV, through radiation with $\lambda < 2040$ Å. On the other hand, recombination occurs wholly through 2-body encounters between an ion and an electron. The degree of ionization is then such that the processes of ionization and of recombination balance each other.

If we now compare the interstellar lines, say, of Ca II with Ca I or Na I, the intensity ratio is, surprisingly, not very different from that in about an F star. How can we account for this? In interstellar gas, as in the stellar atmosphere, the stated processes must balance one another. However, on the one hand, in interstellar matter the electron density N_e is about 10^{16} times smaller than in the stellar atmosphere (5 compared with 2×10^{16} cm^{-3}) and the number of recombinations per ion per second is correspondingly smaller. On the other hand, we have already estimated that the radiation field in interstellar space is diluted by a factor $W \approx 10^{-16}$ compared with that in a stellar atmosphere, so that the number of ionizing processes per second is reduced by about the same factor. In interstellar space both processes therefore proceed about 10^{16} times more slowly than in a stellar atmosphere, but the degree of ionization remains about the same!

As J. S. Plaskett, O. Struve, and others showed, the equivalent widths W_λ of the interstellar lines increase in a fairly regular way with the distance r of the star in whose spectrum they are observed. Conversely, they can therefore be used in order to estimate the distance of suitable objects. Moreover, interstellar lines show Doppler displacements which correspond to about half the differential galactic rotation given by equation (23.21) for the star concerned. This is evidence that the interstellar gas occupies the space between the stars in the galactic plane more or less uniformly. Later observations by T. Dunham, W. S. Adams, and others, with better dispersion, have shown, however, that the interstellar calcium lines are often composed of several components, which can be associated with the various spiral arms of the Galaxy traversed by the light path. After allowing for local differences of ionization and excitation, it is found that the abundance distribution of the chemical elements in the interstellar gas and in diffuse gaseous nebulae is everywhere about the same as in the Sun. It is the same

Figure 24.6. The southern Milky Way. This picture shows the "Coal Sack," a local dark nebula in the constellation of the Southern Cross. On the far left, lying in the Milky Way, is α Centauri, the brightest star in the southern sky; the brightest globular cluster, ω Centauri, is seen in the upper left-hand corner about equidistant from the top and left-hand edges.

as that given by our analysis of the spectra of stars of the spiral arm and disk populations of our Galaxy (Table 19.1).

Spectroscopic observations in the optical, ultraviolet and radio regions have revealed to us [cf. equation (24.8) and Table 24.1] a large number of interstellar molecules. From the intensity of the H_2 bands in the extreme ultraviolet we can now estimate that about half the interstellar hydrogen is in atomic form and half is in molecular form.

The continuous absorption and reddening of starlight by *cosmic dust* discloses quite other aspects of interstellar material. Against the background of bright star clouds, especially in the southern Milky Way, we can even with the naked eye discern dark clouds such as the well-known "Coal Sack" in the Southern Cross (Crux), the "dark nebula" in Ophiuchus, etc. E. E. Barnard, F. Ross, M. Wolf, and others have obtained very beautiful photographs with relatively small, wide-angle cameras (Figure 24.6). They show a strong concentration of the dark clouds in the galactic plane. The well-known "splitting" of the Milky Way is obviously caused by an elongated dark cloud. Photographs of distant galaxies give a still clearer picture of the association of dark nebulae with spiral arms. Max Wolf first estimated the distances of certain dark nebulae with the aid of the diagram bearing his name: On a plate with photometric standards one counts the number of stars $A(m)$ in the apparent-magnitude range $m - \frac{1}{2}$ to $m + \frac{1}{2}$ per square degree within the region of the dark cloud and in one or more neighboring comparison fields. Were all stars to have the same absolute magnitude $\overline{M}$, then a dark cloud that causes an absorption of Δm magnitudes in the range of distance r_1 to r_2, that is to say in the range of reduced (that is, "absorption free") distance moduli $(m - \overline{M})_1 = 5 \log r_1/10$ to $(m - \overline{M})_2 = 5 \log r_2/10$, will diminish the star-numbers $A(m)$ in the manner shown schematically in Figure 24.7. Because of the actual dispersion of absolute magnitudes, the accuracy of the method is low, but it suffices to show that many of the conspicuous dark clouds are no more than a few hundred parsecs away. Probably even the great complexes in Taurus and in Ophiuchus are connected with each other past the Sun.

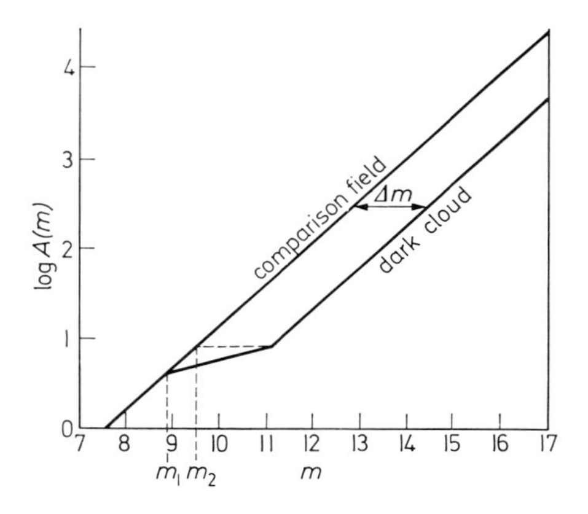

Figure 24.7
Wolf diagram for determining the distances of galactic dark clouds. The number $A(m)$ of stars per square degree in the magnitude range $m - \frac{1}{2}$ to $m + \frac{1}{2}$ is plotted as a function of m. The front and back faces of the cloud correspond to the mean stellar magnitudes m_1 and m_2, while the absorption of the cloud is Δm stellar magnitudes.

Bright diffuse nebulae with a continuous spectrum, the so-called *reflection nebulae* such as those that envelope the Pleiades, occur where bright stars with temperatures below some 30 000 K light up clouds of dust. On the plates we can often see directly the passage from dark clouds to bright nebulae.

In our Milky Way, as also in distant galaxies, the form of the dark nebulae gives the impression that structures of a few parsecs across have been deployed over a distance of a hundred parsecs or more in the direction of the spiral arm.

Although the absorption of starlight in extended, and often not sharply delimited, regions of dark nebulosity is readily noticed, it was not until 1930 that the knowledge of a *general interstellar absorption and reddening* became widespread. Such absorption has a large effect in the photometric measurement of great distances.

If a star of absolute magnitude M is at a distance $r = 1/p$ pc, then in the absence of interstellar absorption its apparent magnitude m would be given, according to equation (14.2), by what we now call, more precisely, the *true distance modulus*

$$(m - M)_0 = 5(\log r - 1), \tag{24.9a}$$

which is then simply a measure of distance. If on its way to us the light suffers an absorption of γ mag/pc, or altogether γr mag, we then obtain as the difference between the apparent magnitude that is actually measured and the absolute magnitude of the star the *apparent distance modulus*

$$m - M = 5(\log r - 1) + \gamma r. \tag{24.9b}$$

In order to bring out the significance of these important formulas, in Figure 24.8 we have illustrated the relation between the apparent distance modulus $m - M$ and the distance r, or the true distance modulus $(m - M)_0$, for $\gamma = 0$ (zero absorption) and absorption of $\gamma = 0.5$, 1.0, 2 mag/1000 pc. With absorption of 1 to 2 mag/kpc our view is practically cut off, as in a fog, at a few thousand parsecs.

R. Trümpler in 1930 gained the first well-founded ideas of the value of γ for average interstellar absorption in the galactic plane by comparing the ways in which angular diameter and luminosity fall off with distance in the case of open clusters. He was thus able to obtain a direct relation between geometric and photometric measurements of distance. Just as important was Trümpler's discovery that the absorption is always accompanied by a reddening of the starlight. Also more recent spectrophotometric investigations have confirmed that the interstellar reddening (change of color index) is always proportional to the interstellar absorption (change of apparent magnitude), which is to say that *the wavelength dependence of the interstellar absorption coefficient* is everywhere the same.

Away from discernable dark clouds, on average in the galactic plane we can reckon on absorption (visual) of $\gamma \approx 0.3$ mag/kpc; if we do not exclude the dark clouds we find up to 1 to 2 mag/kpc. As to the distribution of

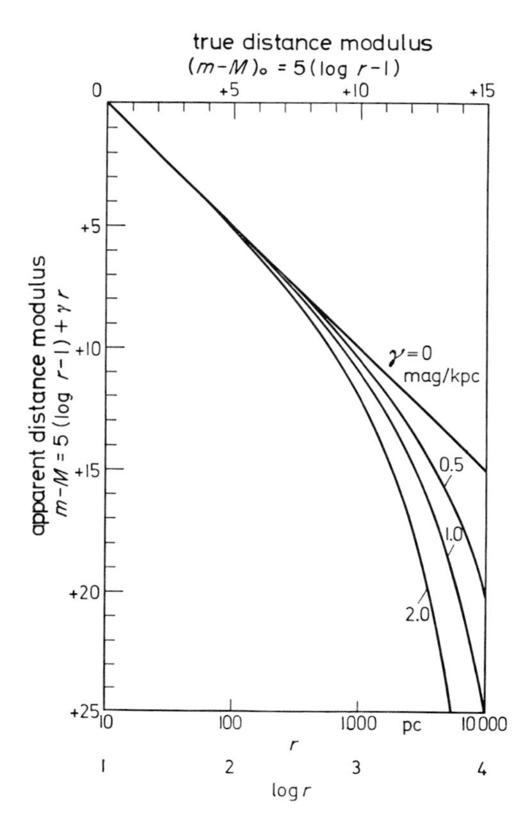

Figure 24.8
Relation between apparent distance modulus $m - M$ and distance r (parsecs) for stars with no interstellar absorption ($\gamma = 0$) compared with that for an (assumed uniform) interstellar absorption of $\gamma = 0.5$, 1 and 2 mag/kpc, respectively. The *true* distance modulus $(m - M)_0$ corresponds to the true distance, that is, to $\gamma = 0$.

absorbing material perpendicular to the galactic plane, that is, the dependence of γ upon galactic latitude b^{II}, we gain a picture from E. Hubble's discovery of the "zone of avoidance" (1934). Hubble studied the distribution over the sky of distant galaxies brighter than a given limiting magnitude. In the galactic polar caps their number per square degree $N(m)$ is almost constant. From 30° to 40° galactic latitude $N(m)$ falls off more and more steeply toward the galactic equator, in such a way that in the vicinity of this equator there exists a zone almost devoid of galaxies. Hence Hubble concluded that the absorbing material in the Galaxy forms a flat disk, in the middle plane of which we find ourselves with the result that extra-galactic objects suffer visual absorption of about $0^{\text{m}}.2$ cosec b^{II}. Observations of stars in our galactic neighborhood then show further that the entire half-value thickness of the absorbing layer in this neighborhood is about 300 pc. As was found later, this corresponds to about the thickness of the hydrogen gas.

Spectrophotometric measurements showed that in the photographic and visual region the dependence of the interstellar absorption or of γ upon the wavelength λ is well represented by the law $\propto 1/\lambda$. Hence, as well as from the photometry of objects of known color, we obtain as the relation on the average between, for instance, the weakening of the V magnitude A_V and the increase of the color index $B-V$, the so-called color excess $E_{B-V} = \Delta(B-V)$

$$A_V = (3.0 \pm 0.2)E_{B-V}. \tag{24.10}$$

The distribution of interstellar dust in the Milky Way is so irregular (cf. Figure 24.6) that for a particular star cluster or other object it is best to determine the *interstellar absorption* A_V directly. This can be achieved, for example, if we measure a color index (usually $B-V$) for stars (for example, bright B stars) whose absorption-free color index is known from nearby examples and then deduce the absorption from the color excess.

Since the color excesses for two color indices, for example $U-B$ and $B-V$, are proportional to each other ($\propto \lambda_{\text{eff}}^{-1}$), interstellar reddening displaces a star in the two-color diagram (Figure 15.5) along a straight *reddening line,* whose direction is indicated in the figure. If, for example, we know that a star belongs to the main sequence, we can return to the "main-sequence line" from the measured color indices $U-B$ and $B-V$ along a reddening line of the slope marked and read off both color excesses and the unreddened color indices of the star. From the color excess E_{B-V}, we also obtain immediately, using equation (24.10), the amount of (visual) interstellar absorption. This technique, which naturally permits several variants, is one of the most important aids in modern stellar astronomy.

In 1949 W. A. Hiltner and J. S. Hall made the astonishing observation that the light from distant stars is partially *linearly polarized* and that the degree of polarization grows about in proportion to the interstellar reddening E_{B-V} and so to the interstellar absorption A_V. The electric vector of the light waves (perpendicular to the conventional plane of polarization) vibrates preferentially parallel to the galactic plane.

If we denote the intensity of the light vibrating parallel and perpendicular to the plane of polarization by $I_{\|}$ and $I_{\perp}$, respectively, then the

$$\textit{degree of polarization } P = \frac{I_{\|} - I_{\perp}}{I_{\|} + I_{\perp}}. \tag{24.11}$$

Frequently one uses also the "polarization" Δm_P calculated in magnitudes:

$$\Delta m_P = 2.5 \log I_{\|}/I_{\perp} \approx 2.17P \quad (\text{for } P \ll 1), \tag{24.12}$$

as is easily verified.

The largest values of the degree of polarization are in the region of a few percent; as a function of wavelength, it shows a flat maximum at about 5500 Å. The interstellar polarization is correlated with the interstellar reddening E_{B-V} and absorption A_V; for example

$$\Delta m_P \leq 0.063 A_V. \tag{24.13}$$

The interstellar polarization indicates that the particles, which also give rise to the interstellar absorption and reddening, are *anisotropic,* that is, needle shaped or plate shaped, and partially aligned. L. Davis Jr. and J. L.

Greenstein attribute the orientation of the particles to a galactic magnetic field of about 5×10^{-6} to 2×10^{-5} G. The particles gyrate in this, keeping the axis of their principal moment of inertia parallel to the magnetic field lines, while the remaining components of motion are braked.

Figure 24.9, taken from data by D. S. Mathewson and V. L. Ford (1970), shows the preferred direction of the electric vector and the degree of polarization in percent for about 7000 stars. Other things being equal, the strongest polarization is obviously obtained where the lines of force run at right angles to the line of sight.

The *Zeeman effect in the 21-cm line* offers another possible way of measuring the interstellar magnetic field. The technique of measurement, which is very difficult, represents a direct extension to the decimeter-wave region of Babcock's procedure for measuring weak magnetic fields on the Sun.

Then we shall see that the nonthermal *synchrotron radiation* of many radio sources is polarized. Such measurements give information about the magnetic field at the point of origin of the radiation. However, the plane of polarization of the radio radiation is then rotated by its passage through an interstellar plasma with a magnetic field. This is just the familiar *Faraday effect,* whose theory shows that the rotation ψ of the plane of polarization is proportional to λ^2 times the so-called *rotation measure* RM, which is independent of wavelength:

$$\psi = \lambda^2 \text{ RM}, \text{ where RM} = 0.81 \int_0^L N_e B_{\parallel} \, ds. \qquad (24.14\text{a\&b})$$

Here ψ is in radians, the wavelength λ in meters, the electron density N_e per cubic centimeter, the longitudinal component of the magnetic field $B_{\parallel}$ in microgauss (10^{-6} G) and the distance L of the radio source and its increment ds in parsecs. From measurements at several wavelengths it is possible to find from equation (24.14a) the initial position of the plane of polarization and the rotation measure RM. If the electron density N_e "en route" is known, then equation (24.14b) can be used to obtain a correspondingly weighted mean value of $B_{\parallel}$.

In summary, there are available to us the following ways of investigating the interstellar magnetic field:

Method	Medium involved	Evidence about:
(1) Polarization of starlight	dust	$B_{\perp}$
(2) Zeeman effect in 21-cm line	neutral hydrogen	$B_{\parallel}$
(3) Synchrotron radiation	relativistic electrons	$B_{\perp}$
(4) Faraday rotation	thermal electrons	$B_{\parallel}$

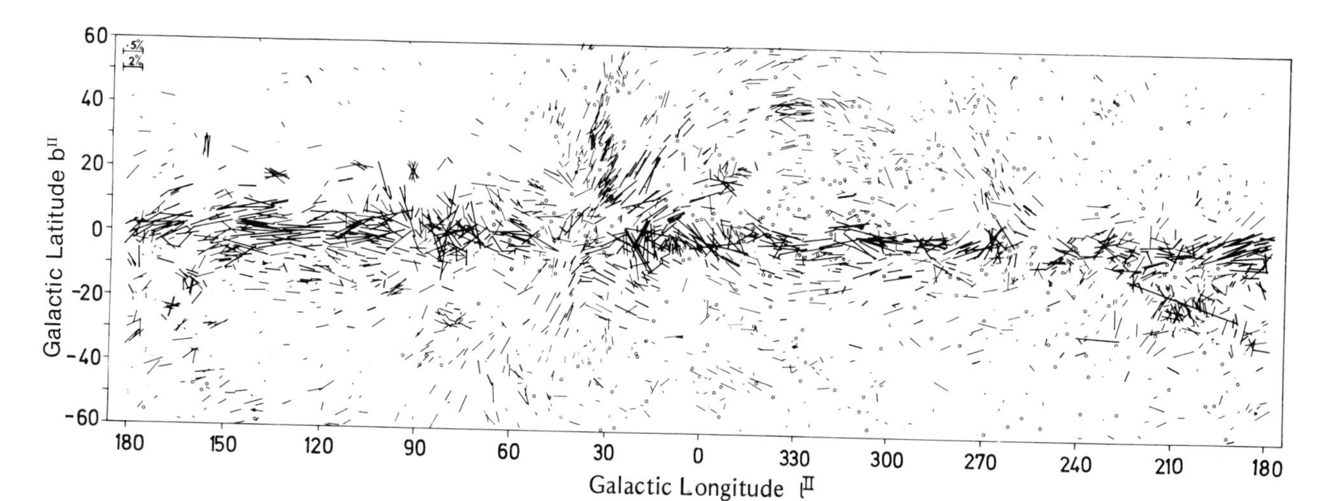

Figure 24.9. Interstellar polarization, displayed in galactic coordinates (after D. S. Mathewson and V. L. Ford, 1970). Each line is centered on a star and shows the direction of the electric vector of the optical polarization. The length of the line gives the value of the degree of polarization P; small circles denote stars with $P < 0.08$ percent. The scales for the degree of polarization (upper left-hand corner) are to be understood as follows: the upper scale is for stars with $P < 0.6$ percent (thin lines), the lower for stars with $P \geqslant 0.6$ percent (thick lines). Roughly speaking, we may regard our picture as analogous to the well-known experiment with iron-filings, which are scattered on a sheet of paper overlying a magnet.

Methods (1) to (4) yield different mean values, which correspond to the different spatial distribution of the indicated media and are not comparable without further information. It is therefore not surprising that attempts to construct a model of the galactic magnetic field have not yet led to a definite result.

After this excursion concerning interstellar magnetic fields, which will be very important with respect to our later discussion on the origin, propagation and storage of synchrotron electrons and cosmic ray particles in galaxies, let us turn back again to interstellar absorption in the optical region and raise the question of the physical nature of the interstellar dust.

Besides the known interstellar lines, whose narrowness makes them conspicuous in stellar spectra, P. W. Merrill discovered in 1934 several broad interstellar absorption bands. The strongest lies at λ 4430 Å with a total half width of 26 Å. Altogether about a dozen such bands are known. Because of their width, they can surely be caused only by colloidal particles or by large molecules. All detailed attempts at explanation still have a very speculative character. People have considered small metallic spheres with plasma oscillations or silicate grains (garnets) with transitions into the unfilled electron shells of the Fe^{+++} ion. An entirely different proposal attributes the interstellar bands to porphyrin molecules, noted for their great stability.

Eventually it may be possible to build a successful picture of the composition and structure of the interstellar dust which explains the observations presented above. The theory of the scattering and absorption of light by colloidal particles, which was begun by K. Schwarzschild and developed further by P. Debye, G. Mie, H. C. van de Hulst, and others, shows first of all:

Coarse particles (sand) $\gg \lambda$ absorb and scatter independently of the wavelength λ; according to Rayleigh's law, very fine particles (molecules, in particular) $\ll \lambda$ do so as λ^{-4}. Interstellar grains must therefore have diameters of the order of the wavelength of visible light, $\sim$ 6000 Å or 0.6 μm; their mass is then about 10^{-13} g. According to van de Hulst, the average density of interstellar dust amounts to about 1.3×10^{-26} g/cm^3, that is, only about 1 percent of all interstellar matter. The interstellar polarization indicates that the particles are anisotropic and partly aligned. Apart from that, however, theories with metallic or dielectric particles, with graphite platelets and hydride-flakes, with and without surface mantles are advocated with equal enthusiasm: we do not know. The anisotropic structure of the interstellar grains supports the view that they should not be regarded as counterparts of interplanetary meteor dust (powder), but according to J. Oort they should be ascribed to condensation (smoke) from the interstellar gas. Nevertheless, the latter idea has the difficulty that then the reverse process of destruction must also be exhibited, since otherwise we should not understand why at any rate not all the heavy atoms are condensed.

In conclusion, we try to construct a preliminary picture of the disposition and significance of interstellar matter in the Galaxy: The gas and dust form a flat disk, of which the thickness from $r = 3$ kpc to 10 kpc is about 200 pc. Within this disk, the gas and dust are concentrated toward the spiral arms; along the arms they form clouds, which are further subdivided in finer structures. In general along the arms the field lines of the interstellar magnetic field are frozen into the highly conducting plasma. The OB associations of bright blue stars and the younger galactic clusters are mostly embedded in diffuse nebulae, which we must regard as condensations of the interstellar matter. We shall see that these groups of stars have been formed out of the interstellar material. We shall support and extend these ideas in Sections 27–29 with the help of recent research on distant galaxies.

25. Internal Constitution and Energy Generation of Stars

As long ago as 1913 H. N. Russell was convinced as to the significance of his diagram for the study of stellar evolution. However, the interpretation of the diagram, and therewith a theory of stellar evolution based on observation, first became possible in conjunction with the study of the internal constitution of the stars. The older work of J. H. Lane (1870), A. Ritter (1878–89), R. Emden (whose *Gaskugeln* appeared in 1907), and others could be based essentially only upon classical thermodynamics. A. S. Eddington then succeeded in combining their results with the theory of radiative equilibrium and with the Bohr theory of atomic structure, which had meanwhile been developed. His book *The internal constitution of the stars* (Cambridge, 1926) formed the prelude to the whole development of modern astrophysics. We can grasp the basic ideas of Eddington's theory with only a very modest use of mathematics:

25.1 Hydrostatic equilibrium

In a star, consider a volume element having base 1 cm² and height dr, where r is the distance from the center of the star. Its mass $\rho(r) \times 1 \times dr$, where $\rho(r)$ is the density, is under the gravitational attraction of all the mass M_r inside a sphere of radius r. We have

$$M_r = \int_0^r \rho(r) 4\pi r^2 \, dr \quad \text{or} \quad dM_r/dr = 4\pi r^2 \rho(r). \tag{25.1}$$

According to the Newtonian law of attraction, this mass produces at its surface a gravitational acceleration GM_r/r^2, where G is the gravitation constant. Therefore the pressure changes over the depth of our volume element by the amount

$$-dP = \rho(r)dr \cdot GM_r/r^2 \tag{25.2}$$

that is force/cm² = mass/cm² × acceleration.

The *hydrostatic equation* for the star is therefore

$$dP/dr = -\rho(r)GM_r/r^2. \tag{25.3}$$

In most stars, P is practically equal to the *gas pressure* P_g; the radiation pressure P_r has also to be taken into account explicitly only in very hot and massive stars, so that $P = P_g + P_r$. Using equation (25.3) we can easily estimate the pressure at, for example, the center of the Sun ($P_{c\odot}$) if on the right-hand side we substitute for $\rho(r)$ the mean density $\bar{\rho}_\odot = 1.4$ g/cm³ and for M_r half the solar mass $M_\odot = 2 \times 10^{33}$ g. Then

$$P_{c\odot} \approx \tfrac{1}{2}\,\bar{\rho}_\odot GM_\odot/R_\odot = 1.3 \times 10^{15}\ \text{dyn/cm}^2 \text{ or } 1.3 \times 10^9\ \text{atm}. \tag{25.4}$$

A more exact calculation shows that equation (25.4) holds good exactly for a homogeneous sphere, and generally, assuming only that $\rho(r)$ increases monotonically inward, it gives a least value for P_c. More appropriate models give for the Sun a value of the central pressure about a hundred times larger.

25.2 Equation of state of the material

The relation, at each position, between pressure P, density ρ, and the temperature (absolute) T, which we must now introduce as the third parameter describing the state of the material, is expressed by the *equation of state*. Following Eddington, we start with the equation of state of an ideal gas

$$P_g = \rho \mathcal{R} T/\mu \tag{25.5}$$

where $\mathcal{R} = 8.317 \times 10^7$ erg/K mol is the universal gas constant and μ is the mean molecular weight. We may employ equation (25.5) so long as the interaction between neighboring particles is sufficiently small compared with their thermal (kinetic) energy. On the Earth, we are accustomed to this no longer holding good at densities of about $\rho \geqslant 0.5$ to 1 g/cm³. In stars this limit is raised a great deal by *ionization*. In particular, the commonest elements H and He are completely ionized in all stars below a relatively small depth.

The *mean molecular weight* μ for complete ionization is equal to the atomic weight divided by the number of all the particles, that is, nuclei plus electrons. Thus we have for

	Hydrogen	Helium	Heavy elements	
	$\mu = \frac{1}{2}$	$\frac{4}{3}$	$\sim 2.$	(25.6)

We can then readily calculate μ for any given mixture of chemical elements.

Using equation (25.5) we can estimate the mean temperature inside the Sun taking $\bar{P}_g = \frac{1}{2} P_{c\odot}$ according to equation (25.4), $p = \bar{\rho}_\odot$, and $\mu = \mu_H = 0.5$, obtaining

$$\bar{T}_\odot \approx 6 \times 10^6 \text{ K}. \tag{25.7}$$

25.3 Temperature distribution and energy transfer in stellar interiors

In order to calculate the temperature distribution $T(r)$ in the interior of the star, we must examine the nature of the energy transport. Inefficient transfer leads to a steep temperature gradient; efficient transfer leads to a low gradient. (Let anyone who questions this, use his fingers to hold, with one end in a flame, first a matchstick, and then a metal nail!)

Following K. Schwarzschild's investigations of the solar atmosphere, A. S. Eddington first considered *energy transfer by radiation,* that is, radiative equilibrium (see Section 18, also in regard to notation).

From the intensity of the radiation I we calculate the flux of radiation πF by multiplying by $\cos\theta$ and integrating over all directions. Denoting now the element of optical depth corresponding to mean mass-absorption coefficient κ by $d\tau = -\kappa\rho\, dr$ we obtain from equation (18.4) for the net flux of the total radiation

$$\pi F = -\int_0^\pi \frac{dI}{\kappa\rho\, dr} \cos^2\theta\, 2\pi \sin\theta\, d\theta. \tag{25.8}$$

In the stellar interior the radiation field is almost isotropic, so that on the right we can take out the integration over θ; it is

$$\int_0^\pi \cos^2\theta\, 2\pi\sin\theta\, d\theta = 4\pi/3. \tag{25.9}$$

Furthermore, we can here at once write for I the value given by the Stefan–Boltzmann law as a function of T, that is

$$\pi I = \sigma T^4. \tag{25.10}$$

We often write the radiation constant σ in the form $\sigma = ac/4$, where c is the light speed; with a so defined, the energy density of the blackbody radiation field is then simply aT^4 erg/cm^3.

Thus we first obtain, when written in full,

$$\pi F = -\frac{4}{3}\pi \frac{d}{\kappa\rho\, dr}\left(\frac{ac}{4\pi} T^4\right) \tag{25.11}$$

and the total radiative energy that flows outward per second across the spherical surface of radius r becomes $L_r = 4\pi r^2 \cdot \pi F$ or

$$L_r = -4\pi r^2 \frac{4}{3} ac \frac{T^3}{\kappa\rho} \frac{dT}{dr}. \tag{25.12}$$

At the surface of the star $r = R$, the quantity L_r becomes the observable luminosity L; see also equation (14.5). At temperatures of about 6×10^6 K the maximum of the Planck radiation curve $B_\lambda(T)$ is from equation (11.23) around $\lambda_{max} = 5$ Å, that is, in the X-ray region. Here the absorption coefficient is determined by the bound-free and free-free transitions of the atomic states that have not yet been fully "ionized away." In the mixture of elements such as that in the Sun and in stars of population I and the disk population, these are high stages of ionization of the more abundant heavy elements like O, Ne,. . . . In "metal-poor" subdwarfs we must also take account of the contributions made by H and He.

Besides energy transport by radiation, under the conditions discussed on page 193 in the case of the Sun, for example, a hydrogen- and helium-convection zone may also occur in the stellar interior. Other convection zones may arise also in connection with the nuclear generation of energy. According to L. Biermann, in all such convection zones (apart from boundary regions) energy transport by convection (the ascent of hot material and the descent of cooled material) exceeds by far that by radiation. The connection between the temperature T and the pressure P is then determined by the familiar adiabatic relation

$$T \propto P^{1-1/\gamma} \tag{25.13}$$

where $\gamma = c_p/c_v$ is the ratio of the specific heats at constant pressure and at constant volume. By logarithmic differentiation with respect to r we derive

$$\frac{1}{T}\frac{dT}{dr} = \left(1 - \frac{1}{\gamma}\right)\frac{1}{P}\frac{dP}{dr} \tag{25.14}$$

and thence the temperature gradient in a convection zone

$$\frac{dT}{dr} = \left(1 - \frac{1}{\gamma}\right)\frac{T}{P}\frac{dP}{dr}. \tag{25.15}$$

As we have already explained in Section 20, the results of the calculation of convective energy transfer are even today still very unsatisfactory in quantitative respects.

25.4 Energy generation in stellar interiors by nuclear reactions

Making use of the knowledge that, since the formation of the Earth some 4.5×10^9 years ago, the luminosity of the Sun has not changed significantly, J. Perrin and A. S. Eddington realized even in 1919–1920 that the hitherto considered mechanical or radioactive energy sources did not nearly suffice for supplying the radiation of the Sun. So they came to conjecture that in the interiors of the Sun and stars *nuclear energy* must be generated by the conversion or, as we also say, the "burning" of hydrogen into helium.

Using Einstein's relation (1905)

$$\Delta E = \Delta m\, c^2 \tag{25.16}$$

the total energy ΔE liberated by the union of four hydrogen atoms to give one helium atom, $4H^1 \rightarrow He^4$, can easily be calculated from the mass difference Δm between the initial and final states. The mass of four hydrogen atoms amounts to 4×1.008145 units of atomic weight (each equal to 1.660×10^{-24} g), while that of the helium atom is 4.00387 units. The difference corresponds to

$$\begin{aligned} 0.02871 \text{ units of atomic weight} &\approx 4.768 \times 10^{-26}\ \text{g} \\ &\approx 4.288 \times 10^{-5}\ \text{erg} \approx 26.72\ \text{MeV}. \end{aligned} \tag{25.17}$$

If we suppose that, for example, one solar mass of pure hydrogen were converted into helium, we should obtain a quantity of energy of 1.27×10^{52} erg, which could supply the present rate of energy output of the Sun $L_\odot$ for a life span of 105 billion years (billion = 10^9).

The rapid advance of nuclear physics in the 1930s enabled H. Bethe and others in 1938 to work out the nuclear reactions that are possible at temperatures of about 10^6 to 10^8 K in solar material and in other mixtures of chemical elements. Experimental investigations, particularly those of W. A. Fowler, of reaction cross sections at low proton energies have contributed very significantly to our knowledge of energy generation in stars. We collect the important reactions in accordance with the usual scheme

$$\text{initial nucleus (reacts with \dots, gives out \dots) final nucleus.} \tag{25.18}$$

We use the notation

$$\begin{array}{llll} \mathrm{p} & \text{proton} & \alpha & \mathrm{He}^{++} \text{ particle} \\ \beta^+ & \text{positron} & \mathrm{e}^- & \text{electron} \\ \gamma & \text{radiation quantum} & \nu & \text{neutrino.} \end{array} \tag{25.19}$$

By combination with a thermal electron e^-, an emitted positron β^+ immediately produces two γ quanta.

Because of their minute cross sections, neutrinos escape from stellar interiors without suffering collisions. We shall therefore write on the right-hand side following the equations of each reaction the "usable" energy liberated and in brackets the energy escaping in the form of neutrino radiation. The slowest reaction, which determines the time required for the reaction chain, is printed in heavy type.

We begin with the proton-proton or pp chain:

$$\mathbf{H^1(p,\beta^+\nu)D^2} \to \mathrm{D^2(p,\gamma)He^3} \begin{cases} \nearrow \mathrm{He^3(He^3,2p)He^4} \dots \\ \searrow \mathrm{He^3(\alpha,\gamma)Be^7} \begin{matrix} \nearrow \dots \\ \to \dots \end{matrix} \end{cases}$$

$$\left.\begin{array}{ll} \dots\dots\dots\dots\dots\dots\dots\dots\dots\dots & 26.21 + (0.51)\ \text{MeV} \\ \dots \mathrm{Be^7(e^-,\nu)Li^7} \to \mathrm{Li^7(p,\gamma)2He^4} & 25.92 + (0.80)\ \text{MeV} \\ \dots \mathrm{Be^7(p,\gamma)B^8} \to \mathrm{Be^8} + \beta^+ + \nu \to 2\mathrm{He^4} & 19.5\ \ + (7.2)\ \ \text{MeV} \end{array}\right\}. \tag{25.20}$$

As the temperature increases the lower branches become significant.

A second possibility for the conversion of hydrogen into helium is offered by a reaction cycle which H. Bethe and C. F. v. Weizäcker had studied earlier (1938). In it the elements, C, N, O take part and essentially determine the speed of the process, but at the end of the process they are quantitatively reinstated.

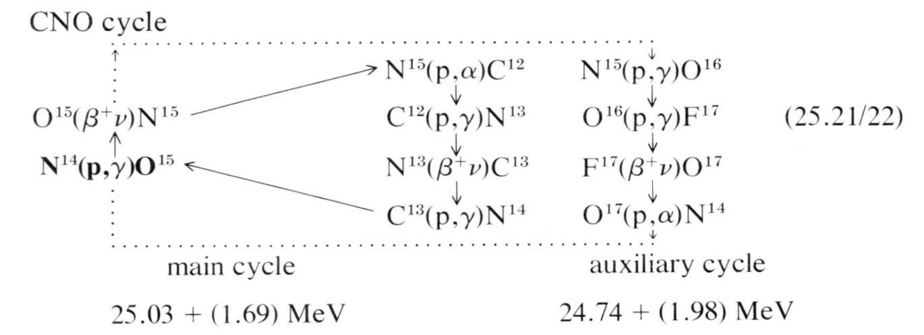

The energy generation comes principally from the cycle written on the left. In the Sun, for example, the cycle on the right is performed about 2200 times less often.

As we must remark here, in a steady state the relative abundances of the participating isotopes are determined by the speed of the individual reactions as in a state of "radioactive equilibrium." At temperatures of about 10^7 to 10^8 K the mass originally present as C, N, and O is transformed by the CNO cycle mostly into N^{14}. If the original matter has cosmic abundances (Table 19.1), the nitrogen abundance increases in these reactions by a factor of about 15. The abundance ratio of the carbon isotopes $C^{12} : C^{13}$, which amounts to ~ 90 in terrestrial and solar matter, adjusts itself to a much smaller value of ~ 4. This is an important indicator of (possibly earlier) hydrogen burning in the CNO cycle.

Again following W. A. Fowler, in Figure 25.1 the mean energy generation in erg/g s by "hydrogen burning" is shown on a logarithmic scale for stars with central temperatures T_c from 5 to 50 × 10^6 K and, somewhat

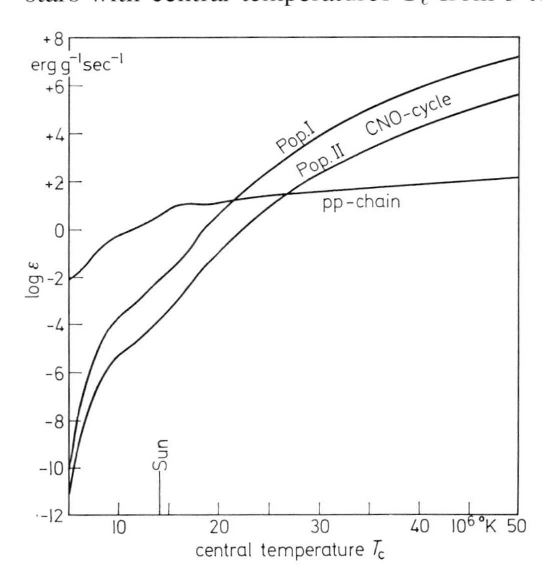

Figure 25.1
Rate of energy generation ε in stars of various central temperatures T_c, for a central density ρ_c = 100 g/cm³. The pp chain dominates in cooler stars such as the Sun, and the CNO cycle dominates in the hotter stars. (After W. A. Fowler, 1959.)

arbitrarily, a central density $\rho_c = 100\ \mathrm{g/cm^3}$. The age of the stars is assumed to be 4.5×10^9 years, so that the curves apply to the older population I and disk stars, and to younger population II stars. The composition by mass assumed for the former is

$$\begin{array}{llll} \mathrm{H} & \mathrm{C} & \mathrm{N} & \mathrm{O} \\ 50 & 0.3 & 0.1 & 1.2 \text{ percent (the remainder is He);} \end{array} \tag{25.23}$$

for population II the abundance of CNO is reduced by a factor 25. The most important result is: Cool stars with central temperatures up to $T_c = 21 \times 10^6$ K for population I or 27×10^6 K for population II (in particular also the Sun with $T_c \approx 13 \times 10^6$ K) produce their energy by the pp chain; hotter stars (in the upper part of the main sequence) produce it by the CNO cycle.

In the interior of a star, suppose that a considerable part of the hydrogen has been used up and that, as a result of contraction, the temperature has risen to more than 10^8 K. As E. J. Öpik and E. E. Salpeter remarked in 1951–1952, *helium burning* then sets in first to give C^{12} in accordance with the triple-α process:

This begins with the slightly endothermic combination of two helium nuclei

$$\mathrm{He}^4 + \mathrm{He}^4 + 95\ \mathrm{keV} = \mathrm{Be}^8 + \gamma. \tag{25.24}$$

The Be^8 is present in thermal equilibrium, though with a minute concentration, and is then further built up by the reaction

$$\mathrm{Be}^8(\alpha,\gamma)\mathrm{C}^{12} \text{ with an energy-release of 7.3 MeV per } \mathrm{C}^{12} \text{ nucleus.} \tag{25.25}$$

Reckoned per He atom, this burning yields only 2.46 MeV, that is, about 10 percent of the energy released in forming the atom. Starting from here, the following nuclei, having mass numbers that are multiples of four, can be built up by further (α,γ) reactions with comparable energy-production:

$$\mathrm{C}^{12}(\alpha,\gamma)\mathrm{O}^{16}\ 7.15\ \mathrm{MeV};\ \mathrm{O}^{16}(\alpha,\gamma)\mathrm{Ne}^{20}\ 4.75\ \mathrm{MeV};\ \mathrm{Ne}^{20}(\alpha,\gamma)\mathrm{Mg}^{24}\ 9.31\ \mathrm{MeV}. \tag{25.26}$$

Ne^{20} can also lead to the reaction-sequence

$$\mathrm{Ne}^{20}(\mathrm{p},\gamma)\mathrm{Na}^{21} \rightarrow \mathrm{Na}^{21}(\beta^+\nu)\mathrm{Ne}^{21} \rightarrow \mathrm{Ne}^{21}(\alpha,\mathrm{n})\mathrm{Mg}^{24}. \tag{25.27}$$

The last reaction appears suitable for providing neutrons (n) for the synthesis of heavy elements, which, because of the strong coulomb fields, can obviously not be achieved under any circumstances by charged particles.

The application to the theory of the internal constitution of the stars is basically entirely simple: Let ϵ be the energy generated per second per gramme of stellar matter by all the nuclear reactions proceeding at the values of the temperature T and density ρ concerned. Therefore in a spherical shell r to $r + dr$ energy $\rho\epsilon\ 4\pi r^2 dr$ erg/s is generated, and the energy flux L_r [see equation(25.2)] increases in accordance with the equation

$$dL_r/dr = 4\pi r^2 \rho\epsilon. \tag{25.28}$$

25.5 Summary: the fundamental equations of the theory of the internal constitution of the stars and general consequences

For a better grasp, we bring together the equations of the theory of the internal constitution of the stars from sections 25.1 to 25.4; today their solution is normally obtained with the use of a large electronic computer. To these are added the equation of state (section 25.2) and the (in practice very complicated) equations that relate the parameters of the material, ϵ and κ or γ, to two of the state parameters P, T, ρ. As is important for what follows, all these relations depend essentially upon the chemical composition of the matter.

Hydrostatic equilibrium under the influence of self-gravitation (25.3)

$$dP/dr = -\rho G M_r/r^2$$

and (25.1)

$$dM_r/dr = 4\pi r^2 \rho$$

energy generation (25.28)

$$dL_r/dr = 4\pi r^2 \rho \epsilon$$

energy transfer — radiation (25.12)

$$\frac{dT}{dr} = -\frac{3}{4ac}\frac{\kappa\rho}{T^3}\frac{L_r}{4\pi r^2}$$

energy transfer — convection adiabatic (25.15)

$$\frac{dT}{dr} = \left(1 - \frac{1}{\gamma}\right)\frac{T}{P}\frac{dP}{dr}. \qquad (25.29\text{a–e})$$

Finally, our problem is fully determined by the boundary conditions:

(a) In the center of the star, we must obviously have

$$\text{for } r = 0\text{: } M_r = 0 \text{ and } L_r = 0. \qquad (25.30)$$

(b) At the surface of the star, the equations for the stellar interior must in principle pass over into those of the theory of stellar atmospheres already discussed.

So long as we are interested *only* in the internal structure, we can use the above equations up to $r = R$, say, where $T \to 0$. We can make some general consequences of the theory intuitively clear; obviously we can also deduce them by formal calculations.

Consider a mass of gas M furnished with given energy sources. This system, which at first forms an indefinite picture as regards its arrangement in space, we now consolidate in imagination into a star with mass M and luminosity L. Provided a stable configuration is indeed possible, this will adjust itself to a definite radius R. On the other hand, L is related to the radius R and the effective temperature T_{eff} by

$$\begin{aligned} L &= 4\pi R^2 \sigma T_{\text{eff}}^4 \\ \text{luminosity} &= \text{stellar surface} \times \text{total flux}. \end{aligned} \qquad (25.31)$$

So the effective temperature of our star is fixed. Thus for stars having the same structure and composition (which we must not overlook), so-called homologous stars, a unique relation between mass M, luminosity L, radius R (or effective temperature T_{eff}) is satisfied

$$\varphi(M,L,T_{\text{eff}}) = 0. \tag{25.32}$$

A. S. Eddington discovered such a relation in 1924. According to his calculations, the dependence of the function φ on T_{eff} is so weak that he was able to speak simply of the *mass-luminosity relation*. At first, the agreement with observation (Figure 16.2) seemed to be really good; later many exceptions were found, but in the light of the general theory these are not at all unexpected.

Furthermore, we take into account the fact that, from the theory of nuclear energy generation, ϵ is determined as a function of the state parameters (for example, T and ρ). This produces a further relation between the three quantities M, L, T_{eff}. For stars in a steady state, having the same structure, we have therefore an equation of the form

$$\Phi(L,T_{\text{eff}}) = 0. \tag{25.33}$$

This implies that in the HR diagram, or in the color-magnitude diagram, these stars lie on a definite line. This assertion is often called the Russell–Vogt theorem. In fact, we shall become acquainted with a line of this kind in the color-magnitude diagram which forms the so-called *zero-age main sequence*. On the other hand, the presence of red giants and of supergiants then shows that already at least one further parameter comes into play. As we shall see, this is the chemical composition of the star, which changes with the age. Simultaneously the internal structure of the star changes as well and our previous considerations are no longer directly applicable.

26. Color-magnitude Diagrams of Galactic and Globular Clusters and Stellar Evolution

Our current ideas about the origin, evolution, and fate of stars originate in the study of color-magnitude diagrams of stellar clusters. Here we have groups of equidistant stars. From the magnitudes and colors measured photoelectrically with high accuracy, the subtraction of a common distance modulus and a single correction for interstellar absorption and reddening leads to the values of the (most commonly employed)

true absolute magnitude $M_{V,0}$

and true color index $(B - V)_0$.

As we saw, the color-magnitude diagram is basically equivalent to the HR diagram. However, while for very faint stars, down to about 22^{m}, it is still possible accurately to measure color indices, it is no longer possible to photograph spectra that can be classified.

To fix the ideas we once again briefly summarize the important properties of both sorts of cluster:

26.1 Galactic or open clusters

Their distances are determined by comparison of proper motions μ and radial velocities V [see equation (23.7)], or by using the method of spectroscopic parallaxes, or by photometric comparison of stars in the lower part of the main sequence with corresponding stars of a standard cluster (usually the Hyades) or with stars of our neighborhood. Galactic clusters (Figure 23.5) contain from a few dozen to several hundred stars, their diameters are of the order of 1.5 to 20 pc; they are always found (Figure 23.7) in or near the spiral arms of the Galaxy and, like other objects belonging to population I, they move in nearly circular paths about the galactic center. We know some 400 galactic clusters; taking account of regions of the Galaxy that are very remote or obscured by dark clouds, we estimate the total number to be about 20 000. Closely related to the galactic clusters are the looser moving clusters, as well as the OB and T associations mentioned in Section 23.

Loose clusters, in which the star density is not much greater than in the region round about, will disperse even on kinematic grounds after about one galactic revolution ($\sim 2.5 \times 10^8$ years). More compact clusters will be gradually dissipated by the gravitational fields of gas and star clouds. All the same, detailed calculations allow the Pleiades, for example, which is still a fairly compact cluster, a life of some 10^9 years.

26.2 Globular clusters

In our galactic system we know something over 100 globular clusters. We have already discussed at length the photometric determination of their distances with the help of cluster variables or RR Lyrae stars. A typical globular cluster (Figure 23.4) contains within a region of about 40 pc diameter several hundred thousand stars, so the mean star density is about 10 times larger than in galactic clusters. Toward the center of the cluster the star density increases so strongly that the night sky must be really bright! The absolute magnitude of a globular cluster is around -8. Analogously to known results in the kinetic theory of gases, we can estimate the total mass of a globular cluster from the dispersion in the radial velocities of its stars. In this way we get values of a few times $10^5 M_\odot$. Since the globular clusters travel round the galactic center in elongated elliptic paths, they pass through the galactic plane about every 10^8 years. Because of their compact structure, the resulting "jolt" has in general no significant effect.

After these preliminary remarks, we turn to the color-magnitude diagrams, first of galactic and then of globular clusters.

26.3 Color-magnitude diagrams of galactic clusters

We mention briefly the pioneering work of R. Trümpler in the 1930s on the HR diagrams, etc., of galactic clusters. In the following we rely upon the more recent investigations made photoelectrically, or at any rate using photoelectric luminosity scales, by H. L. Johnson, W. W. Morgan, A. R. Sandage, O. J. Eggen, M. Walker, and others. In deciding if any particular star is a member of a nearby cluster, we are guided by proper motions and maybe radial velocities; progress is here naturally not so rapid.

In Figures 26.1a and b we show first the actual observational material for Praesepe and NGC 188. Then in Figure 26.2 [after A. Sandage and (in part) O. Eggen] there is a collection of the color-magnitude diagrams (M_V versus $B-V$) of the galactic clusters NGC 2362, h + χ Persei, Pleiades, M 11, NGC 7789, Hyades, NGC 3680, M 67 and NGC 188. The lower parts of the main sequences (up to about the Sun G 2) can be brought into coincidence

Figure 26.1a. Color-magnitude diagram of Praesepe (after H. L. Johnson 1952). Apparent magnitude V plotted against $B-V$. Distance modulus for this cluster is $6^m.2 \pm 0.1$. The stars lying about 1 magnitude above the main sequence are most probably binaries. The intrinsic scatter of magnitudes on the main sequence is less than $\pm 0^m.03$.

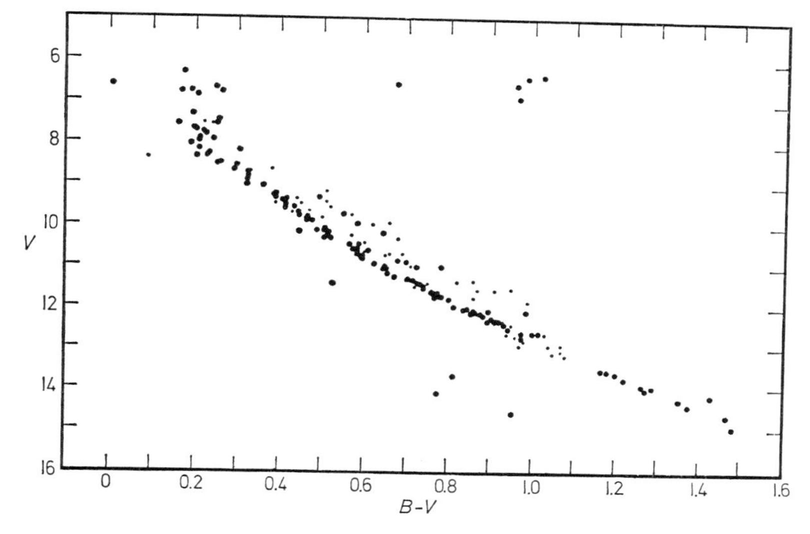

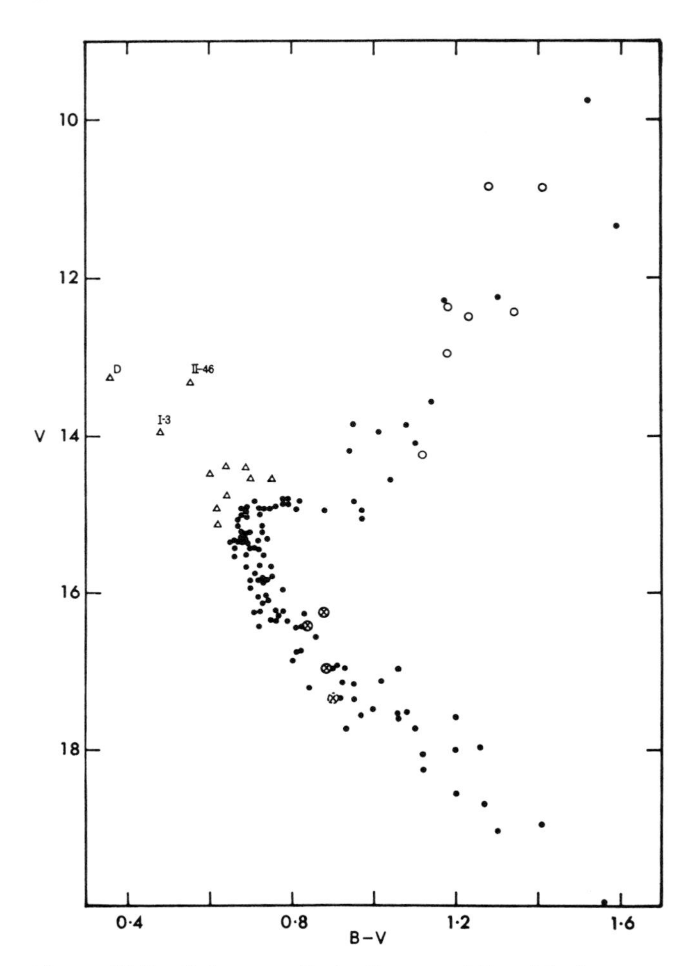

Figure 26.1b. Color-magnitude diagram of the oldest galactic cluster NGC 188 (after O. J. Eggen and A. Sandage, 1969). Uncorrected values of apparent magnitude V and color index $B-V$. From the two-color diagram and the position of the main sequence one obtains interstellar reddening, absorption, and the true distance modulus $(m - M)_0 = 10.85$ mag.

● and ○ recent and older measurements

Δ possible "blue stragglers," arising from evolution in close binary systems

⊗ four eclipsing variables (drawn in as single stars)

without any forcing. On the other hand, higher up sooner (in h + χ Persei at the O and B stars) or later (in Praesepe at about the A stars) the sequence turns off to the right. At about the absolute magnitude of the turn-off (the so-called "knee"), we find on the right-hand side at large (positive) $B-V$ some red giant stars. In NGC 188, M 67 . . . the passage from the main

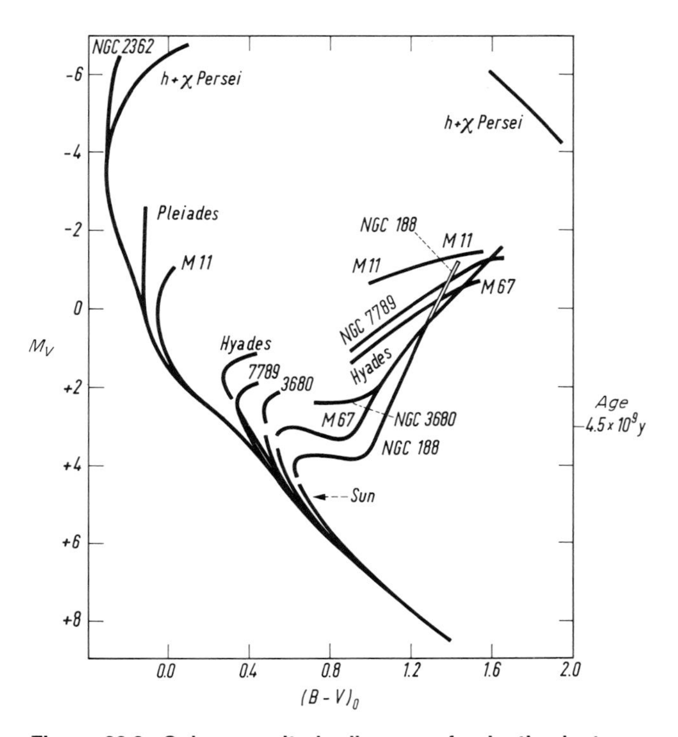

Figure 26.2. Color-magnitude diagram of galactic clusters (after A. R. Sandage and O. J. Eggen, 1969, and A. R. Sandage 1957). The absolute magnitude M_V is plotted against the color index $(B-V)_0$; both are corrected for interstellar absorption and reddening. The point where the main sequence turns off to the right (the "knee") gives the age of the star cluster. While the youngest clusters NGC 2362 and h + χ Persei are only a few million years old, the oldest cluster NGC 188 has an age of from 8 to 10×10^9 years; this is the same as the age of the galactic disk. The position of the knee for 4.5×10^9 years, corresponding to the age of the Sun, is marked on the right-hand scale. The Sun itself still lies on the (almost) "unevolved" main sequence.

sequence to the red-giant branch takes place continuously. We shall call attention to further details in connection with the theory of stellar evolution.

26.4 Color-magnitude diagrams of globular clusters

The structure of the color-magnitude diagrams of globular clusters remained unclear until 1952 when, under the leadership of W. Baade of the Mt. Wilson and Palomar Observatories, a group of young astronomers, A. R. Sandage, H. C. Arp, W. A. Baum, and others set about determining the main sequence—which even in favorable cases begins somewhere in the

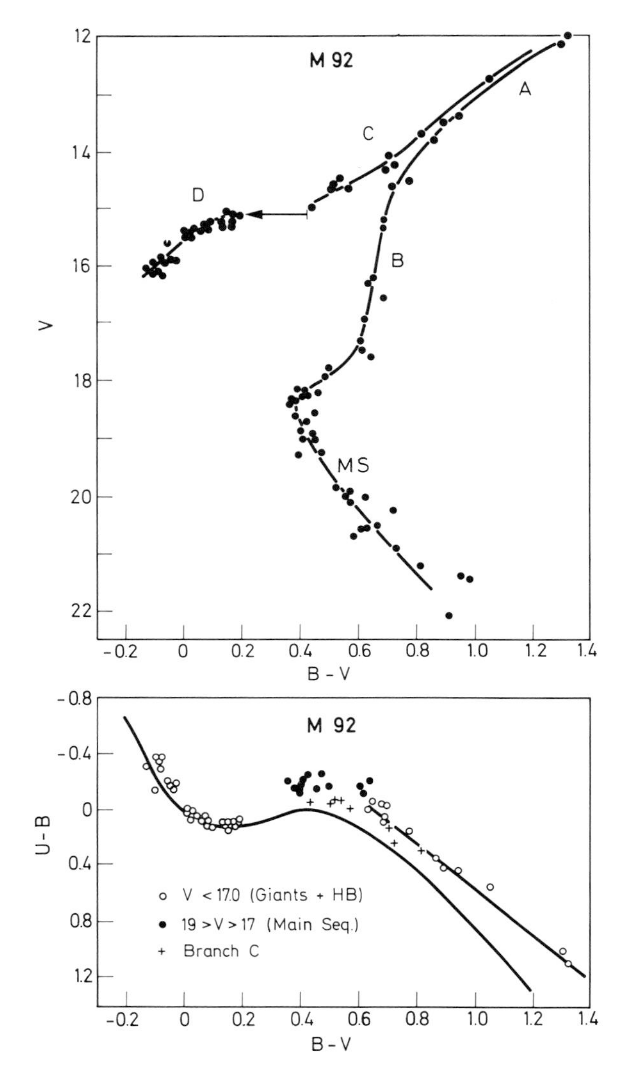

Figure 26.3. Color-magnitude diagram of the globular cluster M 92. Apparent magnitude V and color index $B-V$ are plotted, without the (very small) corrections for interstellar absorption and reddening. A: giant branch; B: sub-giant branch; C: asymptotic branch; D: blue horizontal branch; MS: main sequence. The cluster variables belong in the gap on the horizontal branch at $V = 15$. The apparent visual distance modulus is $m - M = 14.63$. The two-color diagram is added below. The curved line shows the two-color diagram for the Hyades. The line on the right shows its displacement for extremely metal-poor stars.

region 19^m to 21^m. In this way, a comparison with stars of our neighborhood and with those of galactic clusters was possible for the first time. Figure 26.3 (after A. R. Sandage, 1970) shows for example the color-magnitude diagram of Messier 92. Above the main sequence, which runs from the faintest stars presently observable at $V = 22$ up to the "knee" at $V = 18.4$, are joined on first the subgiants B and then the red giants A. The asymptotic sequence C—clearly separated from the giants—leads down to the left from this tip. To this sequence is added the horizontal branch D, with a well-defined gap in which the cluster variables (not marked in Figure 26.3) are situated. The color-magnitude diagrams of other globular clusters show also that *all* stars in this region are variable. The red part of the horizontal branch, to the right of the variable star region, is missing in M 92 and other metal-poor clusters.

In Figure 26.4 we present a composite diagram of the globular clusters M 3, M 13, M 15, and M 92 which have recently been particularly carefully studied by A. Sandage and his colleagues; in this case careful determinations of the distance moduli and the reddening have been used to give absolute magnitude M_V and true color index $(B-V)_0$. The main sequences can be brought into coincidence with no further effort. The subgiant and giant branches are certainly similar, but nonetheless differ significantly from one another, corresponding, as we shall see, to the metal abundance.

Figure 26.4. Color-magnitude diagrams of the globular clusters M 3, M 13, M 15, and M 92 (after A. Sandage 1970). The absolute magnitude M_V is plotted against the true color index $(B-V)_0$.

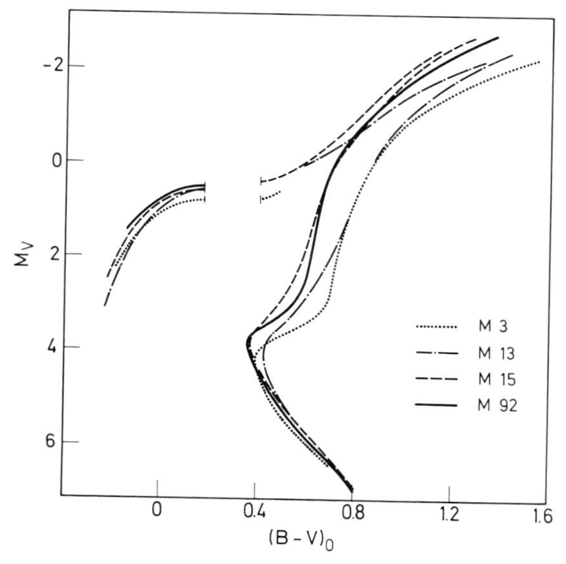

Along with many other questions, we shall compare diagrams 26.2 and 26.4 for galactic and globular clusters when we come to discuss theoretical results on the evolution of stars.

26.5 Nuclear evolution of the stars and interpretation of the color-magnitude diagrams of galactic and globular clusters

After these preliminaries we shall first consider, as an example, the energy balance of the Sun. Its central temperature (calculated according to Eddington's theory) $T_c = 13 \times 10^6$ K has obviously adjusted itself (see Figure 25.1) so that the pp process is responsible for the energy generation.

Provisionally treating the Sun's material as homogeneous, its mass $M_\odot = 1.983 \times 10^{33}$ g consists of about 70 percent hydrogen. The complete conversion of this into helium by the pp process (25.20) would produce 0.86×10^{52} erg. At its present luminosity $L_\odot = 3.84 \times 10^{33}$ erg/s, the Sun would therefore consume 10 percent of its hydrogen—which would mean a noticeable change in its properties—in 7×10^9 years. Since the time when the Earth acquired its solid crust 4.6×10^9 years ago, the Sun can in fact have scarcely changed.

How is it then with the energy balance of other main-sequence stars? Their central temperatures T_c rise from lower values at the cool end of the main sequence up to about 35×10^6 K in B0 stars, etc. Somewhat above the Sun, therefore, the CNO cycle takes over the energy generation (Figure 25.1), without any essential change in its efficiency [equations (25.20) and (25.21)]. From known values of the masses $M/M_\odot$ and luminosities $L/L_\odot$ (that is, the energy production) of main-sequence stars we now easily calculate the time in which 10 percent of the hydrogen is consumed, which we call for short the evolution time t_E (Table 26.1)

$$t_E = 7 \times 10^9 \frac{M/M_\odot}{L/L_\odot} \text{ years.} \tag{26.1}$$

Table 26.1. Stars of the main sequence and their evolution times.

Spectral type	Effective temperature T_e	Mass $M/M_\odot$	Luminosity $L/L_\odot$	Evolution time t_E in years
O 7.5	38 000 K	25	80 000	2×10^6
B 0	33 000 K	16	10 000	1×10^7
B 5	17 000 K	6	600	7×10^7
A 0	9 500 K	3	60	3×10^8
F 0	6 900 K	1.5	6	1.7×10^9
G 0	5 800 K	1	1	7×10^9
K 0	4 800 K	0.8	0.4	14×10^9
M 0	3 900 K	0.5	0.07	50×10^9

Since their time of formation which (to anticipate later results) we cannot put earlier than $\sim 10 \times 10^9$ years ago, main-sequence stars below G0 have used up only a small fraction of their hydrogen. On the other hand, hot stars of early spectral types "burn" away their hydrogen so quickly that they can have come into existence only a relatively short time ago, of the order t_E. The age of the O and B stars is indeed appreciably shorter than the time of revolution of the Galaxy in our neighborhood ($\sim 2.5 \times 10^8$ years); such stars must therefore have been formed in their present surroundings. Before we investigate further the origin of the stars, we first consider their evolution away from the main sequence.

The course of this evolution depends crucially upon whether the material that is changed by nuclear processes in the stellar interior mixes with the rest of the material, or whether it remains in position or remains within the relevant convection zone if one is present. F. Hoyle and M. Schwarzschild were probably the first to show in 1955 that only the latter view leads to an acceptable theory of stellar evolution and also can be made dynamically plausible.

We can illustrate what happens in detail by the example of a star of 5 solar masses (R. Kippenhahn and others 1965). Let this begin its evolution as a fully-mixed main sequence star of type about B5 belonging to population I, that is, with chemical composition (by mass)[1]

$$\text{hydrogen } X : \text{helium } Y : \text{heavy elements } Z = 0.602 : 0.354 : 0.044. \tag{26.2}$$

We show its further evolutionary track in Figure 26.5a in a theoretical color-magnitude diagram taking the effective temperature T_e as abscissa

[1]According to recent spectroscopic studies the abundance ratios $X : Y : Z = 0.70 : 0.28 : 0.014$ appear to be more likely. However, it hardly seems worthwhile repeating the lengthy stellar evolution calculations.

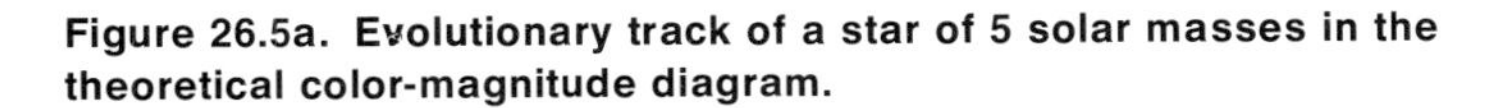
Figure 26.5a. Evolutionary track of a star of 5 solar masses in the theoretical color-magnitude diagram.

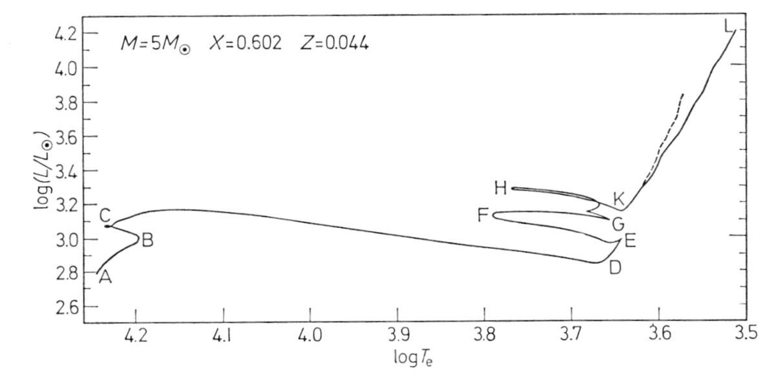

and the luminosity $L/L_{\odot}$ as ordinate and using logarithmic scales. In Figure 26.5b are shown the changes in the stellar interior following the temporal sequence A → B → . . . K → L using a time scale which is different in different parts. Instead of the distance r from the center of the star, M_r/M from equation (25.1) is plotted in order to show what fraction of the stellar mass is in action at the moment. The homogeneous starting model has effective temperature T_e = 17 500K, radius 2.58 $R_{\odot}$ and absolute bolometric magnitude $M_{\text{bol}} = -2.24$. At the center the temperature is $T_c = 26.4 \times 10^6$ K and the pressure is given by log P_c = 16.74 or $P_c = 5.5 \times 10^{10}$ atm. Close to the center we have a hydrogen-burning region, the "nuclear reactor," where hydrogen is burnt to give helium by the operation of the CNO cycle. To this hydrogen-burning region is joined a convection zone within which the reaction products become mixed. This phase of the evolution (A → B → C) lasts 5.6×10^7 years corresponding approximately to our estimated evolution time t_E (Table 26.1). When the core is burnt out, for a short time (C → D → E = 0.3×10^7 years) there is set up a hydrogen-burning zone in the form of a shell. Then at E there arises at the center a $3He^4 \rightarrow C^{12}$ burning region in which at central temperatures now from $T_c \approx$ 130 to 180 × 10^6 K the 3α process [(equations 25.24) and (25.25)] takes over

Figure 26.5b. Time variation of conditions in a stellar interior. The age given on the abscissa refers to the time which has elapsed since the star left the main sequence. Letters *A* to *L* refer to the corresponding letters in Figure 26.5a which label the evolutionary track. The ordinate M_r/M is the fraction of the mass inside a radial distance r. The "cloudy" areas denote convective zones; hatched areas denote regions where nuclear energy generation ϵ exceeds 10^3 erg/g s; dotted areas denote regions where the H (or He) content decreases toward the center. (After R. Kippenhahn, H.-C. Thomas, and A. Weigert, 1965.)

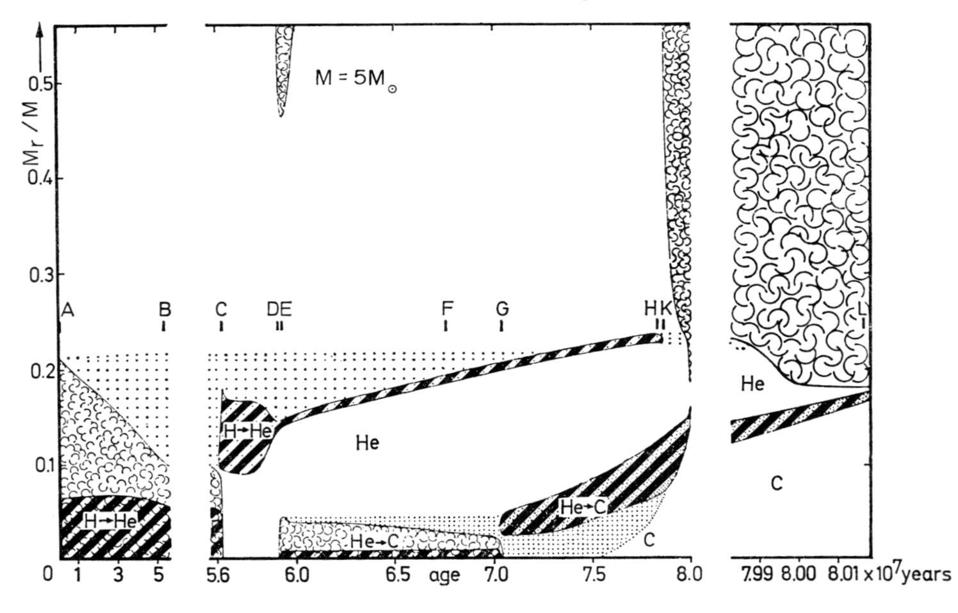

the energy production. When the helium core is also burnt out, a helium-burning zone in the form of a shell is set up. However, from E to K a thin hydrogen-burning shell which is steadily progressing outward also continues to contribute appreciably to the energy generation. From K the star rapidly assumes the status of a red supergiant. All the later stages of evolution during which the star with $T_e \approx 6500$ K several times traverses the region of cepheid instability are run through relatively quickly. After the initial 5.6×10^7 years in the immediate vicinity of the main sequence (A $\rightarrow$ C), another 2.4×10^7 years brings the star to the tip of the "red giant branch." The late stages of evolution which follow this will be described together later.

In Figure 26.6 the evolutionary tracks of (population I) stars of different masses are shown in a theoretical color-magnitude diagram with the logarithm of the luminosity $L/L_\odot$, or the absolute bolometric magnitude M_{bol}, plotted against the logarithm of the effective temperature T_{eff}. The times which the stars need to complete the stages marked 1–2 . . . 10 from the homogeneous initial stage at the beginning of hydrogen burning up to the red giant branch are indicated in Table 26.2.

We can now summarize the most important features of the nuclear evolution of stars in the following way:

An originally homogeneous star forms at its center a hydrogen-burning zone in which the energy is produced by the CNO cycle at higher central temperatures (large stellar masses) and by the pp chain at lower central temperatures (small stellar masses). As Eggen and Sandage (1969) have pointed out, the transition from the pp chain to the CNO cycle is marked in the color-magnitude diagrams of galactic clusters by a gap, whose detailed explanation would however lead us too far from our present discussion.

The stars which are young enough still to be homogeneous are arranged in the (T_{eff}, L) or color-magnitude diagram along a line on which stars of large mass have high luminosity and stars of small mass have low luminosity. This line, which is the locus of the points 1 in the theoretical diagram (Figure 26.6), we term the *initial main sequence* or (not quite accurately)

Table 26.2. Time (in years) which stars of different masses (in solar units) require to traverse the intervals marked 1-2, 2-3, . . . in Figure 26.6.

	Intervals along the evolutionary tracks in Figure 26.6.					
$M/M_\odot$	1 2	 3	 4	 5	 6	 10
15	1.0×10^7	2.3×10^5		7.6×10^4		1.6×10^6
9	2.1×10^7	6.1×10^5	9.1×10^4	1.5×10^5	6.6×10^4	4.0×10^6
5	6.5×10^7	2.2×10^6	1.4×10^6	7.5×10^5	4.9×10^5	1.7×10^7
3	2.2×10^8	1.0×10^7	1.0×10^7	4.5×10^6	4.2×10^6	7.2×10^7
1.5	1.6×10^9	8.1×10^7	3.5×10^8	1.0×10^8	$\geqslant 2 \times 10^8$	
1	7×10^9	2×10^9	1.2×10^9	1.6×10^8	$\geqslant 1 \times 10^9$	

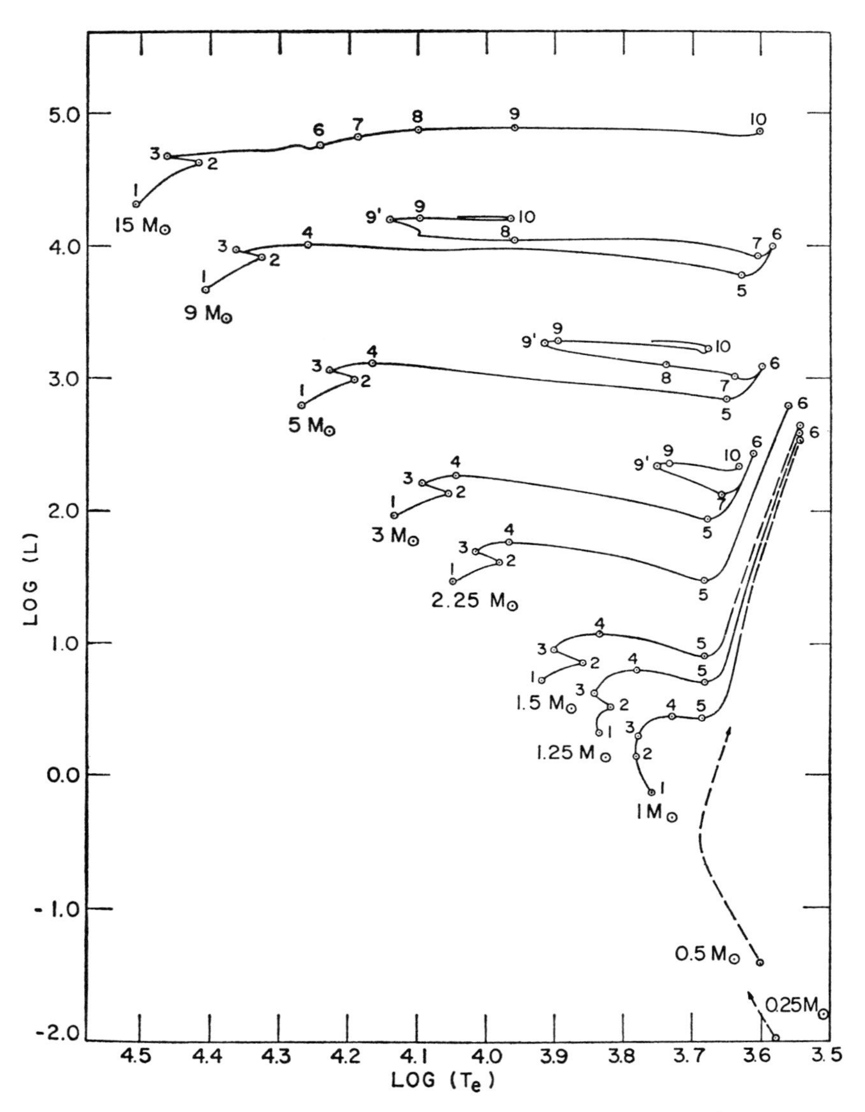

Figure 26.6. Evolutionary tracks for metal-rich population I stars of different masses in the theoretical color-magnitude diagram. Ordinate: luminosity log $L/L_\odot$, or absolute bolometric magnitude M_{bol}; abscissa: effective temperature log T_e. The times in which the paths 1-2-3 . . . are traversed are indicated in Table 26.2. (Reproduced, with permission, from "Stellar evolution within and off the main sequence" by I. Iben, Jr, Annual Review of Astronomy and Astrophysics, Volume 5.

the *Zero Age Main Sequence* (ZAMS). We shall return to the question of its empirical determination.

Stars remain in the immediate vicinity of the main sequence until a considerable fraction of the hydrogen is burned, that is, for a time $\approx t_E$ (Table 26.1). If no mixing takes place between burned and unburned material, the evolutionary track in the (T_{eff}, L) diagram then leads upward and to the right (as M. Schönberg and S. Chandrasekhar calculated as early as 1942), that is, into the region of the red giants. With the help of the times given in Table 26.2 we could now mark the lines in Figure 26.6 along which would lie at time t a group of stars which started at time $t = 0$ on the initial main sequence (points 1). *Isochrones* computed in that way enable us to explain the color-magnitude diagrams of the galactic clusters (Figure 26.2) as an *age sequence* and to refine our earlier age estimates in Table 26.1. h + χ Persei with its extremely luminous blue supergiants, which burn away their hydrogen so extravagantly, is a quite young stellar cluster. The turnoff from the main sequence, the so-called "knee," at $M_V \approx -6$ indicates (Table 26.1) an age of a few million years. The few red supergiants to the right of the upper end of the main sequence are separated from the latter by the so-called "Hertzsprung gap" which has been known empirically for a long time. Becoming less marked, the gap extends down to about F0 III stars. The gap is easily explained by the fact that, for example, in Figure 26.5a (that is, for a star of $5M_\odot$) the portion of track C $\rightarrow$ D lasts for only 0.3 $\times\ 10^7$ years, compared with 2.1×10^7 years for the ensuing red-giant stage or 5.6×10^7 years for the hydrogen-burning phase on the main sequence.

The color-magnitude diagrams of, for example, the Pleiades . . . Praesepe . . . down to NGC 188, in which the main sequence turns off to the giant branch further and further down, indicate ever greater ages. Without question, there is no diagram that turns off significantly lower down than that of NGC 188, that is, there exists a *maximum age* for galactic clusters which, according to the best model calculations at the present time, is about 8 to 10 $\times\ 10^9$ years. This is evidently also the *age of the galactic disk.*

The aggregate of color-magnitude diagrams of galactic clusters in Figure 26.2 up to fairly bright M_V possesses a well-defined envelope, from which the nuclear evolution toward the right begins. This is obviously the initial main sequence, or zero-age main sequence, which has already been demanded by the theory. In its upper part it lies a little below the main sequence of luminosity class V (Figure 15.4) and merges into this at the G stars (Table 26.3).

We can best interpret the familiar color-magnitude diagram of the field stars in our neighborhood as that of a mixture of stars from the remnants of many associations and clusters that have become dispersed in the course of time. The calculated evolution times make it at once understandable that the main sequence conforms closely to the initial main sequence. The (at first sight not easily understandable) bunching together of the yellow and red giants in Russell's giant branch, according to A. R. Sandage, is ascribed to the fact that in this region the evolutionary tracks of the more massive

Table 26.3. Zero Age Main Sequence; after H. L. Johnson, 1964, and (for $B - V \geqslant +1.40$) W. Gliese, 1971.

$B - V$	M_V	$B - V$	M_V	$B - V$	M_V
−0.20	−1.10	+0.40	+3.56	+1.20	+ 7.66
−0.10	+0.50	+0.50	+4.23	+1.40	+ 8.9
0.00	+1.50	+0.60	+4.79	+1.60	+11.5
+0.10	+2.00	+0.70	+5.38	+1.80	+14.0
+0.20	+2.45	+0.80	+5.88	+2.00	+16.5:
+0.30	+2.95	+1.00	+6.78		

and luminous stars from left to right and those of the less massive and fainter stars from the lower main sequence toward the upper right run together as in a funnel (cf. Figure 26.2). Particularly in regard to their masses, the giant stars in our neighborhood may therefore not be treated as a homogeneous group.

We turn now to the color-magnitude diagrams of the globular clusters (Figures 26.3 and 26.4), which actually formed the starting point for the modern theory of stellar evolution. They very much resemble the diagrams of the older galactic clusters with the difference that their giant branch rises more steeply. According to the model calculations of F. Hoyle and M. Schwarzschild (1955), this is a consequence of the fact that, as members of the extreme halo population II, the globular clusters have *very low metal content* (of the order of $\frac{1}{10}$ to $\frac{1}{200}$ times the solar value). This is confirmed by the spectra of their red giants. The opacity and the energy generation are considerably different, as a result. The turnoff of the color-magnitude diagrams (at $M_V \approx 4$, as in NGC 188) shows, according to the most recent model calculations, that *all* globular clusters have an age of $9–12 \times 10^9$ years. The spread of values, which is mainly due to the theory, is probably larger than the actual differences in age of the various globular clusters. As to the absolute value, it is scarcely possible, as suggested also by theoretical difficulties, to guarantee that there is any difference in age between the oldest galactic clusters and the globular clusters. In any case, however, the galactic halo and all the globular clusters must have been formed within a relatively short time, while the formation of young galactic clusters is still being played out today before our very eyes.

In the color-magnitude diagrams of globular clusters (Figures 26.3 and 26.4) we distinguish the red-giant branch (upper right), the asymptotic branch, and finally the horizontal branch which goes down to the B stars of absolute magnitude $M_V \approx +2$. Embedded in the horizontal branch is the "gap" containing the pulsating cluster variables or RR Lyrae stars. The turnaround of the evolutionary track in the red-giant region is ascribed to the sudden ignition of central helium burning (the "helium flash"). Spectroscopic observations show that the chemical composition of the atmosphere (Table 19.3) of, for example, the horizontal branch star HD 161817 (rather to the left of the RR Lyrae gap) can hardly be distinguished from that of the subdwarfs of the population II main sequence.

The stars which are most closely related to the globular clusters are the field stars of the old halo population II. We got to know them earlier as "high velocity stars." They move round the galactic center (Figure 23.11) in elongated orbits of large eccentricity and large inclination and so they mostly have large velocities relative to the Sun and especially a large W component (perpendicular to the galactic plane). Spectroscopic analysis (Table 19.3) shows that in these stars the abundance of all the heavy elements ($Z \geqslant 6$; "metals") relative to hydrogen is reduced by factors of up to about 500 compared to the Sun, while the ratios of the abundances of the heavy elements to one another are the same as in the Sun. This is not true for helium; analyses of several population II planetary nebulae show clearly that the abundance ratio He : H (by number of atoms) was originally about 1 : 10 in populations I *and* II.

Since for practical reasons detailed spectroscopic analyses can be carried out only for relatively few objects, mostly the brighter ones, it is important that it is also possible to determine the metal abundances of stars, at least globally, by means of the two-color diagram. That is, in the cooler, metal-poor halo stars the metallic lines, which become steadily more densely packed as one goes to shorter wavelengths, are so much weaker than in normal metal-rich stars that the color index $U-B$ is considerably displaced toward smaller values; the index $B-V$ on the other hand is displaced by a comparatively small amount. Hence the characteristic line in the two-color diagram is raised in the region $B-V > 0^{\mathrm{m}}.35$ by, at most (that is, for the most metal-poor stars), about $\delta(U-B) \approx 0^{\mathrm{m}}.25$; cf. the two-color diagram of the extremely metal-poor globular cluster M 92 in Figure 26.3. It goes without saying that this effect must be taken into account when determining interstellar reddening from the two-color diagram.

The ultraviolet excess $\delta(U-B)$, which is usually defined more precisely as the difference in $U-B$ compared with stars on the Hyades main sequence with the same $B-V$, also gives a measure of the metal abundance for fainter stars. Still more precise results are obtained from the intermediate band photometry of B. Strömgren.

As we have already noted, the ultraviolet excess or metal abundance of field stars is correlated with their W velocity or the eccentricity and inclination of their galactic orbit. A recent investigation by H. E. Bond (1971), which selected stars not from proper motion catalogs but from objective-prism plates taken with Schmidt telescopes, showed that, while these correlations are certainly very pronounced for extremely metal-poor halo stars, they almost (or completely?) disappear for stars with larger metal abundances. There is therefore a completely continuous transition from the extreme halo population to the disk population of our Galaxy. The color-magnitude diagram of the halo stars roughly fills the space between those of NGC 188 and M 92.

We cannot try to find a cosmogonical significance in these observations until we have come to know (in Section 27) their counterparts in other galaxies, which are in some respects quite different.

26.6 The initial contracting phase of stellar evolution and star formation

We turn back once more to the starting point of our deliberations, and ask, *How do the stars reach the main sequence?* and then, *How and from what did the stars arise?*

The close association in space of young, absolutely bright O and B stars with gas and dust clouds in the spiral arms of our Galaxy and the Andromeda galaxy strongly suggests that quite generally stars are formed in and from cosmic clouds of diffuse matter. The only available energy source they have at first, that is until some nuclear reaction gets under way, is gravitational energy (contraction energy).

It is fitting to recall here that as early as 1846, shortly after the discovery of the law of conservation of energy, J. R. Mayer raised the question of the origin of the radiative energy emitted by, say, the Sun. He considered the fact that a mass of meteorites m falling into the Sun would give up energy of amount, according to equation (6.28),

$$mG\,M/R \tag{26.3}$$

as heat energy, where G is the gravitation constant; M, R denote the mass and radius of the Sun. Since the mass of infalling meteorites is in fact very small, H. von Helmholtz 1854 and Lord Kelvin 1861 were able to show that the contraction of the Sun itself would be a more significant source of gravitational energy.

As we see immediately from equation (26.3), the energy released in the formation of a gas sphere of radius R from initially dispersed material, or in a considerable contraction of a star of radius R, is always of the order of magnitude

$$E_{\text{contr.}} \approx GM^2/R. \tag{26.4}$$

In the case of the Sun, for example, this is about 3.8×10^{48} erg; the content of the Sun in thermal and ionization energy is of the same order. The energy $E_{\text{contr.}}$ could supply the emission from the Sun at its present luminosity $L_\odot = 3.84 \times 10^{33}$ erg/s for only

$$E_{\text{contr.}}/L_\odot \approx 30 \text{ million years.} \tag{26.5}$$

What then is the track in the color-magnitude diagram of a star that is formed by the contraction of initially widely dispersed material? In 1961, C. Hayashi showed that stars with effective temperatures $T_{\text{eff}} < 3$ to 4000 K are essentially convective in structure (cf. Section 25); the greater part of their interior is taken up by an extensive hydrogen convection zone (cf. Section 20). Hayashi was further able to show that such stars are unstable. The line in the theoretical color-magnitude diagram which separates the stable stars (on the left) from the unstable ones (on the right) is called the Hayashi line. To understand what happens, consider a star with prescribed values of mass M, luminosity L, effective temperature T_{eff} and hence also radius R. The structure of its atmosphere is determined by T_{eff} and the gravitational acceleration $g = GM/R^2$; we therefore know in the atmosphere—for example, at optical depth $\tau = 1$—a pair of values of pressure P_a and temperature T_a (and so also a value of density ρ_a). On the other hand (see Section 25), we can also calculate directly the values of P_c, T_c, ρ_c which must prevail in the center of the star if it is built out of an ideal gas in hydrostatic equilibrium. If now, for a star on the *right* of the Hayashi line, we compute inwards from P_a, T_a along an adiabat of the partially ionized hydrogen (and helium), we arrive at pressures and temperatures lower than P_c and T_c, that is, the star would possess, so to speak, a cavity at the center and would have to collapse. Only on the *left* of the Hayashi line, where

Figure 26.7. Initial convective stages of stellar evolution (after I. Iben, Jr, 1965). Stars of more than 1 $M_\odot$ require less than a few times 10^7 years to reach the main sequence.

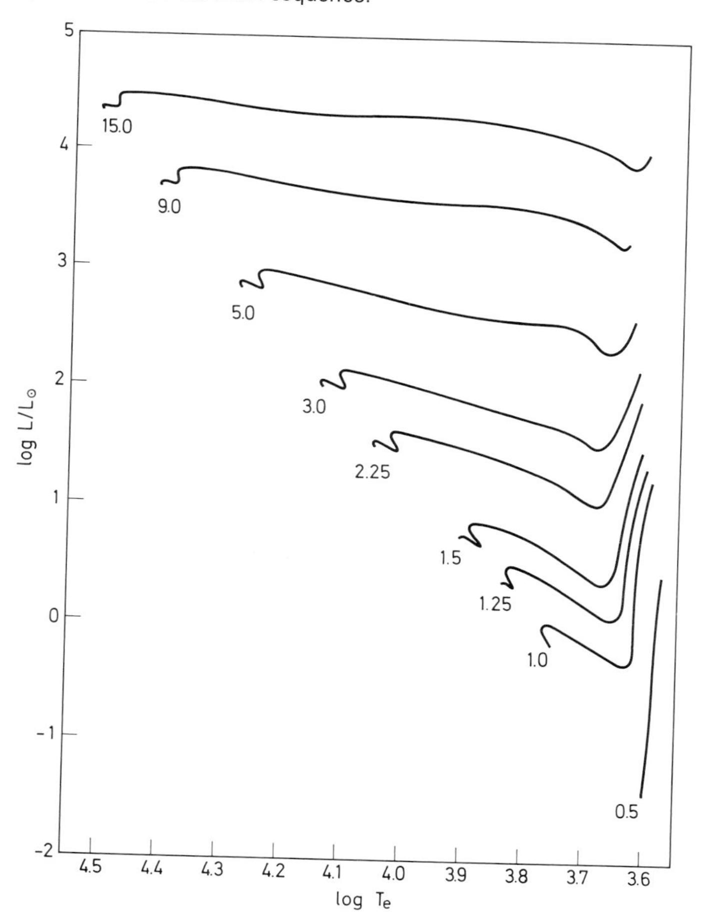

radiative equilibrium can occur, can the star achieve sufficiently high central pressures and temperatures and so find a stable configuration.

Thus a star forming out of interstellar material possesses an extended convective envelope as soon as the hydrogen is partially ionized. In the color-magnitude diagram (Figure 26.7) it moves down immediately to the left of the Hayashi line, with slightly rising temperature, until it reaches (after about 10^7 years for $1M_\odot$) the almost horizontal evolutionary track for radiative equilibrium, previously computed by L. G. Henyey and others. Along this track it meets its energy requirements in the first place from gravitational energy. The familiar nuclear reactions do not ignite until just before the initial main sequence is reached.

We can understand similarly why the evolutionary tracks of the giants in Figure 26.6 climb steeply upward: stars cannot cross the Hayashi line in the opposite direction either.

Toward the upper part of the main sequence the contraction time corresponding to equation (26.5) becomes steadily smaller; a B0 star requires only about 10^5 years. On the other hand, stars of smaller mass form themselves more slowly; an M star of $0.5M_\odot$ requires about 1.5×10^8 years. Contracting masses below a certain value do not at any stage reach the temperature needed for setting nuclear processes in operation. S. S. Kumar has shown (1963) that this limiting mass for population I material is $0.07M_\odot$, and for metal-poor population II material it is $0.09M_\odot$. Smaller masses form cool, wholly degenerate bodies that are naturally called *black dwarfs*. Bodies such as the dark companion of Barnard's star (page 140) with mass about $0.0015M_\odot$ or the large planets, like Jupiter with about $0.001M_\odot$, may have been formed as black dwarfs.

In young galactic clusters, which we recognize by their bright blue stars, M. Walker in fact found to the right of the lower main sequence stars of middle and of late spectral types, which by their T Tauri-variability, Hα emission lines, in some cases fast rotation, etc., are indicated to be young stars in the course of formation. For example, Figure 26.8 shows following M. Walker (1956) the color-magnitude diagram of the cluster NGC 2264; according to the evidence of the brightest stars, this has an age of only 3×10^6 years. In agreement with the theory stars below spectral type A0 ($< 3M_\odot$) have therefore not yet reached the standard main sequence.

Dynamical considerations (see Section 29.1) indicate that only relatively large masses of gas, say 10^2 to $10^4M_\odot$, can become unstable and condense into stars. In fact as long ago as 1947, V. A. Ambartsumian had extracted from the observations the important inference that, at any rate for the most part, stars are formed in times of the order of 10^7 years in groups with masses of about $10^3M_\odot$. These are the OB associations with bright blue stars and the T associations with cooler stars, notably T Tauri variables of small absolute luminosity, the two sorts often occurring together. An OB association, as in Orion or Monoceros, with its enormous bodies of gas, which are ionized by the short-wavelength radiation from the O and B stars immersed within them, first attracts the attention of the observer as an HII region that shines brightly in Hα light. The expansion of the gas heated by

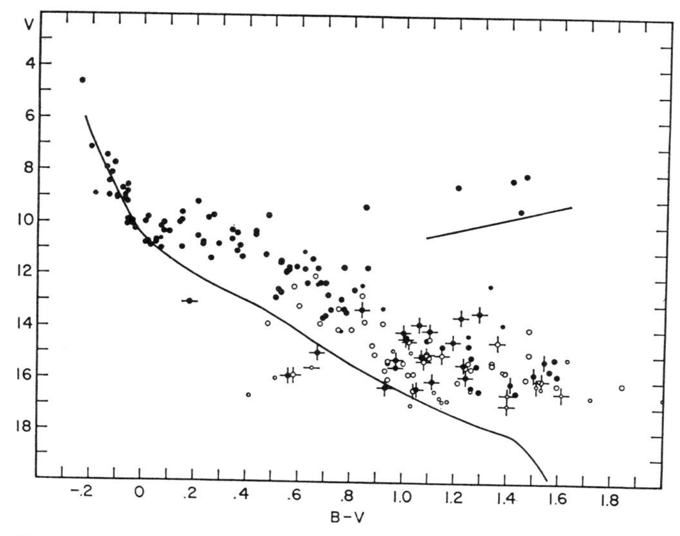

Figure 26.8. Color-magnitude diagram of the very young galactic cluster NGC 2264 (after M. Walker, 1956).
● photoelectric measurements
○ photographic measurements
| variable stars
— stars with Hα emission
The curves indicate the standard (zero-age) main sequence and the giant branch, corrected for uniform interstellar reddening of the cluster. The apparent distance modulus is $9\overset{m}{.}7$, and the distance 800 pc. The lifetime of the cluster (about 3×10^6 years) has been insufficient to enable stars to reach the main sequence below about spectral type A0.

the stars that have already been formed is the probable reason why further stars move away from the center with speeds of the order of 10 km/s. Thus the associations are *not permanent.* The expansion age obtained by extrapolating the stellar motions backward in general agrees with the evolution age of the brightest stars. For the Orion nebula and the Trapezium stars that excite its radiation (and certain similar systems) we find an age of only about 1.5 to 3×10^4 years. In contrast, the Orion cluster itself must be about 100 times older.

In a number of cases, individual stars travel away from their OB association with much larger speeds up to 200 km/s. A. Blaauw sees such a "runaway star" as the surviving component of a fast-revolving double star whose primary component was suddenly dispersed in space as the result of an explosion (possibly as a Type II supernova).

26.7 Evolution in close binary systems

Nuclear evolution in close binary systems may run a quite different course from that in the single stars so far considered. Extremely interesting investigations into this subject have been carried out, observationally by O.

Struve in the 1940s and 1950s and theoretically since 1966 by R. Kippenhahn and A. Weigert and their colleagues.

Let us consider first a binary system whose components are *detached* and study its equipotential surfaces: at a point, which is at a distance r_1 from a mass M_1 and at a distance r_2 from a mass M_2, the gravitational potential [see equations 6.28–30] is $\varphi_g = -G(M_1/r_1 + M_2/r_2)$. If the system rotates with angular velocity ω, the centrifugal acceleration $z\omega^2$ (z = distance from rotation axis) can be represented by an additional potential $\varphi_z = -z^2\omega^2/2$. On a surface $\varphi = \varphi_g + \varphi_z$ = constant a test particle can be moved without doing any work; for this reason the surface of, for example, the ocean or a heavenly body follows an equipotential φ = constant.

In our binary system each component is first surrounded by its own closed equipotential surfaces; then one comes to the first common equipotential surface, the so-called *Roche surface*, which has the shape of an hour glass and surrounds both bodies. Further out all surfaces envelop both bodies.

If the larger mass, M_1, say, now evolves into a giant star, it may overflow the innermost common equipotential surface. We then have a so-called *semidetached* system; gas streams from component 1 to component 2, so that the mass ratio may even be reversed. The details of what happens, that is, what courses the two components would pursue in the color-magnitude diagram, depend on the original masses and orbital radii and, in connection with that, on the stage of its evolution at which component 1 begins to "bubble over." This may occur while hydrogen burning is still confined to the center of the star, or it may not begin until hydrogen burning, or even helium burning, is in progress in a shell source (see Figures 26.5a and b). In these ways binary systems can be formed in which one component is a helium star, which has, to a greater or lesser extent, disposed of its hydrogen-rich envelope, or a white dwarf star (see below). Evolution in close binary systems has also been proposed as an explanation of Wolf–Rayet stars (B. Paczyński) and of Ap and Am stars (E. P. J. van den Heuvel).

Recent calculations of *contact* systems, in which both components fill a common equipotential surface, seem to point to a possible origin for the W Ursae Majoris stars.

26.8 Final stages of stellar evolution

Now that we have seen how stars originate from interstellar material (Section 26.6) and how they evolve further by nuclear burning of hydrogen, then helium, and finally carbon (cf. Section 26.5) the question arises: "How does stellar evolution proceed after exhaustion of all nuclear energy sources and what are its final stages?"

The virial theorem of R. Clausius (1870) forms one of the important foundations for the understanding of the whole problem of stellar evolution (and of many other astrophysical questions). If we restrict ourselves here to

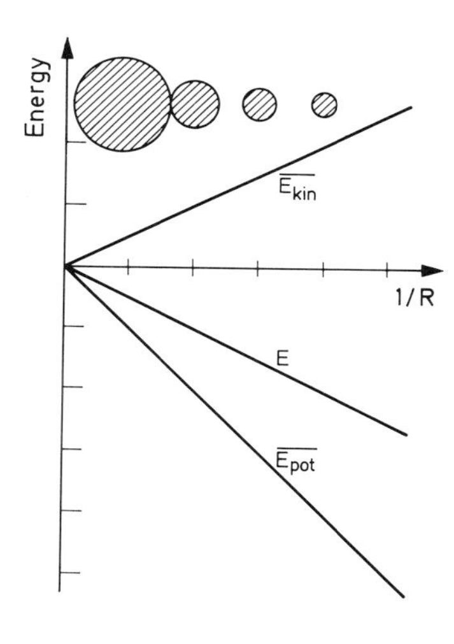

Figure 26.9
The virial theorem for a system of gravitating point masses: $E_{kin} = -\frac{1}{2}E_{pot}$. A contracting star loses potential energy. However, half of this amount is added to its reserves of kinetic energy. According to J. J. Jeans the condition $2E_{kin} < -E_{pot}$ leads to gravitational instability (Section 29.1).

a closed system of point masses, which are bound to one another by forces proportional to $1/r^2$, in particular, gravitation, the virial theorem implies[2] that the total energy E is divided into kinetic energy E_{kin} and potential energy E_{pot} in such a way that, averaged over time [see also equation (6.36)]

$$\bar{E}_{kin} = -\tfrac{1}{2}\bar{E}_{pot} \tag{26.6}$$

or alternatively, since $E = E_{kin} + E_{pot} = \text{constant}$,

$$\bar{E}_{kin} = -\bar{E}. \tag{26.7}$$

Figure 26.9 may demonstrate these important relations still more clearly.

[2]*Proof:*
If we have a system of point masses m_k with position vectors $\boldsymbol{r}_k$ and corresponding momenta $\boldsymbol{p}_k = m_k\dot{\boldsymbol{r}}_k$, then the kinetic energy (see Section 6) is given by

$$2E_{kin} = \sum m_k\dot{\boldsymbol{r}}_k^2 = \sum \boldsymbol{p}_k \cdot \dot{\boldsymbol{r}}_k. \tag{26.a}$$

From this we obtain by integration by parts

$$2E_{kin} = \frac{d}{dt}\sum \boldsymbol{p}_k \cdot \boldsymbol{r}_k - \sum \dot{\boldsymbol{p}}_k \cdot \boldsymbol{r}_k. \tag{26.b}$$

From Newton's equation of motion $\dot{\boldsymbol{p}}_k$ is equal to the force acting and so, since E_{pot} is the path integral of the force,

$$\boldsymbol{p}_k = -\nabla E_{pot}(k)\left(\text{with components } -\frac{\partial E_{pot}(k)}{\partial x_k}, -\frac{\partial E_{pot}(k)}{\partial y_k}, -\frac{\partial E_{pot}(k)}{\partial z_k}\right) \tag{26.c}$$

where $E_{pot}(k)$ is the potential energy of m_k in the field of all the other mass points:

$$E_{pot}(k) = -\sum_{k' \neq k} \frac{Gm_{k'}m_k}{|\boldsymbol{r}_{k'} - \boldsymbol{r}_k|}. \tag{26.d}$$

It is then easy to show by taking the terms in pairs that

$$\sum_k \dot{\boldsymbol{p}}_k \cdot \boldsymbol{r}_k = -\tfrac{1}{2}\sum_{k'}\sum_{\substack{k \\ k' \neq k}} \frac{Gm_{k'}m_k}{r_{kk'}} = E_{pot} \tag{26.e}$$

where $r_{kk'} = |\boldsymbol{r}_{k'} - \boldsymbol{r}_k|$, E_{pot} is the total potential energy of the system and the ½ allows for repeated terms in the summation. If we now average (26.b) over a sufficiently long time, the mean value of the first term on the right-hand side vanishes and we obtain the *virial theorem:*

$$2\bar{E}_{kin} = -\bar{E}_{pot}. \tag{26.f}$$

We now illustrate them by a simple estimate, which we had already discussed in another way in connection with equation (25.7). The potential energy[3] of a star, which we shall treat as a homogeneous sphere, is (per unit mass) $E_{\text{pot}} = -\frac{3}{5}GM/R$. If we assume a monatomic gas of molecular weight μ, the only kinetic energy available is the thermal energy per unit mass $E_{\text{kin}} = \frac{3}{2}\mathcal{R}T/\mu$ ($\mathcal{R}$ = gas constant). We can now deduce immediately from the virial theorem (26.f) (see footnote 1), together with known data for main sequence stars (so long as radiation pressure plays no important role), that temperatures of from 10^6 to 10^7 K must prevail in their interiors.

Now back to our problems of stellar evolution! After exhaustion of a nuclear energy source—for example, hydrogen burning—a star makes use of its gravitational energy, that is, it contracts, until its central temperature has risen (Figure 26.9) sufficiently for the next energy source—in our case, helium burning—to ignite. In these evolutionary phases of very high luminosity an essential part is played not only by the nuclear energy but also by the gravitational energy which is released by the ever-increasing concentration of the burned material toward the center of the star. It would be interesting to follow these processes in detail with the help of our color-magnitude diagram. However, we return here at once to our question: "What happens after the exhaustion of all nuclear energy sources?"

One possibility, which we shall investigate first, is offered by the lowly existence of a white dwarf star (see p. 135), which then slowly cools to invisibility, at constant radius.

As R. H. Fowler discovered in 1926, the material (more exactly, the electrons) in these stars is, apart from a very thin atmospheric layer, degenerate in the sense of Fermi–Dirac statistics. The meaning of this is as follows. In classical or Maxwell–Boltzmann statistics, one derives, for example, the velocity distribution of the electrons or the equation of state of the gas by calculating the distribution of particles in phase space according to the rules of probability theory. (The six-dimensional element $\Delta\Omega$ of phase space is formed from the product of the familiar volume element $\Delta V = \Delta x\ \Delta y\ \Delta z$ in position space and the element in momentum space $\Delta p_x\ \Delta p_y\ \Delta p_z$). However, for an electron gas, for example, the applicability of this procedure is restricted by the Pauli principle, which requires that one quantum state or one quantum cell in phase space of size $\Delta\Omega = h^3$ (h = Planck's constant) may be occupied by at most one electron of each spin direction, that is, by two electrons altogether. Fermi–Dirac statistics take account of this. Now, in the interior of a white dwarf star the material is so compressed due to enormous pressure and a comparatively low temperature that all cells h^3 in phase space, up to a certain threshold energy E_0 or maximum momentum p_0, are completely filled, that is, are occupied by two

[3]We calculate the potential energy of a homogeneous sphere of mass M, radius R, density ρ as follows: If we add a shell of mass dM_r to a sphere of radius r and mass $M_r = (4\pi/3)\rho r^3$, the gain in potential energy is $dE_{\text{pot}} = -GM_r dM_r/r$. Using $r = (3M_r/4\pi\rho)^{1/3}$ we obtain immediately $-E_{\text{pot}} = \int_0^R GM_r dM_r/r = G(3/4\pi\rho)^{-1/3}\,\frac{3}{5}M^{5/3} = \frac{3}{5}GM^2/R$.

electrons. This state is described as complete Fermi–Dirac degeneracy of the gas.

The equation of state of the degenerate electron gas can easily be calculated (the proton gas which is also present does not become degenerate until much higher densities; its pressure can be neglected):

Let there be n electrons/cm^3 in a volume V, and so altogether nV electrons. In momentum space p_x, p_y, p_z (Figure 26.10) these electrons uniformly fill a sphere whose radius is the maximum momentum p_0 (or the threshold energy, the so-called Fermi energy $E_0 = p_0^2/2m$). We therefore have a volume $\frac{4\pi}{3} p_0^3 V$ in phase space and with two electrons per phase cell of size h^3 we obtain the relation

$$nV = \frac{2}{h^3} V \frac{4\pi}{3} p_0^3 \quad \text{or} \quad p_0 = \sqrt{2mE_0} = h\left(\frac{3n}{8\pi}\right)^{1/3}. \tag{26.8}$$

Now the pressure, just as in an ideal gas, is given by the familiar formula

$$P = \tfrac{2}{3} n\bar{E}, \tag{26.9}$$

where $\bar{E}$ denotes the mean energy per electron. However, we can immediately find the relation between $\bar{E}$ and E_0 with the help of Figure 26.10. It is in fact (using $E = p^2/2m$)

$$\bar{E} = \int_0^{p_0} E\, 4\pi p^2\, dp \bigg/ \int_0^{p_0} 4\pi p^2\, dp = \frac{3}{5}\frac{p_0^2}{2m} = \frac{3}{5} E_0 \tag{26.10}$$

and so we obtain the equation of state of a completely degenerate electron gas:

$$P = \frac{2}{5} nE_0 = \frac{8\pi}{15}\frac{h^2}{m}\left(\frac{3n}{8\pi}\right)^{5/3}. \tag{26.11}$$

Here there is no further mention at all of temperature; that is precisely the characteristic property of degeneracy. (It is easy to verify that the pressure of the proton gas, which is equally dense but not yet degenerate, is very much smaller than that of the degenerate electron gas.)

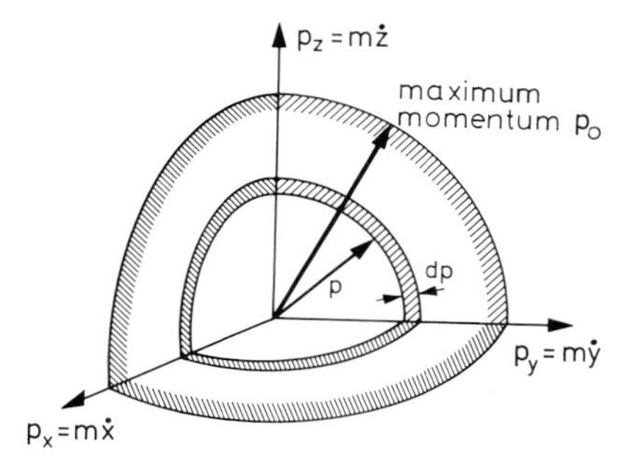

Figure 26.10
Degenerate electron gas (Fermi–Dirac statistics). In momentum space, of which only an octant is shown, the electrons uniformly fill a sphere of radius p_0, the so-called Fermi sphere.

The relation between n and the density (in g/cm^3) is most easily expressed in terms of the mass μ_E (in atomic mass units) which "belongs" to one electron. Then:

$$n = \rho/m_H\mu_E, \tag{26.12}$$

where $\mu_E = 1$ for hydrogen, 2 for helium, etc.

An estimate, $\rho \approx 10^6$ g/cm^3, for the density in a white dwarf may illustrate these relations: from equation (26.12) we then have $n \approx 10^{30}$ electrons/cm^3 and from equation (26.11) a pressure of the degenerate gas of $P \approx 10^{17}$ atm, a million times higher than, for example, in the center of the Sun. For an ideal gas to have the same pressure, it would need to have a temperature of $\sim 2 \times 10^9$ K.

We can use the equation of state $P \propto \rho^{5/3}$ together with our earlier estimate of the pressure in the interior of a star of mass M and radius R, that is [see equation 25.4)] $P \propto \rho GM/R$, and the trivial relation $\rho \propto M/R^3$, to obtain approximately[4] a mass-radius relation for white dwarf stars, viz.:

$$R \propto M^{-1/3} \tag{26.13}$$

that is, the radius decreases with increasing mass.

Using this, we can also write down the luminosity $L = 4\pi R^2 \sigma T^4{}_{\text{eff}}$. According to this, white dwarf stars of mass M should be ordered by mass in the theoretical color-magnitude diagram and lie on lines

$$L \propto M^{-2/3} T^4_{\text{eff}}. \tag{26.14}$$

Comparison with observations yields the at first surprising result that *all* white dwarfs have masses of about 0.6 $M_\odot$.

However, our whole scheme of approximation still needs completion and correction, although mostly only in fine detail. With increasing density the energy of the electrons first increases like $\rho^{2/3}$, until, for $E \gtrsim mc^2$ (that is, in massive and therefore denser stars), the relativistic change of mass of the electrons becomes noticeable. In the case of fully relativistic degeneracy ($\bar{E} \gg mc^2$), a repetition of our calculations taking account of special relativity yields an equation of state

$$P \propto \rho^{4/3} \tag{26.15}$$

that is, the material is more easily compressed. As S. Chandrasekhar realized in 1931, with increasing relativistic degeneracy the radius of a

[4]We neglect here the facts that the outermost part of the star is not degenerate and that the innermost part is relativistically degenerate (see later).

white dwarf star tends to the limiting value zero at a finite limiting mass of $1.4M_\odot$ (with a slight dependence on chemical composition); that is, white dwarf stars are stable, or can exist, only for masses $M \leqslant 1.4M_\odot$.

Before we further discuss the origin and evolution of the white dwarfs, we should say a few things about their spectra and the chemical composition of their atmospheres.

About 80 percent of white dwarfs show in their spectra almost nothing except strongly pressure-broadened Balmer lines of hydrogen; they are classified as spectral type DA. Clearly they cannot have any hydrogen in their interior, or there would immediately be a nuclear explosion. The spectral type DB is characterized by equally strongly broadened lines of helium. Then there are stars which are clearly cooler because they have metallic lines. The very small ratio of metals to H + He in these stars suggests that in degenerate stars hydrogen is transferred outward and heavy elements inward, probably by diffusion. The great abundance of helium (and possibly carbon) means that the interior of white dwarfs is composed of material processed by nuclear burning. By what evolutionary path have they traveled?

The white dwarfs cannot have started as stars of $0.6M_\odot$, since according to Table 26.1 these could scarcely have even begun to evolve in the lifetime of the Galaxy. Their ancestors must have been much more massive stars which have ejected a significant fraction of their mass "along the way." In fact, the evolutionary paths of such stars lead for short periods to such high luminosities that for strongly ionized material the radiative acceleration alone can almost (or completely?) balance the gravitational acceleration, and so material can easily be ejected. The remnant always still has a very high temperature and blows off an envelope, with smaller velocities, over a longer period of time: this forms a *planetary nebula*. Support for the origin of white dwarfs from the central stars of planetary nebulae comes from the observation of V. Weidemann (1971) that the birth rates of white dwarfs (estimated from their space density) and of planetary nebulae (estimated from their number and life time) nearly agree (2 to $3 \times 10^{-12}/\mathrm{pc}^3$ yr).

We have already (in Section 26.7) dealt with the origin of white dwarfs as a result of evolution in close binary systems. There are also indications that novae have something to do with white dwarfs.

But what happens to stars whose masses at the end of nuclear evolution are still larger than the white dwarf limiting mass of $1.4M_\odot$? Collapse to significantly higher densities is the only way left to form a stable star. Eventually the electrons and protons are so strongly squeezed together that they unite to form neutrons—the inverse of the β-decay process. The possibility of such neutron stars was investigated by L. Landau as early as 1932 and by J. R. Oppenheimer and G. M. Volkoff in 1939. The (degenerate) neutron material is very similar to that in heavy atomic nuclei since (apart from the electrostatic Coulomb forces, which are unimportant here) the forces of interaction between protons and between neutrons are equally strong. On this basis we make an elementary estimate, in which we treat the neutron star as, so to speak, a giant atomic nucleus.

An atomic nucleus, for example Fe^{56}, has a mass $M_{Fe} = 56 \times 1.67 \times 10^{-24}$ or 9.3×10^{-23} g and a radius of about 5.7×10^{-13} cm, and so has a mean density of $\rho \approx 1.2 \times 10^{14}$ g/cm³. If a star of for example $1 M_{\odot}$ were compressed to this density, it would shrink to a radius of about $5.7 \times 10^{-13} (M_{\odot}/M_{Fe})^{1/3}$ or 16 km! However, one should not take too literally the comparison between atomic nuclei and neutron stars, since the latter certainly cannot possess an electric charge. While in atomic nuclei there are roughly equal numbers of protons and neutrons, in stars the neutrons take precedence by far, since a proton + electron are "squashed together" to make a neutron. The transition from a neutron fluid of density about 10^{14} g/cm³ to the exterior ($\rho \rightarrow 0$) places before the theoretician a series of very interesting problems.

If a star initially rotated like the Sun with a period of 25 days, then, assuming complete conservation of angular momentum, its period would change to about 10^{-3} s (1 ms)! (Even so, the surface would be nowhere near the velocity of light.)

If the star originally had a reasonably ordered field of about 5 G, its field lines would be pressed together in about the same proportion as the cross-sectional area of the star, and the neutron star would contain an enormous magnetic field of about 10^{10} G.

In 1967 A. Hewish and his colleagues first discovered such neutron stars, in a quite unexpected way, in the *pulsars*.

Using the Cambridge radio telescope, they noticed signals in the meter-wave band which repeated completely regularly with a period of 1.337 s. Observations of the apparent motion on the sky showed that they were not dealing with terrestrial perturbations, but with a cosmic object, which was designated CP 1919, that is, Cambridge Pulsar with RA 19h 19m. Now about 60 such objects are known, with periods between 3.74 and 0.0331 s. The signals, which last for about 10 percent of the period, arrive consistently somewhat later at low frequencies than at high frequencies, because the propagation speed of electromagnetic signals in interstellar plasma decreases toward lower frequencies. The delay is proportional to the number of electrons in a centimeter squared column from the observer to the pulsar, the so-called

$$\text{dispersion measure } DM = \int N_e \, dl \qquad (26.16)$$

where it is customary to measure the electron density N_e in cm^{-3} and the distance l in parsecs. Using $N_e \approx 0.05$ cm^{-3} as a rough mean value (see below), we can reverse the process and estimate the distances of the pulsars; it is found that about half are within about 2 kpc.

Since the radio-frequency radiation from the pulsar is polarized (in a complicated way), it is possible to measure the Faraday rotation of the plane of polarization as a function of frequency and so determine the integral

$$\int HN_e \, dl \tag{26.17}$$

which is essentially the so-called rotation measure. By comparison with the dispersion measure (26.16) we obtain a mean value for the interstellar magnetic field of $H \approx 0.5$ to 2.8×10^{-6} G, in good agreement with other determinations.

It was a great step forward when in 1968–1969 the pulsar NP 0532 with the shortest known period $P = 0.0331$ s was identified optically with the central star of the Crab nebula, RA 5h 31m 31.46s, $\delta + 21°58'54''.8$ (1950). Observations through a rotating sectored disk tuned to the period P soon succeeded in showing that this star emits the same kind of "pulses" in the optical and even in the X-ray regions. The pulse shape is the same from X rays to meter waves; at longer wavelengths the pulses are broadened by rapid local fluctuations in the interstellar electron density N_e, the interstellar scintillation. The extraordinarily high accuracy of pulsar observations soon made it possible to confirm that the periods P of all pulsars are increasing. As a concrete measure of their rate of slowing down we give the time T within which the period would be doubled if it changed at the fixed rate dP/dt: The Crab pulsar NP 0532 changes fastest, with $T = 2500$ years; for the slower pulsars the times range up to $T \approx 3 \times 10^7$ years. This points to a pulsar lifetime of 10^7 to 10^8 years, which we will confirm by other means.

The extraordinarily short periods and the regularity of the pulses, especially in the case of NP 0532, leave no alternative to the idea of a neutron star, rotating with period P, which emits, like a light house, a beam of cone angle about 20°, and does so over the gigantic range of frequency from X rays to meter waves. The requirement that the stellar surface at any rate cannot rotate at the speed of light already means a significant upper limit for the star's radius.

The emission mechanism of pulsars has not yet been explained in detail, but it is believed that a rapidly rotating neutron star with a huge magnetic field blows plasma outward; initially this plasma "corotates" and so, at a cylinder of radius $cP/2\pi$ (= 1580 km for the Crab pulsar, for example), reaches the speed of light c. There the conditions for the emission of a cone of radiation are at any rate favorable. Why the Crab pulsar emits a weaker secondary pulse ("interpulse") between two main pulses, while other pulsars do not, is not yet satisfactorily understood.

Before we pursue these considerations any further, let us remember that the Crab nebula has played a decisive role in the development of astrophysics on several earlier occasions. W. Baade and R. Minkowski had in 1942 already made a detailed study of this very interesting object, M 1 = NGC

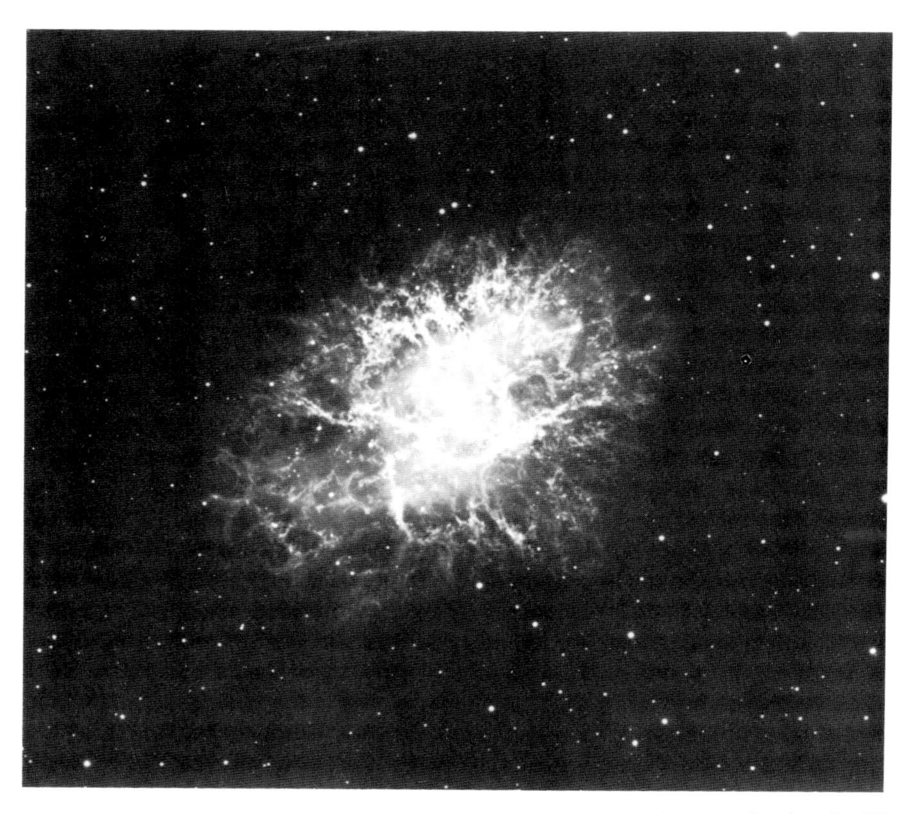

Figure 26.11. The Crab nebula, Messier 1 = NGC 1952. Red photograph taken by W. Baade on the 200-in Hale telescope. In 1949 J. G. Bolton successfully made the first optical identification of a radio source by proving the identity of Taurus A with the Crab nebula. The homogeneous inner part emits continuous synchrotron radiation, the outer part—the "legs" of the Crab—a nebular spectrum, particularly the red hydrogen line Hα. The Crab nebula originated in a supernova explosion in 1054 A.D. Its remnant forms the pulsar NP 0532, a neutron star, which "ticks" from X-rays to meter-waves with the period $P = 0.0331$ s, at present the shortest known.

1952, which has $m_{\text{pg}} = 9.0$ (Figure 26.11 shows one of the particularly fine Hα plates taken by Baade). It consists of an inner, almost amorphous region $3'\!.2 \times 5'\!.9$ with a continuous spectrum and an envelope, whose bizarre filaments (the "legs" of the crab) shine predominantly in Hα. As we can infer from East Asian records, the Crab nebula is the envelope of a supernova, which reached maximum brightness $m_V = -5$ or $M_V = -18$ in the year 1054 A.D. Comparison of the radial expansion speeds of about 1000 to 1500 km/s with the corresponding outward proper motions confirms the date of the explosion and in particular gives the distance of the Crab nebula as between 1.5 and 2 kpc.

Following I.S. Shklovsky's bold proposal in 1953, we explain the continuum radiation of the Crab nebula from radio waves to X rays as synchrotron radiation (see Section 28) from energetic electrons in a magnetic field whose order of magnitude is 10^{-4} G. This is supported by the following arguments: If the radiation were thermal, the electron temperature T_e would have to be higher than the highest radio brightness temperature T_ν, that is, $\geqslant 10^9$ K, while $T_e < 10^4$ K in all gaseous nebulae. Furthermore, the amorphous core shows *only* a continuum in the optical region and no lines at all. However an unambiguous argument in favor of the synchrotron theory is given by the verification of the polarization of the continuum which it demands (Figure 28.1). For the optical spectral range, this was obtained by Dombrowsky and Vashakidze in 1953–1954, and later with better resolution and accuracy by Oort, Walraven, and Baade. Later, in spite of difficulties caused by the Faraday effect, polarization measurements in the centimeter and decimeter range were also secured. A penetrating discussion of the lifetime of relativistic electrons in the Crab nebula led Oort to the view that even today, almost a thousand years after the supernova explosion, such electrons are being continually produced.

Soon after the discovery of the Crab pulsar it became clear that its rotation was the ultimate energy source for the Crab nebula and that enormous quantities of superthermal particles were produced in connection with the rotation and braking of the surrounding plasma and magnetic field. The production of synchrotron radiation requires electrons of about 10^7 eV for the radio region and of about 10^{14} eV for the X-ray region. There is no doubt that the corresponding positively charged atomic nuclei, that is, cosmic rays, would be accelerated at the same time. Theoretical considerations do not exclude the possibility that even cosmic rays of the highest energy so far observed, about 10^{20} eV, may be produced in this way.

A comparison of the energy emitted per unit time in the form of superthermal particles and the rotational energy of the Crab pulsar confirms our earlier estimate of its lifetime.

The remnants of the supernovae of 1572 (Tycho Brahe) and of 1604 (Kepler) can also be observed both radio astronomically and optically; Cassiopeia A, the strongest radio source in the northern sky, may also be a supernova remnant.

Then Hα pictures show a larger number of ring-shaped or circular nebulae that also emit radio radiation, like the well-known Cygnus Loop (the Veil nebula) or IC 443 (Figure 26.12). Finally, the north galactic radio "spur," which has been a puzzle for a long time, extending from near the galactic center almost to the galactic north pole, as well as certain other similar structures, appears to be part of a gigantic ring on the sky. All such radio sources might likewise be connected with old supernovae. Measurements in 1972–1973 from the UHURU satellite have revealed that most of them are also X-ray sources in the keV region.

Altogether the story of the Crab nebula teaches us that a (supernova) instability can occur in the evolution of a massive star, as a result of which a part of the mass collapses to become a neutron star while another part is flung out into space. At the present time the exact conditions for the

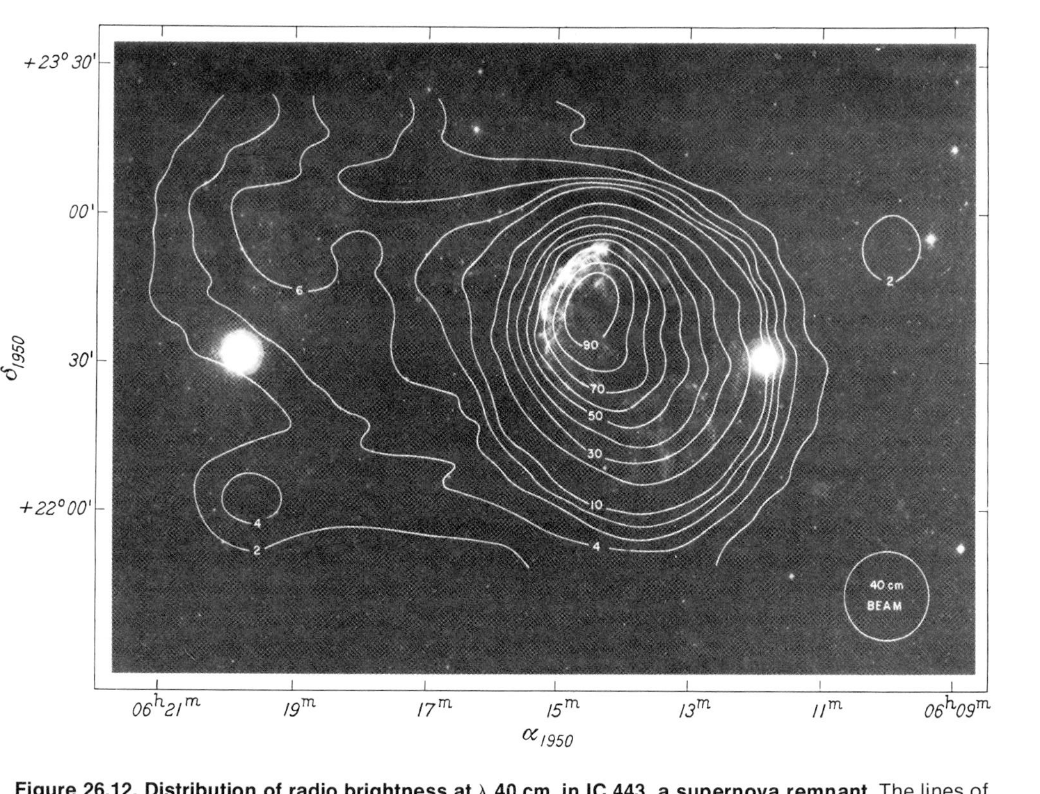

Figure 26.12. Distribution of radio brightness at λ 40 cm, in IC 443, a supernova remnant. The lines of equal radiation temperature are marked in units of 0.95 K. The radio map is shown superposed on a photograph taken from the Mount Wilson and Palomar Sky Survey.

occurrence of such an instability are not yet known. Possible "explosives" are the nuclear energy, which is released, for example, by the fusion of the remaining C^{12} nuclei into still heavier atoms, and the considerably greater gravitational energy.

Related to the pulsars are the pulsating X-ray sources discovered by satellites at wavelengths from 1 to 10 Å. Her X-1 could be identified with the eclipsing variable HZ Her discovered by C. Hoffmeister in the thirties. The X-ray signal shows first a period of about $1^s.24$, which can be ascribed to the rotation of a neutron star. This period is modulated by the Doppler shift caused by orbital motion with a period of 1.7 days. The latter period recurs in the eclipses of the X-ray source and in the strengthening of the optical radiation which always occurs if the side of the optical component which is turned toward us is illuminated by the X-ray source. A further period of 35 days is probably connected with the precessional motion of the axis of the neutron star. The mass of the neutron star can be estimated as from 0.3 to 1 solar mass. Several more pulsating X-ray sources like Her X-1 are known. Cyg X-1 is even (with considerable reservations) attributed to a black hole (see below).

What further possibilities exist and what role is played by rotation (because of conservation of angular momentum) we do not yet know. Equally, it is not yet clear why some neutron stars radiate mainly in the radio region and others mainly in the X-ray region.

The general theory of relativity has indicated—first of all as a theoretical possibility—an even more extreme contraction than in neutron stars. If, for example, one solar mass is shrunk to a radius < 2.9 km, the relativistic bending of radiation prevents any further energy from escaping in the form of radiation or matter; we obtain a *black hole*. We shall be able to give a better description of these remarkable structures—to which several astrophysicists ascribe great cosmological significance—when we discuss relativity theory in Section 30.

To summarize, we review the following evolutionary possibilities:

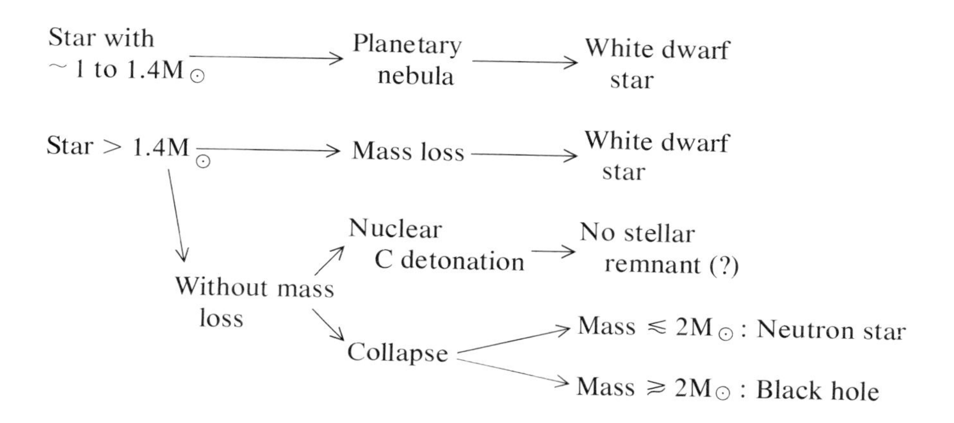

26.9 Stellar statistics and stellar evolution and rates of formation of stars

By statistical studies of stars with known parallaxes, during the 1920s J. Kapteyn, P. J. van Rhijn, and others derived the luminosity function for stars in our neighborhood (Section 23). Here we restrict ourselves to stars still on the main sequence and in Figure 26.13 we plot the luminosity function $\Phi(M_V)$ = number of main-sequence stars with absolute magnitudes $M_V - \frac{1}{4}$ to $M_V + \frac{1}{4}$ per cubic parsec in the solar neighborhood.[5] The fact that from $M_V \approx 3.5$ this number falls off rapidly toward brighter magnitudes was related by E. E. Salpeter (1955) to the fact that stars fainter than absolute magnitude 3.5 have been accumulating since the formation of the Galaxy $T_0 \approx 10^{10}$ years ago, without significant change. Brighter stars, on the other hand, after the lapse of about the evolution time t_E (Table 26.1) from the epoch of their formation, move away from the main sequence and ultimately, after a total time which is certainly small compared with T_0, they become white dwarfs, etc. Thus we can readily calculate the *initial luminosity function* $\Psi(M_V)$ giving the number of stars *that have been formed* since the start of the Galaxy in the magnitude interval $M_V \pm \frac{1}{4}$, per

[5] $\Phi(M_V)$ is often defined instead for the magnitude interval $M_V \pm \frac{1}{2}$.

Figure 26.13. Luminosity function Φ (M_v) and initial luminosity function Ψ (M_v) of main sequence stars in the solar neighborhood. The functions Φ and Ψ give respectively the present number of stars per cubic parsec in the magnitude range $M_v - \frac{1}{4}$ to $M_v + \frac{1}{4}$, and the number of stars born in this range since the formation of the Galaxy.

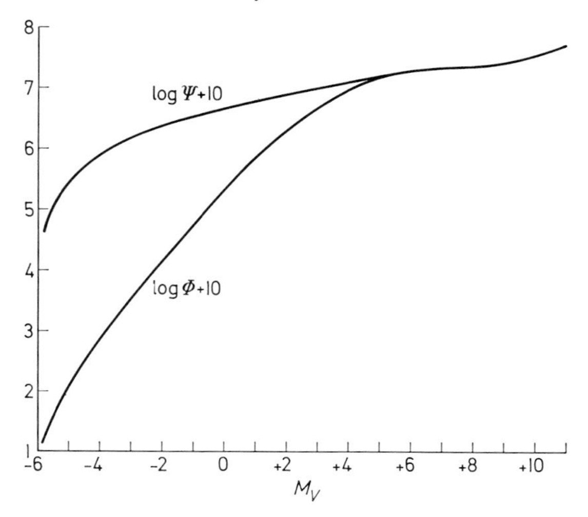

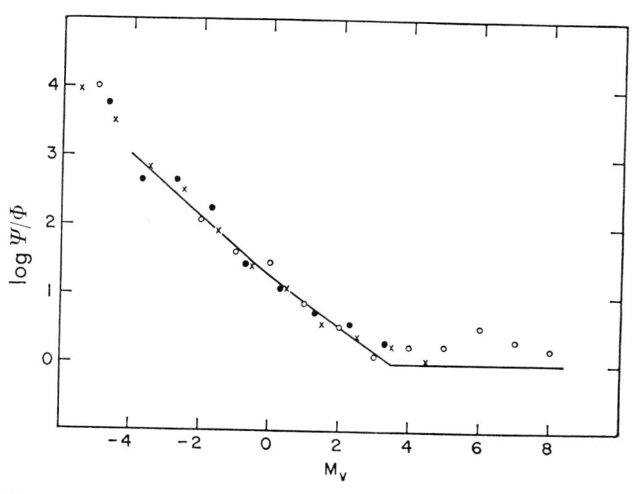

Figure 26.14. Comparison of observed luminosity functions of young galactic clusters NGC 6530 (dots), NGC 2264 (circles), and the Orion Nebula cluster (crosses) with E. E. Salpeter's zero age luminosity function. The ordinate is the logarithm of the ratio of the observed luminosity function of the cluster (points) or of the zero age luminosity function $\Psi(M_v)$ (line), to the observed luminosity function $\Phi(M_v)$ for main sequence stars in the vicinity of the Sun. The luminosity functions of the clusters have been adjusted for the star density in each cluster relative to that of the solar neighborhood. (After M. Walker, 1957.)

cubic parsec (external conditions being assumed to remain the same). For brighter stars we have

$$\Psi(M_V) = \Phi(M_V)\,\frac{T_0}{t_E(M_V)}; \tag{26.18}$$

for fainter stars $\Psi(M_V)$ goes over continuously into $\Phi(M_V)$. In Figure 26.13 we have included the initial luminosity function $\Psi(M_V)$ following A. R. Sandage, who has extended the calculations made by E. E. Salpeter.

If our ideas are correct, then the luminosity function of young galactic clusters must correspond to the initial luminosity function Ψ and not to the Φ of our neighborhood. In Figure 26.14 M. Walker has compared the luminosity functions of three very young clusters, which he has studied, with the calculated function Ψ, derived from the luminosity function Φ for our neighborhood. The differences between the total stellar densities in clusters and in our neighborhood are taken into account by appropriate displacements of the logarithmic ordinate scale. The shapes of the curves show remarkable agreement. This supports the view that the resolution of an original body of gas into stars proceeds throughout according to the same initial luminosity function $\Psi(M_V)$.

The difference $\Psi(M_V) - \Phi(M_V)$, summed over all M_V, represents those stars that have evolved off the main sequence at any time since the origin of the Galaxy. By far the majority of these stars must now be white dwarfs. Actually the space density of white dwarfs calculated from Figure 26.13 agrees well with the observed value, certainly as well as could be expected in view of the uncertainty of the data.

The foregoing data and discussion refer, more precisely, to the disk and spiral-arm population I of the Milky Way. What about the metal-poor halo population II? Of prime interest is the luminosity function of those stars which stem from the early days of the galactic system, that is, the main-sequence stars beneath the "knee" in the color-magnitude diagram, with $M_V > 4$. We can best avoid the selection effects, which are so dangerous in determining such a statistic, by comparing differentially the luminosity functions of the halo and of the disk (plus spiral-arm) populations. To do this, we make use of the very homogeneous observational material for stars within 25 pc, for which Sir R. Woolley, S. B. Pocock, E. A. Epps, and R. Flinn (1971) have gathered together all the data, in particular their galactic orbital elements (generalized eccentricity e and inclination i) and two-color diagrams (for the recognition of metal-poor stars by their ultraviolet excess). Brighter stars ($M_V < 4$) are lacking, as is to be expected, while, within the statistical errors, the luminosity function of the still present stars of the extreme halo population II ($e \geqslant 0.3$, $i > 0.5$) cannot be distinguished from that of the population I stars.

A. Sandage has investigated the luminosity function, including giants, of the globular cluster M 3. Apart from a maximum, or peak, at $M_V \approx 0$, which is populated by cluster-variables, the luminosity function of the globular cluster differs little from that of our neighborhood. Therefore Sandage filled in the region fainter than $M_V = +6$, where the stars are too faint to be observed, in accordance with the van Rhijn luminosity function. Then we have for the entire cluster:

	Number	Mass		
Luminous stars	588 000	1.75×10^5	$2.45 \times 10^5\ M_\odot$	(26.19)
White dwarfs	48 500	0.70×10^5		

While half the luminosity of the cluster comes from stars brighter than $M_V = -0.14$, half the mass is reached only at $M_V = +11.28$. The overall ratio of mass to luminosity is $M/L \approx 0.8$ in solar units in excellent agreement with other studies (including dynamical studies) of M 92.

After we have become familiar with other galaxies in the following Sections 27 and 28, we shall look more closely at the fundamental problems of the origin and evolution of our Milky Way System, its stellar populations and the abundance distribution of the chemical elements.

27. Galaxies

The advance into cosmic space beyond the confines of the Milky Way into the realm of distant galaxies—or extra-galactic nebulae, as they used to be called—and the beginnings from that of a *cosmology* founded upon observation must for all time belong to the greatest achievements of our century.

We mentioned earlier (page 231) the classic catalog of Messier (M) 1784 and also, arising out of the work of W. and J. Herschel, the *New General Catalogue* (NGC) of 1890 by J. L. E. Dreyer and his *Index Catalogue* (I.C.) of 1895 and 1910, the notations of all of which are still in use today. The Andromeda nebula (Figure 27.1), which is, apart from the Magellanic Clouds in the southern sky, the brightest galaxy and which S. Marius had already observed in 1612, has, for example, the catalog numbers M 31 and

Figure 27.1. Andromeda galaxy M 31 (NGC 224) and its physical companions, the elliptical galaxies M 32 (NGC 221) and NGC 205 (lower left). M 31 has an inclination of 11°.7, and a distance of 670 kpc. Mount Wilson and Palomar Observatories, 48-in Schmidt camera.

NGC 224. In modern times we mention first the Shapley–Ames catalog of 1932, which surveyed the whole sky uniformly and recorded 1249 galaxies brighter than 13^m. Out of that grew the *Reference Catalogue of Bright Galaxies* by G. and A. de Vaucouleurs (1964), which covers the whole sky and lists 2599 galaxies brighter than 14^m. Unsurpassed illustrative material is given in *The Hubble Atlas of Galaxies* by A. Sandage, Mt. Wilson and Palomar Observatories, 1961.

In the 1920s the cosmic status of the "spiral nebulae" was the subject of strenuous debate amongst astronomers. Here we can only just mention the important contributions made by H. Shapley, H. D. Curtis, K. Lundmark, and others.

In 1924 E. Hubble succeeded in partially resolving the outer parts of the Andromeda galaxy M 31 and of some other galaxies (to use the modern terminology) into stars, and in identifying various objects with known absolute magnitudes, which could be used as a basis for photometric distance determination:

(1) *Cepheids*. The light curves of these (Figure 27.2), with periods of from 10 to 48 days, were first interpreted naturally in terms of the "general" period-luminosity relation. About 1952 W. Baade realized, partly as a result of discrepancies regarding the absolute magnitude of red giant stars, that the "classical" cepheids of population I, in particular therefore the long-period cepheids of the Andromeda galaxy, are about $1^m.5$ *brighter* than the corresponding cepheids of population II. This then required an increase of extragalactic distances by a factor of about 2. Nowadays the relation between period P (in days) and mean abso-

Figure 27.2. Light curves of four Cepheids in the Andromeda galaxy M 31, after E. Hubble, 1929. The abscissas give the time in days, and the ordinates the photographic magnitude m_{pg}.

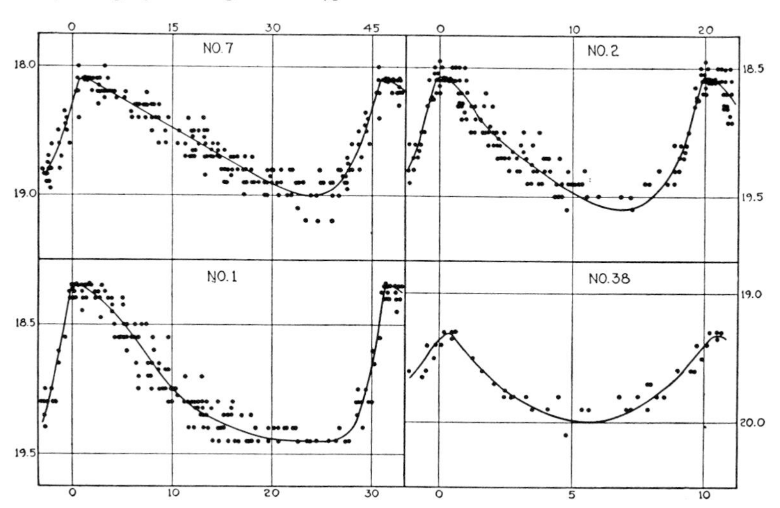

lute magnitude M_V, together with small corrections which depend on the color $B-V$ or on the pulsation amplitude of the star, is well determined between $\log P = 0.4$, $M_{<V>} = -2.65$ and $\log P = 2.1$, $M_{<V>} = -7.07$. In this way we can penetrate into space to a distance modulus of about $m - M = 28$.

(2) *Novae.* Hubble was able to detect and make use of novae, whose light curves fully resemble those of galactic novae. The very much brighter S Andromedae with $m_{V,\max} \approx 8$ observed by Hartwig in 1855 was later recognized to be a supernova.

(3) *Brightest nonvariable stars* with $M_V \approx -10$ or, better, the brightest *globular clusters* with $M_B \approx -9.8$. Again certain corrections were subsequently needed, after a fraction of the "brightest stars" turned out to be groups of stars of H II regions.

(4) *H II regions.* After more careful calibration, the angular diameters of bright H II regions, determined from Hα plates, proved to be a remarkable aid to measuring greater distances. In this way distance moduli $m - M \approx 32$ can be reached.

Here we cannot enter into the questions (very difficult in their details) of the exact determination of the absolute magnitude of various sorts of objects nor of the equally important improvement of the magnitude scale for fainter stars.

After taking account of a small amount of interstellar absorption, we now find for the Andromeda galaxy M 31, our nearest neighbor in space, a true distance modulus $(m - M)_0 = 24.12$ or a distance of 670 kpc or 2.2 million light years. (In his pioneering work in 1926 E. Hubble had assigned a distance of 263 kpc.)

W. Baade first succeeded in 1944 in resolving the central part of the Andromeda galaxy and of the neighboring smaller galaxies M 32 and NGC 205 (Figure 27.1) by using the utmost refinement of photographic technique. Here the brightest stars are red giants with $M_V = -3$, while the brighter blue stars of the spiral-arm population I are absent.

As a result of E. Hubble's investigations it was definitely established that galaxies like M 31 and others are broadly similar to our Milky Way Galaxy. What can we say then about the distribution of the galaxies in the heavens?

As we saw, the almost galaxy-free zone around the galactic equator between about $b = \pm 20°$, the so-called "zone of avoidance" is produced by a thin layer of absorbing material in the equatorial plane of the Milky Way.

Fairly commonly we observe *groups* of two, three . . . galaxies that obviously belong together physically. The Milky Way along with the Andromeda galaxy M 31, its two (physical) companions (Figure 27.1), and about 20 other galaxies form the so-called *local group*. Again in higher galactic latitudes, where galactic obscuring clouds are absent, the galaxies show a nonuniform distribution in the sky. As Max Wolf had already noted in a number of cases, they form *clusters of galaxies* like, for example, the Coma cluster (in Coma Berenices) at about 10^8 pc = 100 Mpc (megaparsec) distance which includes several thousand galaxies in a region of about 3 Mpc diameter. The local group, together with a series of similar groups, appears to belong to the Virgo cluster, whose center is about 15 Mpc away.

Now consider more closely the manifold forms of galaxies. E. Hubble was able to arrange these in a scheme of classification (Figure 27.3) which, with improvements (already incorporated in the figure), also forms the basis for the *Hubble Atlas of Galaxies* and at the same time is more exactly specified by the *Atlas*. As in the case of the Harvard sequence of spectral types, it is clear in advance that such a purely descriptive scheme need by no means represent an evolutionary sequence.

The elliptical galaxies E0 to E7 have rotationally symmetric form with indications of some further structure. The observed (apparent) ellipticity is naturally that determined by the projection of the (true) spheroid on the sky. The statistics of the apparent ellipticities show that the true ellipticities of the E galaxies are fairly uniformly distributed. The surface brightness decreases uniformly from the center toward the outside.

From the ellipticals there is a steady transition to the *spiral galaxies*. These are all more strongly flattened. However, here we have a fork in the sequence.

The *normal spirals* (S) have a central bulge from which the spiral arms spring more or less symmetrically. In the case of the *barred spirals* (SB) a straight "bar" comes out of the central bulge, and to its ends the spiral arms are attached almost perpendicularly.

Between the spiral arms in normal, as well as in barred, spirals there are great numbers of stars so that in photometric recordings the arms do not stand out at all prominently.

The sequence of types S0 . . . Sc or SB0 . . . SBc is characterized by the property that the *central region*—also called the nucleus or the lens of the spiral galaxy—becomes relatively smaller, while the windings of the *spiral arms* become more open, as we proceed along the sequence. For example, the Andromeda galaxy M 31 (Figure 27.1) is a typical Sb spiral; our Milky Way is on the boundary between Sb and Sc.

Figure 27.3. Classification of galaxies.

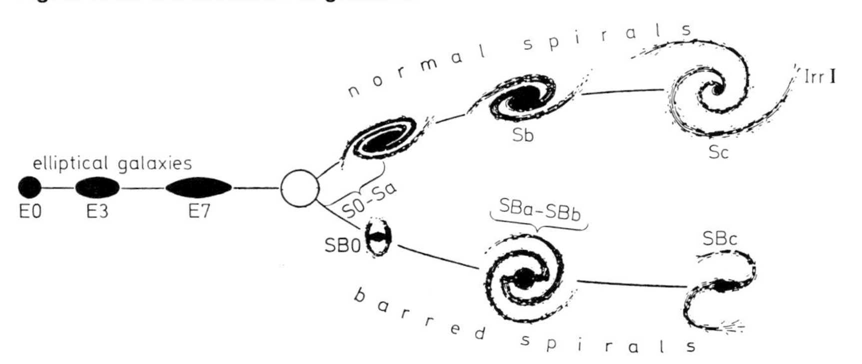

In the center of many galaxies there is a clearly delimited, apparently starlike nucleus[1], whose angular diameter is, for example, for M 31 $1''.6 \times 2''.8$, corresponding to a size of about 5.4×9.4 pc.

The *Hubble Atlas* further divides the S and SB spirals into two subclasses according as the arms spring directly out of the nucleus, or out of the ends of the bar (suffix s), or come tangentially out of an inner ring (suffix r).

To the Sc galaxies (like M 33 in the local group) are joined continuously the irregular galaxies Irr I (not shown in Figure 27.3). These relatively rare systems show in the first place no rotational symmetry and no well-defined spiral arms, etc. The best-known examples are our neighbors in the local group, the Large Magellanic Cloud LMC (Figure 27.4) and the Small Magellanic Cloud SMC. These are situated in the southern heavens at distances of about 55 and 63 kpc. More exact investigation has shown that the irregular luminosity distribution of the Irr I systems is only brought out by the bright blue stars and neighboring gas nebulae that feature strongly in the blue plates. In agreement with radio-astronomical 21-cm measurements of the location and velocity distribution of the hydrogen, the substrate of fainter red stars shows that the main body of, for instance, the Magellanic

[1]In the literature the central region (or lens) of a galaxy is often described as the "nucleus." This leads to confusion and misunderstanding. In the future, the term *nucleus* should be reserved for the much smaller, but—as we shall see—extremely important bright condensation in the center of a galaxy.

Figure 27.4. Large Magellanic Cloud LMC. The numerous H II regions stand out clearly in this Hα photograph.

Clouds exhibits a much more regular, flattened form and considerable rotation.

The Irr II systems, which only superficially resemble the Irr I, such as M 82 have recently been recognized as galaxies in the nucleus of which an explosion of unimaginable violence is occurring.

As must be stressed once again, in the Hubble sequence the galaxies are classified only according to their *form*. Considered physically, as B. Lindblad has emphasized, we clearly have an ordering according to increasing angular momentum (flattening).

The study of the distances of individual galaxies and of their membership in clusters of galaxies has shown that, for any one Hubble type, their absolute magnitudes and diameters are spread over a wide range.

For spirals, barred spirals, and irregular galaxies, van den Bergh (1960) found that the luminosity increased with increasing development of spiral arms and analogous structures. On this basis he founded a luminosity classification, analogous to that of stars, which for this range of Hubble types permits a unique calibration according to absolute magnitude:

	Luminosity class	Mean absolute magnitude M_{phot}
I	Supergiant	−20.2
II	Bright giant	−19.4
III	Normal giant	−18.2
IV	Subgiant	−17.3:
V	Dwarf galaxy	−15::

While the luminous systems are more abundant among spirals, the dwarf galaxies by contrast predominate amongst the irregulars.

Figure 27.5. Color-magnitude diagram of Draco System. This dwarf galaxy is at a distance of 99 kpc and has a diameter of 1.4 kpc. The system is very similar to a metal-poor globular cluster, but is about 10 times larger. (After W. Baade and H. Swope, 1961.)

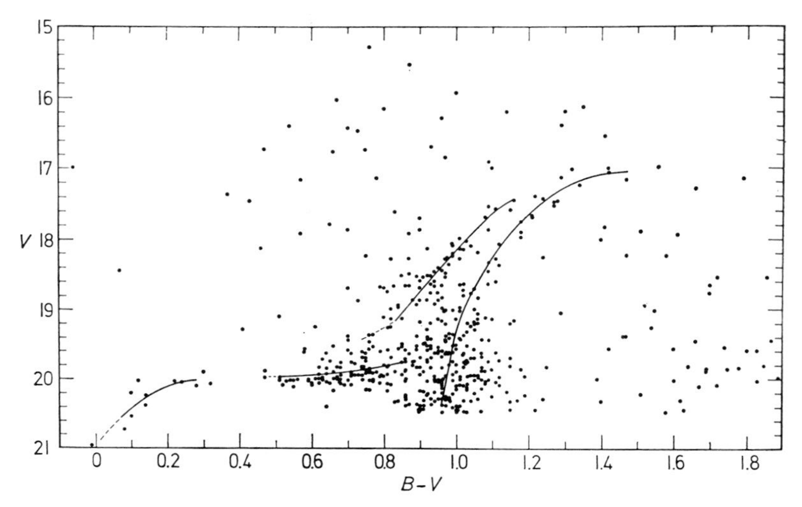

Figure 27.6 The dwarf irregular galaxy IC 1613 (Ir V; M_V = 14.8). Red plate (103aE emulsion + RG2 filter) taken by W. Baade with the Hale telescope in 1953. Several H II regions can be seen in the lower right. The visual luminosity of IC 1613 is some 30 times smaller than that of the Large Magellanic Cloud.

For elliptical galaxies, the surface brightness at the center of the image (proportional to the number of stars along the line of sight) is a measure of the luminosity. This shows in particular that the brightest E galaxy in a cluster (the so-called "first ranked E galaxy") always has, within very close limits, the same absolute magnitude $M_B = -21.7$, similar to the brightest spiral galaxies. There is a range of luminosity of a good 5 powers of 10 from the giant elliptical galaxies through the normal ellipticals to the faintest dwarf E galaxies. There is no sharp distinction between the latter and the spheroidal dwarf galaxies like the Draco system, with $M_V = -8.6$, which can most easily be compared to a very large globular cluster, a comparison which extends to its color-magnitude diagram (Figure 27.5).

In a similar way we contrast the Large Magellanic Cloud, which has $M_V = -18.5$ (Ir or SBc III–IV) (Figure 27.4), with the dwarf irregular galaxy IC 1613 which also belongs to the local group and has $M_V = -14.8$ (Ir V) (Figure 27.6). The latter also has a series of bright OB stars and H II regions.

The small dwarf galaxies often appear to be physically linked to neighboring giant galaxies. As well as the Magellanic Clouds, the systems in

Sculptor, Draco, and Ursa Minor, all closer to us than 100 kpc, may be "satellites" of our Milky Way system. The E galaxies NGC 205 and M 32, with $M_V = -16.4$ (Figure 27.1), have long been known to be companions of the Andromeda galaxy M 31. Recently, S. van den Bergh (1972) has discovered near the Andromeda galaxy another three dwarf spheroidal galaxies, And I–III, with absolute magnitudes $M_V \approx -11$ and diameters of from 0.5 to 0.9 kpc; they closely resemble, for example, "our" Sculptor system. The recognition of such mini-galaxies is already very difficult in the local group, and is hopeless outside it.

With regard to the *spectra of galaxies,* in the first place the Fraunhofer absorption lines establish that the galaxies are composed mainly of stars. Emission lines of H II regions are observed in the arms of spiral galaxies and even more markedly in the irregular systems. Some quite different origin must obviously be found for the very broad emission lines in the nuclei of the so-called Seyfert galaxies and of the Irr II systems like M 82 as well as in those of many giant ellipticals (see Section 28).

The Doppler shifts of the absorption and emission lines (it is better to measure the latter) give information about the radial velocity and the rotation of a galaxy.

In the Andromeda galaxy M 31, radial velocity measurements have been carried out for many emission objects (H II regions) by H. W. Babcock (1939), N. U. Mayall (1950), and, more recently, V. C. Rubin and W. K. Ford (1970). On the whole, these are confirmed by 21-cm radio measurements.

For the center of the galaxy, one obtains to begin with a radial velocity of −300 km/s (in round figures). Since the galactic coordinates of Andromeda are $l^{\mathrm{II}} = 121°$, $b^{\mathrm{II}} = -21°$, this velocity is for the most part a reflection of our orbital motion in the Galaxy. The portion of the radial velocity which remains after subtracting −300 km/s is symmetrical about the nucleus; we can at once convert it to a rotation speed (Figure 27.7) by assuming circular orbits and taking account of the inclination of about 77° of the rotation axis to the line of sight.

The rotation speed $V(R)$ as a function of distance R from the center (1′ corresponds to 0.2 kpc) must first increase as we go inward from the outside. (In particular, sufficiently far outside all attracting mass, Keplerian motion would correspond to a relation $V \propto R^{-1/2}$.) After going through a flat maximum of ~ 270 km/s at $R \approx 10$ kpc, V decreases inward to a deep minimum at $R \approx 2$ kpc. Further in, V increases again and reaches a second maximum of 225 km/s at $R = 0.4$ kpc. With the help of an electronic image converter, A. Lallemand, M. Duchesne, and M. F. Walker (1960) carried out measurements in the region of the star-like nucleus of M 31 which showed that the speed goes through a second deep minimum at $R \approx 20$ pc and then through a sharp maximum of about 87 km/s at a distance $R = 7$ pc (2″.2) from the center before finally falling to zero.

By analogy with the earlier calculations for our Milky Way system, we first obtain from Kepler's third law (equation 23.26; $V^2 = GM/R$) the mass

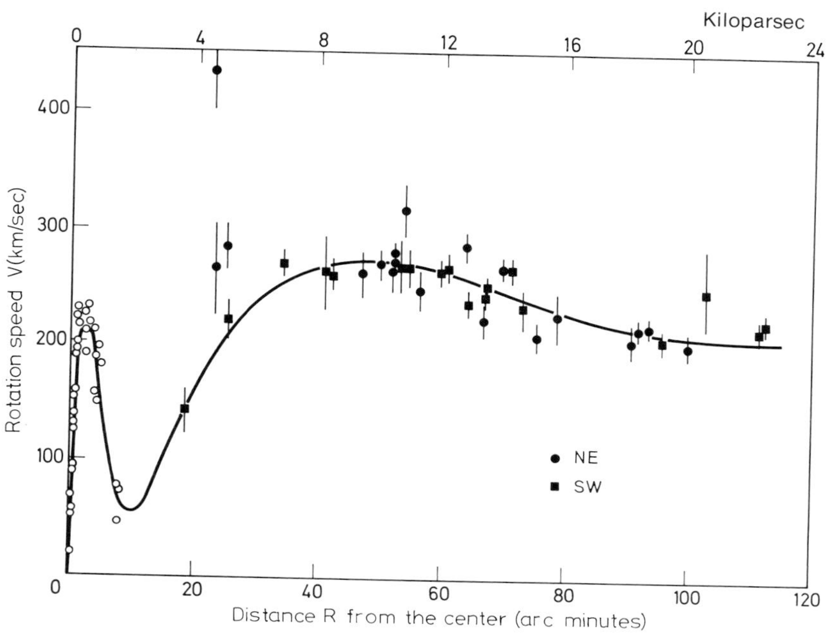

Figure 27.7. Rotation speed V for OB associations in the Andromeda galaxy M 31 as a function of the distance R from the center ($1' \approx 0.2$ kpc) after V. C. Rubin and W. K. Ford (1970). The points denote measurements using the Hα line, open circles measurements using [N II] lines. It is not possible to show on this diagram the maximum in the nucleus of the galaxy, where $V \approx 87$ km/s at $R = 7$ pc.

M of M 31 within 24 kpc as 1.85×10^{11} solar masses; the total mass may lie between 2 and $3 \times 10^{11} M_\odot$. According to 21-cm observations only about 1.5 to 6 percent of the mass is in the form of neutral atomic hydrogen.

To determine the mass distribution, we construct models using nested homogeneous ellipsoids or disks, similar to those which M. Schmidt used for the Milky Way. For these models we calculate the potential field and hence the distribution of rotation speed $V(R)$. In fitting this to the observations (Figure 27.7), we obtain essentially the surface density of the mass distribution, that is, the mass contained in a cylinder of cross section 1 pc^2 perpendicular to the plane of symmetry. This increases inward and reaches a maximum of $\sim 400 M_\odot/\text{pc}^2$ at $R = 5$ kpc. At ~ 2 kpc from the center the surface density falls almost to zero; no model seems to be able to reproduce properly the corresponding minimum in the rotation speed. The "lens" lying further in, within ~ 0.4 kpc, has a mass of $\sim 6 \times 10^9 M_\odot$. In the actual nucleus of the galaxy, that is, in a region of only 7.4 pc radius, about 1.3×10^7 solar masses are crowded together; the mass density is about 10^4 times that in our neighborhood. The mass-to-light ratio, known only to order of magnitude, shows that the nucleus of M 31 mainly consists of stars. For the whole Andromeda galaxy the mass-to-light ratio is about 12 (in solar units).

If we compare M 31 with our Milky Way system, we recognize a very great resemblance, which we shall not go through in detail.

For elliptical galaxies rotation fades into the background in comparison with the random motions of the stars in their common gravitational field. In this case it is only possible to estimate the total mass of the galaxy; this is done by determining the mean squared speed from the width of the spectral lines and using the virial theorem (Figure 26.9), which states that (averaged over time) the kinetic energy of the galaxy equals $-\frac{1}{2} \times$ its potential energy. People have also tried to use the motions of double galaxies and of galaxies in clusters (again using the virial theorem) in order to obtain an estimate of their masses.

In Table 27.1 we have gathered together some information about (extreme) giant and dwarf galaxies of different Hubble types. The numbers can only give an order-of-magnitude idea, or act as points of reference.

Table 27.1. Absolute magnitude *Mv*, color index $B - V$, mass M, mass-to-light ratio M/L (in solar units), and hydrogen content (percent by mass) of galaxies of different Hubble types. "Giants" and "Dwarfs" denote galaxies at the extremes of the magnitude range.

	E		Sb		Irr	
	Giants	Dwarfs	Giants	Dwarfs	Giants	Dwarfs
M_v	-23.3	-10 to -8.6	-22.5	-19	-18.5	-14.8
$B - V$	$+1.0$		$+0.9$ to 0.7		$+0.3$	
M	$\sim 10^{12}$	$\sim 10^{6}$	3×10^{11}	10^{10}	6×10^{9}	4×10^{8}
M/L	~ 10		~ 5		~ 1	
Percent H I	< 0.01		1 to 5		~ 30	

The luminosity function of galaxies (Figure 27.8; in contrast to Figure 26.13, *not* logarithmic) gives information about the frequency of galaxies of different absolute magnitudes (or luminosities). The curve found by Hubble in 1936 is today largely of historical interest. The luminosity functions derived by E. Holmberg (1950) from nearby galaxies and by F. Zwicky (1957) from clusters of galaxies agree well up to $M_p \approx -12$. The dwarf galaxies in any case form a considerable fraction by *number* of the popula-

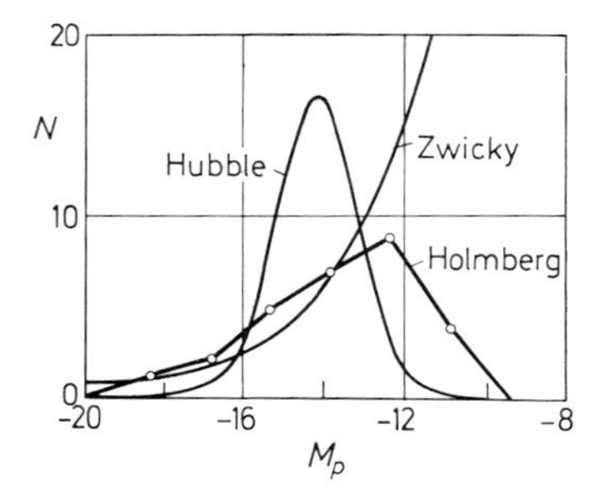

Figure 27.8
Luminosity function of galaxies according to E. Hubble, 1936, E. Holmberg, 1950, and F. Zwicky, 1957.

tion of the universe, although they make scarcely any contribution in luminosity and mass.

The frequency distribution of Hubble *types* is *not* the same in different clusters of galaxies.

As in the star clusters of our Milky Way, we should like to construct color-magnitude diagrams for the distant galaxies and then study different kinds of stars and stellar populations in order to gain insight into the origin and evolution of galaxies. In practice, we first come up against a problem of shortage of photons; the construction of larger telescopes in the southern hemisphere, and especially the development of ever improving image converters and the expansion of the accessible spectral regions justifies great hopes for the coming decades.

Here we shall concern ourselves first with more or less "normal" galaxies in the optical spectral region, about 3000 to 10000 Å. Then in Section 28 we shall discuss radio astronomy and the related questions of high-energy astronomy and the activity of galaxies, particularly of their nuclei. Only in Section 29 will we, very cautiously, venture to approach some theoretical problems.

Even in the nearby Andromeda galaxy we can observe single stars only down to absolute magnitude about −3.5. Only in the still closer Magellanic Clouds has it been possible, quite recently, to analyze with satisfactory dispersion the spectra at least of a few extreme supergiants and gaseous nebulae. In general we must be content with integrated spectra, particularly of the central regions.

The classification of the spectra of galaxies is based mainly on photographs of the central regions or of the whole galaxy with a dispersion of only 100 to 400 Å/mm. It aims to give information about the stellar populations of the system and to relate their classification to the Hubble sequence.

Since the spectrum of a galaxy consists of the superposition of many stellar spectra (composite spectrum), other things being equal, in the short-wavelength region hot blue stars will dominate the spectrum, and in the long-wavelength region cooler red stars will do so. W. W. Morgan and N. U. Mayall (1957) restricted themselves principally to the interval λ 3850–4100 Å:

(1) *A systems* have broad Balmer lines; in the range λ 3850–4100 the spectrum corresponds to spectral type A, near λ 4340 to that of an F8 star. A typical representative is the Irr I galaxy NGC 4449 which is similar to the Magellanic clouds. The Hubble types Irr I, Sc, SBc all come in here.

(2) *F systems* correspond in the violet to spectral type F, and near λ 4340 to G. A typical example is the Sc galaxy M 33 = NGC 598; the Sb spirals also occur here.

(3) *K systems* produce a spectrum that can be interpreted as the superposition of the spectra of normal (that is, not metal-poor) G8 to early M giants (CN criterion) with fainter F8 to G5 stars. Earlier spectral types

make no appreciable contribution. The prototype is the Andromeda galaxy M 31 = NGC 224. Here one can also see that the spectra of the central region and of the disk are not significantly different. Besides the large Sb and Sa spirals, the corresponding barred spirals, giant ellipticals like the well-known radio-galaxy M 87 = NGC 4486 or NGC 4636 as well as "dust-free" Sb and Sa systems are all included. Between the main types A–F–K one can insert the intermediate types AF and FG.

To summarize once again, Figure 27.9 shows which regions in the HR diagram according to Morgan and Mayall make essential contributions to the spectrum in the case of A systems and in the case of K systems.

Closely linked with the spectral type is the *color index* of the galaxies. Since spirals seen edge on show absorption and reddening by their own interstellar material, on the side presented to us, we must correct for these (as well as for galactic absorption and reddening). According to E. Holmberg (1964) this true color index $C^* = (m_{\mathrm{pg}} - m_{\mathrm{pv}})^*$ shows a close correlation with the Hubble type and with the spectral type of galaxies, independent of their luminosity class. On the average C^* is 0.12 for Irr, 0.41 for Sb, and 0.77 for E galaxies, corresponding qualitatively to their types according to Morgan and Mayall or to the Hertzsprung–Russell diagram (Figure 27.9).

We shall attempt to bring some order into the at first rather bewildering diversity of observations of our Milky Way and of other galaxies by making use of the concept of *stellar populations*.

In 1944, under the astronomically convenient conditions of the war-time blackout, W. Baade succeeded in using the 100-in telescope to resolve into stars the central region of the Andromeda galaxy M 31 and also its elliptical companions M 32 and NGC 205 (Figure 27.1). Here the brightest stars

Figure 27.9. Hertzsprung–Russell diagram. This diagram shows the regions principally contributing to the light of (1) A system NGC 4449 (Irr I) (2) K system NGC 224 = M 31 (Sb). As in the case of the Milky Way, the cooler parts of the main sequence do not contribute significantly.

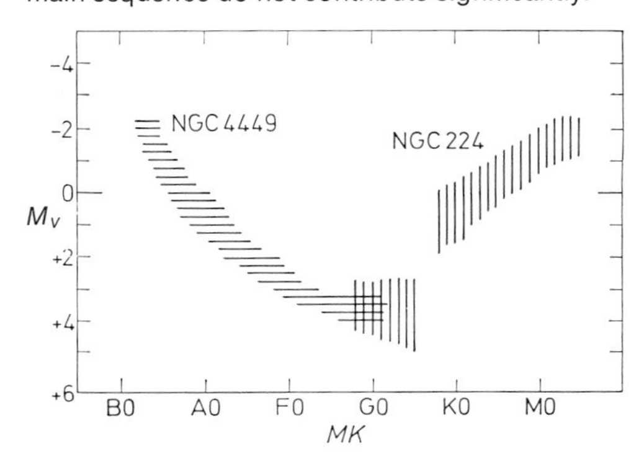

proved to be not the blue OB stars as in the spiral arms but *red giants*, just as in the globular clusters of the Milky Way system. Baade realized that different galaxies, or parts of galaxies, are inhabited by different *stellar populations* with quite different color-magnitude diagrams.

In the first place he distinguished:

(1) *Population I*, with a color-magnitude diagram similar to that of our neighborhood (Figure 15.2). The brightest stars are blue OB stars with $M_V \approx -7$.
(2) *Population II*, with a color-magnitude diagram similar to that of the globular clusters (for example, M 92, Figure 26.3). In this case the brightest stars are red giants.

He showed further that the stellar populations, or parts of them, also differ in respect of *stellar dynamics* and *stellar statistics*, that population I is composed of *young*, and population II of *old*, stars and that a part of the latter consists of the halo population II metal-poor stars.

In the course of time a clarification and refinement of the classification scheme of stellar populations became necessary. This was worked out at a conference on stellar populations in Rome in 1957 by J. H. Oort and others. Table 27.2 gives a review, with some recent additions, of the new classification of stellar populations, which are now characterized primarily by the *spatial distribution* and *orbits* of their stars in the galaxy. *Age* and *metal abundance* are certainly related to this one-dimensional sequence as further parameters, but they are not uniquely prescribed by it. In more detailed investigations they would need to be specified in much greater detail. After some remarks on ways of determining empirically the last-named parameters, we shall proceed to study the three main populations in more detail.

The *age* of a star cluster or of an aggregate of stars with a common origin can (as we have seen in Section 26) be obtained from the color-magnitude diagram by comparison with theoretical isochrones. However we should keep in mind that the *chemical composition* of the stars enters into these calculations. Furthermore, it has generally been assumed up to now that in the course of evolution no *mass loss* occurs, and that there is no *mixing* between original material and material which has undergone nuclear processing (except in the convection zones, whose mixing-length theory is still rather uncertain). The fact that the latter assumptions are still very problematic, especially in the region of the giant branch, cannot leave unaffected the precision of age determinations.

Our main information about the *chemical composition* or *metal content* of stars comes from quantitative analysis of high-dispersion spectra. This method is wearisome and mostly confined to stars brighter than about 8^m; however, it does yield the most detailed information.

We must expect *a priori* that in some stars material from the "nuclear reactor" in the center somehow succeeds in reaching the atmosphere, so that it is possible to observe directly traces of hydrogen, helium, and/or carbon burning. Here the helium stars, the carbon stars, the Ba II and the S

Table 27.2 Stellar populations, especially in the Galaxy.

Population:	(Old) halo population II	Disk population	Extreme or spiral arm population I
Transition groups:	Intermediate population II	Older population I	
Typical members:	Subdwarfs	"Normal" stars of disk and nucleus	Interstellar matter
	Globular clusters	Planetary nebulae Novae	Galactic clusters and associations
	RR Lyrae variables with $P > 0^d.4$		
	Bright red giants		Bright blue OB stars
Mean distance from galactic plane:	2000	400	120 pc
Mean velocity $\bar{W}$ perpendicular to galactic plane:	75	18	8 km/s
Concentration towards galactic center:	strong	considerable	weak
Metal abundance (relative to Sun):	1/500 to 1	mainly ~1; ⅓ to 3	~1
Age of the stars and time when they were formed:	Old stars In about 10^8 years at beginning of galactic system ~10^{10} years ago	Mainly old stars From time of formation of halo to the present	Young stars Within the last 10^8 or 10^9 years or so

stars are relevant, among others. We shall not concern ourselves with such objects here, interesting though they are.

With this restriction we learn from the data gathered in Tables 19.1–19.3 and from many other stellar analyses that the *relative* abundances of the heavy elements (atomic number $Z \geqslant 6$) are everywhere the *same* for stars of the *spiral arm, disk,* and *halo populations* of our Milky Way system. Except in extremely *metal-poor* stars of halo population II and at the other extreme in *metal-strong* (super-metal-rich) stars, there is no perceptible difference between the *relative* abundance ratios of the heavy elements in stars and in the Sun. Furthermore, as has been noted already, the possibility of a *two*-dimensional classification of "normal" stars shows that their spectra are determined completely by T_{eff} and g, while differences in chemical composition play *no* role.

Let us be permitted one—almost *too* trivial—remark on the effect of *errors of analysis:* If analyses of a uniform ensemble contain a variety of errors, the ensemble may seem to be nonuniform. However, if a nonuniform ensemble is analyzed with significant errors, it will hardly appear to be uniform!

It therefore appears to be justifiable to speak simply of *the* abundance of the heavy elements or—as one usually says—*the metal abundance.* The abundance of these elements is first referred to hydrogen and then the star is compared with the Sun (as standard). We therefore have

$$\epsilon = (M/H)_{\text{star}}/(M/H)_{\odot}.$$

It is also common to write $\log \epsilon = [M/H]$.

Using broadband *UBV* photometry, it is possible to obtain the metal abundance of fainter stars, though without the possibility of picking out abundance anomalies of individual elements. As a result of the crowding of the Fraunhofer lines in the short wavelength region of the spectrum ($\lambda <$ 4500 Å) in stars of intermediate and later spectral types (cf. Figures 12.4 and 18.5) their intensity I_λ^L in that region is depressed in metal-rich stars and augmented in metal-poor stars (relative to normal ones). In metal-poor stars therefore the color index $(U-B)_0$ is shifted significantly toward negative values while the color index $(B-V)_0$ is only shifted slightly. The metal-poor stars of intermediate and later spectral types therefore form a sequence in the two-color diagram (Figure 15.5) which is above that for normal stars, as we have already seen in the example of the globular cluster M 92 (Figure 26.3). Unfortunately, the connection between this ultraviolet excess $\delta(U-B)$—whose dependence on $(B-V)_0$ is of lesser importance—and the metal abundance ϵ can be established only by detailed analyses of the spectra of selected stars at high dispersion.

Before using the two-color diagram the color indices must be corrected for interstellar reddening. Since the slope of the reddening line (Figure 15.5) is known, it is possible to construct "metal indices" which are almost independent of interstellar reddening. Thus S. van den Bergh uses the

$$\text{metal index } Q = U-B - 0.72(B-V). \qquad (27.1)$$

Within the framework of his intermediate-band photometry B. Strömgren introduced a corresponding index Δm_1. Naturally, one must exercise a certain caution in interpreting such metal indices; for example, Am or Ap stars would not be distinguished, without further investigation, from stars with altered metal abundance. Besides their use for faint stars, the great significance of metal indices lies not least in the fact that they make it possible to obtain at least a rough measure of the metal abundance for the *composite spectra* of very distant globular clusters, galaxies, and possibly parts of these.

In recent years the so-called *scanner technique* has brought about a further refinement in the analysis of such composite spectra, not only with regard to metal abundance but beyond that to the whole Hertzsprung–

Russell diagram or to stellar populations, for example in galaxies. In this technique narrow spectral regions are selected by means of a slit in the focal plane of a spectrograph instead of by a color filter and are measured photoelectrically. For example, in the range from 3000 to 10700 Å, H. Spinrad and B. J. Taylor (1969–1971) have normally used 36 regions of 16 to 32 Å in width, corresponding to important lines and bands, and in this way have made measurements on the one hand for several galaxies, including M 31, and on the other hand for numerous stars with known spectral types, luminosity classes, metal abundances, etc. Then they try to calculate what mixture of stars or stellar populations will correctly reproduce all 36 narrowband magnitudes. The obvious question, of how much detail one can trust in such an analysis, requires much further discussion.

After these, perhaps rather extended, preliminaries, let us return to the stellar populations, of our Milky Way system (Table 27.2) and then of other stellar systems. In doing this, it seems appropriate to start with the two extreme main populations, the *spiral arm population I* and the *halo population II,* and then to consider the *disk population*. The transition populations require no further special discussion.

The *extreme* or *spiral arm* population I is characterized by its bright ($M_V \approx -6$) blue O and B stars in *young* galactic clusters and associations (Figure 26.2). These formations are all associated with interstellar matter (often compressed), out of which they have clearly been formed, some of them only a short time ago. Our earliest ancestors could have observed the formation of the bright blue Orion stars from their observatories in the trees! As is shown by Table 19.1 and many other stellar analyses, the chemical composition of all these formations cannot be distinguished from that of the Sun. It is true that sometimes in strongly compressed gas clouds—as recent infrared and radio studies have shown—a significant fraction of the condensable material, especially the heavy elements, separates out as *cosmic dust* or "smoke" from the gas, that is, hydrogen, helium, etc. However, in the process of *star formation* the solid and gaseous fractions are clearly united and so the original mixture of elements is reconstituted.

As we have already seen (Section 23), the galactic orbits of the young stars of population I and of the interstellar matter are almost circular. The theory of the spiral arms themselves will be thoroughly discussed in Section 29.2.

The *halo population II*. As we have seen, the halo is an almost spherical subsystem of our Milky Way, with a radius of 10 to 20 kpc. It comprises the *globular clusters* and numerous field stars which show up first as *high velocity stars* through their large space velocities of up to ~ 300 km/s relative to the Sun. If their velocity components U and V in the galactic plane are plotted in a Bottlinger diagram (Figure 23.11), it can be seen that these stars revolve round the galactic center in elongated elliptical orbits, some of which are even retrograde. The large velocity component W perpendicular to the galactic plane shows that, in contrast to the almost coplanar orbits of the disk population and population I, the orbital inclinations are distributed almost at random. The mass of the halo may be estimated to be about 15 percent of the mass of the whole system, and so the present orbits of the halo stars are determined essentially by the gravitational field of the disk.

The shape of the halo shows at once that its *angular momentum per unit mass*—more precisely, its component $h = VR$ in the direction of the rotation axis of the galactic system—is small. We compare it with that of the Sun and with an average value which J. H. Oort (1970) has calculated for the whole Milky Way system:

	Halo	Whole Milky Way System	Sun		
$h =$	$\leqslant 50$	170	250	km/s × 10 kpc.	(27.2)

These values are clearly of significance in connection with the origin of the Milky Way system (Section 29).

The halo population II—globular clusters and field stars—is recognized by its color-magnitude diagram, which corresponds to that of globular clusters (Figures 26.3 and 26.4). In this case there are *no* bright OB stars, but rather the brightest stars are *red giants* with $M_V \approx -3$. Then there are the characteristic cluster variables, the RR Lyrae stars with periods greater than $0^d.4$. To the classical Cepheids (like δ Cep) of population I there correspond here the rarer W Virginis stars with periods of about 14 to 20 days.

All stars and star clusters of the halo population are more or less metal deficient. Among the high velocity stars, the K2 giant HD 122563 with a metal abundance $\epsilon \approx 1/500$ and the Sun-like G0 main sequence star HD 140283 with $\epsilon \approx 1/200$ must belong to the most deficient stars. Metal abundances of about $1/10$ (relative to the Sun) are rather common.

The metal-poor main sequence stars of the halo population II are also described as *subdwarfs*. This has the following, purely historical background: Since the Fraunhofer lines of metal deficient stars are weaker than those of normal stars of the same T_{eff} and g, and since on the other hand the metal lines become weaker on going along the main sequence to earlier spectral types, the metal deficient stars were originally classified too "early" in comparison with normal stars of the same effective temperature T_{eff}. For example, HD 140283 was originally described as A2, and later as sd F5, while its effective temperature $T_{\text{eff}} \approx 5940$ K, corresponds to about G0 V. For this reason, such stars came to lie *below* the main sequence in the HR diagram and were "subdwarfs." However, if one uses for example the effective temperature and the bolometric magnitude or luminosity as coordinates of the diagram, the subdwarfs lie almost *on* the usual main sequence.

More recently, extensive observational material has been obtained for the halo population II: (1) about their *stellar dynamics* using radial velocities and proper motions, and (2) with regard to *metal abundances*, through more analyses of high dispersion spectra, proper classification of low dispersion spectra, and, especially for fainter stars, two-color diagrams and the metal indices Δm_1, Q, and the like.

One then naturally asks about a correlation between the metal abundance and the elements of the galactic orbit or equivalent variables. One finds to begin with that there is a strong correlation between very low metal abundance and large eccentricity e and inclination i of the galactic orbit or large velocity component $|W|$ perpendicular to the galactic plane. However, this statement requires qualification: In the older studies, which were based on proper motion catalogs, those metal-poor stars whose orbits are similar to those of the disk stars were necessarily discriminated against. This "injustice" was first redressed, to some extent, by the work of H. E. Bond (1971) who used a Schmidt telescope with an objective prism to classify low-dispersion spectra. This showed that for larger metal abundances, from about $1/10$ of the solar value, and for orbital eccentricities $e < 0.5$ or velocity components $|W| < 50$ km/s the correlation becomes much weaker and possibly completely disappears. We therefore have, at any rate with respect to both metal abundance and galactic orbits, a rather numerous "in-between population," which forms a continuous transition from the extreme halo population II to the disk population (see below). Whether it is expedient to define a special *intermediate population II* remains to be seen.

For globular clusters, the *two-color diagrams* and similarly the *metal indices* show first that M 92 (Figure 26.3) is about as metal poor as the extreme subdwarfs (metal abundance $\approx 1/200$), while other globular clusters form a continuous transition to smaller metal deficiencies. With increasing metal abundance the *color magnitude diagram* (Figure 26.4) changes from the M 92 type to that of the old galactic cluster NGC 188 with normal metal abundance. In the process the giant branch moves downward and its upper end terminates at earlier types, while the horizontal branch steadily shrinks.

If the position of the globular clusters in the sky is considered, it is discovered that in general the metal abundance increases toward the galactic center. Some globular clusters near the center appear—just from the color indices—to have only a small underabundance of metals.

The color-magnitude diagrams of *globular clusters* (Figures 26.3 and 26.4) are our primary source of information about the *age* of the halo population II. By comparison of these diagrams with theoretical evolutionary tracks, F. Hoyle and M. Schwarzschild drew the conclusion as early as 1955 (cf. Section 26) that the halo population II, with an age of about 10^{10} years, forms by far the oldest component of the galactic system. More recently, the color-magnitude diagrams of globular clusters have become remarkably accurate and complete. Using them and after weighing up all the uncertainties, A. Sandage (1970) arrived at an age of the globular cluster system of 9 to 12×10^9 years. By establishing very precisely the position of the "knee" on the $(B-V)_0$ axis of the color-magnitude diagram of globular clusters and high-velocity stars, Sandage was further able to show that in each case the age of *all* members of halo population II must be the *same*, within narrow limits. On the other hand, because of the uncertainties in some of the assumptions made in calculating stellar models, one should certainly not overestimate the precision of the *absolute* age determination.

Superficially, the *disk population* of the Milky Way system consists of stars of roughly normal metal abundance which have the familiar C–M diagram of our neighborhood (Figures 15.2 and 15.3) and roughly circular galactic orbits. A more exact study is needed to reveal its relation to halo population II and to spiral arm population I and other important information about the evolution of the galactic system.

Our information about the *age* of the disk population comes in the first place from the C–M diagrams of the *galactic clusters* (Figures 26.1b and 26.2). For the oldest, NGC 188, a recent investigation by A. Sandage and O. J. Eggen (1969) gave an age of 8 to 10 $\times$ 10^9 years; M 67 is only a little younger. There are also many field stars of the disk population, with well-known absolute magnitudes and color indices, which fit into the color-magnitude diagram of the oldest galactic clusters; however, the subgiants never lie significantly below those of NGC 188. Although the maximum age of the disk population, for the reasons already explained above, may perhaps not be known so exactly as has been believed from time to time, nonetheless there can be no doubt that it is well defined. The value given is admittedly 1 to 2 billion years less than the age given above for the globular clusters, but this difference may already be very close to the present error bounds. There are younger clusters in great numbers (see, for example, Figure 26.2); the formation of the disk population therefore occupies the whole space of time between the halo population II and the spiral arm population I. As we shall discuss more precisely later, associations and clusters are continually forming out of interstellar matter in the spiral arms. Some of these groups "dissolve" and so continually replace the evolutionary losses of the disk population.

A more careful study of the metal abundances of the disk stars then showed that they certainly differed only slightly from each other but that nevertheless there was a scatter within a factor of 3 to (at most) 5. In particular, intermediate-band photometric studies by B. Strömgren and by G. Gustafsson and P. E. Nissen (1972) have shown that the metal abundance of the *Hyades* is about 3 times larger than that of the Sun. The metal abundances of 60 population I stars of spectral types F1 to F5 lay between these values. The *Pleiades* proves to be metal poor with respect to the Hyades by a factor of 2. A connection between metal abundance and *age* is not to be found. In accord with our foregoing discussions, the relatively metal-poor stars clearly form a transition to the halo population II, which may perhaps be described as intermediate population II. But what about the *metal-rich* stars? (As remarked already, the description "super-metal-rich" seems to us to be thoroughly pretentious for an effect which barely lies above the presently achievable error bounds.) By means of multicolor photometry, M. Grenon (1972–1973) succeeded in finding for a larger number of such stars a correlation between *metal overabundance* and *galactic orbit* (in the Bottlinger diagram), and, what is more, in the sense that the metal abundance in the galactic disk decreases steadily outward. Furthermore, observations by H. Spinrad and others of other disk galaxies had already suggested an effect of this kind.

At this point we conclude for the moment the discussion of our Milky Way system and turn to the other galaxies, working our way from right to left in the Hubble diagram (Figure 27.3) and starting with the *giant galaxies* which are comparable to our system. After that we investigate the dwarf galaxies.

Among the *irregular galaxies,* our nearest neighbors, the Magellanic Clouds LMC and SMC, have been thoroughly investigated. Their dominant characteristic is an extensive *population I,* with gaseous nebulae, blue OB stars, etc. However, there is also an *old population* with red giants, etc. Furthermore, we know many *globular clusters* and at least the upper parts of their color-magnitude diagrams. To some extent they are the same as in the Milky Way system, but some contain, among other things, bright blue stars whose nature is still very mysterious. The chemical composition can be determined for some *gaseous nebulae* and several of the extremely bright *supergiants,* such as HD 33579 (B. Wolf, 1972). With regard both to the relative abundances of the heavy elements and to the metal abundance M/H, they can*not* be distinguished (within the errors) from the galactic population I. We still know extremely little about the *old* stars of the Magellanic Clouds.

As we would expect, the *disk* or *spiral galaxies,* like M 31, resemble the Milky Way system in many ways. With regard to the *globular clusters* in M 31, S. van den Bergh was able to use color indices to verify that they are metal poor, although perhaps slightly less so than ours. Then it was possible to observe that in the inner regions of some galaxies the nitrogen lines of the gaseous nebulae are anomalously strong. This could be caused by stellar evolution according to the CNO cycle being particularly active here. It is certainly well known that in this process after a sufficiently long time almost all the CNO group has been converted to N^{14}.

The *elliptical galaxies* consist mainly of *old* stars, as is shown by their brightest stars being *red giants*. Just occasionally W. Baade noticed small filaments of dark matter in which bright blue stars have recently started to shine. Our main question is naturally directed to the *metal abundance* of the E galaxies. We discuss this forthwith, with respect both to the *luminosity* of the galaxy and to the *mass* (which is roughly proportional to it). R. D. McClure and S. van den Bergh (1968) determined for a large number of galaxies (and globular clusters) a color index C^* for the region 3800–4500 Å and the metal index Q which we have already mentioned [equation (27.1)]; both are to a great extent independent of interstellar reddening. Recent narrow- and intermediate-band photometric studies by H. Spinrad and B. J. Taylor and by S. M. Faber (1972–1973) confirm their results, a sample of which is shown in Table 27.3.

Table 27.3. Relation between absolute magnitude M_B, the color index C^* for the wavelength range λ3800–4500 Å (constructed so as to be independent of interstellar reddening) and the corresponding metal index Q [from equation (27.1)] for some E and S0 galaxies. After McClure and van den Bergh (1968).

Galaxy	M_B	C^*	Q
NGC 185, dE	−13.7	0.85	−0.31 (metal poor)
NGC 3245, S0	−18.7	1.12	−0.16
NGC 4472, E	−20.8	1.25	−0.06 (metal rich)

In agreement with spectroscopic observations, we discover that the *giant* E galaxies have normal or (perhaps) somewhat enhanced metal abundances, but that with decreasing luminosity the metal content also decreases, to values that approach those of metal-poor globular clusters. In the *dwarf* E galaxies we are clearly concerned (almost?) entirely with metal-poor population II.

Since, as is well known, there are no dwarf spiral galaxies, it only remains to discuss the question of the metal abundance in the *irregular dwarf galaxies* such as IC 1613. W. Baade's Hα photograph (Figure 27.6) allowed several H II regions to be discovered, and so members of a population I. However we do not know in what proportion this is present relative to an old stellar population which is certainly also present.

Still very puzzling are the two *blue dwarf galaxies*, studied by L. Searle and W. L. W. Sargent (1972), in which they find on the one hand an underabundance of O and Ne and on the other hand a numerous population I with blue stars. Are we dealing with young objects? Do they have anything to do with the blue globular clusters?

To close this Section we discuss again the question of the abundance ratio H:He of the two lightest elements, hydrogen and helium, in the different stellar populations. As will become apparent, this question is very important for cosmology. For the *spiral arm population I* and the *disk population* the ratio H:He is ascertained mainly from analyses of B star spectra; in O stars departures from thermodynamic equilibrium may create difficulties. The analyses of the *interstellar gas* are immediately comparable with those of young stars. In that case H:He is obtained from recombination lines in gaseous nebulae or H II regions, for example the Orion nebula, and from the radio transitions between energy levels of very high quantum number (Section 24). All these methods give (cf. Table 19.1) H:He $\approx$ 10. The same value is obtained for many planetary nebulae (Table 19.1), although these certainly represent a rather advanced stage of stellar evolution. It is clear—although in no way obvious—that in the formation of their envelopes it is mainly unburned matter from the outermost layers of the central star that is blown off. In the case of the halo population II the determination of H:He runs up against the serious difficulty that at the lower temperatures of their main sequence and giant stars helium lines can in no circumstances be expected. We are therefore directed to the planetary nebulae, and perhaps other late stages of evolution, where there is however the perpetual danger that material burned in nuclear reactions may have succeeded in reaching the surface. The most thorough investigation is by O'Dell, Peimbert, and Kinman of the planetary nebula K 648 which, as a member of the globular cluster M 15 = NGC 7078, certainly belongs to population II. This showed that even for reduced metal abundance the ratio H:He remained at about 10, just as in the stars of population I and the disk. These and similar observations, together with the theory of the color-magnitude diagram of globular clusters, indicate that the two lightest elements H and He were formed in a different way from the heavy elements with $Z \geqslant 6$.

28. Radio Emission from Galaxies, Galactic Nuclei, and Cosmic Rays and High-energy Astronomy

Besides the well-known "optical window" from the ozone limit at about 3000 Å to the infrared at about 22 μ the Earth's atmosphere has a second range of transparency from $\lambda \approx 1$ mm to as far as the onset of ionospheric reflection at $\lambda \approx 30$ m. Since K. G. Jansky's discovery (1931) radio astronomy has exploited this spectral range.

In Section 9 (pp. 100–101) we have described radio telescopes and radio interferometers, for the precise measurement of the position and intensity distribution of cosmic radio sources. By combining accurately calibrated measurements at various wavelengths λ, or frequencies $\nu = c/\lambda$, we obtain *radio spectra*. In recent times it has also become possible to measure the *polarization* of radiation in radio frequencies from cosmic sources. In Section 20 we became acquainted with the radio emission from the Sun. This is composed of the thermal radiation of the quiet Sun produced by free-free transitions in the plasma of the corona at 1 to 2×10^6 K, and also of the nonthermal radiation of the disturbed Sun, which, along with other phenomena of solar activity, is produced by plasma oscillations, synchrotron radiation (see below), etc.

28.1 Radio radiation of the Galaxy and galactic sources

As regards the radio radiation of the Galaxy, we have already discussed the thermal part in Section 24. We are primarily concerned with the line emission of neutral hydrogen (H I regions), at $\lambda = 21.105$ cm or $\nu = 1420.40$ MHz, as well as of the OH radical and many other molecules. We should remark in addition that in some radio sources the intensity ratios, for example of the OH lines, do *not* correspond to thermal population of the energy levels. Rather, some lines appear to be strengthened by stimulated emission, that is, the *maser effect,* as a result of overpopulation of the higher energy level. What process led to this unusual population of the energy levels is not yet clear.

Then we also know about the continuum of free-free radiation of ionized hydrogen, etc., especially in the H II regions and gaseous nebulae near the galactic plane and in planetary nebulae. We recognize such thermal continua by the fact that in the case of emission by an optically thin layer the intensity, or the flux per unit frequency interval S_ν, is nearly independent of ν. The radiation temperature T_ν according to the Rayleigh–Jeans law[1] behaves approximately as $T_\nu \approx \nu^{-2}$. In contrast to this, the radio emission of the Galaxy coming from all over the sky, as well as that of the strong radio sources, shows a spectrum expressed approximately by

$$I_\nu \approx \nu^{-0.7} \quad \text{or} \quad T_\nu \approx \nu^{-2.7}. \tag{28.1}$$

Even taking account of absorption and self-absorption, this cannot be ascribed to thermal emission. Again, the radiation temperatures of more than 10^5 K measured at long wavelengths can scarcely be interpreted as thermal. So H. Alfvén and N. Herlofson in 1950 appealed to the mechanism of *synchrotron radiation,* or magnetobremsstrahlung, for the explanation of the nonthermal radio continua. These notions were then further developed by Shklovsky, Ginzburg, Oort, and others. It was known to physicists that relativistic electrons (that is, electrons moving with nearly the speed of light, whose energy E appreciably exceeds their rest energy $m_0c^2 = 0.511$ MeV), that describe a circular path in the magnetic field of a synchrotron, emit intense continuum radiation in their direction of motion, the spectrum of which extends into the far ultraviolet. This continuum differs from that of free-free radiation or bremsstrahlung in that the acceleration of the electrons results, not from atomic electric fields, but from a macroscopic magnetic field H.

V. V. Vladimirsky and J. Schwinger developed the theory of synchrotron radiation in 1948–1949. It rests upon the following consideration. According to the laws of relativistic kinematics, an electron circulating with almost the speed of light emits radiation within a narrow cone of angle $\alpha \approx m_0c^2/E$ (Figure 28.1). Like a lighthouse beam, this sweeps over the observer in rapid succession so that, taking account of the relativistic Doppler effect, he receives a succession of radiation flashes, each of duration Δt, say. The spectral resolution or, mathematically speaking, the Fourier analysis of the radiation gives a continuous spectrum, whose maximum lies at the circular frequency $\sim 1/\Delta t$. More exact calculation gives for the frequency of the maximum ν_m [Hz] the formula

$$\nu_m = 4.6 \times 10^{-6} H_\perp E_{\text{eV}}^2 \tag{28.2}$$

where $H_\perp$ is the component of the magnetic field (in gauss) perpendicular to the direction of motion of the electron and E_{eV} is its energy in electron-

[1]Equation (11.25): $S_\nu = 2\nu^2 kT_\nu/c^2$.

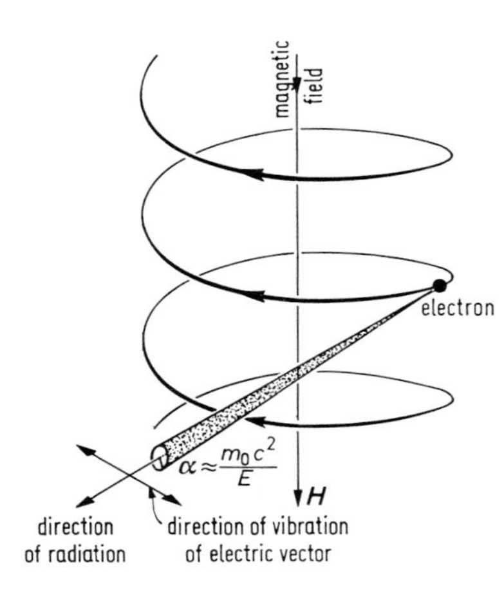

Figure 28.1
Synchrotron radiation of a relativistic electron in a magnetic field *H*.

volts. If we work with a mean galactic magnetic field of $H_{\perp} \approx 5 \times 10^{-6}$ G, the following electron energies correspond to the various frequencies ν_m or wavelengths λ_m shown:

$$\begin{array}{llllll} \lambda_m = 3000\ \text{Å} & 30\ \mu & 3\ \text{mm} & 30\ \text{cm} & 30\ \text{m} & \\ \nu_m = 10^{15} & 10^{13} & 10^{11} & 10^{9} & 10^{7}\ \text{Hz} & \\ E\ = 6.6 \times 10^{12} & 6.6 \times 10^{11} & 6.6 \times 10^{10} & 6.6 \times 10^{9} & 6.6 \times 10^{8}\ \text{eV}. & \end{array} \tag{28.3}$$

Even for the production of radiation in the radio-frequency range, and certainly in the visible, electrons must be available with energies in the region of those of cosmic rays. If we can represent the energy distribution of the electrons by a power law

$$N(E)\, dE = \text{constant}\ E^{-\gamma}\, dE \tag{28.4}$$

then the intensity of the synchrotron radiation is given by

$$I_\nu \sim H^{(\gamma+1)/2} \nu^{-(\gamma-1)/2}. \tag{28.5}$$

In 1939 when G. Reber investigated the distribution of radio radiation over the sky at $\nu = 167$ MHz or $\lambda = 1.8$ m, making use of his radio telescope of modest angular resolving power, he noted at once its concentration toward the galactic plane and toward the galactic center.

Radio astronomy, and also astronomy with X rays and γ rays, presents us with the possibility of learning something about the occurrence of highly energetic electrons in cosmic systems. This is of great significance since the original distribution of directions of the charged particles arriving at the Earth cannot be inferred, because of their complicated deflections by the terrestrial and interplanetary magnetic fields. An obstacle is that we still know little about the magnetic field H that also appears in equation (28.5). At present we can still not determine how the galactic emissivity in radio frequencies derived from the observations is shared between the factor $\sim H^{1.7}$ and the factor N_e, the density of relativistic electrons in the relevant energy range.

In addition to the nonthermal radiation from the galactic disk (Figure 28.2) (which joins up with the thermal radiation of ionized hydrogen at decimeter and centimeter wavelengths) one observes, also in the meter-wavelength region, another component which is distributed relatively uniformly over the whole sky. Following J. E. Baldwin, this has been ascribed to the galactic halo in which there ought to be present (dynamically independent of the stars): (1) relativistic electrons, (2) a magnetic field, (3) a plasma, in which the magnetic field is "suspended." In order to explain the smallness of its concentration toward the galactic plane, we should ascribe to the plasma either macroscopic velocities (like those of high-velocity stars) of some hundreds of kilometers per second, or corresponding thermal velocities, that is, a temperature of from 10^5 to 10^6 K.

More recent measurements with ever greater angular resolving power have traced an ever greater proportion of the "halo radiation" to galactic and, still more, extra-galactic radio sources (see below) that are densely distributed over the sky. By now it is questionable as to whether between the many sources anything at all is left over for the galactic radio halo!

The determination of the spatial distribution of radio sources in our galaxy is made particularly difficult and uncertain by the fact that we ourselves are *in* the system. The construction of more and more powerful radio interferometers has brought us in recent years a much more direct knowledge of the distribution of radio emission in *neighboring galaxies*.

We are already familiar with the most important types of *galactic*, mainly *nonthermal, radio sources:*

(1) Supernova remnants; on the one hand the *pulsars* and their surrounding nebulae (such as the Crab nebula, M1 = Tau A, cf. Figure 26.11) and on the other hand the extended *ring-shaped* or *circular nebulae* (like the well-known Cygnus Loop (the Veil nebula) or IC 443).
(2) Flare stars; red dwarf stars, on whose surfaces eruptions (flares) occur at irregular intervals, which differ from the analogous phenomena on the Sun only in their much larger energy output.
(3) Very recently it has been possible to detect a few other stars as radio emitters, for example α Scorpii = Antares, or more precisely its blue component B.

28.2 Extra-galactic radio sources

In 1946 J. S. Hey and his collaborators discovered the first radio source Cygnus A because of its intensity fluctuations. Simultaneous observations at stations far removed from each other showed later on that these fluctuations are to be interpreted as scintillation produced in the ionosphere, chiefly in the F2 layer at a height of about 200 km. Like optical scintillation, which occurs only for stars but not for planets (which have a larger angular diameter), radio scintillation occurs only in the case of sources of sufficiently small angular diameter. For the so-called quasi-stellar radio sources (see below), some of which have angular diameters less than a second of arc, another sort of scintillation has recently been discovered, which is produced in the interplanetary plasma. Using larger and larger radio telescopes, the *radio interferometer* invented by M. Ryle and in particular the technique of *aperture synthesis* invented by him in 1962, a series of research groups have discovered more and more radio sources and have measured their position, angular diameter or intensity distribution, and radio spectrum. These studies will be most effectively supplemented by means of the recent advances in *infrared* astronomy, which enable the gaps in the spectrum from the near infrared to the millimeter region to be almost closed.

For quantitative discussion, we must first bring order into the jumble of notation and units, which has arisen from the fact that physicists and stellar spectroscopists began from the classical work of G. Kirchoff, M. Planck, and others and as a result favored the cgs system with basic units centimeter, gram, second. The radio astronomers were linked to the usage of high-frequency techniques and to the MKS system with units meter, kilogram, second. The frequency unit in many European countries is now officially the Hertz = s^{-1}; in some countries the clumsy notation c/s = cycle per second is still preferred. As well as these notations there are still the physically quite nonsensical "DIN 1970" standards, which should be abolished as soon as possible. Table 28.1 gives a review of the *definitions* and *units* used to describe on the one hand extended sources and on the other unresolved (point) sources.

For the description of extended sources the so-called *brightness temperature* T (or T_b) is commonly used instead of intensity or brightness. Here one asks what temperature T a blackbody must have in order to radiate at frequency ν, according to the Rayleigh–Jeans' law (sufficiently accurate in the radio range), precisely the radiation intensity I_ν. From equation (11.25)—written directly in MKS units—we have the relation

$$I_\nu = 3.075 \times 10^{-28} \nu_{\mathrm{MHz}}^2 \, T \mathrm{W/m^2\ Hz\ sr}. \tag{28.6}$$

The spectral energy distribution of radio sources is usually approximated by a power law

$$I_\nu \quad \text{or} \quad S_\nu \approx \nu^\alpha \tag{28.7}$$

where α is called the *spectral index* (some authors prefer to define $-\alpha$ as the spectral index). For the synchrotron radiation of the Milky Way and many galaxies α lies in the range -0.7 ± 0.1. For thermal radiation from H II regions $\alpha \approx 0$ in optically thin layers and $\alpha = +2$ in optically thick layers.

After these somewhat tedious explanations, we turn to the exciting problems of radio emission from galaxies. Next to the origin and dynamics of the spiral arms (see Section 29), the present focus of interest is in the investigation of *galactic nuclei*. We distinguish the (normally) lens-shaped and relatively dense *central* region of a galaxy from the much smaller,

Table 28.1. Basic definitions and units of optical and radio astronomy.

	Optical astronomy	Radio astronomy	DIN 1970 (Lighting technology)	cgs units	MKS units
Extended sources	Intensity I_ν	Brightness (distribution)	Spectral radiation density	erg/s cm^2 Hz sr	W/m^2 Hz sr
Unresolved (point) sources	Net flux S_ν or πF_ν	Flux density (or flux)	Spectral specific emission	erg/s cm^2 Hz	W/m^2 Hz*
				10^3 cgs units = 1 MKS unit	

*In radio astronomy $10^{-26} Wm^{-2}Hz^{-1}$ is often denoted by 1 f.u. (flux unit) or 1 Jy (Jansky).

compact, actual *nucleus,* only 0.1 to 10 pc in size. The galactic nuclei are the seat of an "activity" which varies strongly from galaxy to galaxy and also with time; the significance of this was pointed out by V. A. Ambarzumian as early as 1954, in several papers scarcely noticed at the time. The activity manifests itself in the release of huge amounts of energy in the form of superthermal particles and photons. Here, clearly, is the most important source of synchrotron electrons and cosmic rays, of nonthermal radio waves, X rays, and γ rays and of nonthermal long wavelength infrared radiation, etc.

In what follows we discuss—in order of increasing activity—first the (more or less) *normal galaxies,* of which our Milky Way is one. Then we turn to the *Seyfert galaxies,* to actual *radio galaxies* and to *quasars.* In these we are dealing with energy transformation on such a gigantic scale that we may well ask whether the known laws of physics remain adequate for its explanation.

28.2.1 Normal galaxies

In 1950 M. Ryle, F. G. Smith, and B. Elsmore were able to show that several of the well-known brighter galaxies emit radio radiation of about the strength expected from their similarity to our Milky Way. Soon afterward R. Hanbury Brown and C. Hazard succeeded in determining, in rough outline, the radio brightness distribution of the Andromeda galaxy M 31; the two galaxies proved to be very similar in the radio region as well.

As already indicated, the investigation of the brightness distribution, and therewith the origin of the radio emission in galaxies, entered an entirely new phase with the use of *aperture synthesis.* By means of this technique G. G. Pooley (1969) in Cambridge first studied the Andromeda galaxy M 31 at 408 and 1407 MHz. Then, using the Dutch radio telescope at Westerbork, which came into existence on the initiative of J. H. Oort, D. S. Mathewson, P. C. van der Kruit, and W. N. Brouw (1971) made a thorough study of the Sc galaxy M 51 = NGC 5194 and its companion, the Ir galaxy NGC 5195, in the continuum at 1415 MHz (λ 21 cm). In that study they achieved a resolution (whole half-power beam width) of 24″ in right ascension and 32″ in declination, corresponding to about 450 pc. With computer reduction, the observations can be displayed either as isophotes or as a genuine radio picture. Then, using the same instrument, M 33, NGC 4258 (P. C. van der Kruit, J. H. Oort, and D. S. Mathewson, 1972) as well as several normal and Seyfert galaxies (P. C. van der Kruit, 1971) were investigated. As a representative of the Ir galaxies, the Large Magellanic Cloud LMC is well known. The angular resolution of galactic nuclei is—so far as the nonthermal radiation is concerned—naturally left to long baseline interferometry or to infrared astronomy. We try to give an account of the very detailed investigations, summarizing them as far as is possible.

For convenience, we distinguish several components (developed to different extents in different galaxies) of the nonthermal, that is, synchrotron, radiation, which we relate to the *core,* to the *central region,* and to the *disk* (uniformly distributed) as well as to the *spiral arms.*

For a long time it was generally accepted that the Milky Way, and similar galaxies as well, had an extended, slightly flattened *radio halo* (not to be confused with the halo of population II stars!), which must then consist of plasma with synchrotron electrons and a magnetic field. With improved resolution, this radiation proves more and more to be from discrete sources, partly in the galaxy concerned and partly in the background. Probably there is in general no radio halo (although it played an important role in several theories of cosmic rays!).

(1) *Spiral arms.* As is seen most clearly in M 51, the ridge line, that is, the maximum brightness of the spiral arm, does *not* agree precisely with that on blue-sensitive plates, that is, the maximum star density. Rather, the *radio* arm lies along the inner side of the optical arm in the region of the *dust* and *gas* arm marked by dark clouds and (on Hα-plates) H II regions. D. S. Mathewson *et al.,* following W. W. Roberts, attribute this to the fact that, according to the *density-wave theory* of spiral arms (cf. Section 29), a compression of the gas and the magnetic field lines is to be expected here. The larger density of synchrotron electrons thus caused, and the strengthening of the magnetic field, would produce an increase in radio emission. The intensity of the radio arms decreases outward. In M 51 the decrease is more gradual in the arm at whose end—about $4'.5 \approx 5.2$ kpc from the nucleus—is situated the strongly emitting Ir galaxy NGC 5195. An estimate of the number of supernovae exploding each year and of their remnants shows further that—contrary to the commonly held view—supernovae scarcely represent a sufficient source for the synchrotron electrons or for the nonthermal radiation of the spiral arm. We shall return presently to the question of their origin.

Before that, we must briefly refer to the quite astonishing observations which P. C. van der Kruit, J. H. Oort, and D. S. Mathewson (1972) have made of NGC 4258 using the same technique. The blue plates in the *Hubble Atlas* at first gave the impression of a rather normal Sb galaxy. On the other hand, the radio picture at 1415 MHz showed, besides the two "usual" optical spiral arms of the blue plates, two *further* radio arms, which emerge from the central region and then run between 5 and 15 kpc further, almost in a straight line. These arms first obtained an optical counterpart with the use of interference-filter pictures in Hα, which show, just as in the radio picture, an unusually smooth run of brightness along this arm. The radio and Hα pictures together with corresponding radial velocity measurements can now be interpreted in terms of a model in which about 18×10^6 years ago two plasma clouds were flung out of the nucleus of the galaxy in opposite directions with velocities of approximately 800 to 1600 km/s. These masses (about 10^7 to $10^8 M_\odot$) would have compressed the gas of the disk and so produced the "extra" radio arms. The entire formation will then in the course of one rotation of the galaxy or less, that is, $\sim 10^8$ years, be gripped by the differential rotation and so more and more assume the appearance of normal spiral arms. What role do such processes play in the evolution of galaxies with regard to the origin of new spiral arms or the replacement of old "ground-down" ones? To this question too we can return only in the next Section.

(2) *Disk component.* The brightness of this component, which can naturally only be caught sight of between the spiral arms, shows a smooth run as a function of distance from the center. Its connection with the density of the neutral and ionized gas certainly needs further elucidation.

(3) *Nuclear component.* To begin with, it is evident that even optically similar galaxies show great differences in regard to the radio brightness of the central region and of the nucleus. Of the galaxies we have discussed, M 51 yields the largest flux density and our galaxy a much smaller one; finally M 33 and the LMC show no measurable nuclear component at all.

The ratio of the three components of the synchrotron radiation should give us information about their interrelation and ultimately about the origin of the synchrotron electrons. The observations that on the one hand a significant correlation exists between the nuclear component and the disk plus spiral arm components and that on the other hand supernovae (at any rate in M 51) are not the main source of the synchrotron electrons suggest that their source should be sought in the nuclear region. Also, one has the impression that the intensification of synchrotron radiation in normal spiral arms (for example, M 51) as well as in the extra radio arms of NGC 4258 arises from compression of previously existing plasma together with magnetic field and synchrotron electrons. It appears conceivable (but is not proved) that high-energy particles may even stream directly out of the nucleus along magnetic field lines into the spiral arms.

Optical observation shows in the center of most galaxies a small *nucleus* of high surface brightness. In the Andromeda galaxy M 31—where it was discovered already by Hubble—its dimensions are $1''.6 \times 2''.8$ or 5.4×9.4 pc (diameter). The spectrum shows that the nucleus consists mainly of normal stars of late spectral type. The brightness, as well as the internal motions, leads to a total mass of 10^7 to 10^8 solar masses and thus to a star density of about $2 \times 10^5/\text{pc}^3$. Similar nuclei are also observed optically in the local group, for example, in the E galaxies NGC 205 and M 32—the companions of M 31—and in the Sc galaxy M 33.

However the true significance of galactic nuclei was first revealed by *radio astronomy.* Even in the category of "normal" galaxies many—though not all—proved to be strong nonthermal radio sources. These show in principle the same phenomena that we shall meet in increased measure in Seyfert and N galaxies, quasi-stellar objects, etc. Besides compact components, whose diameters amount only to 0.1 to about 10 pc, there are extended central components out to a few hundred parsecs. Some galaxies even possess both. All these objects show, commonly in the radio and sometimes in the optical, *brightness variations* on a time scale of a few months.

Of special interest to us is the central region of our own galaxy. It is hidden from optical observations by the dense dark clouds in the Scorpius–Sagittarius region. Only recently, using high resolution radio antennas at centimeter wavelengths and as a result of the spectacular evolution of infrared astronomy, has progress been made.

Using the isophotal representation of the central region of the Galaxy (Figure 28.2) at 8.0 GHz (λ 3.75 cm), which D. Downes, A. Maxwell, and M. L. Meeks (1966) recorded with a 120-ft paraboloid, we recognize at the center the radio source *Sagittarius A* with a (half-power) diameter of $3\overset{'}{.}5$, which corresponds to 12 pc at the distance of the galactic center of $\sim$ 10 kpc. The position relative to many other formations leaves no doubt that Sag A represents the center or *nucleus* of our galaxy. Even at higher resolution (to $\sim 1'$) no further structure can be detected in the radio. The radio spectrum is *nonthermal* for $\nu > 2$ GHz with a spectral index $\alpha = -0.7$. In the surrounding region are found numerous radio sources, whose spectra ($I_\nu \approx$ constant) shows that we are dealing with thermal free-free emission from H II regions. The whole central region is covered by an extended, flattened thermal source, whose diameter along the equator corresponds to about $60'$ or 170 pc. Furthermore, *absorption* by ionized hydrogen depresses the long wavelength continuum of Sag A for $\nu < 2$ GHz.

Infrared astronomy brought completely new kinds of information about the nucleus of the Galaxy. E. E. Becklin and G. Neugebauer studied the spectral region λ1.65 to 19.5 μm in 1968–1969, F. J. Low, D. E. Kleinmann, F. F. Forbes, and H. H. Anmann the region 5 to 1500 μm in 1969. First of all, the former group found an infrared source which agreed precisely with the radio source Sag A in position and size ($\approx 3\overset{'}{.}5$). On the other hand the size and the (almost extinction-free) surface brightness at 2.2 μm also fits the nucleus of the Andromeda galaxy M 31. We may reasonably assume that both galactic nuclei are composed of *stars* and have the same spectral energy distribution at shorter wavelengths as well. Then we can determine, to begin with, the extinction in the near infrared. If we extrapolate rather further, we find toward the galactic center a visual extinction of 27^{m}!

Then it was shown that in the middle of Sag A an even more compact infrared source stands out against the background, with a size of only about $16''$ or 0.8 pc. Finally, there sits within that another *point source,* $\approx 0\overset{''}{.}02$, possibly a star.

However, the greatest surprise of infrared astronomy was that the spectral energy distribution S_ν of the nucleus of our galaxy, and also of other galaxies (little studied up to the present), shows a maximum within the range $\sim$ 3 μm to 1000 μm (= 1 mm) at $\sim$ 70 μm; this maximum rises above the background by about 3 powers of 10 in our galaxy and by about 5 powers of 10 in other galaxies and makes a decisive contribution to the

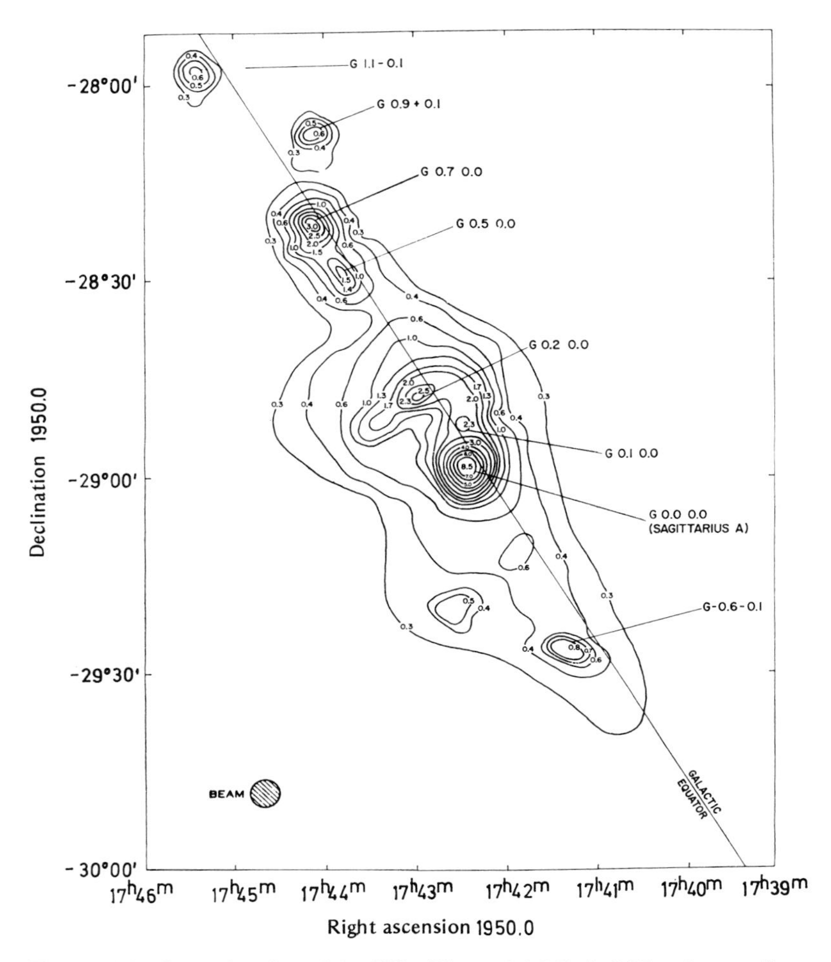

Figure 28.2. Central region of the Milky Way at 8.0 GHz (λ 3.75 cm) according to D. Downes, A. Maxwell, and M. Meeks (1966). The numbers on the isophotes correspond to the antenna temperature. The radio sources G are denoted by their galactic coordinates l^{II}, b^{II}. Sagittarius A is the nonthermal central source of our Galaxy. The remaining sources and also the extended sources ($l^{\text{II}} \approx \pm 0°.5$) are thermal. The "true" equator apparently lies 2′ south of the position accepted at the time that the new galactic coordinates were fixed. The hatched circle of 4′.2 diameter gives the resolving power, that is, the half-power width of the antenna's pencil beam. For a distance to the galactic center of 10 kpc, 1 arc minute corresponds to 2.91 pc.

luminosity. The origin of this component of the radiation is at present not yet clear.

Observations of radio galaxies (see below) suggest that the emission of nonthermal radio and infrared radiation in the center of the galactic system is connected with explosion(s) in its nucleus. In fact, the discovery of the *high-velocity H I clouds* (which can *not* be explained by rotation) led P. C.

van der Kruit in Leiden (1970–1971) to the idea that there may have been flung out of the galactic center $\approx 6 \times 10^6$ years ago about a million solar masses (essentially in the form of hydrogen) at $\approx$ 130 km/s and even earlier, 12×10^6 years ago to within a few million years, a mass some 5 to 10 times larger, at $\approx$ 6000 km/s. While the former is still "in flight," the latter, which flew out at an angle of 25° to 30° to the galactic plane, has fallen back into the plane and has given rise to the expansion motions observed in the inner parts of the Galaxy, for example the 3-kpc arm.

We postpone all questions of energy and concern ourselves first with "active" galaxies.

28.2.2 Seyfert galaxies

The activity of the nucleus can be more clearly understood, both optically and radio astronomically, in this class of galaxies, mainly spirals, which were first discovered optically in 1943 by C. K. Seyfert. The "Seyfert nuclei" show a strong emission spectrum, in which the Balmer lines of hydrogen as well as of He I and II and perhaps other permitted transitions are always distinguished by their large widths. For example, in NGC 4151 the half width of $H\beta$ corresponds to a Doppler broadening of about 6000 km/s. On the other hand, the *forbidden* lines—mostly of high excitation—of N II, OI–III, Ne III and V, S II, A IV, Fe III, IV, X, and XIV have the *same* width in some Seyfert galaxies but in others are much *narrower* (for example, in NGC 4151 $\approx$ 450 km/s). Interpretation of the spectra leads to the conclusion that—in the same way as in diffuse and planetary nebulae—plasma with about 10^5 electrons/cm^3 is stimulated to emit by the presence of hot stars ($T_{\text{eff}} \approx$ 20 000K). This plasma occupies only about 1 percent of the nucleus, and moreover the highest ionization states are explained by using only small amounts of gas at much higher temperatures. According to D. E. Osterbrock (1971) the chemical composition of the plasma cannot be distinguished from that of the Sun and of normal stars. The origin of the Doppler widths of the Balmer lines may not yet be explained. Probably we are dealing with *gas streaming* at a few 100 km/s; however, it is also possible that the originally narrower lines become broadened by *scattering* by free electrons (like the Fraunhofer lines in the solar corona).

More recently it has been shown that all Seyfert galaxies emit strong nonthermal radio radiation, that is, *synchrotron radiation*. The interferometric determination of radio source positions led now to extremely interesting results. According to P. C. van der Kruit (1971) some Seyfert galaxies have a radio source in the nucleus. Others have two emitting regions placed symmetrically with respect to the core and at first sight resemble radio galaxies (see below), but in the case of Seyfert galaxies the double source apparently lies in the plane of the galaxy. Finally NGC 4736, for example, has both kinds of source. In some cases the radio sources could be resolved by very long baseline interferometry down to $\sim 10^{-3}$ arc seconds; in this way one obtained, for example, for NGC 1275 = 3C84 three formations fitting inside one another, with (half-power) diameters

	300″	0″.02	0″.001
or	80 kpc	5 pc	0.3 pc.

Variations of Seyfert nuclei on a time scale of months also leads to the conclusion of finer structure.

The phenomena already described in "normal" galaxies give the impression of a somewhat fragmentary or atrophied counterpart to Seyfert galaxies. We therefore come to the view that all or in any case most bright galaxies succumb to the "Seyfert disease" from time to time. Since about 1 percent of galaxies show the Seyfert phenomenon, its total duration (possibly with interruptions) may amount to 10^8 years, to order of magnitude.

The classification criterion due to Seyfert (1943) is not the only one possible. Besides that one describes as *N galaxies* (W. W. Morgan 1958) objects which (optically) have a bright, star-like nucleus with a weak, hazy envelope. Probably all Seyfert galaxies are also N type, but the reverse is certainly not true. Related to the N type, but still further condensed, is the class of *compact galaxies* (F. Zwicky 1963) which are defined as having within them a region whose surface brightness is greater than 20^m per □″. Finally one denotes as *Markarian galaxies* those which, for example on low dispersion spectra, reveal a significant ultraviolet excess (compared with normal objects). The relationship of the criterion chosen by B. E. Markarian (1967) to that due to Seyfert is obvious. It must remain a task for the future to find a classification which represents at a glance and in the closest possible connection with the observations both the permanent and the temporary features of galaxies.

28.2.3 Radio galaxies

By 1954 the measurements of the positions of several of the stronger radio sources had attained such accuracy that W. Baade and R. Minkowski succeeded in identifying them with optical objects. In particular, the second strongest radio source in the northern sky, Cygnus A, could be associated with a remarkably faint object of photographic magnitude 17^m9. Besides a weak continuum and the Hα line, it shows forbidden lines of [OI and III], [NII], [NeIII and V] . . ., with a redshift (Section 30) corresponding to 16 830 km/s. Thus we have to do with an extragalactic object at a distance about 170 Mpc. The optical picture shows two "nuclei" about 2″ apart. Baade and Minkowski first interpreted this as two colliding galaxies. In the collision the stars would be only a little affected; the gas, on the other hand, would be swept out of both galaxies and excited into a state of radio emission. Subsequent investigation has not been able to endorse this picture.

Measurements by R. Hanbury Brown, R. C. Jennison, and M. K. Das

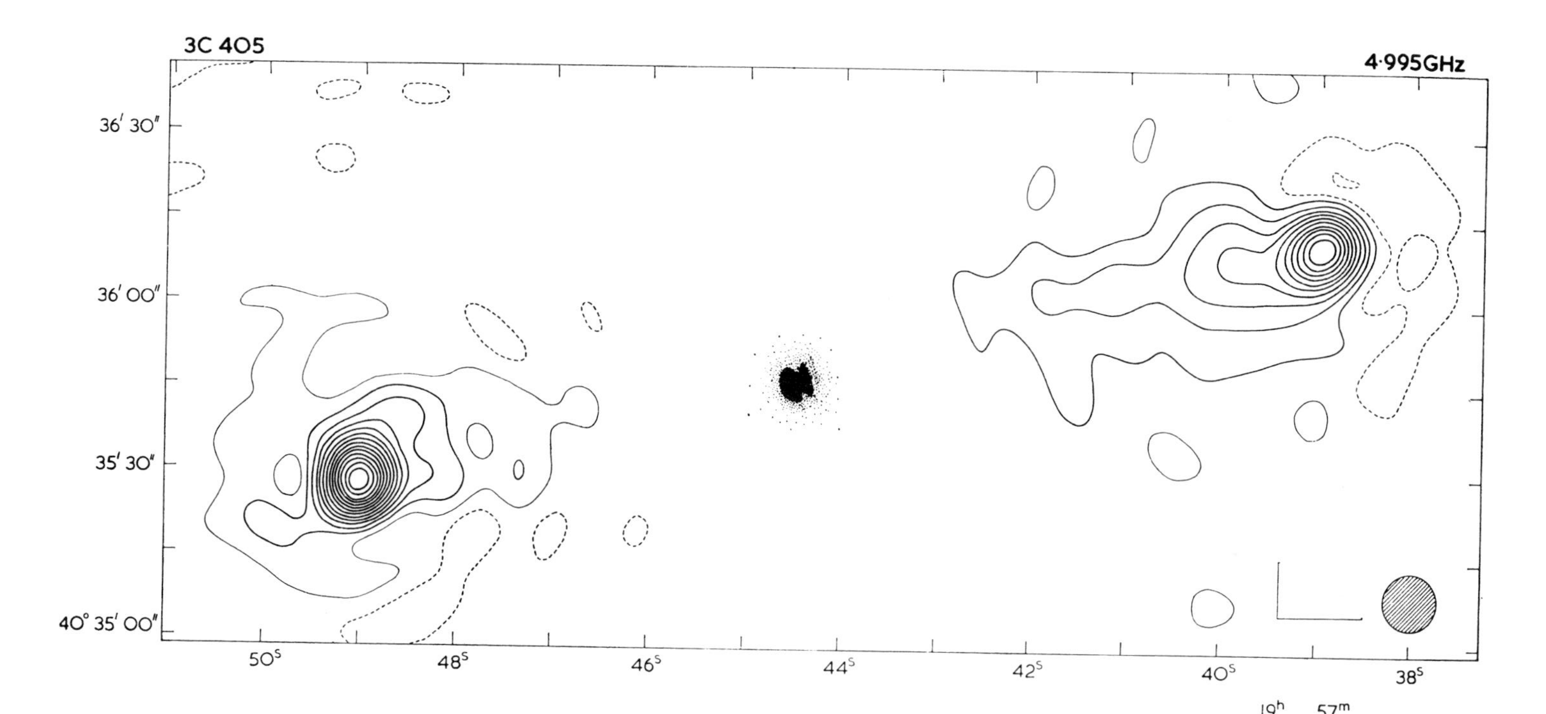

Figure 28.3 Cygnus A at 5 GHz (λ 6 cm). Radio isophotes at intervals of 4800 K in radiation temperature (S. Mitton and M. Ryle, 1969). The optical shape of the galaxy (W. Baade and R. Minkowski, 1954) is indicated. The hatched circle marks the antenna beam size. The scales in RA and δ are chosen so that this appears as a circle; each arm of the L-shaped scale is 10″ long.

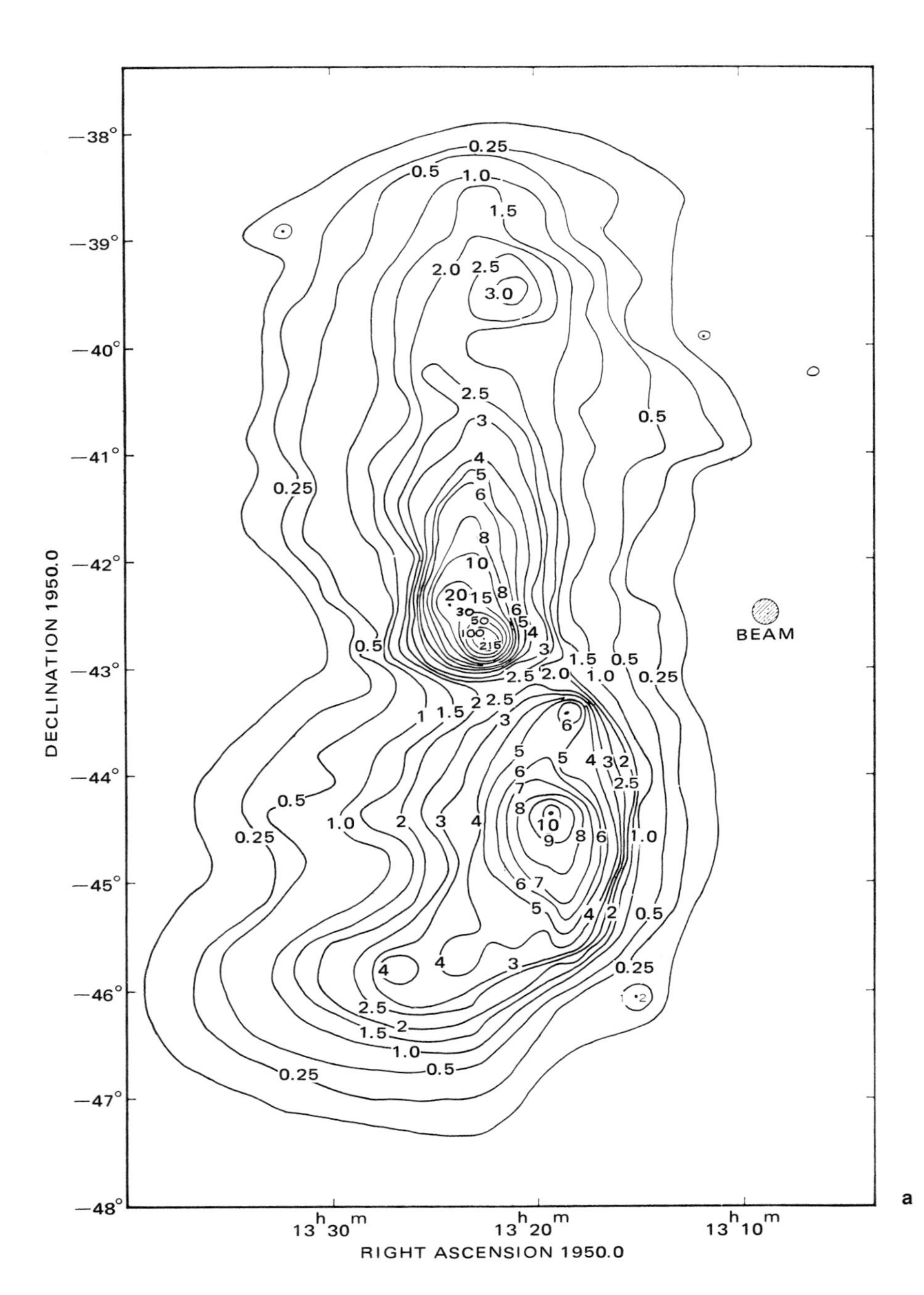
DECLINATION 1950.0
RIGHT ASCENSION 1950.0
−38°
−39°
−40°
−41°
−42°
−43°
−44°
−45°
−46°
−47°
−48°
13h30m
13h20m
13h10m
BEAM
a

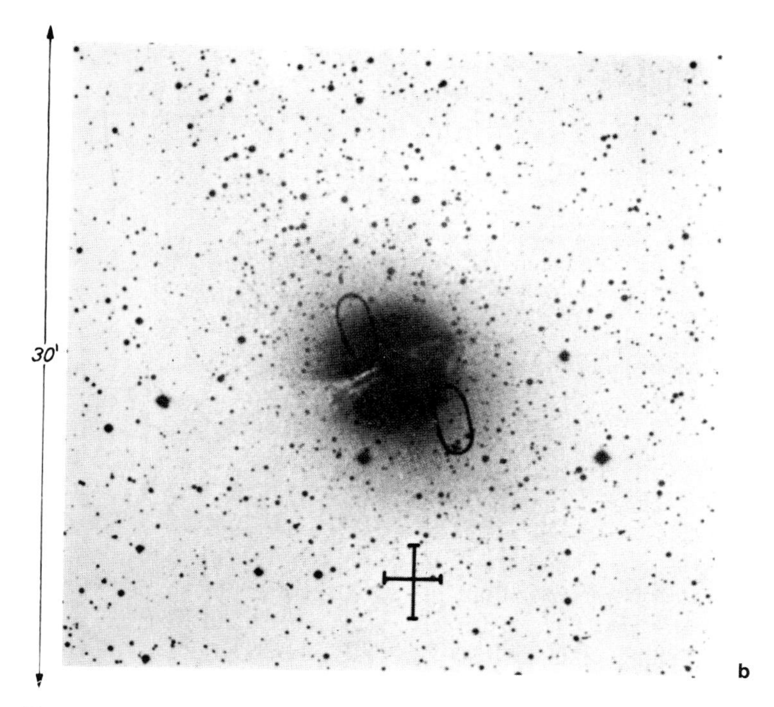

Figure 28.4 Radio source Centaurus A (NGC 5128). a—Isophotes of the radio continuum at 1420 MHz (λ 21 cm) by B. F. C. Cooper, R. M. Price, and D. J. Cole, 1965. Radiation temperatures are given in Kelvin. The circle (right) of 14′ diameter denotes the resolving power (beamwidth at half intensity) of the aerial. b—The EO galaxy NGC 5128 (RA $13^h\ 22^m\ 31\overset{s}{.}6$, $\delta -42°\ 45\overset{\prime}{.}4$, 1950) has been identified with the unresolved central source in Figure 28.4a. According to interferometer measurements by P. Maltby (1961) this galaxy itself consists of two sources, which are almost symmetrically placed with respect to the remarkable absorbing lane which surrounds the galaxy. There is no *optical* indication of either the narrow (younger) or wider (older) pair of radio sources. The sole indication of the explosion responsible for flinging the radio sources out of the core is a pair of weak extensions just discernible in Figure 28.4b, and approximately along the line joining the two inner radio sources. This line is itself nearly parallel to the rotation axis of the galaxy.

Gupta using the great correlation interferometer at Jodrell Bank Observatory showed in 1953 that the radio emission of Cygnus A does not arise from the galaxy itself but from two components placed almost symmetrically with respect to it. Cygnus A is the prototype of a *radio galaxy;* in some objects the galaxy itself also emits radio radiation.

In Figure 28.3 we show a contour map which S. Mitton and M. Ryle (1969) have obtained with the Cambridge aperture synthesis radio telescope at 5 GHz (λ 6 cm) with a resolution of $6\overset{\prime\prime}{.}5 \times 9''$ (in RA, δ, respectively). The

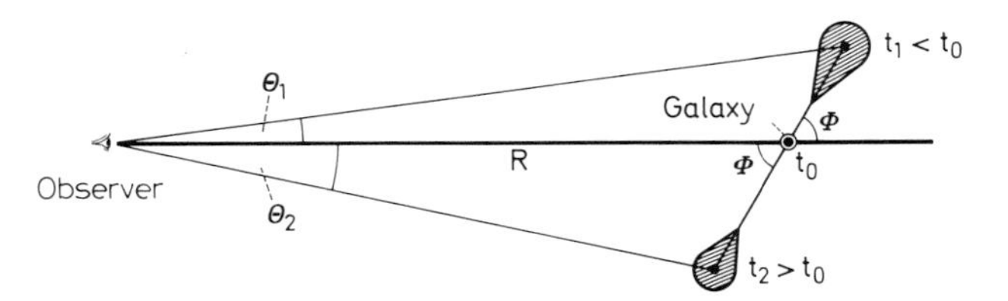

Figure 28.5. Radio galaxy. At time $t = 0$ two plasma clouds (and magnetic fields) were flung out from the galaxy in opposite directions with velocity v at an angle Φ to the line of sight. Because of the finite speed of light, we see—compared to the galaxy—an earlier state of the cloud 1 moving away from us and a later state of the cloud 2 moving toward us. For this reason, cloud 2 also appears rather further from the galaxy than cloud 1.

optical shape of the galaxy is indicated.[2] The observations of Cyg A and of many similar radio galaxies suggest that two plasma clouds were flung out of the central galaxy, approximately along its axis, by a gigantic explosion. Their present separation amounts to $2'$ or ≈ 90 kpc (perpendicular to the line of sight); their diameters are about 20 kpc. The elongated shape of both components indicates that lines of force of a magnetic field were dragged out of the galaxy along with the plasma. The outer boundaries of both plasma clouds are conspicuously sharp. Taking account of the finite resolution, the main decrease in brightness occurs in $< 1''$. According to D. S. de Young and W. I. Axford (1967) the ejected plasma is here separated from an *intergalactic medium* by a shock wave.

The radio galaxy Centaurus A = NGC 5128, which is much closer to us, shows itself in the first place in the optical picture (Figure 28.4b) to be an elliptical galaxy. Quite generally, in fact, among the radio galaxies the Hubble type E predominates, with a strong preference for giant galaxies. At right angles to the axis NGC 5128 is girdled by a rather turbulent band of dark material (the one in Cyg A simulates two galaxies). In exposures with high contrast, we can trace faint appendages along the rotation axis out to about 40 kpc in both directions. On the radio map (Figure 28.4) we recognize at once two pairs of plasma clouds, whose distances from the galaxy amount to about 400 000 and about 13 000 light years. In this galaxy, then, (at least) two explosions must have taken place, at times at *least* the corresponding number of years ago (since the expansion velocity must be $< c$). The energy radiated in the meantime can only stem from the explosions.

We now try to discover the *age, energy content,* and *power output* of radio galaxies. From these we can then hope to obtain at least an estimate for the density of the intergalactic medium.

[2]More recent measurements by P. J. Hargrave and M. Ryle (1974) at 5 and 15 GHz show also radio emission from the galaxy and more details, which may be significant for the further development of the theory.

Following M. Ryle and M. S. Longair (1967) we start with a simple kinematic consideration (Figure 28.5). A galaxy has ejected at time $t = 0$ two plasma clouds with the same speed v at an angle Φ to the line of sight. If we observe at time $t = t_0$, then the cloud 1 moving *away* from us—from which the light takes longer to reach us—appears to us as it was at an earlier time t_1, while the state of the cloud 2 moving toward us corresponds to a later time $t_2 > t_0$. At the moment of observation cloud 1 has covered a distance $t_1 v$ while cloud 2 has covered the greater distance $t_2 v$. We therefore have

$$t_0 - t_1 = t_1 v \cos \Phi / c \qquad t_2 - t_0 = t_2 v \cos \Phi / c \tag{28.8a}$$

and for the angular distances θ_1 and θ_2 of the two clouds from the galaxy

$$\theta_1 = t_1 v \sin \Phi / R \qquad \theta_2 = t_2 v \sin \Phi / R. \tag{28.8b}$$

Here R denotes the distance of the galaxy, which is obtained from its red shift (Section 30).

From these four equations we can calculate t_0, t_1, t_2, and v/c without further ado, if we make a plausible assumption for the angle Φ. Since the axes of the galaxies will be randomly distributed, the Φ ranges 0°–60° and 60°–90° are equally probable. The reader may easily convince himself that no great error is made by putting $\Phi \approx 60°$ throughout.

If we return first to the radio source Cygnus A, we can deduce from the data already given that the two components are traveling away from their parent galaxy at present at speeds of $v/c = 0.1$ to 0.2 (depending on the assumed angle Φ). Their age amounts to $t_0 \approx 5 \times 10^5$ years. The energy which is at present contained in the synchrotron electrons and magnetic fields of the two components can be estimated by means of the theory of synchrotron radiation. First of all, according to equations (28.4) and (28.5) the spectral index gives information about the velocity distribution of the electrons. Then one can calculate the emission per unit volume as a function of the number of electrons per cubic centimeter *and* of the strength of the magnetic field, or its energy density. Finally, making the plausible assumption that the energy densities of the synchrotron electrons plus protons and of the magnetic field are approximately equal, one can obtain the total energy density, or more precisely a lower limit for it. In this way we learn, for example, that the radio components of Cyg A possess at present an energy content of $\sim 4 \times 10^{58}$ erg. On the other hand, by integrating over the entire radio spectrum we calculate the power output as $\sim 5 \times 10^{44}$ erg/s. This energy loss could be provided from the supply of 4×10^{58} erg for 0.8×10^{14} s $\approx 3 \times 10^6$ years, which agrees well with the dating of Ryle and Longair.

More extensive observational material teaches us that the explosions of different radio galaxies occurred from 10^4 to 3×10^6 years ago. During this time the velocities of the components went down from $\sim c$ to $\sim 0.1c$ and their brightness at 1407 MHz decreased by about 3 powers of 10. Their initial energy content can be estimated (with considerable uncertainty) to be $\sim 10^{62}$ erg. As we remarked in connection with de Young and Axford, the sharp outer boundary of the two plasma clouds is clearly the result of the formation of a *shock wave*. In that case, however, the "ram pressure" in the intergalactic medium of density ρ_0, that is, $\rho_0 v^2/2$, must be of the same order of magnitude as the internal pressure or energy density in the components immediately behind the shock wave. If one uses the earlier approximate values for Cyg A, one obtains, following S. Mitton and M. Ryle

$$\rho_0 \approx 10^{-28} \text{ g/cm}^3. \tag{28.9}$$

This numerical value for the *density of the intergalactic medium*—which we shall later compare with other estimates—is of the greatest significance for the construction of cosmological world models (Section 30).

28.2.4 Exploding galaxies; M 82 and NGC 1275

We obtain more precise information about the explosion in the nucleus of a galaxy in the few cases that we can also observe optically and spectroscopically. In 1963 C. R. Lynds and A. R. Sandage were the first to succeed in doing this, for the galaxy M 82.

Going by its membership of the M 81 group, the galaxy M 82 is at a distance of about 3 Mpc; since 1961 it has been known to be a radio source. Superficially it has the appearance of an irregular galaxy. Actually we have to do with a flattened galaxy out of whose nucleus, as spectra and photographs in Hα show (Figure 28.6), enormous masses of hydrogen are shooting out in both directions along the axis. The speed of each mass is proportional to its distance from the nucleus. We can trace these *filaments* on both sides out to a distance of about 4000 pc from the nucleus of the galaxy; there the speed is about 2700 km/s. Hence we easily calculate that the explosion took place 1.5 million years ago. From the intensity of Hα we can infer the mass of the ejected filaments to be $\lesssim 5.6 \times 10^6$ solar masses, and the mean density to be about 10 protons/cm^3. The total kinetic energy of the moving bodies of gas comes to $\lesssim 2.4 \times 10^{55}$ erg.

For comparison we can easily verify that 5.6×10^6 Suns in a time interval of 1.5 million years would radiate about 4 percent of this amount of energy.

The synchrotron emission of M 82 integrated from the region of radio waves into the optical region for the 1.5 million years since the explosion (assuming the emission to remain constant in time) amounts to 9×10^{55} erg. This amount of energy must then have been stored up in synchrotron electrons with energies up to about 5×10^{12} eV. According to Lynds and Sandage, this appears to be possible, if the magnetic field is less than 2×10^{-6} G. The total amount of energy liberated in the explosion can be only tentatively estimated at 10^{56} to 10^{58} erg. The present emission in the radiofrequency range [from 10^7 to 10^{11} Hz; see equation (28.3)] from radio galaxies like Cygnus A is about 10^5 times stronger than that from M 82 (5.1×10^{44} as compared with 4.2×10^{39} erg/s).

It has been shown recently that the Hα radiation of M 82 is partially *polarized*. If this polarization were attributed to the fact that the radiation emitted in the nucleus reached us via scattering by free electrons, the density estimates would need to be correspondingly modified.

In 1970 R. Lynds observed similar filaments in the galaxy NGC 1275 in the Perseus cluster, which had already been identified in 1954 by W. Baade and R. Minkowski with the strong radio source Perseus A. As already mentioned, its inner region shows all the characteristics of a Seyfert galaxy. Over and above that, however, one descries numerous elongated (apparently unpolarized) Hα filaments out to about 140″ from the center; these have a certain superficial resemblance to those of the Crab nebula. Clearly NGC 1275 = Per A is also a radio galaxy which is in the middle of exploding. While in the cases of M 82 and NGC 1275 we are concerned with comparatively mild explosions, in the following section we shall discover in

Figure 28.6. Radio source M 82. Hα photograph (negative) taken by A. R. Sandage (1964) with the Mount Palomar 200-in telescope. The hydrogen filaments were flung out to about 4000 pc on either side of the disk of the galaxy by an explosion which occurred about 1.5 million years ago.

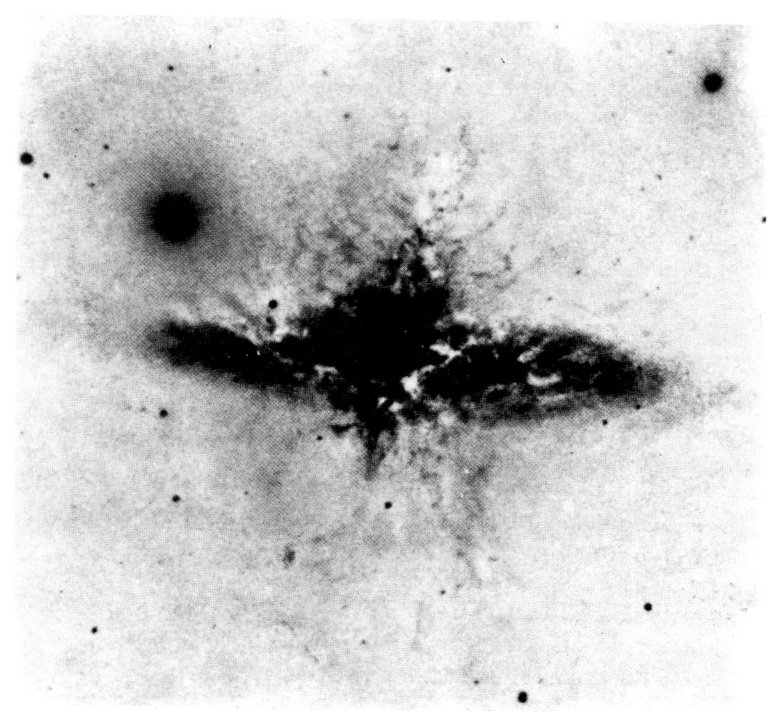

the *quasars* the actual early stages in the life of radio galaxies of the types of Cyg A, etc.

28.2.5 Quasars and quasi-stellar objects

It was discovered in the sixties that a number of not particularly weak 3C sources (3rd Cambridge catalog of discrete radio sources) were to be identified with optical objects which even on 200-in plates could not be distinguished from normal stars. The diameters determined optically and by radio astronomy were $<1''$ and led first to the realization that the new kind of object possessed in both regions of the spectrum an extraordinarily large surface brightness.

In 1962–1963 M. Schmidt investigated their spectra; they show a *continuum* and strong *emission lines,* similar to those of the already known radio galaxies. The new feature was the enormous red shifts which—interpreted in the sense of the Hubble relation (Section 30)—showed that one has to do with very distant, blue compact galaxies. Their visual absolute magnitude exceeds that of normal giant galaxies by up to 5^m, that is, a factor of about 100. Their radio power (for example, integrated from 10^7 to 10^{11} Hz) corresponds roughly to that of Cyg A. At this stage of the investigation the designation *quasar* or quasi-stellar radio source was invented for them.

However, in 1965 A. Sandage realized that there are many more *quasi-stellar galaxies,* which with their compact structure, high surface brightness and blue color cannot be distinguished optically from the quasi-stellar radio sources, but emit *no* radio radiation (or are at most very weak radio sources). Now, therefore, we often distinguish between QSR = quasi-stellar radio source or QSG = quasi-stellar galaxy and QSO = quasi-stellar object (without radio emission).

Because we may well say nowadays that we are dealing with the same kind of compact galaxy, of which at any time a small fraction is suffering from the "radio disease," the terminology seems to be changing, in the sense that one speaks of quasars or quasi-stellar galaxies (objects) in all cases and adds as required whether or not they are strong radio emitters.

If we consider the two-color diagram, $U - B$ against $B - V$, it turns out (A. Sandage 1971) that *quasars* (with or without radio emission), *Seyfert nuclei, N-galaxies,* and *blue compact galaxies* are arranged from left to right along a narrow strip, roughly along the blackbody line (cf. Figure 15.5), and so can be clearly distinguished from stars (apart from white dwarfs or strongly reddened objects). Such a diagram also emphasizes the close relationship of the objects we have mentioned.

Observations with high resolution radio interferometers show further that many quasars consist of two components (like, for example, Cyg A) and/or a central source. If we return to the considerations of M. Ryle and M. S. Longair, we find that for quasars the two plasma balls are still moving almost at the speed of light (naturally we must in this case use the complete formulas of relativistic kinematics) and that the times since the explosion in the central galaxies are here in the range 10^2 to 10^5 years. At least within certain limits—still to be discussed—we can therefore view quasars as the youthful stages of radio galaxies of the Cyg A type. Their nuclei, on the other hand—as we have already remarked—show a relationship with the much "gentler" ones of the Seyfert galaxies.

Let us then try to learn at least something of the certainly very "unconventional" physical conditions in quasars. There is no doubt that their radio spectrum is to be explained as *synchrotron radiation*. In some cases, as has been shown, *self-absorption* must also be taken into account. That even the optical continuum, with its ultraviolet excess (with respect to normal galaxies), was attributable to the synchrotron mechanism first became apparent in observations of the giant elliptical galaxy M 87 = NGC 4486, the radio source Virgo A. Out of its center is shooting a "jet," which emits bluish, polarized light which can only be synchrotron radiation. For the production of a continuum in the optical region there must be available synchrotron electrons of in any case up to 5×10^{12} eV, and certainly also heavy particles of comparable energy. Obviously then we here also have before us sources of *cosmic rays*.

The *total energy* of synchrotron electrons, related heavy particles, and magnetic field which is available to a quasar can again be estimated under the assumption of approximate equipartition and comes to about 10^{62} erg (some authors regard even values of up to 10^{64} erg as possible). We ought to form a clear idea of the size of this quantity of energy, which can be liberated in a galactic explosion: it corresponds to the *nuclear energy* (0.007 mc^2) of about 8×10^9 solar masses or—as an extreme possibility—to the total *rest mass* energy (mc^2) of 6×10^7 solar masses!

The *power output,* integrated over the optical, infrared and radio regimes, reaches values of up to 10^{46} erg/s, which can therefore be covered by the aforementioned supply even in the most extreme cases.

Concerning the spatial extension of quasars, the best modern information comes from very long baseline interferometry at centimeter and decimeter wavelengths, where the resolving power of about 10^{-3} arc seconds far exceeds that of all optical instruments.

As an example, let us consider the quasar 3C 273 which, because of its proximity, has been thoroughly studied. From its red shift, $z = 0.158$ (cf. Section 30), a distance of $\sim$ 500 Mpc is deduced; 1″ then corresponds to $\sim$ 2.4 kpc. In the first place, the optical and radio pictures agree in showing a thin "jet"—called 3C 273A—stretching radially out of the actual quasar to a distance of $\sim$ 20″. Since it cannot have been moving faster than the speed of light c its age must amount to at *least* 1.5×10^5 years and may be of the order of 10^6 years. The actual quasar consists of a sort of halo of size $\sim 0\overset{''}{.}022 \approx 50$ pc (component B); within that is situated component C with a size of $\sim 0\overset{''}{.}002 \approx 5$ pc and the still smaller, at most partially resolved component D, whose size is $\leqslant 0\overset{''}{.}0004 \approx 1$ pc. At frequencies $> 10^4$ MHz, the smallest component D makes the largest contribution to the radiated flux. However, toward longer wavelengths the flux falls off rapidly, first in

D then in C and finally in B, as a result of synchrotron self-absorption, particularly in the denser regions.

The surprisingly small size of the central component D in 3C 273, and in other quasars. which is clearly to be thought of as the actual source of their energy and "activity," is confirmed by the time variability of the optical and radio radiation. Their characteristic time scales are of the order of weeks to years, which means that the extent of the emitting regions can amount at most to a few tenths of a light year.

Our knowledge of quasar *masses* is still very modest: With some uncertainty as regards the excitation mechanism, we can estimate the mass of emitting gas from the absolute intensities of the optical emission lines and find, to order of magnitude, 10^6 solar masses. At the present time we know nothing of the total mass; we may surmise that it corresponds roughly to the 10^7 to 10^8 solar masses of quiet galactic nuclei.

It is of the utmost interest to ask of what the emitting gas clouds—probably therefore the outermost parts—of quasars are composed. We cannot discuss here the very difficult questions relating to the excitation processes of the emission lines. However, if we merely in the first place notice which chemical elements appear, through whatever lines, it turns out that they are all those elements whose abundance in the Sun is larger than or equal to that of iron. With great surprise we conclude from that that in quasars—just as we found earlier for Seyfert nuclei—we have to do with fairly normal cosmic material.

This conclusion is confirmed by the absorption lines which have been found in the spectra of many quasars. These lines are mostly *violet* shifted relative to the emission lines and arise in gas clouds which are flung out of the quasar with velocities of from a few hundred kilometers per second to about one third of the speed of light. According to Y. W. Tung Chan and E. M. Burbidge (1971) these quasar absorption spectra also imply a chemical composition which is not at all distinguishable from that of the usual cosmic material.

The observations of quasars, whose outlines have here merely been sketched in, allow us to recognize that there is obviously a continuous transition from them to the activity (only quantitatively less prodigious) of radio galaxies, Seyfert nuclei, N galaxies, etc. Recent observations at the Hale Observatories, and their statistical exploitation, make it in fact as good as certain that the apparently star-like quasars are nothing other than the *nuclei* (about 5^m "too bright") of galaxies, whose outer regions are for the most part too faint to be photographed even with the 200-in telescope.

The main question now concerns the origin of the quantities of energy ($\sim 10^{62}$ erg) which have come to light in quasars, radio galaxies, etc. We have already estimated that *nuclear* energy is in no case sufficient. Even if—as an extreme possibility—the entire rest-mass energy Mc^2 of a mass M can be made "useable," we need, as we have seen, $\sim 6 \times 10^7$ solar masses, that is, a significant part of the mass of a galactic nucleus. The sole process known within the framework of present-day physics which could release a significant fraction of the relativistic rest-mass energy Mc^2 of a cosmic mass with $M \approx 10^8$ solar masses is the gain of gravitational potential energy as a result of contraction, that is, gravitational collapse. If, for example, we form a homogeneous sphere of radius R (this is to be understood only as a crude estimate!) from initially dispersed material, the resulting gravitational energy V is:

$$V = \frac{3}{5}\frac{GM^2}{R}. \tag{28.10}$$

Other forms of energy will be of the same order of magnitude. If we now require that V should be about Mc^2, then—as we shall see in Section 30—R is of the same order as the *Schwarzschild radius*, that is, the collapsed material must take up a structure which is at least similar to that of a *black hole*. For 6×10^7 solar masses this radius would be $r_s \approx 170 \times 10^6$ km ≈ 1 AU. However, in quasars we are probably dealing not with a single homogeneous object but with a group of several. It seems premature, however, further to elaborate our entirely provisional estimates.

On the other hand we should mention briefly that some astronomers—especially in view of the huge energy requirements of quasars—have argued against the use of Hubble's red-shift law for determining their distances. They try rather to explain the large red shifts of quasars in some other way and to interpret quasars as a *local* phenomenon. This hypothesis contradicts the empirical fact that in every respect (energy output and content; separation of outer components; membership of clusters of galaxies) there is a continuous transition from quasars to radio, Seyfert, and N galaxies. Whether there are some quite other kinds of local objects, which could be mistaken for quasars, may remain undecided.

28.3 Cosmic rays

The cosmic rays may be contrasted with the nonthermal radio radiation as a second form of nonthermal radiation. We treat first the *primary* cosmic rays, which are investigated *before* their interaction with the material of the Earth's atmosphere by means of balloons and rockets and, by means of artificial satellites and space probes, even before they have been influenced by the geomagnetic field. We shall discuss later the remaining effects, which are of solar origin.

We consider first the *nucleon component,* consisting of protons, α particles (He^{++}), and heavy nuclei (that is, fully ionized atoms). With the use of nuclear emulsions, suitable arrangements of counters, or tracks in solids it is possible to determine their atomic number Z and so also to obtain their *abundance distribution.* By and large this is the same as the cosmic (or solar) abundance distribution, with the strange difference that the light elements Li, Be, B, which are extremely rare in stars, have in cosmic rays almost the same abundance as the subsequent heavy elements. The nuclei Li, Be, B originate in cosmic rays by *spallation,* that is, as a result of the destruction of heavy nuclei, particularly Fe, by energetic protons or α particles. The reaction cross sections for such spallation processes can be measured using large accelerators—the proton synchrotron at CERN reaches $\sim$ 25 GeV; they are of the same order as the geometrical nuclear cross sections. From the number ratio of the lightest and the heavy nuclei in cosmic rays it is possible to calculate that they have traversed an amount of matter of 4 to 6 g/cm^2. However, this number should be considered more as an upper limit. P. H. Fowler *et al.* (1967) have in fact proved that there are in cosmic rays nuclei up to the end of the periodic system ($Z > 80$) and perhaps even beyond. Such "dense" nuclei, and further all low-energy nuclei, can have traversed only a *small* quantity of material. Making appropriate assumptions, on the one hand, about the distribution of the traversed layers of the interstellar medium and using, on the other hand, measured or calculated cross sections for the production and annihilation of energetic nuclei, M. F. M. Shapiro *et al.* (1972) have deduced the chemical composition of cosmic rays at their place of origin. We have already included the results in Table 19.1 and noted that the abundance distribution of the elements in cosmic rays (at their place of origin) is, within reasonable error bounds, the same as the cosmic abundance distribution in the Sun and other normal stars. Before we further consider these conclusions, as unexpected as they are important, we must first concern ourselves with the *energies* of the cosmic ray particles. Their penetrating power—even into deep mines—already showed that we have here to do with energies which far exceed all normal bounds.

The *energy distribution* of cosmic ray particles—we restrict ourselves at first to the nucleon component—is obtained at the smaller energy values by magnetic and electric deflection experiments and at large energies by measuring the ionization energy which *extensive air showers* (Auger showers) deposit in the Earth's atmosphere. Figure 28.7 shows the result of such measurements over the enormous energy range $10^8 < E < 10^{20}$ eV.

We cannot here go more deeply into the complicated events which accompany the impact of the primary cosmic rays on the Earth's atmosphere. However we must discuss in more detail the deflection of the charged cosmic ray particles by *magnetic fields,* to wit, the *geomagnetic* field, the *interplanetary* magnetic field of the Sun, and the *interstellar* magnetic field of the Galaxy.

A particle with charge e and momentum $p = mv$ in a magnetic field H moves in a plane perpendicular to the field on a (Larmor) circle whose radius r_H is given by[3]

[3]If we call the particle velocity v and the light speed c, the rest mass m_0 and the relativistic mass $m = m_0(1 - v^2/c^2)^{-1/2}$, then the momentum of the particle $p = mv$ and the energy $E = mc^2$. For motion perpendicular to the field we have centrifugal force = Lorentz force, or $mv^2/r_H = evH/c$ which is $mv = p = eHr_H/c$.

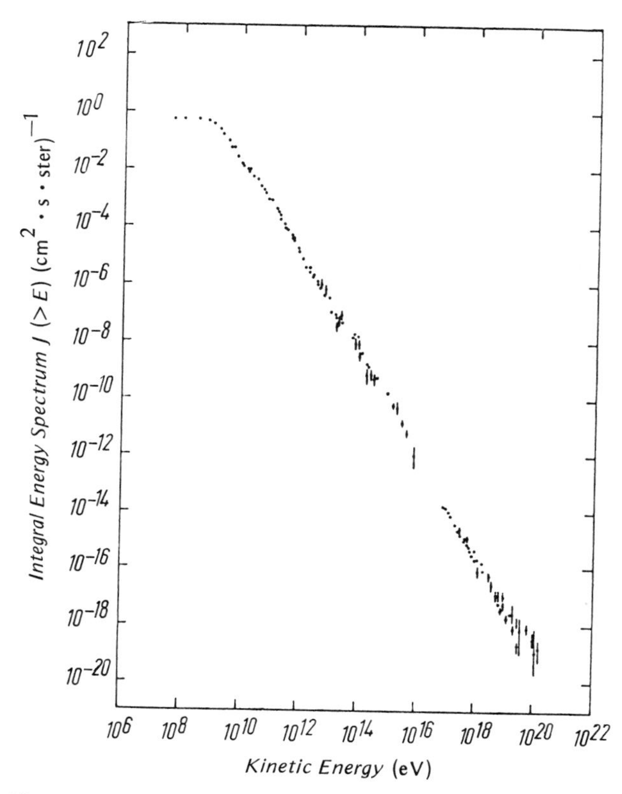

Figure 28.7. Integral energy spectrum of cosmic rays. The number $J(>E)$ of particles with energy greater than E passing through 1 square centimeter per unit solid angle per second is plotted as a function of E in electronvolts. For $E > 10^{10}$ eV the influence of the interplanetary magnetic field is unimportant. From that point to the highest energy measured in air showers, $\sim 10^{20}$ eV, the energy spectrum can be represented by the interpolation formula:

$$J(>E) = 1 \times 10^{16} E^{-1.74 \pm 0.1}$$

$$pc = eHr_H. \tag{28.11}$$

The so-called magnetic "rigidity" Hr_{H} (G cm) is thus exactly connected with the momentum p. For relativistic particles ($E \gg m_0c^2$, or 0.51 MeV for electrons, 931 MeV for protons; $v \approx c$) we have $E \approx cp$ and therefore

$$E \approx eHr_H. \tag{28.12}$$

If we calculate E in units of

$$1 \text{ eV} = \text{electron charge } e \times \frac{1}{300} \text{ electrostatic units of potential}$$

we get for singly charged particles

$$r_H = \frac{E_{eV}}{300} \frac{1}{H} \quad [\text{cm}] \tag{28.13}$$

or for our purposes more usefully in parsecs and GeV (gigaelectronvolts) = 10^9 eV

$$r_{H,\text{pc}} = 1.08 \times 10^{-12}\, E_{\text{GeV}}/H. \tag{28.14}$$

The general form of path of a charged particle in a magnetic field is given by combining the circular motion with a translation along the field lines which is not affected by the field. Thus the particles move in spirals along the lines of magnetic force.

The *geomagnetic* field has the effect that relatively low-energy charged particles can reach the Earth's surface only in a certain region around the geomagnetic pole. This is the latitude effect discovered by J. Clay. As can easily be estimated, using r_H = radius of Earth, H = 1 G, the effect vanishes for energies greater than about 100 GeV.

Then, as the cosmic rays penetrate the planetary system, they begin to interact with the *solar wind* and with the *interplanetary* magnetic field of $\sim 10^{-6}$ G which is streaming outward along with the wind plasma. It is therefore not surprising that the intensity of the cosmic rays within the planetary system displays a dependence on the 27-day cycle of the synodic rotation of the Sun and on the 11-year cycle of solar activity.

In the *galactic* magnetic field of 5×10^{-6} G the Larmor radius for a singly charged particle of energy E is given by

$$\left.\begin{array}{llllll} E & = \begin{cases} 10^9 \\ 1 \end{cases} & \begin{array}{l} 10^{12} \\ 10^3 \end{array} & \begin{array}{l} 10^{15} \\ 10^6 \end{array} & \begin{array}{l} 10^{18}\ \text{eV} \\ 10^9\ \text{GeV} \end{array} \\ r_H & = 2 \times 10^{-7} & 2 \times 10^{-4} & 0.2 & 200\ \text{pc.} \\ & \quad (0.045\ \text{AU}) & & & \end{array}\right\} \tag{28.15}$$

On account of the deflection of the particles by the galactic magnetic field (and for smaller energies by the interplanetary and terrestrial fields as well), it is not to be wondered at that, within the accuracy of the measurements, the distribution of directions of the particles is isotropic. This makes it all the more important to learn something about their origin through radio astronomy and γ-ray astronomy.

In at least one case we can observe the production of cosmic rays at close quarters since S. E. Forbush was able to show in 1946 that in great eruptions (flares) the Sun emits cosmic-ray particles up to some GeV. We cannot here go further into these and other investigations by A. Ehmert, J. A. Simpson, P. Meyer, and others who principally elucidated the propagation of cosmic rays in the interplanetary plasma and magnetic field. Concerning the physical mechanism of the acceleration of charged particles in the magnetic plasma of the solar chromosphere or corona to energies of 10^9 to 10^{10} eV we are still ignorant. Whether an induction effect as in a betatron is crucial, or whether the particles are squeezed between two magnetic "mirrors" (possibly shockfronts) and thereby accelerated (E. Fermi) we cannot yet discern. One thing is, however, clear and definite, that the acceleration of particles to high energies in the Sun as in the giant radio sources takes place in a highly *turbulent plasma with a magnetic field present*.

Let us next discuss the cosmic rays in our galaxy. From the abundances of certain isotopes which arise from spallation processes in meteorites, we know that the intensity of cosmic rays has remained practically constant for at least 10^8 years.

We then compare estimates of energy densities of various sorts in our neighborhood, the values being erg/cm^3:

Cosmic rays (center of gravity of energy spectrum $\sim$ 7 GeV)	1×10^{-12}
Thermal radiation, that is, total starlight	0.7×10^{-12}
Kinetic energy of interstellar matter $\rho v^2/2$ ($\rho \approx 2$ proton masses/cm^3, $v \approx 7$ km/s)	0.8×10^{-12}
Galactic magnetic field $H^2/8\pi$ ($H \approx 5 \times 10^{-6}$ G)	1×10^{-12}

(28.16)

According to the second law of thermodynamics the possibility seems to be excluded that in the Galaxy as much energy should be supplied in the form of extremely nonthermal cosmic rays as in the form of thermal radiation. In fact the paths of the charged cosmic ray particles become wound up in the galactic magnetic field G, that is, their motion at right angles to G is in (Larmor) circles, whose center performs a translational motion parallel to G. The particles are stored in the Galaxy in screw-shaped orbits of this kind. According to equation (28.15), one would expect that this mechanism operates up to energies of $\sim 10^{17}$ eV. Assuming a mean density for the interstellar medium of $\sim 2 \times 10^{-24}$ g/cm^3, we deduce from the quantity of material traversed, ~ 4 g/cm^2, a path length for the particles of ~ 600 kpc or 2×10^6 light years, and so a lifetime of $\sim 2 \times 10^6$ years. Compared to undeflected particles, for example photons, which leave the galactic disk on average after a path of ~ 300 pc, this means then that cosmic rays are more abundant by a factor of about 2×10^3. The fact that the storage is about the same for different nuclear charges Z shows that the end of the "flight path" of a cosmic ray particle is due in general not to a collision with another nucleus but to its escape from the Galaxy.

To the energy densities in equation (28.16) there corresponds in each case a pressure of about the same magnitude (erg/cm^3 = dyn/cm^2). The equality as regards order of magnitude of the magnetic and turbulence pressures in the interstellar medium appears to be plausible on magnetohydrodynamic grounds. The approximate equality of the cosmic-ray pressure with these might be understood to mean that cosmic rays accumulate in the Galaxy, being retained by its magnetic field, until their pressure suffices for them to leak away into surrounding space, probably taking with them a certain amount of interstellar material and its magnetic field.

Since the average time spent by a cosmic-ray particle in the Galaxy is only $\sim 2 \times 10^6$ years, while the intensity of the cosmic rays has remained fairly constant for more than about 10^8 years, the supply must be replenished. One would have to look for their sources in highly turbulent plasma with magnetic fields, of which the energy density $H^2/8\pi$ might adjust itself to match approximately the kinetic energy density $\rho v^2/2$. These are however also the strong nonthermal radio emitters. The author emphasized this relationship as long ago as 1949.

It was reinforced by recent measurements of the *electron component* of cosmic rays, discovered by P. Meyer *et al.* in 1961. Their energy distribution, in the range 1 to 10 GeV that is important for the production of galactic synchrotron radiation, corresponds to a power law [see equation (28.4)] with exponent $\gamma = 2.1$, which according to equation (28.5) would lead to a radio spectrum $I_\nu \propto \nu^{-0.55}$, in satisfactory agreement with the observed $I_\nu \propto \nu^{-0.7}$.

For a long time it was uncertain whether the electron component of cosmic rays originated indirectly through the decay chain: π meson $\rightarrow$ μ meson (penetrating component) $\rightarrow$ electron or was directly accelerated together with the nucleons of a plasma. In the first case roughly equal numbers of electrons e^- and positrons e^+ should be formed while in the second case the electrons e^- should by far predominate. The measured ratio $e^-/e^+ \approx 10$ indicates that the cosmic ray or synchrotron electrons in the Galaxy originated in the same sources as the nucleon component.

What celestial bodies are these sources? We have already seen that the Sun produces cosmic rays and synchrotron radiation. However the flares of the Sun and even the flare stars make altogether only a tiny contribution to the galactic radiation. I. S. Shklovsky, V. L. Ginzburg, and others have therefore drawn attention to the significance of *supernovae*. It has been learned recently that their (possible) remnants, the *pulsars,* accelerate considerable amounts of material to high energies. Although the contribution to cosmic rays of supernovae and their remnants—concerning which we can only make rather uncertain estimates—could be of the right order of magnitude, it has very recently become steadily more likely, in the light of radio investigations, that we should look for important sources of synchrotron electrons and cosmic rays in the nucleus of our galaxy as well as in the nuclei of more distant galaxies and even quasars. (It is still unclear how the number ratio of electrons to protons in cosmic rays, $\sim$ 1:50, has come about.)

In this way, too, we might explain the puzzle of the most energetic component of cosmic rays, up to $\sim 10^{20}$ eV. Since the Larmor radius at 10^{18} eV already corresponds to the thickness of the galactic disk, and at 10^{20} eV corresponds to the dimensions of the whole system, such particles could under no circumstances be stored. They are therefore, in the Galaxy alone, at a disadvantage relative to the less energetic particles by a factor $\sim 2 \times 10^3$. Then we know that the energy spectrum of the solar component is steeper than that of the galactic radiation. All this leads to the concept that in the energy spectrum of cosmic rays at *high* energies the contribution of distant galactic nuclei and quasars steadily gains the upper hand. Furthermore, the extreme isotropy of that component could otherwise scarcely be understood.

Important contributions to this area of research are expected from *γ-ray astronomy*. However, it seems premature to summarize the results available at present.

On the other hand we should briefly indicate the possibilities and existing results of *neutrino astronomy*. We now distinguish—according to their origin in meson or electron processes—between μ and e neutrinos as well as their corresponding antineutrinos. So far only the e neutrinos have achieved astronomical significance. We have already pointed out that several percent of the energy produced at the center of a star escapes directly into space as neutrinos. The extraordinarily small reaction cross section of the neutrino with all kinds of material allows these to run through a star, indeed even the whole universe, without a single collision. Neutrino astronomy can therefore give us direct information about the energy-producing core of the Sun.

(The neutrino radiation of the whole universe together with its evolutionary history and related matters on a large scale (Section 30) are at present still very interesting problems for the future.)

Neutrinos can be detected by their nuclear reaction with Cl^{37}, in which A^{37} is formed, which then decays with a half life of $\sim$ 35 days. The Auger electrons resulting from this decay are then counted. Since 1964 R. Davis, Jr., has used this method to carry out measurements to detect *solar* neutrinos.

In order to minimize perturbations due to cosmic rays, the apparatus was erected at a depth of 1.48 km in the Homestake gold mine in South Dakota. A tank containing 378 000 liters of tetrachloroethylene C_2Cl_4 (normally used as a chemical cleaning fluid) serves as "receiver." After intervals of 2 or 3 half lives the A^{37} is flushed out with helium and the decay electrons are counted. At the time of writing (June 1972)[4] the result corresponded to $< 1 \times 10^{-36}$ neutrino processes per second per Cl^{37} atom.

According to theory, on the other hand, it would be expected that neutrinos in the energy range to which the Cl^{37} apparatus is sensitive would be formed, essentially as a result of the pp process by decay of B^8 [equation (25.20), last line]. However the neutrino production calculated in this way is about 9 times larger than the measured upper limit.

Many varied speculations have appeared concerning the origin of this discrepancy. Plausible suppositions might be either that the theory of convection in the interior of the Sun is still too rough and ready or that we do not sufficiently understand the sequence of reactions in the pp chain.

[4]Translator's note (July 1976): Very recently, some higher values have been reported; if confirmed, these would reduce, but not remove, the difference between observation and theory.

29. Galactic Evolution

The formation and evolution of galaxies is still full of puzzles and unsolved problems. Even a superficial glance through A. Sandage's *Hubble Atlas of Galaxies* (1961) and even more through H. Arp's *Atlas of Peculiar Galaxies* (1966) might serve as a "dreadful warning" to overhasty theoreticians. We therefore restrict ourselves here primarily to the description of several mechanisms which might play an important role in the life of galaxies:

(1) The *Jeans' gravitational instability* and the origin of star clusters and stars.
(2) The dynamics of *spiral arms,* in particular the density-wave theory.
(3) Formation of a *galactic disk* by collapse.
(4) Remarks on the *origin* of galaxies, insofar as this problem does not belong to the realm of cosmology (Section 30). To these will be added some reflections on the following:
(5) The significance of the abundance distribution of the elements and the *chemical (nuclear) evolution of galaxies.*

29.1 Jeans' gravitational instability and the formation of star clusters and stars

In Section 26 we have already indicated that, and in what way, stars in young galactic clusters and associations form out of interstellar matter. Now we seek to answer the far more widely ranging question of the conditions under which a mass of gas which is distributed in space becomes *unstable,* so that it collapses under the influence of its own weight. There then follows, as we shall see, a further splitting and the formation of single stars. Our question is answered by the criterion of gravitational instability discovered by J. Jeans (1902, 1928). We shall be content with an estimate which brings out the essential point, and we begin with a roughly homogeneous sphere of radius R, density ρ, and mass

$$M = \frac{4\pi}{3}\rho R^3. \tag{29.1}$$

If this is in equilibrium, then, according to the virial theorem (6.36) or (26.6), the ratio of twice the kinetic energy $2E_{\text{kin}}$ to the negative potential energy $-E_{\text{pot}}$ is one. If, on the other hand, E_{kin} (that is, the pressure in the interior) is too small or $-E_{\text{pot}}$ is too large, then gravitational instability sets in and the mass collapses.

As we saw in Section 26, E_{kin} equals the thermal energy of the atoms, etc.; we obtain E_{pot} by a simple integration. Then we have at the limit of stability:

$$\frac{2E_{\text{kin}}}{-E_{\text{pot}}} = \frac{M3\mathcal{R}T/\mu}{3GM^2/5R} = 1. \tag{29.2}$$

As usual, G signifies the constant of gravitation, $\mathcal{R}$ the gas constant, T the temperature of the mass of gas, and μ the mean molecular weight. If there are also turbulent motions in the gas we could simply replace $3\mathcal{R}T/\mu$ by the mean square velocity $\langle v^2 \rangle$, obtained, for example, from the Doppler effect in the 21-cm line. Thus a mass of gas can only collapse if its *radius* is *smaller* than

$$R = \frac{1}{5}\frac{GM}{\mathcal{R}T/\mu} \quad \text{or} \quad \frac{3}{5}\frac{GM}{\langle v^2 \rangle}. \tag{29.3}$$

If we also use equation (29.1), it follows at once that its *mass* must be *larger* than

$$M = \frac{5^{3/2}}{(4\pi/3)^{1/2}}(\mathcal{R}T/\mu G)^{3/2}\rho^{-1/2}. \tag{29.4}$$

The value of the constant is 5.46.

On closer inspection, our result is not yet very satisfactory. In practice we would not start off with a gaseous *sphere*, but would have to consider a more or less inhomogeneous larger mass of gas. Following W. H. McCrea (1957) we can then carry out a calculation in which we imagine ourselves exerting a pressure on the surface of the sphere. As a result, its collapse is encouraged and may start somewhat earlier. Then we would have to discuss whether the collapse was more nearly adiabatic or isothermal. However, all these refinements effect only moderate changes in the constants of equations (29.3) and (29.4).

As the most important application, we investigate an instability in the interstellar gas, whose density might be $\rho \approx 10^{-24}$ g/cm³ and whose temperature (in neutral regions, including turbulence) might be $T \approx 10^4$ K. According to equation (29.4) these figures give

$$M \geqslant 6 \times 10^7 M_\odot \tag{29.5}$$

That is, to begin with in a galaxy only a structure of the order of magnitude of a whole star cluster can form by gravitational instability. Only if, for example, its density increased by a factor $\sim 10^4$ (corresponding to $\sim 10^4$ atoms/cm³), the turbulence died away and the temperature dropped to ~ 10 K could the formation of single stars begin. With regard to many other problems as well, such as the origin of galaxies, of planetary systems, etc., we note that: the larger the density ρ of the original matter, the smaller are the celestial bodies formed therefrom.

29.2 Dynamics of spiral arms and density wave theory

Most galaxies which possess a rotating disk also have, embedded therein, spiral arms. S0 galaxies are an exception; then we recall also that there are no dwarf spiral galaxies.

The naïve idea that a spiral arm is always composed of the *same* stars, gas clouds, etc., founders on the realization that such a formation would be destroyed by the differential rotation in the course of a few rotations of the disk, that is, some 10^8 years. As recent measurements of the interstellar magnetic field show, even magnetohydrodynamic forces are insufficient for stabilization. A mechanism for the continuous renewal of material spiral arms could not be found.

On the other hand, many years ago B. Lindblad tried to attribute the maintenance of the spiral structure over long periods of time to a sectorial system of *density waves*. The spiral arms of a galaxy should then consist of different stars at different times, just as, for example, the crest of an ocean wave is formed from continuously changing particles of water. To prove this, B. Lindblad and his colleagues carried out extensive calculations of the orbits of *single stars* in the mean gravitational field (potential field) of a galactic disk. These investigations met with little approval, probably because Lindblad believed for a long time that the spiral arms were "leading," that is, moving with the concave side facing forward. In many spiral galaxies, however, for example using dark clouds, it is possible to distinguish unambiguously what is "in front" and what is "behind" and one then sees that in fact the spiral arms "trail."

Only in 1964 did C. C. Lin and others resuscitate the density wave theory of the spiral structure, this time in the form of a *continuum theory*. If we have in the disk of a galaxy—whose differential rotation will be described using a suitable potential field—a place where, for example, the gas density is larger, this will produce a change (dip) in the potential field. This in turn influences the velocities of the stars and so the mass density of the "star gas." In order that the potential field, gas density, and star density should now be consistent with each other, a rather complicated system of differential equations must be satisfied, which describes the structure and extension of density waves in the disk of the galaxy. From the multiplicity of possible solutions C. C. Lin and his co-workers now select one, which corresponds to a quasi-stationary spiral structure (QSSS hypothesis). This rotates rigidly, so to speak, with a constant angular velocity Ω_p. If this constant is fixed, we can calculate the form of the potential minimum which then forms the basis of the spiral arm. For our galaxy, the observations give $\Omega_p \approx 125$ km/s/10 kpc, that is, in our neighborhood the density wave is rotating at about *half* the speed of the stars, etc. According to W. W. Roberts (1969) the observable spiral arms are now arranged (Figure 29.1) in such a way that, as we said, the interstellar *gas* is primarily streaming into the density wave from its concave side (at ~ 125 km/s near us). As a result of the potential minimum, the gas suffers a compression, which reveals

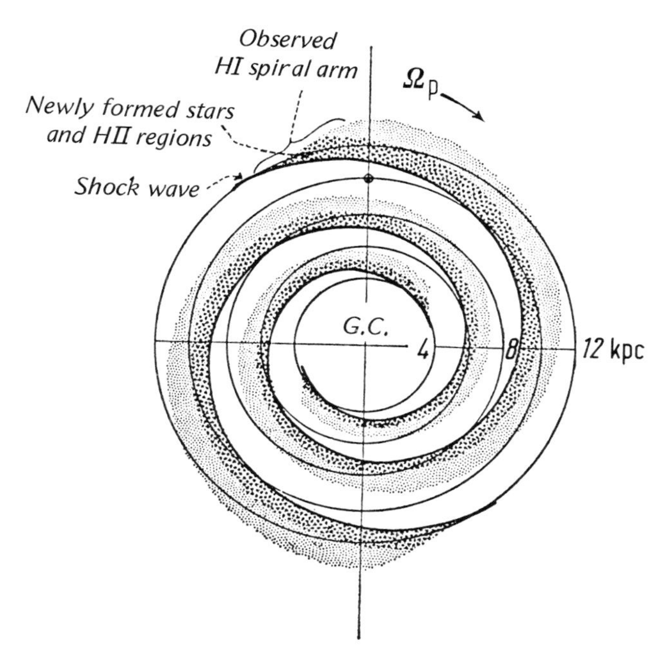

Figure 29.1. Spiral structure of our Galaxy (⊙Sun). According to C. C. Lin *et al.* the quasi-stationary spiral density wave (two arms) rotates with the angular speed $\Omega_p \approx 125$ km s^{-1}/10 kpc. It is therefore continuously overtaken by the matter (gas, stars, . . .) which is rotating about twice as fast and is compressed in the region of the potential trough of the density wave. In the gas there are therefore formed dark clouds, strengthened synchrotron radiation, and a shock wave (heavy line). In the "shocked" gas are formed in the course of $\sim 10^7$ years young bright stars and H II regions. At a larger distance from the shock are older stars. The greater part of the potential trough is filled with neutral hydrogen (H I; 21-cm radiation). [After W. W. Roberts (1969).]

itself mainly by the condensing out of interstellar dust. W. Baade had already emphasized a long time ago that the dust arms form the basis of the spiral structure. We have already mentioned the recent pictures of M 51, which reveal the corresponding strengthening of the radio emission as a result of the compression of the magnetic field and the synchrotron electrons.

The compression of the gas leads further (Figure 29.1) to the formation of a spiral shock wave, which on its part favors Jeans' instabilities and so the creation of young stars and H II regions. On the convex side of the shock front one does in fact observe a narrow band of bright blue stars and H II regions. Next to that there is a vague, wider band of older stars and star clusters; finally, the old disk population is distributed almost uniformly.

Quantitatively, Lin estimated the amplitudes of the gravitational field and of the gas and star densities to within about 5 percent.

The density wave theory of spiral structure has without doubt been very stimulating, particularly for the interpretation of the 21-cm observations. No important discrepancies between theory and observation have yet emerged. On the other hand, we see immediately that the theory in its present form is still very incomplete. We might simply draw attention to a few questions, without going into the still provisional proposals for their solution.

(1) Why does one observe only a quasi-stationary density wave? What determines its "pattern speed" Ω_p?
(2) What is the situation with regard to the original excitation and to the damping of density waves? As far as the excitation goes, people have thought about the tidal effect of nearby galaxies (for example, the Magellanic Clouds) or about Jeans' instabilities in the outer part of the spiral arms.
(3) Under what circumstances does a galaxy *not* develop spiral structure? Why are there no dwarf spirals? We cannot consider here either the particular problem of the barred spirals.

In the face of this list of desires, a totally different kind of mathematical technique gains in interest: using a large computer, several research groups (R. H. Miller, K. H. Prendergast, and W. J. Quirk 1968; F. Hohl 1970 and others) have simulated the motions of some 10^5 stars, under the influence of their mutual gravitation, as an N-body problem and have represented pictorially numerous sequential stages of such a system. The significance of this technique lies not least in the fact that it is now possible to *experiment*, as it were, with different kinds of galaxies. In the cases investigated so far a spiral structure is formed from a homogeneous disk in the course of less than one rotation. It does not have a quasi-stationary character, but changes and renews itself continuously.

As to why it does that and what bearing such simulation experiments have on Lin's theory, the computer is silent. We may however conjecture that, although the QSSS hypothesis may not be strictly valid, it does have the significance of a suggestive approximation.

29.3 Shapes of galaxies and formation of a galactic disk through collapse

Since the directions of the rotation axes will be randomly distributed, it is possible to calculate the distribution of true axial ratios q (= minor/major semiaxis) from the frequency distribution of the apparent axial ratios of galaxies of different Hubble types as measured on photographic plates. In continuation of the classical work of E. Hubble, an investigation by A. Sandage, K. C. Freeman, and N. R. Stokes (1970) gave the result that the true axial ratios of elliptical galaxies occupy the range from $q = 1$ to $q \simeq 0.3$, while *all* galaxies of types S0, SB0 (without spiral arms) and Sa, Sb, Sc (with spiral arms) exhibit axial ratios in the narrow range $q = 0.25 \pm 0.06$.

The two groups are also distinguished in a characteristic way by the radial distribution of *surface brightness* $I(r)$.

According to G. de Vaucouleurs and I. R. King the brightness distribution in E galaxies and in globular clusters, which are in many respects similar (apart from mass), may be represented (with certain restrictions; see below) by an empirical formula of the type

$$\log I(r)/I_0 = -(r/r_k)^{1/4} \quad \text{or} \quad I(r)/I_0 = \left(1 + \left(\frac{r}{r_c}\right)^2\right)^{-1} \tag{29.6}$$

This type of brightness distribution and the corresponding density distribution occurs because for large star density the *relaxation time,* in which a close approximation to a Maxwellian velocity distribution for the stars is achieved by close encounters, is relatively short. On the other hand, however, in the outer parts of globular clusters and dwarf E galaxies stars with high energy, which travel too far from the center, are continually "plucked off" by the tidal forces of nearby giant galaxies. The clusters and dwarf galaxies are therefore cut off at a certain "tidal radius" r_t; as a result, equations like equation (29.6) are no longer applicable for $r \gtrsim r_t/10$. Finally, the estimates, here merely sketched in, clearly do not take account of the strong mass concentration in the nuclei of giant E galaxies.

On the other hand, in the outer parts of almost all *disk* galaxies one observes a brightness distribution

$$I(r) = I_0 e^{-r/r_0}. \tag{29.7}$$

In addition to this exponential disk some galaxies contain in their inner parts another positive or negative term with a more spheroidal distribution, similar to the E galaxies. If one treats the exponential part of the galactic disk by itself, then, according to K. C. Freeman (1970), there follows a series of remarkable laws:

For 28 out of 36 galaxies investigated I_0 has, in the system of blue magnitudes B (there are still too few measurements in other spectral regions), almost the *same* value of 21.65 ± 0.30 mag/□″; this was true for galaxies of all types from S0 to Ir. (The extreme dwarf galaxy IC 1613 has lower surface brightness; some galaxies up to $\sim 3^{\mathrm{m}}.5$ brighter have mainly a strong spherical component.) Since—as we cannot here justify in every detail—the mass to light ratio M/L within all galaxies from type S0 to Ir is rather uniformly about equal to 12 (in solar units; L in photometric B system), we may conclude that in all these galaxies there is also a corresponding law for the radial distribution of surface (mass) *density* $\mu(r)$

$$\mu(r) = \mu_0 e^{-r/r_0} \tag{29.8}$$

with a uniform value for the central surface density μ_0.

The values of the length scale r_0 scatter between 1 and 5 kpc for the earlier types SO to about Sbc; in galaxies of later types values of ~ 2 to 1 kpc are preferred. Galaxies with anomalously large central surface density are for the most part particularly small.

Measurements of the Doppler widths and shifts of the Fraunhofer lines in the spectra of galaxies (cf. also Section 27) show primarily that in *elliptical* systems direct and retrograde stellar orbits occur with comparable frequency, just as in the halo of our galaxy. Clearly, such more or less spherical systems have low angular momentum. In the *disks* on the other hand—we may as well restrict ourselves here mainly to the exponential disks—every star circles in the same direction round the center. That is, the galactic disk systems have high angular momentum, as is to be expected from their flattening.

It is now an obvious question whether and how an (exponential) galactic disk could have formed out of a more or less spherical cloud?

As the simplest model of such a protogalaxy, let us consider a gaseous sphere which is rotating rigidly with a constant angular velocity $\Omega = 2\pi/T$ (T = rotation period). Let $M(r)$ be the mass within a sphere of radius r, R be the radius, and $M = M(R)$ be the total mass. Then the ratio of centrifugal force to gravity perpendicular to the rotation axis (Figure 29.2) is

$$\frac{\text{centrifugal force}}{\text{gravity}} = \frac{\Omega^2 r \sin\theta}{GM(r)\sin\theta/r^2} = \left[\frac{(\Omega r^2)^2}{GM(r)}\right]\frac{1}{r}, \tag{29.9}$$

where G again denotes the constant of gravitation.

Now let us, in thought, allow our sphere to collapse, at first radially, in such a way that each mass element conserves its angular momentum. Thus in particular for a spherical shell of radius r the angular momentum per unit mass Ωr^2 and the enclosed mass $M(r)$ remain fixed, and the square bracket in equation (29.9) undergoes no change. However, because of the factor $1/r$ the ratio of centrifugal force to gravity increases until equilibrium is reached for the components of the two forces perpendicular to the axis. The component of gravity parallel to the rotation axis remains unaltered, however, and so the protogalaxy must flatten, that is, collapse along the rotation axis. The energy liberated in this way is mostly dissipated; thus we obtain finally a thin disk. For there to be overall centrifugal balance in this disk (gravitation = centrifugal force), there must naturally result a radial distribution of mass density; the rotation will in general no longer be uniform.

We cannot follow these more complicated processes in detail, but we may expect that during them every mass element conserves its *angular momentum*. Thus the fraction dM/M of the mass whose angular momentum per unit mass lies in the range h to $h + dh$ must be the same in the final disk as in the protogalaxy. In place of $dM(h)/Mdh$ we could naturally equally well consider, after integration over h, the fraction $M(h)/M$ of the

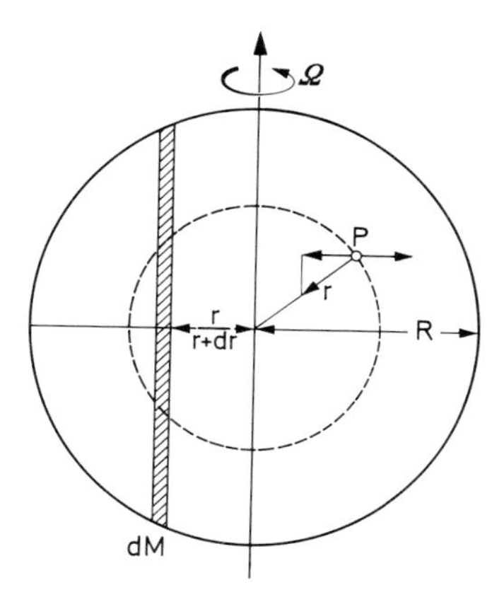

Figure 29.2
Model of a protogalaxy.

mass in which the angular momentum per unit mass is smaller than a given value h.

To calculate this ratio, we take, in the absence of more precise information, as the simplest model of a protogalaxy a homogeneous sphere of radius R, density ρ, and mass M. According to equation (29.9), since

$$M(r) = \frac{4\pi}{3}\rho r^3 \quad \text{and} \quad M = \frac{4\pi}{3}\rho R^3 \tag{29.10}$$

we can arrange the angular velocity Ω in such a way that there is centrifugal balance everywhere in the equatorial plane by taking

$$\Omega^2 = \frac{4\pi}{3}G\rho = GM/R^3. \tag{29.11}$$

The angular momentum per unit mass as a function of distance r from the axis is

$$h(r) = r^2\Omega \tag{29.12}$$

and the total angular momentum (moment of inertia times Ω) is

$$H = \tfrac{2}{5}MR^2\Omega = \tfrac{2}{5}(GM^3R)^{1/2}. \tag{29.13}$$

To calculate the distribution of angular momentum, we cut out of the sphere a cylinder parallel to the axis with radii r and $r + dr$. Its mass is then (cf. Figure 29.2, on the left)

$$dM = \rho 2R\left(1 - \frac{r^2}{R^2}\right)^{1/2} 2\pi r\, dr, \tag{29.14}$$

and its angular momentum is

$$h\, dM = r^2\Omega\, dM = \rho\Omega 2R\left(1 - \frac{r^2}{R^2}\right)^{1/2} 2\pi r^3\, dr. \tag{29.15}$$

According to equation (29.12), $dh = \Omega\, 2r\, dr$. Then using equation (29.10) we easily obtain finally the fraction dM/M of the mass whose angular momentum per unit mass lies in the range h to $h + dh$. If for simplicity we further refer h to its maximum value ΩR^2 and write

$$h/\Omega R^2 = x \tag{29.16}$$

we then have

$$dM/M = \tfrac{3}{2}(1 - x)^{1/2}\, dx. \tag{29.17}$$

By integration (with respect to $1 - x$) we hence obtain the fraction of the mass $M(h)$ or $M(x)$ in which the angular momentum per unit mass is less than x:

$$\frac{M(x)}{M} = \frac{3}{2}\int_0^x (1 - x)^{1/2}\, dx = 1 - (1 - x)^{3/2}. \tag{29.18}$$

For comparison, we have for the *whole* sphere from equations (29.13) and (29.16)

$$\bar{x} = H/M = \tfrac{2}{5}. \tag{29.19}$$

We have represented the two distribution functions dM/Mdx and $M(x)/M$ graphically in Figure 29.3. On the lower abscissa scale we have further, and with more physical significance, referred h not to its maximum value ΩR^2 but to the mean value for the whole sphere $\tfrac{2}{5}\Omega R^2$, writing $x' = x/\bar{x} = 5x/2$.

D. J. Crampin and F. Hoyle (1964), and later J. H. Oort (1970), had already shown that some Sb and Sc galaxies, including our own, do in fact show angular momentum distributions which follow equations (29.17) and (29.18) and which indicate an origin from a uniformly rotating, homogeneous, spherical protogalaxy. The previously mentioned study by K. C. Freeman (1970) of exponential disks, whose surface density μ (mass per unit area) follows the distribution function (29.8), takes us considerably further.

From the distribution of surface density μ one obtains immediately the total mass M_E of an exponential disk

$$M_E = 2\pi\mu_0 r_0^2. \tag{29.20}$$

Note in passing that, for example, 80 percent of the total mass is within $r \leqslant 3r_0$.

Then it is possible to calculate the angular velocity of rotation $\Omega(r)$ and so the distribution of angular momentum per unit mass $h_E(r)$. Along with this one finds also the fraction of the mass $M_E(h_E)/M_E$ in which the angular momentum per unit mass is $\leqslant h_E$. Furthermore, one calculates by numerical integration the total angular momentum

$$H_E = 1.109(GM_E^3 r_0)^{1/2}, \tag{29.21}$$

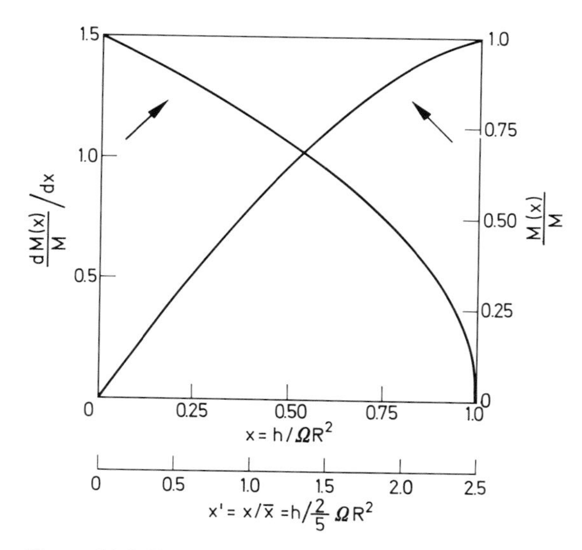

Figure 29.3. The fraction of mass dM $(x)/M$ (left) for which x lies between x and x + dx and the fraction of mass M $(x)/M$ (right) in which x is less than the value of the abscissa. These are plotted as functions of x, the angular momentum h per unit mass referred to its maximum value $h_{max} = \Omega R^2$, and of $x' = x/\bar{x}$, which is h referred to its mean value for the whole mass. During the formation of a disk galaxy out of the spherical protogalaxy, with conservation of angular momentum for each mass element, $dM(x)/Mdx$ or $M(x/\bar{x})/M$ should remain unchanged as functions of $x/\bar{x}$.

so that h can be referred to its mean value H_E/M_E, writing

$$x'_E = h_E/1.109(GM_E r_0)^{1/2}. \tag{29.22}$$

If the function $M_E(x'_E)/M_E$ for exponential disk galaxies is now compared with the function $M(x')/M$ calculated for our "theoretical" protogalaxy [equation (29.18) and/or Figure 29.3, right], then according to K. C. Freeman there is excellent agreement. This means that such galaxies could in fact have been formed by collapse from a spherical protogalaxy.

It is also interesting to compare the scale-length r_0 of the disk according to equation (29.7) or (29.8) with the radius R of our hypothetical protogalaxy for $M_E = M$ and $H_E = H$. From equations (29.21) and (29.13) one obtains immediately

$$r_{0E}/R = 0.13. \tag{29.23}$$

Thus even the radial contraction of the protogalaxy is very considerable. For this reason, because of angular momentum conservation, the velocities in the protogalaxies must have been significantly smaller than the present-day rotation speeds in the disk galaxies.

The observation that exponential disk galaxies are characterised by a *single* scale length r_0 and have consistently the *same* central surface brightness I_0 (in the B system) and surface density μ_0 leads to two further interesting points. First, the total luminosity should behave like $I_0 r_0^2$. In fact K. C. Freeman finds for disk galaxies (without a large spheroidal contribution)

$$\text{absolute magnitude } M_B = -16.93 - 5 \log \mathrm{r}_0, \tag{29.24}$$

where r_0 is measured in kiloparsecs.

Furthermore, one sees at once that the total angular momentum of a galactic disk should depend only on its mass M_E, and indeed one obtains from equations (29.21) and (29.20)

$$H_E \propto M_E^{7/4}. \tag{29.25}$$

This relation should also be reasonably well satisfied for galaxies in general, so long as their spheroidal component is not too large.

We can estimate the time scale required for the collapse of a galactic disk from two points of view: In a homogeneous spherical galaxy a point mass performs harmonic oscillations. The time τ for which the plunge from the surface (radius R) to the center lasts is thus equal to one quarter of the period of that oscillation or of the rotation period of a circular orbit with radius R. Thus by equation (29.11) we have

$$\tau_1 = \frac{\pi}{2} \sqrt{(R^3/GM)}. \tag{29.26a}$$

If the galaxy is already strongly condensed, then τ will be about equal to half the period of a Keplerian ellipse with semimajor axis $R/2$, that is,

$$\tau_2 = \frac{\pi}{2\sqrt{2}} \sqrt{(R^3/GM)}. \tag{29.26b}$$

With, for example, $M = 1.5 \times 10^{11} M_\odot$ and $R = 25$ kpc one obtains

$$\tau_1 = 2.5 \times 10^8 \text{ years } \tau_2 = 1.8 \times 10^8 \text{ years}. \tag{29.27}$$

Both times are naturally of the order of the galactic rotation period.

29.4 Intergalactic matter and the formation of galaxies

With regard to the further question of the *formation* of galaxies, we are first of all interested in the mean density of matter in the universe (1) in the form of galaxies and (2) in the form of intergalactic matter—mainly gas.

J. H. Oort (1958) calculated from the known masses and distances of *galaxies* that they contributed on average a mass density of

$$\rho_{\mathrm{gal}} = 3 \times 10^{-31} \text{ g/cm}^3. \tag{29.28}$$

What is the contribution of an *intergalactic medium* or gas, so far detected only rather indirectly?

As we shall see in Section 30, plausible cosmological models require a mean matter density of

$$\rho_{\text{cosm}} \approx 2 \times 10^{-29}\ \text{g/cm}^3 \tag{29.29}$$

(with an uncertainty of a factor of 2 either way).

From the braking of the plasma clouds in Cygnus A, S. Mitton and M. Ryle (1969), and others, disclosed that the surrounding matter had a density of

$$\rho_{\text{Cyg A}} \approx 10^{-28}\ \text{g/cm}^3, \tag{29.30}$$

which may well lie *above* the average since Cyg A belongs to a rich cluster of galaxies.

From consideration of intergalactic X rays R. A. Sunyaev (1968) discovered that any plasma must have

$$\rho \leqslant 0.7 \times 10^{-29}\ \text{g/cm}^3 \tag{29.31}$$

and a temperature of order 10^6 K.

The application of the virial theorem to the internal motions in clusters of galaxies or in the local group leads to similar values for the density of the intergalactic matter. In what follows we shall adopt for calculations the currently most probable value of

$$\rho_{\text{intergal}} \approx 10^{-29}\ \text{g/cm}^3. \tag{29.32}$$

Starting from the consideration that in a turbulent medium the three velocity components could certainly not produce compression everywhere, J. H. Oort estimates that of the order of $1/16$ of the originally available material would finally combine into galaxies and he interprets the ratio $\rho_{\text{gal}}/\rho_{\text{intergal}}$ from equations (29.28) and (29.32) in this sense.

Now, what was the prevalent mean density at the time of formation of the galaxies? We should direct our attention in the first place to the giant galaxies, like our Milky Way system, since markedly dwarf galaxies could very well have been formed as a secondary effect. Since J. H. Oort (1970) and others have shown that angular momentum exchange between "completed" galaxies cannot have played any essential role, we must require that the protogalaxy already possessed the angular momentum of the present-day system. The angular momentum of galaxies must therefore stem from the turbulence in the intergalactic medium of that time. If we demand further that a protogalaxy can acquire the requisite angular momentum, we are led back, at least in essence, to the model of the previous section. A present-day galaxy with a scale-length $r_0 \approx 2.5$ kpc must therefore, according to equation (29.23), have started out as a sphere of radius $R \approx 20$ kpc. If the galaxy contains $\sim 1.5 \times 10^{11}$ solar masses (corresponding to the Milky Way or other giant galaxies), then the mean density of the sphere was

$$\rho_s \approx 3 \times 10^{-25}\ \text{g/cm}^3. \tag{29.33}$$

Following J. H. Oort, we must therefore set the formation of the protogalaxies, and of the galaxies as well, back in an earlier era of the expansion of the universe (Section 30), in which the mean density of matter was still $\sim 3 \times 10^{-25}/10^{-29} \approx 3 \times 10^4$ times larger than it is today and the radius of the universe was some 30 times smaller.

In this way we can also understand the origin of giant galaxies in terms of the Jeans' gravitational instability. If we substitute a density $\rho = 3 \times 10^{-25}$ g/cm^3 and a temperature $T \approx 10^6$ K into equation (29.4), we find in fact that in such a gas unstable regions with masses of order 10^{11} solar masses are formed preferentially.

The theory of gravitational instability certainly explains in rough outline the division of intergalactic matter into galaxies, but does not explain its more detailed properties. If the masses of galaxies were generally to follow equation (29.4), so that $M \propto \rho^{-1/2}$, then according to equation (29.11) their masses M, and therewith also their luminosities L, would have to be proportional to their characteristic scale (for example, r_0), while according to K. C. Freeman [equation (29.24)] $L \propto r_0^2$.

It might be more significant to consider first the origin of the angular momentum per unit mass from the turbulence of the interstellar medium. According to K. C. Freeman the angular momentum per unit mass for disk galaxies (H_E/M_E, from equations (29.21) and (29.20) with μ_0 constant) is proportional to $r_0^{3/2}$. This relation is naturally equivalent to the law $L \propto r_0^2$ just mentioned for the luminosity. On the other hand, if we tentatively use the theory of homogeneous, isotropic turbulence, this says that to a turbulent element of size r_0 there belongs a characteristic velocity whose order of magnitude is $v_0 \propto r_0^{1/3}$, and so the corresponding angular momentum per unit mass is $r_0 v_0 \propto r_0^{4/3}$. As is easily verified, this leads to a distribution of masses and so of luminosities $L \propto r_0^{5/3}$, which might be in sufficient agreement with the empirical relation $L \propto r_0^2$.

The formation of *clusters* of galaxies has clearly nothing to do with that of the galaxies themselves; at any rate we know of no theoretical clue for such a connection. The formation of clusters of galaxies at an earlier time might be based on the same tendency for gravitating masses to coagulate which we observe (on a smaller scale) in the computer experiments which were done to explain spiral structure, but which, as here, we do not yet correctly understand.

29.5 Dynamical and chemical evolution of galaxies and origin of the abundance distribution of the elements

We first briefly summarize our previous investigations of the origin and evolution of galaxies from the point of view of mechanics and enlarge on a few points.

We saw that we probably have to begin at an earlier stage of the universe whose density was still about 3×10^4 times that of present-day intergalactic matter ($\approx 10^{-29}$ g/cm^3). Under these circumstances single gas clouds could be formed by gravitational instability whose masses correspond roughly to those of giant galaxies of 10^{11} to 10^{12} solar masses.

Some obtained from the turbulence of the gas as much angular momentum as they were able to absorb. During its contraction, such a protogalaxy must have formed a disk, by collapse, within a few hundred million years. The structure of this disk, at least in its outer parts, is astonishingly similar in all galaxies from S0 to Ir. It appears that what distinguishes the Hubble types of the series S0, Sa, Sb, Sc from each other physically is in the first place (or exclusively?) the fraction of the total mass which is still present in the form of gas—mainly H I. It varies from less than 1 percent in the uniform S0 disks to ~ 20 percent in irregular galaxies. The formation of spiral arms is clearly intimately connected with the gas of the disk. The density wave theory certainly explains a good number of the observed characteristics but may hardly be considered as confirmed so long as a series of important theoretical problems is still unresolved.

The giant E galaxies clearly emerged from protogalaxies with less angular momentum. While there is still no actual theory of the division of the originally turbulent gas into clouds with different amounts of angular momentum, yet no fundamental difficulties appear to stand in the way of our assumption.

The formation of dwarf galaxies can *not* be traced back to gravitational instability in the primaeval gas; they must rather be formed specifically in association with the giant galaxies. In the local group, in fact, it is possible to associate almost every one of the dwarf galaxies with one of the giant galaxies.

Our Milky Way system, in which we can study things exactly, does not follow completely either the scheme of the disk galaxies or that of the elliptical galaxies. Rather, as we saw, it possesses a strongly flattened disk *and* a much more weakly flattened halo.

After this brief introduction to the *dynamical* aspects of the evolution of galaxies we turn to the *nuclear* problems. We have already laid down the empirical foundations for this discussion in Section 28 in connection with the fundamental idea of stellar populations.

In connection with the relativistic theory of the expanding universe, G. Lemaître and G. Gamow had developed in the 1940s the idea that the mixture of chemical elements—at that time regarded as completely universal—had its origin in the "Big Bang." In due course nuclear physicists pointed out that the construction of heavy elements in the manner suggested by Gamow must come to an end as early as mass number $A = 5$; on the other hand, it was discovered that high-velocity stars and subdwarfs are, to a varying degree, metal poor.

E. M. and G. R. Burbidge, W. A. Fowler, and F. Hoyle (1957) then made the first attempt at a joint solution of the two problems of the abundance distribution of the elements and of the evolution of galaxies; this is normally referred to briefly as the B^2FH theory. Here for the first time was an attempt to apply nuclear physics to astrophysics on a wider basis. In particular, W. A. Fowler's measurements of nuclear reaction cross sections at low energies, in connection with stellar energy generation, placed this branch of research for the first time on a secure foundation. This should not be forgotten when in what follows we criticize the astronomical part of the B^2FH theory.

In astronomical respects, the theory commences with a universe which emerges from the Big Bang consisting of hydrogen and helium (10 : 1) and perhaps traces of heavier elements (here we straight away make recent models of the universe our starting point).

A galaxy starts its life as a roughly spherical mass in which the first stars are at once formed. In their interiors nuclear processes now take place (see below) which lead to the construction of heavier elements. In many cases the result of evolution is to lead to the continuous or explosive (supernovae) ejection of stellar material into the interstellar medium. In this way a second generation of metal-richer stars can form, and so on. In the E galaxies this process may come to an end sooner or later, according to mass; in the disk galaxies the formation of heavy elements must be essentially finished by the time the disk collapses.

We now assemble, in outline only, the nuclear processes taken into account by B^2FH:

(1) We have already discussed in Section 26 the processes which are important for energy generation in stellar interiors: At $\sim 10^7$ K the nuclear burning of *hydrogen* to helium commences, first by means of the pp or fusion process and at somewhat higher temperatures predominantly by the CNO cycle. Above $\sim 10^8$ K *helium* is then burnt to carbon and at still higher temperatures *carbon* is itself burnt.

(2) At ~ 1 to 2.5×10^8 K the "*alpha nuclei*" C^{12}, O^{16}, Ne^{20} are formed. Further reactions then yield in particular *neutrons* for the ensuing formation processes.

Besides these were considered:

(3) The construction of heavy alpha nuclei up to Ca^{40} at $\sim 10^9$ K, using α particles; the *α process*.

(4) The *e process* generates the elements of the iron group, V, Cr, Mn, Fe, Co, and Ni in thermal *e*quilibrium at $\sim 4 \times 10^9$ K and a proton to neutron ratio of ~ 300.

Because of the strong Coulomb repulsion, heavy nuclei are inaccessible to charged particles. The construction of heavy elements can therefore result only from *neutron processes*. The following are distinguished:

(5) The *s process* consists of neutron capture (in light or Fe elements) at a rate *s*low compared to that of the concurrent β decays. The s process generates, for example, Sr, Zr, Ba, Pb and in general the stable nuclides

at the foot of the valley in the energy surface.[1] That the s process plays a role in bringing about, for example, the solar abundance distribution, was already recognized by G. Gamow from the fact that the product of abundance times the reaction cross section for ~ 25 keV neutrons is a smooth function of mass number for the nuclides considered.

(6) The term *r process* is used for the corresponding neutron captures which occur *r*apidly in comparison to the concurrent β decays. This process produces the neutron-rich isotopes of the heavy nuclei; B²FH propose that it is responsible in particular for the formation of the radioactive elements, for example U^{235} and U^{238} (at the expense of the iron group).

(7) The *p process* produces the neutron-poor or proton-rich isotopes of the heavy elements in a hydrogen (*p*roton)-rich medium at $\sim 2.5 \times 10^9$ K.

While there is no doubt that the reactions (1) responsible for the energy generation and evolution of stars occur and some assistance from the s process seems to be well documented, this cannot be said of the other processes without further elaboration.

In addition to (3), (4), and (6), J. W. Truran, W. D. Arnett and others (1971) have recently carried out extensive calculations of *explosive nucleosynthesis* with the idea of applying them to shock waves, supernovae, etc. The calculations are concerned with processes which take place at temperatures of the order of 10^9 K and on time scales which in some cases are considerably less than 1 s! Quite apart from the related hydrodynamic problems, so many disposable constants are needed for quantitative calculations of the r process as well as of explosive nucleosynthesis that a comparison of theory and observation becomes rather problematical.

More promising may be a proposal developed by J. P. Amiet and H. D. Zeh (1968) from considerations due to H. E. Suess and J. H. D. Jensen. It starts from a state—whose astrophysical "residence" is not yet explained—with a density $\rho \approx 2 \times 10^{10}$ g/cm³ and temperature $T \approx 5 \times 10^9$ K. Under such conditions, free electrons become, so to speak, pressed into the nuclei and the "valley" of stable nuclei in the N–Z plane (N = number of neutrons, Z = number of protons or nuclear charge, $N + Z = A$ is the mass number) is shifted to the side where the neutron-rich nuclei are (Figure 29.4). If the pressure decreases, there are formed from this neutron-rich "protomatter" by β decay (upper left) the neutron-poorer nuclei which are stable under normal circumstances; however, there may also (more rarely) remain relatively neutron-richer stable nuclei. The abundance maxima of the heavy nuclei (Figure 29.4) lie at the neutron magic numbers (that is, closed neutron shells) of the neutron-rich protomatter and *not* at the

[1]If one plots the binding energies of atomic nuclei on a plane with coordinates N = neutron number and Z = nuclear charge or proton number, these energy surfaces permit a clear representation of nuclear reactions, including their energetics. The stable nuclei are located near the foot of the valley in the energy surface (see also Figure 29.4).

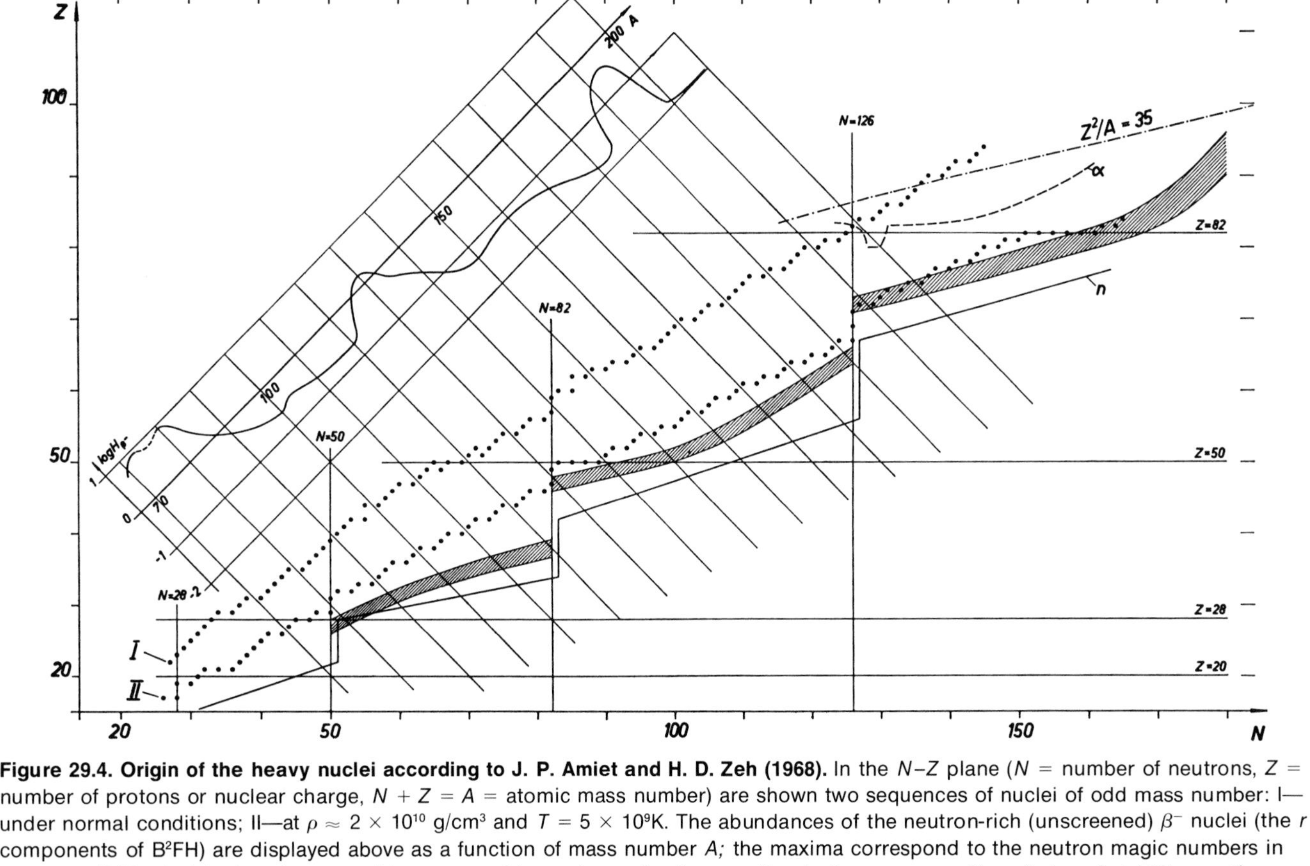

Figure 29.4. Origin of the heavy nuclei according to J. P. Amiet and H. D. Zeh (1968). In the N–Z plane (N = number of neutrons, Z = number of protons or nuclear charge, $N + Z = A$ = atomic mass number) are shown two sequences of nuclei of odd mass number: I—under normal conditions; II—at $\rho \approx 2 \times 10^{10}$ g/cm³ and $T = 5 \times 10^9$K. The abundances of the neutron-rich (unscreened) β^- nuclei (the r components of B²FH) are displayed above as a function of mass number A; the maxima correspond to the neutron magic numbers in sequence II. The hatched surfaces would correspond to nucleosynthesis according to the r process. Boundaries of stability: continuous line stands for neutron emission, dashed line stands for α decay, dot–dashed line stands for fission.

present-day magic nuclei. Many details clearly suggest that the s process with neutrons of ~ 25 keV operated first on the relaxed protomatter and brought about, so to speak, an adjustment of the abundance distribution.

A much more *uniform* origin for the heavy elements (from about Fe) than is adopted in the B^2FH theory is also suggested by empirically discovered laws for the abundance distribution of the elements in meteorites and the Sun, laws already discovered in part by H. E. Suess in the 1940s. These show a completely homogeneous behavior of nuclides, which would be attributed partly to the r and partly to the s process. This appears quite incomprehensible in the framework of the B^2FH theory.

After this excursion into nuclear physics, let us return to our problem of the origin of galaxies and of the element abundances found in them.

Following B^2FH we first start with a protogalaxy of almost pure hydrogen (+10 percent He) and discuss the further proposal that the heavy elements have been formed in conjunction with the appearance and sometimes explosive disappearance of perhaps several generations of stars. This process must have come to a close earlier in the dwarf E galaxies than in the giant galaxies. It seems then all the more surprising that in all sufficiently large galaxies—in particular independent both of the formation of a disk and of the mass fraction still present in the form of interstellar hydrogen—almost the same fraction of the original material would be changed into heavy elements, *and* that from the first beginnings of the formation of the heavy elements up to the achievement of their present abundances their mixing ratio would always be reproduced. In our galaxy we can trace the time of formation of the heavy elements more precisely: Since the oldest galactic clusters have practically normal composition, the formation must already have been completed when the disk was consolidated. According to equation (29.27), therefore, the formation of the heavy elements lasted *at most* a few times 10^8 years. Then however according to Table 26.1 only stars with more than 2 to 5 solar masses can have had anything to do with it. If these had been available in the requisite quantity, and if also the formation of the halo stars had corresponded even approximately to the initial luminosity function (Figures 26.13 to 26.14), there would have to be still present today a host of (long-lived) metal-poor dwarf stars, which in no way corresponds to observation. Because of this J. W. Truran and A. G. W. Cameron (1971) made the ad hoc hypothesis that "at the beginning" almost all the stars formed were more massive than $5M_\odot$. In the meantime, however, it became possible (see p. 302) to determine the luminosity function of the old population II, at least up to the "knee" in the color-magnitude diagram. It shows no kind of difference from that of the disk population; a sudden rise in the region of stars which are no longer present today seems extremely unlikely. Furthermore, this does not explain why in the explosion of ever more recent supernovae—with increasing metal abundance—always the *same* mixture of elements should be ejected.

In this connection we might point out how small in general is the turnover of nuclear energy of the Milky Way and similar galaxies in their present state. In equation (26.1) we have already connected the hydrogen consumption of a heavenly body with its mass-to-light ratio. If we put the latter ≈ 10, we find that, in its *present* state, our galaxy has consumed only 1 to 2 percent of its hydrogen in $\sim 10^{10}$ years. The production of the majority of the heavy elements within about the first 10^8 years by means of stellar evolution, but equally by any other mechanism, implies that at that time our galaxy had a much greater luminosity than at present.

We know that in all cases the heavy elements are formed within about 10^8 years, at the beginning of the evolution of the Galaxy, in its then enormously bright nucleus, by "mass production" so to speak. What then is more natural than to propose that during the formation of the halo a *quasar*[2] formed with a violent gravitational collapse at the center? In this way too one could understand the temporary occurrence of the enormous pressures and temperatures which according to Amiet and Zeh are required to produce the neutron-rich "protomatter." Then, as we saw, observation shows in every case that in quasars, Seyfert nuclei, etc. a mixture of elements is found which is astonishingly similar to that in normal stars. With respect to the dynamics, we must suppose, as does not seem unreasonable, that the matter spurting out of the quasar mixes in the halo with hydrogen and helium in various proportions and that at the collapse of the disk a rather uniform mixture is formed straightaway, with somewhat larger metal abundance in the central region than toward the rim.

Furthermore, it appears plausible to explain in passing the relation between metal content and mass of the E galaxies, in that the dwarf galaxies could only form substantially weaker quasars than the giant E galaxies, in agreement with radio observations.

The knowledge that by far the majority of all heavy elements were formed within a short time at the beginning of the Galaxy enables us to tie down the cosmic time scale (see also Section 30) from considerations of nuclear physics: Let us take two radio-active nuclides, which would *not* be produced later by any parent elements, and which decay with very different half lives, for example the uranium isotopes 235 and 238 with $T_{1/2} = 7.1 \times 10^8$ and 4.5×10^9 years, respectively. Their present abundance ratio is 1 : 138. In earlier times it must have been larger. If we work back to an initial value of the ratio rather larger than 1 we find that the isotopes must have been formed at a time 7 to (at most) 8×10^9 years ago. A similar calculation

[2]The differences between the various conceptions of the evolution of the heavy elements at the beginning of the evolution of a galaxy are perhaps not *so* large as they might appear at first sight. Because we cannot get away with the hypothesis of "normal" stellar evolution, we require an initial stellar population of a kind which has never actually been observed. On the other hand, we know that quasars must have a detailed structure, but we can still say next to nothing definite about it.

for the ratio Th^{232}/U^{238} confirms this estimate. These times are certainly somewhat shorter than the most probable age, $\sim 9 \times 10^9$ years, of the Galaxy as found from the color-magnitude diagrams of the globular clusters and the oldest galactic clusters (9–12 and 8–10 $\times 10^9$ years). However, we would do well not to forget that a series of rather uncertain assumptions still enters into the theory of color-magnitude diagrams (and of isochrones) which lies at the basis of "astronomical" age determination:

(1) The constancy of stellar masses is assumed, while recently indications of mass loss are increasing in number in various places.
(2) It is assumed that no kind of mixing occurs outside convection zones. Here also observation shows exceptions, which we still do not understand.
(3) The mixing-length theory of convection stands hydrodynamically on very insecure foundations.
(4) The neutrino emission from the Sun.

We therefore regard it as entirely reasonable to extend somewhat further the error bounds of the astronomical age determinations and to fix $\sim 8 \times 10^9$ years as a probable value for the age of the Galaxy and therewith of the whole universe of galaxies.

Although the new hypothesis of a *mass production* of the heavy elements has without doubt much to be said for it, there nonetheless still remains, just as in the B^2FH theory, the unanswered question as to why nearly the *same* fraction of the original material was transformed into heavy elements in *all* large galaxies, such as our Milky Way system, M 31, the Magellanic Clouds, etc. (but *not* in the dwarf E galaxies). This clearly depends on the still equally open question as to what fraction of the mass of a galaxy collapses to form a quasar.

At any rate we may register as one of the most important discoveries of modern astrophysics the fact that quasar-like explosions in the centers of galaxies also decisively influence their other evolution, as we see directly on the one hand in the Leiden studies of the inner regions of the Galaxy and on the other in the work on M 82 and NGC 1275. V. A. Ambarzumian has emphasized for several decades the great cosmogonical and cosmological significance of the nuclei of galaxies. In his theoretical interpretation he even goes a step further than the theory described above. He holds that the appearance and explosion of a compact galaxy is a kind of *primary process,* which it is beyond the capabilities of present-day physics to understand. Then, in place of the single Big Bang of the cosmological models of the Lemaître type (see Section 30) there appear many similar processes of galactic size. In this way, admittedly, the principal difficulties of each theory of a "beginning" are in reality merely redistributed.

30. Cosmology

Five years after the measurement of the distances of remote galaxies, E. Hubble in 1929 achieved a second discovery of stupendous significance: *The red shift of the lines in the spectra of distant galaxies increases in proportion to their distance.* We write the red shift

$$z = \frac{\Delta\lambda}{\lambda_0} = \frac{\lambda - \lambda_0}{\lambda_0} \tag{30.1}$$

where λ_0 is the laboratory wavelength and λ the measured wavelength. If r is the distance and if we interpret the red shift as a Doppler shift,[1] then we have for the speed of recession $v = dr/dt$ of the galaxies the relation

$$v = c\,\Delta\lambda/\lambda_0 = H_0 r. \tag{30.2}$$

Apparent exceptions for nearby galaxies—the Andromeda galaxy, for example, is *approaching* us at 300 km/s—are easily explained as the reflection of the rotation of our galaxy. Recent measurements of the 21-cm hydrogen line have given the best confirmation of the nondependence of the effect on wavelength assumed in equation (30.2).

For the Hubble constant H_0 Hubble himself in 1929 obtained the value 530 km/s/Mpc. Following on W. Baade's revision of the cosmic distance scale (distinguishing between cepheids of populations I and II), in 1958 A. R. Sandage computed the most probable value as 75 km/s/Mpc. A further revision of the extra-galactic distance scale led in 1972–1973 to 50–60 km/s/Mpc. Other recent discussions yielded values up to 100 km/s/Mpc. In the calculations that follow we normally use

$$H_0 = 75 \pm 25 \text{ km/s/Mpc}. \tag{30.3}$$

Conversely equation (30.2) is frequently applied, for want of anything better, in order to derive a distance r from the measured red shift in the spectrum of a galaxy that is no longer telescopically resolvable. In the present state of the subject, one must then state what value has been used for H_0.

One can first interpret the relation (30.2) quite naively by saying that an expansion of the universe from a relatively small volume began at a time T_0 years ago. If a particular galaxy, that we now find to be at distance r, received for the expansion the velocity v, it requires in order to travel the distance r the time

$$T_0 = r/v = 1/H_0 \tag{30.4}$$

which is the same for all galaxies. This so-called "Hubble time" T_0 is thus the reciprocal of the Hubble constant $1/H_0$. If as usual we reckon H_0 in

[1] We restrict ourselves here to $z \ll 1$; otherwise we should have to use relativistic calculations.

kilometers per second per megaparsec, and T_0 in years, then we have (1 Mpc = 3.084×10^{19} km; 1 year = 3.156×10^7 s)

$$T_0\,[\text{years}] = \frac{978 \times 10^9}{H_0\,[\text{km/s/Mpc}]}. \tag{30.5}$$

While Hubble's older value for H_0 led to a value 1.86×10^9 years for the age of the universe that was far too short by comparison with the ages of globular clusters, etc., we obtain from the more recent value of 75 km/s/Mpc

$$T_0 = 13 \times 10^9 \text{ years} \tag{30.6}$$

with an estimated uncertainty of $\pm 5 \times 10^9$ years.

The kinematics of the expanding universe expressed by equation (30.2) seems at first sight to imply a reversion to heliocentric ideas. However, this is not so. If we write equation (30.2) as a vector relation

$$\boldsymbol{v} = H_0 \boldsymbol{r} \tag{30.7}$$

where the origin of the coordinate system lies in our galaxy, then, seen from another galaxy that has position $\boldsymbol{r}_0$ and velocity $\boldsymbol{v}_0$ relative to ours (so that $\boldsymbol{v}_0 = H_0 \boldsymbol{r}_0$) we have

$$\boldsymbol{v} - \boldsymbol{v}_0 = H_0(\boldsymbol{r} - \boldsymbol{r}_0). \tag{30.8}$$

Thus the universe expanding in accordance with equation (30.7) presents exactly the same appearance to observers in different galaxies. Therefore our *kinematic world model is homogeneous and isotropic*. We can show that equation (30.7) represents the *unique* field of motion that satisfies this condition, so long as we require the motion to be irrotational (curl $\boldsymbol{v} = 0$).

E. A. Milne and W. H. McCrea (1934) extended this at first purely kinematic model so as to make it a *Newtonian cosmology*. They investigated the motions of a medium (the "gas" of galaxies) that can take place in accordance with newtonian mechanics if one demands throughout *homogeneity, isotropy,* and *irrotational* motion[2].

Consider at time t a galaxy at distance $R(t)$, then according to Newton's law of gravitation this is attracted by the mass within the sphere of radius R given by $M = (4\pi/3)R^3\rho(t)$, where $\rho(t)$ is the mass-density at the instant considered. Therefore the equation of motion[3] of this galaxy is

[2]One shows without difficulty that the requirement of isotropy everywhere implies homogeneity, but not conversely. World models with curl $\boldsymbol{v} \neq 0$ have been studied, but they must remain outside our present scope.

[3]Taking explicit account of pressure—here ignored for the sake of simplicity—would not affect the result. The convergence problem in the case of infinitely extended systems, connected with the slow falloff of Newton's inverse square law of attraction, for a long time held up the development of Newtonian cosmology. Here we have somewhat loosely passed it over. An exact formulation is given, for example, in the article by O. Heckmann and E. Schücking, 1959, in the Handbuch der Physik.

$$\frac{d^2R}{dt^2} + \frac{GM}{R^2} = 0, \quad \text{where } M = \frac{4}{3}\pi R^3 \rho(t) = \text{constant}. \tag{30.9}$$

If one multiplies by $\dot{R} = dR/dt$, then it is possible to integrate without more ado and obtain the *energy equation*

$$\frac{1}{2}\left(\frac{dR}{dt}\right)^2 - \frac{GM}{R} = h \tag{30.10}$$

where h is a constant, or

$$\frac{\dot{R}^2}{R^2} - \frac{8\pi}{3} G\rho(t) + \frac{kc^2}{R^2} = 0, \tag{30.10a}$$

in which we have written $-h = kc^2/2$ in anticipation of later comparison with relativistic calculations.

It is easy to convince oneself that at a definite time $t = t_0$ the same red-shift law would be observed from each galaxy within our sphere. If we denote all variables that relate to the present time $t = t_0$ by a subscript $_0$, the

$$\textit{Hubble constant } H_0 = \dot{R}_0/R_0. \tag{30.11}$$

For a complete characterization of a model universe, we need, besides H_0, a second variable that describes the inward acceleration, due to the mass $M = (4\pi/3)R_0^3\rho_0$, which is working against the expansion of the universe. This is the so-called

deceleration parameter $q_0 =$

$$-\left(\frac{\ddot{R}_0}{R_0}\right) \Big/ \left(\frac{\dot{R}_0}{R_0}\right)^2 = -\frac{\ddot{R}_0}{R_0 H_0^2} = \frac{4\pi G\rho_0}{3H_0^2} \tag{30.12}$$

using equation (30.9). It relates the acceleration $\ddot{R}_0$ to a uniform acceleration which would lead to the observed velocity R_0H_0 at distance R_0 in the Hubble time $T_0 = H_0^{-1}$, starting from zero velocity.

The solution of our equations leads to world models which, from a starting point (singularity) of infinitely great density, either expand monotonically (total energy $Mh \geqslant 0$) or oscillate periodically between $R = 0$ and an $R_{\max}$ (if $h < 0$). Static models are not possible within the framework of equation (30.9).

Generally speaking, Newtonian cosmology extends the world picture of the purely kinematic cosmology, which we discussed first, in that it treats the Hubble constant H as a function of the time t. In a periodic universe, for example, an era with red shifts would be followed by one with blue shifts, and conversely. Rather than discuss the choice between the numerous world models of Newtonian theory, we examine its basic difficulties and their resolution within the framework of relativistic cosmology.[4]

[4]Our short introduction will not replace a textbook of relativity theory. We wish only to indicate its significance for astronomy and cosmology. The few formulas presented will serve only to give some impression of its theoretical structure.

The theory of special relativity developed by A. Einstein in 1905 starts out from the result of the Michelson experiment in that it requires the propagation of a light wave in different coordinate systems, which may be in relative translational motion, to present the same appearance, that is, satisfy the same equation. If the light traverses an element of path length $dr = (dx^2 + dy^2 + dz^2)^{1/2}$ in a time element dt with vacuum light speed c, we have

$$dx^2 + dy^2 + dz^2 - c^2\,dt^2 = 0. \tag{30.13}$$

We may treat the quantity on the left as the line element ds^2 of a four-dimensional space with three spatial coordinates x, y, z and the fourth coordinate ct (= light path) or, following H. Poincaré and H. Minkowski, more "intuitively" as ict [where $i = \sqrt{(-1)}$]. Then, somewhat more generally, we require that in passing from one Cartesian coordinate system to another, which is in translational motion relative to the first, the four-dimensional line element ds^2 shall remain invariant, where

$$ds^2 = dx^2 + dy^2 + dz^2 - c^2\,dt^2 \quad \text{or} \quad ds^2 = dx^2 + dy^2 + dz^2 + d(ict)^2. \tag{30.14}$$

Such a transformation obviously cannot be restricted to the spatial coordinates $x, y, z \rightarrow x', y', z'$, but must also transform the time along with these, $x, y, z, t \rightarrow x', y', z', t'$. As we see at once from the second form of equation (30.14), this so-called *Lorentz transformation* is none other than a rotation in the four-dimensional space x, y, z, ict, in which by definition the line element ds is invariant. The essential advance made by special relativity as compared with Newtonian theory consists in the fact that it takes account of the exceptional status of the (vacuum) velocity of light from the outset. Corresponding to this, its further development leads to the result that no material motion and no signal of any kind can exceed the speed $c = 3 \times 10^{10}$ cm/s. Thus our Newtonian world models can be trusted only so far as no speed occurs that exceeds the light speed, that is, $v < c$ and $z < 1$.

In regard to the whole of physics, special relativity further requires the invariance of all natural laws under Lorentz transformation, that is, that the laws be physically independent of motions of translation.

Should it not then be possible so to formulate natural laws that they are invariant under *arbitrary* coordinate transformations? In 1916 A. Einstein had the brilliant idea in his theory of *general relativity* of combining this requirement with a theory of *gravitation*. Within the framework of classical theory, the equality of gravitational and inertial mass, independently of the kind of matter (or in modern terms, of the kind of elementary particles) was indeed a feature to be wondered at. Newton himself, then Bessel, and later Eötvös had verified it experimentally with increasing accuracy. But what did it mean? Einstein promoted the experience that in a freely falling reference system (a lift) the gravitational force mg appears to be cancelled by the inertial force $m\ddot{z}$ to the status of a fundamental postulate. That is to say, gravitational force and inertial force are in the last resort *the same*. We can transform away these forces by a local transformation of the four-dimensional Cartesian coordinate system with the Euclidean metric[5] (30.14). Conversely the coefficients g_{ik} of the Riemannian metric $ds^2 = \Sigma_{i,k} g_{ik}\, dx^i\, dx^k$ with an arbitrary coordinate system determine, as can be shown, the gravitational field and the inertial field that are in operation in the whole of the space described by the metric.

According to K. Schwarzschild (1916) the gravitational field of a mass M, for example of the Sun, and so also the theory of planetary motions can be represented by using the metric

$$ds^2 = dr^2/(1 - r_s/r) + r^2\,(d\theta^2 + \sin^2\theta\, d\phi^2) - (1 - r_s/r)c^2\, dt^2, \quad (30.15)$$

in which, as usual, r, θ, ϕ denote spatial polar coordinates and t is the time.

The constant of integration

$$r_s = 2GM/c^2 \quad (30.16)$$

is called the *gravitational radius* of the mass M; for $M = 1$ solar mass, for example, $r_s = 2.9$ km. Its size determines the departures from the Euclidean metric of empty space. The planets, or any kind of test particle, for example photons, follow *geodesics* (that is, shortest paths) in the space (30.15).

In the domain of the planetary system, the predictions of general relativity differ very little from those of Newtonian mechanics and gravitation theory. The tests which can distinguish between the two theories therefore require very precise measurements. The present observational situation is the following:

(1) *Equality of gravitational and inertial mass.* In his classical experiment to test the equality of gravitational and inertial mass (the Earth's gravity and the centrifugal force of the Earth's rotation were used) R. v. Eötvös achieved an accuracy of 10^{-9} even in 1922. More recently V. B. Braginsky and V. N. Rudenko (1970) have brought this down to $\sim 10^{-12}$.

(2) *Deflection of light.* The light of a star that (as observed during a total solar eclipse) is seen at an apparent distance of R solar radii from the center of the solar disk should suffer a deflection by the Sun's gravitational field of $1.75/R$ seconds of arc. The extremely difficult measurements give for the value of the constant between $2''.2$ and $1''.75$.

Since 1969 entirely new possibilities have been opened up in the centimeter wavelength range by the very long baseline interferometer whose angular precision is about 3×10^{-4} arc seconds. Fortunately the two bright quasi-stellar radio sources 3C273 and 3C279 are every year either occulted by or close to the Sun and so the Einstein bending of light can be determined from measurements of their separation from each other. The theoretical value is confirmed to within an error of about ± 3 percent.

[5]The form (30.14) differs from that of the Euclidean metric of ordinary three-dimensional space in that one term is negative. Such a metric ds^2 is called indefinite or pseudo-Euclidean.

(3) *Delay of radar signals.* I. Shapiro (1964) discovered that, according to general relativity, a radar signal which passes close to the Sun should experience a delay of the order of 2×10^{-4} s. Such radar reflections could be obtained from Mercury and Venus (passive) as well as from the space probes Mariner VI and VII (active : that is, the incoming signal triggers a transmitter). In this way the predictions of the theory have been confirmed to within a few percent.

(4) *Gravitational red shift.* A photon $h\nu$ that traverses, for example, the potential difference Sun–Earth $GM_\odot/R_\odot$, must show a red shift compared with a laboratory light source expressed by

$$-\Delta(h\nu) = \frac{GM_\odot}{R_\odot}\frac{h\nu}{c^2}. \tag{30.17}$$

This is formally equivalent to a Doppler effect $-c\ \Delta\nu/\nu = c\Delta\lambda/\lambda = 0.64$ km/s. The observations provide qualitative confirmation, but it is at present not possible completely to isolate the gravitational red shift from the Doppler effect of motions in the solar atmosphere. Also the red shift in the spectra of white dwarf stars (like the companion of Sirius) cannot be measured with sufficient accuracy. An experiment with the exceedingly sharp γ lines of the Mössbauer effect in the gravitational field of the Earth is capable of considerably higher accuracy. Using the recoil-less γ line of Fe^{57}, R. V. Pound and G. A. Rebka in 1960 were able to verify within 10 percent the frequency shift of only $\Delta\nu/\nu = 2.5 \times 10^{-15}$ which is that calculated for an altitude difference of 22.6 m. In 1965, with an improved experiment, R. V. Pound and J. L. Snider achieved a precision of ± 1 percent.

(5) *Perihelion advance of the planets.* The very small advance of the perihelion of Mercury calculated by Einstein agrees well with the old computations of Leverrier. The reduction by G. H. Clemence and R. L. Duncombe ($\sim$ 1956), using electronic computation, of a vast amount of observational material gave with considerably improved accuracy:

Planet		Mercury	Venus	Earth
Perihelion advance	Observed	$43''.11 \pm 0.45$	$8''.4 \pm 4.8$	$5''.0 \pm 1.2$
per century	Calculated	$43''.03$	$8''.6$	$3''.8$

(30.18)

Since the perihelion advance is a second-order effect, while the deflection of light rays, the delay of radar signals, and the red shift are only first-order effects, we may claim that all in all there is a remarkable verification of relativity theory.

The more recent tests, which have achieved precisions thought to be quite impossible only a few years ago, have dismissed from the field almost all competitors of general relativity. At present, there still remains in the running a development proposed by R. H. Dicke, C. Brans, and P. Jordan; however, we cannot go further into it here.

As well as the very small effects described above, the Schwarzschild metric [equation (30.15)] reveals another, much more spectacular possibility, which we have already mentioned several times in connection with the evolution of stars and galaxies: the existence of *black holes*.

In fact, if the radius of our mass M is *less* than its gravitational (or Schwarzschild) radius r_s [equation (30.16)], then within the sphere $r \leqslant r_s$ the coefficients of the spacelike element dr^2 and the time-like element $-c^2\ dt^2$ in equation (30.15) reverse their signs. Although we cannot here prove it in detail, this has the effect that neither matter nor light quanta, and so also no kind of signal, can succeed in escaping from the region $r < r_s$, the so-called black hole, to the outside ($r > r_s$). A black hole—more precisely, of Schwarzschild type—is detectable *only* by means of its gravitational field. We have already pointed out that a black hole is involved as a final stage of certain processes of stellar evolution. With respect to the physics of quasars, the important question is whether or not similar configurations can be produced which may release a significant fraction of the gravitational energy set free by the collapse. The solution to this question was brought nearer by R. P. Kerr's discovery (1963) of a metric which represents a mass *M with angular momentum*.

If one writes equation (30.16) in the form $GM^2/r_s = \frac{1}{2}Mc^2$, one sees that the gravitational collapse of a mass M into a black hole or a related configuration provides the unique possibility of releasing a significant fraction of its rest-mass energy Mc^2. According to equation (25.17), even in the most productive nuclear process, $4H^1 \rightarrow He^4$, only about 0.7 percent of that energy is available.

Before we return to the problems of cosmology let us first think briefly about a phenomenon which has been much discussed in recent years, *gravitational waves*. Soon after the discovery of his gravitational equations A. Einstein deduced from them that a system of moving masses produces gravitational waves which propagate with speed c. If a (polarized) gravitational wave acts on matter, then (at a particular time) it is compressed in one direction perpendicular to the wavepath and expanded in the orthogonal direction; after half a cycle the opposite deformation occurs. The description of a cycle for the "other" polarization is given by a 45° rotation about the direction of propagation.

J. Weber has now been actively concerning himself since 1958 with the detection of gravitational waves from space. As receivers, he uses massive aluminium cylinders, usually 153 cm long and from 61 to 96 cm in diameter (the largest weighs about 3 tons), which are suspended horizontally in such a way as to be as free from vibration as possible and to be shielded from other perturbations. The deformation is detected by means of piezocrystals arranged round the circumference and is amplified electronically. In this way even deformations of order 10^{-14} cm can be measured! J. Weber's aluminium cylinder "perceives" gravitational waves at its resonance frequency of 1661 Hz within a bandwidth of 0.016 Hz; its directional sensitivity is a maximum at right angles to the axis of the cylinder. By using the

Earth's rotation it is then possible to scan at least the horizon with modest angular resolution, as in the first radio antennas of K. G. Jansky.

J. Weber now receives, usually several times per day, short bursts of gravitational radiation at 1661 Hz. It is argued in favor of the reality of these signals that two receivers placed about 1000 km apart show corresponding coincidences and that the signals come preferentially from the direction of the galactic center (or anticenter, but that would make rather little sense).

Theoreticians immediately raised the objection against the reality of the gravitational radiation "measured" by J. Weber that—even with favorable assumptions about the frequency and angular dependence—its production would require such an enormous expenditure of energy that it could not be generated by any means, even, for example, in the galactic center. Furthermore, one would expect that such an immense emission of energy must be accompanied by some kind of simultaneous radio signals. However, an extensive search, organized from Jodrell Bank Radio Observatory, turned out to be completely negative.[6]

After this digression we return to our problems of relativistic cosmology. As early as 1917 A. Einstein and W. de Sitter constructed special model universes whose curvature was assumed to be independent of time t. Then in 1922–1924 A. Friedman successfully made the important generalization to spaces with time-dependent radius of curvature. His work was unrecognized until about 1927–1930, when G. Lemaître, A. S. Eddington, and others resumed the investigation of expanding universes on the basis of the general theory of relativity.

If from the outset we make the cosmological postulate that the universe must be homogeneous and isotropic throughout, then according to H. P. Robertson and others, we can express the four-dimensional line-element ds in the form (after a suitable choice of units of length and time)

$$ds^2 = dt^2 - R^2(t)\frac{dx^2 + dy^2 + dz^2}{\{1 + \frac{1}{4}k(x^2 + y^2 + z^2)\}^2}. \tag{30.19}$$

The time-dependent function $R(t)$ determines the radius of curvature of the three-dimensional space (t = constant), and it is defined quite analogously to the radius of curvature of a two-dimensional surface. The constant k, which can take the values 0 or ± 1, gives the sign of the spatial curvature, which is everywhere the same for any particular value of t. We have in fact:

(1) $k = 0$ well-known Euclidean space.
(2) $k = +1$ spherical or (otherwise interpreted) elliptical space. This space is closed and has a finite volume.
(3) $k = -1$ hyperbolic space. This space is open.

One can best picture these three spaces or geometries through their two-dimensional analogues (Figure 30.1).

[6]Note added July 1976: Recent experiments at other institutions have failed to confirm J. Weber's observations.

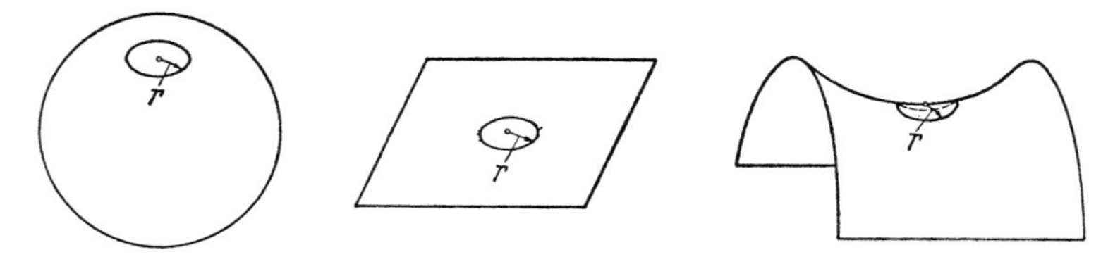

Figure 30.1. Surfaces (i.e. two-dimensional spaces) with curvatures $k > 0$, $k = 0$, $k < 0$ as models of curved space.

Curvature k of surface:	$k > 0$	$k = 0$	$k < 0$
Geometry:	spherical or elliptical	euclidian	hyperbolic (Bolyai-Lobatchewski)
Circumference of circle:	$<2\pi r$	$2\pi r$	$>2\pi r$
Area of circle:	$<\pi r^2$	πr^2	$>\pi r^2$

The further development of relativistic cosmology (which we cannot here review in detail) is accomplished in the following way: The Einstein field equations which relate the g_{ik} of the universe with its material content and the boundary conditions of the problem (see below), reduce for the Robertson metric (30.19) to a differential equation for $R(t)$. For systems of vanishingly small pressure, this proves to be identical with the differential equation (30.10) or (30.10a) of Newtonian cosmology, and so we obtain precisely the same choice of models. However, now *a priori* no velocity exceeding that of light can arise. Relativistic cosmology is the first theory to achieve a self-consistent and contradiction-free description of the universe as a whole.

In relativistic cosmology, just as in Newtonian cosmology, $R(t)$ is primarily fixed by the Hubble constant H_0 and the deceleration parameter q_0 in accordance with equations (30.11) and (30.12). The latter now determines also the type of spatial curvature, that is, the k or the overall character of the universe. There are the following possibilities:

Deceleration parameter	Curvature	Space	Time dependence of $R(t)$	
$0 \leq q_0 < 1/2$	$k = -1$	hyperbolic (open)	monotonically increasing	(30.20)
$q_0 = 1/2$	$k = 0$	Euclidean	monotonically increasing	
$q_0 > 1/2$	$k = +1$	spherical or elliptical (closed)	finite (cycloidal)	

The (present) matter density ρ_0 is again given by equation (30.12).

What observations (partly actual, partly at any rate in principle) are at our disposal in order to discover to which model the actual cosmos corresponds?

We have already discussed [see equation (30.1) ff.] the determination of the Hubble constant H_0 from the red shift z in the spectra of those galaxies whose distance r can still be determined independently (by means of Cepheids, etc.) and the most recent revision by A. Sandage and G. A. Tammann (1972) of the cosmic distance scale.

In the assured region up to $z \approx 0.14$ ($v \approx 42\,000$ km/s) the radio galaxies also follow the relation between z and apparent magnitude V_c (Figure 30.2) for the usual giant galaxies (apart from a negligible difference of 0.3 mag. in the absolute magnitude).

The quasars, with or without radio emission, reach larger values of z. According to R. Lynds (1971) 4C 05.34 has $z = 2.877$; for OH 471 z has even been given as 3.40. However, the quasars, N galaxies, and Seyfert galaxies are scattered in the Hubble diagram (Figure 30.2) between the straight line of the usual giant galaxies and a line up to ~ 5 magnitudes brighter. Some astronomers have seen in this a difficulty for the cosmological interpretation of their red shift z. However, recent observations by A.

Figure 30.2. Hubble diagram for quasars with Δ and without + radio emission, radio galaxies ●, Seyfert and N galaxies etc ⊕. The velocity of recession $v = c\Delta\lambda/\lambda_o$ in km/s is plotted logarithmically against the apparent magnitude V_c, which is corrected in the case of galaxies for interstellar absorption, reddening due to the redshift, etc., but is uncorrected in the case of quasars.

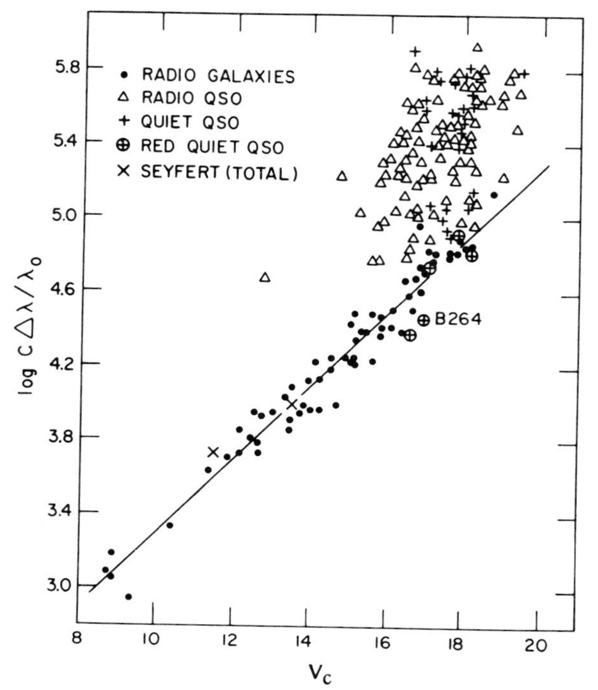

Sandage (1973) of N galaxies and by J. Kristian (1973) of quasars leave no doubt that these structures consist rather of fairly normal giant galaxies with an active nucleus, whose optical emission can exceed that of the rest of the galaxy by 5 or 6 magnitudes, that is, factors up to $\sim$ 200. The actual galaxy is therefore in many cases hard to observe. Further, A. Sandage has pointed out that *no* quasar is fainter than would be expected for its z from the Hubble relation. As a result of this, the cosmological interpretation of the red shift may also be secure for quasars.

It is much more difficult to determine the deceleration parameter q_0 and so the mean density of the universe ρ_0 [equation (30.12)]. Theory shows that for a collection of similar galaxies the connection between the red shift z and their apparent magnitude or their angular diameter depends, for sufficiently large z, on q_0. A. Sandage and others have spent a great deal of effort in trying to obtain an empirical determination of q_0 on that basis. We can say fairly certainly that $q_0 < 0$ (see below) is not consistent with the observations; an "open" universe with $q_0 \leqslant \frac{1}{2}$ is just possible. On the other hand q_0 can under no circumstances be greater than +2.

Further cosmological information is expected from the statistics of apparent magnitudes of galaxies: "How many galaxies per square degree have magnitudes (corrected for galactic absorption etc.) in the range $m \pm \frac{1}{2}$?" In equation (23.1) we have already written down the limiting form which this distribution takes for a Euclidean universe. In this case, too, we can do nothing with the quasars.

Furthermore, as far as the very distant galaxies (of all types) are concerned we must ask whether, in the long time during which the light from them has been on its way to us, they have significantly altered their brightnesses, colors, diameters, etc., as a result of *evolution,* so that a comparison with objects near to us would not be immediately possible.

In further consideration of the evolution of galaxies (Section 29), we investigate a few quantitative aspects of the "world-picture," at least for spatially closed (cycloidal) model universes with $q_0 > \frac{1}{2}$. We must refer the reader to the literature for the mathematical apparatus and for the models with $q_0 \leqslant \frac{1}{2}$.

In Figure 30.3 we depict (schematically) the time dependence $R(t)$ for a world model with $q_0 > \frac{1}{2}$. According to equation (30.11) the Hubble constant H_0 determines the tangent to this curve at the present epoch $t = t_0$. The Hubble time $T_0 = H_0^{-1}$ [equations (30.4) and (30.5)] is the interval between t_0 and the point where this *Hubble line* cuts the t axis (for $q_0 = 0$, moreover, $R(t)$ would be represented *throughout* by this line). The beginning of our epoch of the universe is determined by the intersection of the curve $R(t)$ with the t axis. It is earlier than t_0 by a time interval for which we suggest the name *Friedman time* T_F (Table 30.1). Thus there can be no galaxies, star clusters, etc., whose ages are greater than T_F. Then it is further possible—for given q_0—to calculate for each red shift the corresponding R and so the time interval $t(z)$ which a light or radio signal requires in order to reach us from the object concerned. In this way we

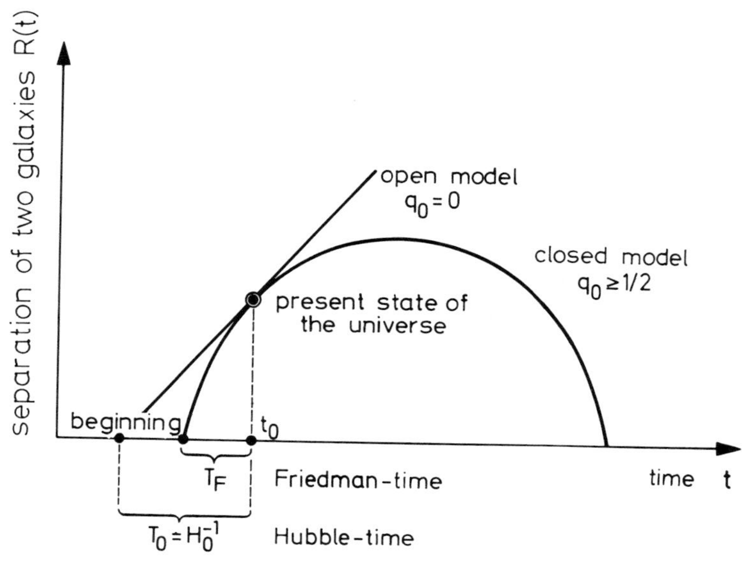

Figure 30.3 Closed world model ($k = +1$; $q_0 > \frac{1}{2}$). The separation of two galaxies is proportional to the scale factor $R(t)$, whose time-dependence is depicted by a cycloid. The small circle represents the present state of the universe ($t = t_0$). The tangent to the $R(t)$ curve at this point would correspond to an open universe ($q_0 = 0$; $k = 1$). The Hubble time $T_0 = H_0^{-1}$ and the Friedman time T_F are also explained (see also Table 30.1).

obtain a lower bound for the age of the furthest galaxies or quasars, $t(z = 3)$ (Table 30.1).

If we try, despite all the uncertainties and doubts, to find, with the help of Table 30.1, *the* model of the universe, we must take into account (with their considerable error bounds) the Hubble constant H_0, the deceleration parameter q_0, and the Friedman time T_F, which fixes the maximum age of

Table 30.1. Hubble time T_0, Friedman time T_F, and $t(z = 3)$ in units of 10^9 years for various values of the Hubble constant H_0 and the deceleration parameter q_0 [after A. Sandage (1961)].

Hubble constant H_0 km/s/Mpc		50	75	100	$t(z = 3)/T_F$
Hubble time $T_0 = H_0^{-1}$ (equal to T_F for	$q_0 = 0$)	19.5	13.0	9.7	0.75
Friedman time T_F	$q_0 = 0.5$	13.0	8.7	6.5	0.87
	$q_0 = 1.0$	11.1	7.4	5.6	0.89
	$q_0 = 2.0$	9.2	6.1	4.6	0.90

the cosmic structure. There are involved the age determinations of (1) globular clusters (2) the oldest galactic cluster (NGC 188), and (3) radioactive elements (uranium, thorium . . .). In regard to the absolute value, we consider the value from (3), $\sim 8 \times 10^9$ years, as the most secure; it may be consistent with the values from (1) and (2) within their rather generous error bounds. Then possible combinations would be $H_0 = 50$ km/s/Mpc and $q_0 \approx 3$ or $H_0 = 75$ km/s/Mpc and $q_0 \approx 0.8$. The first q_0 value seems very high; overall, we may estimate the most probable parameters of our universe to be:

$$\left.\begin{array}{l} H_0 \approx 60 \text{ km/s/Mpc};\ q_0 \simeq 1.5 \text{ to } 2.0 \\ \text{and the corresponding Friedman time} \\ T_F \approx 8.3 \text{ to } 7.7 \times 10^9 \text{ years.} \\ \text{According to equation (30.12) the mean density of} \\ \text{the universe is then} \\ \rho_0 \approx 2.0 \text{ to } 2.6 \times 10^{-29} \text{ g/cm}^3 \end{array}\right\} \qquad (30.21)$$

in good agreement with the empirical values discussed in Section 29. Recently (note added 1976) arguments have been advanced (abundance of deuterium, mass of intergalactic matter, etc.) which argue for a smaller q_0, a larger Friedman time T_F and overall for an open model of the universe. While the suggestion of a production of the heavy elements "in bulk" in the centers of galaxies already seems to be well founded, the discussion of the "correct" world model may still be far from closed.

Our particular interest now turns naturally to the initial stages of cosmic evolution, the *Big Bang*. The fundamental investigations in this area are inextricably linked with the names of G. Lemaître and G. Gamow.

Because of the enormous temperatures in the initial "melt," we must start with the fact that for $kT \geqslant mc^2$ it is energetically possible for every elementary particle to change into another. The theory of the initial stages of the expanding universe therefore shares all the uncertainties of the physics of elementary particles. Present ideas, together with the general theory of relativity, suggest that the universe was initially dominated by hadrons, the strongly interacting particles. After this *hadron era* there followed an equally short *lepton era* (lepton = light elementary particle); then, a few seconds after the beginning of the universe, the *photon* or *radiation era*. In the meantime the universe was cooled from the initial $\sim 1.8 \times 10^{12}$ K (according to R. Hagedorn a higher temperature is not possible because of the production of ever larger numbers of particles) to about 10^9 K (at 5.9×10^9 K we have $kT = m_{\mathrm{el}}c^2$), so that the formation of the chemical elements out of protons and neutrons could begin.

G. Gamow (1948) originally wanted to locate the formation of *all* the heavy elements at this time. However, this idea proved to be untenable, since the building up of nuclei comes to a halt because of an instability as early as mass number $A = 5$. Since it has proved impossible, so far at any rate, to get round this difficulty, we therefore believe that in fact only hydrogen and helium were formed at that stage. Also, a more detailed calculation correctly reproduces the ratio H : He $\approx$ 10 : 1 measured empirically for all objects (still consisting of original material) and valid equally whether they have low or normal metal content.

G. Gamow had already noticed that one could, so to speak, still obtain direct information about that stage of the *primeval fireball*. For soon afterward the interaction between radiation and matter became so slight that the radiation field of the universe expanded *adiabatically*. But L. Boltzmann had already shown that a vacuum radiation field remains black during adiabatic expansion and further that the product T^3 times the volume V of the vacuum remains constant. After the completion of the formation of the atoms H + He, however, the number of their particles $n \times V$ is also conserved. That is, T^3 must decrease like n. From considerations of nuclear physics, Gamow started with $T = 10^9$ K and $n \approx 10^{18}$ cm^{-3} at the element-formation stage. For the present universe, on the other hand, he calculated on average $n \approx 10^{-6}$ cm^{-3}. He therefore concluded that as a result of the 10^{24}-fold expansion in volume the present universe must be filled with blackbody radiation at a temperature of $\sim$ 10 K. In fact, with the techniques of radio astronomy—enormously improved in the meantime—A. A. Penzias and R. W. Wilson and others were able in 1965 to detect a cosmic blackbody radiation field of 2.7 K. That we are actually dealing with the relic of the primeval fireball is confirmed by the fact that the radiation field follows the Planck law from $\lambda \approx 50$ cm to 0.26 cm and is, within the errors of measurement, isotropic and unpolarized. For $\lambda > 50$ cm the galactic radio emission dominates the signal; below $\lambda \approx 0.16$ cm a rapid decrease of the intensity I_ν is expected from the Planck formula (11.23).

After the decoupling of radiation and matter there formed in the cosmos significant density perturbations, which led finally in particular to the formation of the galaxies. The beginning of this era is located at a time $\sim 10^5$ years after the origin of the universe, when the radius of the universe was still $\sim 10^3$ times smaller than it is today. We have already discussed the origin of galaxies in Section 29 and placed it at a considerably later time (radius of the universe $\approx \frac{1}{30}$ of the present). However we scarcely expect the last word to have been spoken about the comparative significance of gravitational instability, turbulence, and angular momentum distribution for the origin of galaxies.

Following this brief review we might add a few more general remarks:

(1) In the construction of our world models we started from the original equations of general relativity or from the virtually identical formulas of Newtonian cosmology. However in 1917 A. Einstein had already extended his gravitational equations by the so-called Λ term, which would mean, for example, an extra term $-\Lambda/3$ on the left-hand side in our equation (30.10a). The purpose of this additional term, which can be introduced with no mathematical inconsistency, was to make it possible to construct a *static* model of the universe, which Einstein believed at that time to be the only sensible model. This point of view became untenable, however, with Hubble's discovery of the red-shift

law. Furthermore, no one has succeeded even recently in developing a physical or astronomical argument for a particular value of Λ. For this reason we shall here—like almost all astrophysicists—start with the assumption $\Lambda = 0$ and follow W. H. McCrea in treating the formal generalization of the gravitational equations with $\Lambda \neq 0$ as a kind of "emergency exit," in case sometime the worst should come to the worst.

(2) An age-old question runs: "What was there *before* the formation of the universe?" To this there is a quite unambiguous answer: Within the framework of the models discussed here, we can obtain no kind of information about whatever conditions or events there may have been longer ago than the Friedman time T_F, since for $T \gtrsim 2 \times 10^{12}$ K there is no longer any structural form and so also no carrier of information.

Are we here dealing with a fundamental naturally imposed limit to our knowledge? Such situations are not unknown in the development of physics. The insight that $c = 3 \times 10^{10}$ cm/s is the greatest possible speed led to *relativity theory*. The further knowledge that $h = 6.62 \times 10^{-27}$ erg s is the smallest action led to *quantum mechanics*. Analogously, we may hope that the knowledge of the fundamental character of the Hubble constant, or of the "age of the universe" $T_0 = H_0^{-1}$ will lead to a *cosmological physics*. For cosmic space-time such a theory could differ considerably from present-day physics, but it would have to include the latter as a limiting case for our own space-time neighborhood.

(3) The world models with $q_0 > \frac{1}{2}$, whose $R(t)$ is represented by a cycloid, have often been referred to in the older literature as *periodic*: it was imagined that another expansion followed the compression to a singularity $R \to 0$. Present-day physics however contains no information about a repulsive force. It is therefore better to speak of *cycloidal* model universes, and we content ourselves with treating our model (30.21) as an approximation to the interpretation of the present epoch.

The model universes which we have discussed so far all start from the *cosmological principle* of spatial homogeneity and isotropy. The distribution on the sky of the very distant galaxies and of the 2.7 K radiation shows that these assumptions are to a good approximation fulfilled. Should we not also require homogeneity of the time scale $-\infty < t < +\infty$? Should not the universe in reality be the same "world without end"? An attempt to make these requirements of many philosophers and theologians quantitative was made by the theory of the *steady-state universe* developed since 1948 by H. Bondi and T. Gold and by F. Hoyle, and then by others. In mechanical respects it can also be treated as an open relativistic model with $k = -1$ and $q_0 = -1$.

This is so to speak a universe for bureaucrats, in which everything at all times shall be regulated in accordance with the same paragraphs! The entire problem of the beginning of the universe is avoided, but as a result there is also no explanation for the 2.7 K radiation. The necessary departure from "ordinary physics"—there are several formulations—is that a mechanism must be postulated that makes possible the continual production of hydrogen in the cosmos, since this is needed for replacing the matter that "expands away" and as fuel for stars. We can thus say that the difficulties which in the *Big-Bang* cosmology are concentrated at $t = 0$ are in the *steady-state* cosmology distributed over the whole time scale. In this respect a middle position is occupied by the proposal advocated by V. A. Ambarzumian that the individual *galaxies* came into the universe as compact nuclei by means of a "superphysical" mechanism.

A. S. Eddington, P. A. M. Dirac, P. Jordan, and others have sought to tackle the problem of "cosmological physics" from a different angle. From the elementary constants of physics on the one hand, e, h, c, m (where it remains undecided whether this is to be the mass of the electron, the proton, or some other elementary particle) and G, and on the other hand the "constants" of cosmology, the time $T_0 \approx 13 \times 10^9$ years $= 4.1 \times 10^{17}$ s and the mean density of matter in the universe $\rho_0 \approx 10^{-30}$ g/cm³, we can form several dimensionless numbers. Factors of the order $2\pi, \ldots$ naturally remain open. In this sense one obtains a set of dimensionless numbers of order of magnitude unity (for example, the Sommerfeld fine-structure constant $\alpha^{-1} = hc/2\pi e^2 = 137$, etc.) and a second set of order 10^{39} to 10^{40}.

[In this connection it may be remembered that the differences in order of magnitude between the strong (nucleon–nucleon), weak (β decay), and electromagnetic interactions—after which gravitation follows—are not yet understood theoretically.]

There thus exist:

(1) The ratio of electrostatic to gravitational attraction between a proton and an electron

$$\frac{e^2}{Gm_{\rm p}m_{\rm e}} = 2.3 \times 10^{39} \tag{30.22}$$

(2) The ratio of the length cT_0 (in a spherical world $\approx$ world radius) to the classical electron radius

$$\frac{cT_0}{e^2/m_{\rm e}c^2} = 4.4 \times 10^{40} \tag{30.23}$$

(3) The number of nucleons in the universe of the order

$$\rho_0 c^3 T_0^3/m_{\rm p} = (1.0 \times 10^{39})^2. \tag{30.24}$$

The same facts show, as we see by combining (30.22)—(30.24), that the so-called deceleration parameter (in the case of zero cosmological constant) $q_0 = 4\pi G\rho_0/3H_0^2$ of relativistic cosmology is of the order of unity. Since the sign of $2q_0 - 1$ is the sign of the world curvature $k = \pm 1$ or 0, this means that the actual universe does not differ too much from a Euclidean one (which is by no means self-evident).

If we consider the equality as regards order of magnitude of the numbers in (30.22) on the one hand and (30.23) and (30.24) on the other as significant, since the age of the universe T_0 occurs in the latter, we can start speculating upon a cosmological time dependence of the elementary constants of physics.

In any case we may suppose that the above relations will play an essential part in any future "cosmological physics." In that connection also the following consideration naturally arises. Both Newtonian cosmology and relativistic cosmology offer us a whole catalog of possible world models. However, why is just our own universe realized with certain definite (dimensionless) numerical constants? We can still give no answer to this. As an example, E. A. Milne made an attempt in "Kinematic Relativity" (1948).

Our presentation of cosmology, the study of the universe as a whole, has so far proceeded from a heuristic viewpoint. It seems appropriate therefore to complete it with a few historical remarks.

H. W. M. Olbers (1826) appears to have been one of the first astronomers to have considered a cosmological problem from an empirical standpoint. Olbers's paradox asserts: Were the universe infinite in time and space and (more or less) uniformly filled with stars, then—in the absence of absorption—the whole sky would radiate with a brightness that would match the mean surface brightness of the stars, and thus about that of the surface of the Sun. That this is not the case cannot depend only on interstellar absorption, since the absorbed energy could not actually disappear. However, a finite age of the universe of about 13×10^9 years would already dispose of Olbers's paradox for any otherwise fairly plausible model (W. B. Bonnor 1963).

The modern development of cosmology started on the one hand from the measurements of radial velocities of spiral nebulae and on the other hand from the theory of general relativity.

In connexion with the older radial velocity measurements by V. M. Slipher (~1912) already in 1924 C. Wirtz had remarked upon their increase with distance, and he had related them to de Sitter's relativistic world model. In 1917 A. Einstein had shown that his field equations of general relativity, when extended to include the Λ term, have a static cosmological solution (Einstein's spherical universe). Also in the same year W. de Sitter found the solution we have mentioned giving an empty expanding universe.

31. Origin of the Solar System: Evolution of the Earth and of Life

From the depths of cosmic space we now return to our planetary system with the old question of its origin. So long ago as 1644, in France René Descartes with his vortex theory was able to advance the daring idea that one could come nearer to the answer, not by handing on traditional myths, but by proper investigation. In Germany even in 1755 I. Kant had to let the first edition of his "Allgemeine Naturgeschichte und Theorie des Himmels" appear anonymously because he feared the (Protestant) theologians. In the book he treated the formation of the planetary system for the first time "nach *Newton*ischen Grundsätzen." Kant started from a rotating, flattened primeval nebula out of which the planets and later the satellites were then formed. A similar hypothesis was at the basis of the somewhat later (independent) account by S. Laplace 1796 in his popular book "Exposition du Système du Monde." We shall not go into details and differences of these historically important beginnings, but we shall summarize once more the most important facts (see Tables 5.1 and 7.1 and Figures 5.5, 7.1, and 31.1) which have to be explained:

(1) The *orbits* of the planets are almost circular and coplanar. Their sense of revolution is the same (direct) and agrees with the sense of rotation of the Sun. The orbital radii (the asteroids are taken together) form approximately a geometrical progression

$$a_n = a_0 k^n \tag{31.1}$$

where $a_0 = 1$ AU, with $n = 0$ for the Earth, and $k \approx 1.77$ (Figure 31.1).

(2) The *rotation* of the planets takes place for the most part in the direct sense (Venus is an exception). For the satellites, we must distinguish between inner satellites, which have belonged to the planets since the beginning, and the outer satellites, which have been captured subsequently, as has recently been confirmed by celestial mechanics calculations. The former—and this should be explained by any cosmogony—have orbits of low eccentricity, small inclination to the equatorial plane, and direct sense, while for the outer satellites the eccentricities and orbital inclinations are considerably larger. For Jupiter only the four Galilean satellites and the innermost fifth satellite, and for Saturn only the satellites out to Titan, are considered to have originally belonged to the planets.

(3) The *terrestrial planets* (Mercury, Venus, Earth, Mars, and the asteroids) have relatively high densities (3.9 to 5.5 g/cm³), while the *major planets* (Jupiter, Saturn, Uranus, and Neptune) have low densities (0.7 to 1.7 g/cm³). Like the Earth, the former consist in the main of metals and rocks, the latter of scarcely modified solar material (hydrogen, helium, hydrides). The terrestrial planets have slower rotation and few satellites, the major planets relatively rapid rotation and—even after deduction of the captured ones—numerous satellites. *Pluto* is physically more closely akin to the terrestrial planets and has clearly obtained its unusual orbital elements only as a result of perturbations. We will accordingly leave it out of consideration in what follows.

(4) The Sun retains within itself 99.87 percent of the *mass* but only 0.54 percent of the *angular momentum* (Σmrv) of the whole system, while

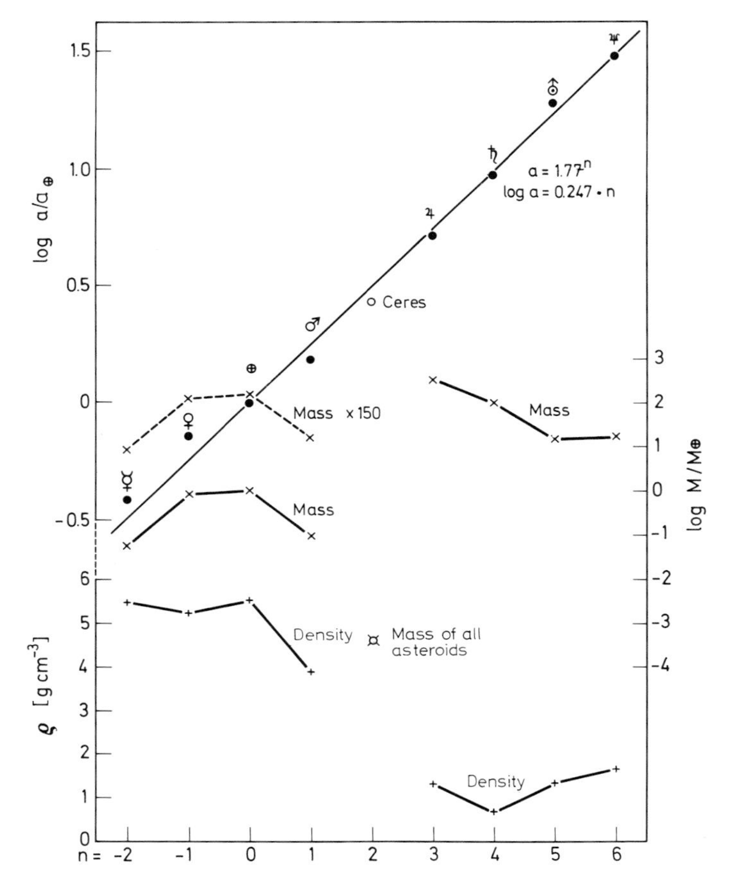

Figure 31.1. Orbital semimajor axes *a*, masses *M*, and mean densities ρ of the planets. The orbital semimajor axis of Ceres and the estimated total mass have been used to represent the asteroids $n = 2$. Pluto, which was possibly captured later, has been omitted. The straight line drawn in the upper half corresponds to the relation $\log a = 0.247\, n$ or $a = 1.77^n$.

conversely the planets (mainly Jupiter and Saturn) have only 0.135 percent of the mass and 99.46 percent of the angular momentum.[1]

While the impressive regularity in the structure of the planetary system speaks for an evolution from within itself—in the sense of the theories of Kant, Laplace, and later v. Weizsäcker, ter Haar, Kuiper, and many others—the paradoxical distribution of angular momentum between the

[1] As we easily calculate, the angular momentum (page 36) of the planetary orbits (♃, ♄) is 3.15×10^{50} g cm²/s, and the angular momentum of the Sun (with some uncertainty, concerning the increase of angular velocity in the interior) is 1.7×10^{48} g cm²/s.

Sun and the planets forms the most important argument for another group (Jeans, Lyttleton, and others) who postulate the interaction of the Sun with a passing star or some such occurrence. We can today put on one side "catastrophe theories" of that kind since—apart from the unlikelihood of such an encounter of two stars—it would be impossible to understand how planets should form from a filament torn out of the Sun by tidal forces.

As regards the cosmogony of the planetary system, discussion of the many older and newer models of the primeval solar system (often very detailed) would prove unfruitful, since the observational basis is in many respects so small and forces the introduction of many hypotheses (often barely recognizable as such).

We now know from radioactive age determinations that the Earth, Moon, and meteorites, and also no doubt the Sun and the entire solar system, were formed 4.55×10^9 years ago within a relatively short time interval of $\sim 30 \times 10^6$ years. The formation of stars from diffuse material is to a great extent comprehensible, both observationally (color-magnitude diagrams, T Tauri stars, etc.—Sections 21 and 26) and theoretically (internal structure and evolution of stars—Section 26). We therefore discuss the origin of the planetary system in this context (A. Poveda and others, 1965).

31.1 Origin of the Sun and the solar system

As we saw, all stars originate by the process in which an interstellar gas cloud of more than about 10^3 solar masses condenses upon itself (during which time some of the atoms combine to form molecules and dust particles) and then fragments into stars. Even the solitary stars, like our Sun, have originally belonged to such multiple systems, associations, or clusters.

Recently R. B. Larson (1969) has made more precise calculations of the formation of a star of one solar mass (admittedly, by necessity, under the assumption of spherical symmetry, that is, zero angular momentum).

The start of the collapse of a gas cloud of radius R is determined by the Jeans criterion of gravitational instability [our equation (29.3); from numerical tests, Larson chooses a constant larger by a factor two]:

$$R \leqslant 0.4 \frac{GM}{\mathcal{R} T/\mu}. \tag{31.2}$$

With $M = M_\odot = 2 \times 10^{33}$ g, a molecular weight $\mu = 2.5$, and an initial temperature estimated to order of magnitude to be $T \approx 10$ K, together with the usual values for the gravitational constant G and the gas constant $\mathcal{R}$, one finds that a cloud of $1 M_\odot$ first becomes unstable when its radius has reached

$$R \approx 1.6 \times 10^{17} \text{ cm} \quad or \quad 11{,}000 \text{ AU} \quad or \quad 0.05 \text{ pc}. \tag{31.3}$$

Its mean density then amounts to

$$\rho \approx 1.1 \times 10^{-19} \text{ g/cm}^3. \tag{31.4}$$

This corresponds precisely to the density in the gaseous nebulae (for example, in Orion) connected with young stellar associations; the interstellar gas has already been compressed by about a factor of 10^4. The collapse of such a cloud to the point of formation of the Sun and solar system occupies a period of some 10^5 years, as may easily be estimated from equations (29.26a and b).

Perhaps we can gain from this some insight into the problem of the *angular momentum* of cosmic masses and in particular of the solar system (see Section 31.3 below). If at the onset of the Jeans instability our mass of gas (31.3 and 31.4) is decoupled (in respect of its angular momentum) from the rest of the Galaxy, it has initially a rotation period of the same order[2] as our galactic rotation period, that is, $\sim 10^9$ years. If the mass of gas contracts to the solar radius $R_\odot = 7 \times 10^{10}$ cm, we obtain, as a result of conservation of angular momentum ($\sim R^2/T$), a rotation period of ~ 0.07 days and an equatorial velocity of ~ 700 km/s.

This is of the same order as the rotation speeds that are observed for young B stars. While stars of later spectral types in general rotate much more slowly, there have been observed in young clusters even stars of later type that rotate at any rate considerably faster than the Sun.

If we could further transfer to the Sun the angular momentum of the planetary orbits (essentially that of Jupiter), its equatorial velocity would increase from 2 km/s to ~ 370 km/s. We thus come to the idea that the large angular momentum of the orbital motion of the planets is none other than a considerable fraction of the angular momentum which *every* structure of $\sim 1 M_\odot$ has originally obtained from the galactic rotation. However, the Sun and similar stars have later given up their *own* angular momentum to the interstellar medium as a result of magnetic coupling through the solar wind (R. Lüst and A. Schlüter 1955), while the orbital angular momentum of the planets has been left behind.

Since the time of I. Kant people have explained the overall structure of the planetary system by the idea that the primeval Sun was surrounded by a flat, primarily gaseous disk—the *solar nebula*—of about the size of the present system. This must clearly have had the same chemical composition as the Sun (Table 19). On the other hand its material can never have been *inside* the Sun since, for example, the Earth and the meteorites still contain about a hundred times more lithium per gram than the Sun, where about 99 percent of it was destroyed in the course of time by nuclear processes. The

[2]More precisely: with a rotation law $v \propto R^{-n}$ ($n = 0$ rigid, $n = 0.5$ Keplerian rotation) the local angular velocity is

$$\omega^* = \tfrac{1}{2} \operatorname{curl} \mathbf{v} = \frac{1-n}{2}\, \omega_{\text{gal}}.$$

solar nebula must therefore have been formed at effectively the same time as the Sun.

That the origin of such a system is no unusual coincidence is shown by the discovery of companions, with masses scarcely more than that of Jupiter, for several stars in our immediate neighborhood (at greater distances the reflection of the orbital motion can no longer be measured), for example, Barnard's star M5V (cf. Section 16). Furthermore, the statistics of mass ratios and of orbital semimajor axes (most probable value ≈ 20 AU) of binary stars support the idea of a substantial frequency of planetary systems in the universe. *When* a planetary or binary system forms may depend on the mass ratio between the primary component and the rest and perhaps also on the angular momentum; nothing more precise is known about it.

We now try to estimate the density ρ in the solar nebula. To start the condensation of, for example, a Jupiter mass, the Jeans condition (29.4) for gravitational instability requires a density of about 10^{-10} g/cm^3, which is, to order of magnitude, in harmony with the pressure $p \approx 10^{-5}$ atm estimated by E. Anders (1972) from physical chemical investigations relating to the origin of the familiar chondrites (see below).

However if larger fragments are to be held together by their self-gravitation against the perturbing effect of the solar tidal forces, the Roche condition (7.10) must be fulfilled. If we denote by ρ the density at distance R from the Sun and by $\rho_\odot$ and $R_\odot$ the density and radius of the Sun itself, the Roche equation (7.10) requires

$$\rho \geqslant 15\rho_\odot (R_\odot / R)^3 = 2.1 \times 10^{-6}/R_{\mathrm{AU}}^3 \text{ g/cm}^3. \tag{31.5}$$

From this one finds $\rho \approx 10^{-7}$ g/cm^3 for the region where, as we shall see, the meteorites are formed, that is, $R \approx 2.8$ AU.

Let us further try to estimate the mass and thickness of the solar nebula. For this we must first take account of the fact that in the terrestrial planets (from Table 19) only about 0.65 percent or $1/150$ of the solar material became condensed and that therefore about equally much mass was originally present in, for example, the regions of the Earth and of Jupiter. This gives in the first instance a lower limit $\geqslant 0.0025 M_\odot$ for the mass of the solar nebula. The original mass of the solar nebula may have been (for example, G. P. Kuiper 1951) about 40 times larger, that is $\sim 0.1 M_\odot$. Still more uncertain are calculations of its thickness; from the inclinations of the planetary orbits we could estimate ~ 0.2 to 1 AU.

While the major planets consist of (almost) unaltered solar material, in the region of the terrestrial planets, of the asteroids and of the meteorites, which were (at least for the most part) formed in their vicinity ($R \approx 2$ to 4 AU), there has taken place a separation of hydrogen, carbon, rare gases, . . . (in astronomical terms, a separation of gas and dust), as well as a fractionation of metals and silicates.

The meteorites (Section 8) contain the most important clues to these events; more recently, the study of lunar rocks has given further information.

31.2 Origin of the meteorites

As we have seen, the earlier view that in the meteorites we had before us "the" cosmic material with "the" cosmic abundances of the elements has long ago given place to a thorough study of their mineralogical, chemical, and isotope structure. In particular, the introduction in the 1940s of neutron activation analysis for determining the abundance of rare elements was an enormously significant step forward.

Figure 31.2 gives a review of the main types of meteorites (left) and of the finer subdivision (right). The iron meteorites are clearly the most strongly differentiated; we might expect the most extensive information from the chondrites. As we saw, these contain *chondrules,* millimeter-sized spheres of various silicates and also of iron (particularly in the enstatites) embedded in a fine-grained matrix of similar composition.

As a starting point for the interpretation of the extraordinary variety of ratios, a very efficient fractionation process must be involved, namely, the condensation and separation of solid bodies out of the gaseous phase of solar composition.

The great achievement of H. C. Urey (1952) was to make fruitful use for the cosmogony of the solar system of the relevant experience and methods of physical chemistry. More recent calculations have been carried out by J. W. Larimer and E. Anders (1967–1968), in which they discussed two limiting cases for the time dependence of the condensation:

(1) Thermodynamic equilibrium. Cooling occurs so slowly that it is possible to form solids in diffusion equilibrium.
(2) Rapid cooling, so that the condensed elements and compounds do not diffuse into each other.

At a pressure of $\sim 10^{-4}$ atm (see below), the condensation of the solar material, under *both* assumptions, takes place essentially in the following steps:

$<$2000 K: nonvolatile compounds of Ca, Al, Mg, Ti,. . . .
1350 – 1200 K: magnesium silicates; Ni, Fe.
1100 – 1000 K: alkali silicates.

By this time $\sim$ 90 percent of the chondritic material has condensed.

680 – 620 K: iron sulphide (FeS, troilite) and other sulphides. Then Pb, Bi, Tl, In. . . .
400 K: Fe_3O_4 is formed from iron and steam.
400 – 250 K: hydrated silicates.

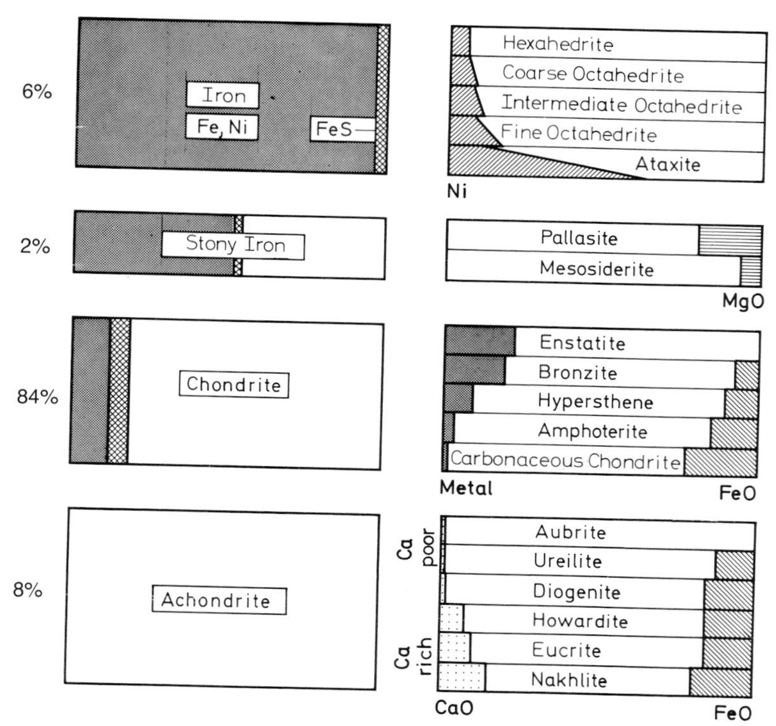

Figure 31.2. Classification of meteorites (E. Anders 1969). The division corresponds primarily (left) to the ratio of metals (shaded) to silicates (white). The finer subdivision (right) takes account of further differences in composition and structure. On the extreme left is given the percentage of the falls which are of each type.

The condensation of many trace elements is more complicated and depends on the assumptions (1) or (2); following the precedent set by H. C. Urey, one may use them as a "cosmic thermometer."

The analysis of the "normal" chondrites (that is, those of the first four subclasses in Figure 31.2) reveals several processes which have occurred for different components (E. Anders distinguishes: early condensate, metal, original dust, remelted dust) at various temperatures from ~1300 K to ~400 K:

(1) Material with a high content of Ca, Al, Ti . . . was—at any rate in the region of formation of the meteorites and terrestrial planets—partially separated at $T \geqslant 1300$ K. It is relatively enriched in the Earth–Moon system but is deficient in the normal and enstatite chondrites.
(2) A metal-silicate fractionation has influenced several types of meteorites and certainly also the terrestrial planets at temperatures of from ~ 1050 to 700 K.
(3) Remelting and outgassing at ~ 600 to 450 K has caused an impoverish-

ment of the volatile elements in meteorites and in the Earth–Moon system.

(4) Accretion of matter took place mainly in a relatively low temperature region at ~ 450 K and ~ 10^{-5} atm. The CI chondrites were formed at ~ 360 K and ~ 2×10^{-6} atm.

We cannot here go into more detail about these results—in particular, details of the reasons for them—especially as it has not yet been possible with the desirable certainty to combine the separate processes known to be important into an astronomical picture. As far as the still rather mysterious reheating process is concerned, we should perhaps think not only about the occasional increases of temperature of the primeval Sun to 8000 or 9000 K required by existing stellar model calculations but also about the observed variability of the T Tauri and RW Aurigae stars and about the related evolutionary phases of the flare stars.

Of particular interest in relation to the origin of the planetary system and especially to the origin of life on the Earth are the rare carbonaceous chondrites. They are divided into three types CI, CII, and CIII according to their (decreasing) content of volatile elements (H, C, S, O . . .). These differences correspond to the relative proportions of the *high*-temperature fraction of chondrules or similar iron particles, which are formed at ~ 1200 K, and on the other hand of the *low*-temperature fraction of their enveloping, fine-grained matrix, whose temperature of formation lay at only ~ 450 to 300 K.

Apart from the volatile elements, the abundance distribution of the chemical elements in this matrix and in the CI meteorites agrees with that in the Sun (Table 19). In this way one can, inter alia, determine chemically or by neutron activation the cosmic abundance of extremely rare elements which even in the Sun can scarcely be detected spectroscopically. Many of these elements are of great importance for understanding nucleosynthesis. The relative abundances of the isotopes are essentially the same in the meteorites, the Earth, and the Sun (where only a few are known precisely). The CI chondrites, which consist entirely of this dark matrix, contain up to 4 percent carbon, mainly in the form of organic compounds. The most obvious comparison is with terrestrial coals or humic acids.

Many investigators were at first tempted to assume a kind of extraterrestrial life. This problem, as well as the closely related question of terrestrial contamination, was unambiguously clarified by the CII chondrite which fell at Murchison in Australia at 11.00 A.M. on September 28, 1969, and whose pieces were collected within a few months. The Murchison meteorite contained, inter alia, primarily amino acids, which play an essential role in terrestrial life forms. However, while in living creatures it is nearly always only the optically laevorotatory L form that appears, in the meteorite the laevo and dextrorotatory molecules are about equally abundant; it contains a racemic mixture. Then the meteorite also contains many other amino acids which do not appear in living organisms. The investigation of the amino acids and also of the (mainly linear) hydrocarbon chains in the carbonaceous chondrites shows that these compounds must already have been formed during the formation of the solar system, but without the assistance of mysterious living creatures.

How can we envisage this event? One possibility is Fischer–Tropsch synthesis, in which hydrocarbons, predominantly of the type C_nH_{2n+2}, are made out of carbon monoxide CO and hydrogen molecules H_2 in the presence of suitable catalysts. It is reasonable to assume that in the solar nebula not all the CO is transformed into CH_4 (as it would have to be in thermal equilibrium below 650 K) since this reaction reaches completion extremely slowly. Then, with the assistance of the Fe_3O_4 and hydrated silicates which are present as catalysts, the remaining CO makes a kind of Fischer–Tropsch synthesis with H_2 at 380–400K. E. Anders and his colleagues have further shown that in the presence of NH_3 (which we know, for example, from the spectrum of Jupiter) there are formed the amino acids, which are of such biological importance, and the many other organic substances found in the carbonaceous chondrites. This idea explains incidentally, inter alia, why the isotope ratio $C^{12} : C^{13}$ in carbonates is somewhat smaller than in the organic compounds mentioned.

We may perhaps assume that the organic molecules discovered by radio astronomers in dense regions of the interstellar medium (Section 24) originated in a similar way to those found in the carbonaceous chondrites.

We should mention that earlier H. C. Urey, S. Miller *et al.* (1953) were already able to synthesize complicated organic molecules by exposing a mixture of CH_4, NH_3, and H_2O—a kind of primeval atmosphere—to a spark discharge or to ultraviolet or γ radiation. However this mechanism does not appear to reproduce, for example, the isotope effect mentioned above.

We shall return later to the exciting question of the origin of life. However, we shall first pursue further the formation of the meteorites and then of the solar system.

A more precise time resolution was made possible by the discovery that many meteorites or, more precisely, parts of meteorites contain not only the noble gases originating in the solar wind but also xenon isotopes which (as is confirmed by terrestrial experiments) could have been formed only by decay of the "extinct" radioactive isotopes I^{129} with a half-life of 16×10^6 years and Pu^{244} with a half-life of 82×10^6 years (fission). These studies, whose details are very complicated, suggest that the formation of all meteorites, including the carbonaceous chondrites, took place within at most $\sim$ 4 million years. Astronomers do not hesitate to connect the production of these short-lived isotopes with a flare-star stage which the Sun ran through briefly before it reached the main sequence. The flares of such stars produce considerably more optical and radio radiation than those of the Sun. We might then also suggest an appreciable production of cosmic ray particles which gave rise to the necessary nuclear reactions.

Then we could determine for many meteorites the so-called irradiation age, that is, the time since the meteorite in question was exposed to cosmic radiation. The primary cosmic ray component penetrates to a depth of about 1 m into the material and in doing so leaves behind it various reaction products, some stable, some radioactive, from whose number ratio we can calculate how long ago the piece considered was irradiated. In many cases the irradiation age gives us the time at which the fragment in question came sufficiently close to the surface as a result of the breaking up of a larger protometeorite (in the asteroid belt). While the irradiation age of the refractory iron meteorites (using a log t scale) lies mostly between 1 and 0.2×10^9 years, the more easily fractured stony meteorites mostly obtained their present size only 10^6 to 3×10^7 years ago. From mineralogical evidence, which allows us to draw conclusions about the force of gravity in the protometeorite, it is estimated that its diameter was 50 to 250 km, and that of the iron core $\sim$ 10 km. Recently metallurgists have succeeded in determining from the Widmannstätter patterns that the cooling rate of the protometeorites was at most 1 to 10 K per million years. From this, using the theory of heat conduction, a confirmation of the above dimensions was obtained.

These findings also provide further support for the origin of the meteorites in the asteroid belt. Only for the carbonaceous chondrites could we consider the formation to be near the orbit of Jupiter or still further out.

31.3 The Earth–Moon System

With regard to the formation of the terrestrial planets or protoplanets—clearly from smaller or larger fragments—let us turn at once to the particularly interesting and informative problem of the origin and evolution of the Earth–Moon system. Here the older studies of astronomers and geophysicists are pushed to one side by the exciting results of the manned and unmanned moon landings (1966–1973). However, our system itself is also, in the true sense of the word, unique: while in all other systems the ratios of mass and (orbital) angular momentum of the satellites to the corresponding quantities for their planets are $<< 1$, the mass ratio Moon : Earth amounts to 1 : 81.3, and the orbital motion of the Moon lays claim to 83 percent of the angular momentum of the whole system.

In order to learn something about earlier configurations of the Earth–Moon system, let us first use terrestrial observations and, following on the classical work of Sir George Darwin (1897), study the phenomenon of tidal friction, which we have already mentioned briefly (p. 43). The two tidal bulges which the Moon continuously drags round, both in the oceans and in the solid body of the Earth, cause a braking of the Earth's rotation. After removing all other effects, the discussion of old eclipses in particular shows an increase in the length of the day of about 0.00164 s per century. The angular momentum lost by the Earth can only be taken up by the orbital motion of the Moon, that is, the period and the orbital radius of the Moon are increasing.

As is easily calculated, this astronomically tiny effect must assume significant proportions in the course of geological time. J. W. Wells and C. T. Scrutton noticed in 1963 that the calcified houses of corals (and other living organisms) which live in seas with strong tides exhibit fine bands which correspond to the periods of the year, the synodic month and the day. While recent corals confirm the familiar astronomical rhythms, the study of fossilized corals showed that, for example, in the mid-Devonian, about 370 million years ago,

$$1 \text{ year} \approx 400 \text{ days and } 1 \text{ synodic month} \approx 30.6 \text{ days},$$

in satisfactory agreement with an extrapolation of present data. However, even the geological data reveals to us only the "recent history" of the Moon; for the more remote past, we are driven to theoretical extrapolation. In addition to the work of G. Darwin, this has been more recently worked out by H. Gerstenkorn (1955) and others, and led to the result that the Moon came closest to the Earth about $(1.4 \pm 0.5) \times 10^9$ years ago and circled round it at a distance of ~ 2.9 Earth radii, causing a tidal wave of quite apocalyptic dimensions (further calculations then led to a "capture theory"; see below). The geological evidence suggests rather unambiguously that this "Gerstenkorn event" never took place. That is not an argument against the theory of tidal friction as such but only against the extrapolatory application of the present frictional constant. In fact, the experts have not yet even agreed whether the main component of the tidal friction is localized in the seas or in the solid body of the Earth. On the other hand, the contribution of the seas to tidal friction might in any case have been less in earlier epochs of the Earth's history than today, because the continents still lay closer together.

Now let us approach our problem of the evolution of the Earth–Moon system from the opposite end and try to evaluate the results of the moon landings in this context.

The whole surface of the Moon is covered with a soil consisting of fine dust and fragments of various sizes, in part baked to form a breccia, the so-called *regolith*. This material was clearly formed by the impact of many large and small meteorites; for the most part it arises from close by, but some has come from very distant impacts. The flight of debris was certainly not hindered by an atmosphere.

The most significant features of the present lunar surface, the maria and craters, were (apart from a few volcanic formations) produced by the impact of meteorites of up to the size of asteroids on the completely solid crust (smaller craters may have been caused by secondary impacts). As can be seen directly from the enormously detailed photographs taken by the astronauts, they were only later largely covered by giant flows of basalt. In Mare Imbrium, 1150 km in diameter and originally ~ 50 km deep, three stages of this lava submersion, which occurred $\sim 3.9 \times 10^9$ years ago, can be recognized. Both the starting points of the lava flows and also the "drowned craters" which here and there project out of the generally flat frozen sheets of lava show that the lava flooding has in most cases nothing to do with the meteorite impact itself. Radioactive age determinations confirm that the lava flows are several hundred million years younger than the meteorite impacts. Small-scale melting as a result of impact plays a subordinate role. The basalt in the maria ($\rho \sim 3.3$ g/cm^3) burst forth rather later through cracks in the rock, whence it is reasonable to conclude that at that time, at least at depths of about 200–400 km, the Moon was still partially molten, at $\sim$ 1300 K. Compared with the rocks of the highlands and of several intermediate regions, the mare basalt displays clear signs of further chemical differentiation; in particular, it is significantly poorer in Al_2O_3.

If we compare (see Figure 7.4) the number of craters of different diameters in the highlands, on the floors of maria of different ages, and in the larger craters themselves it is apparent that the intensity of the cosmic bombardment decreased by *several powers of ten* in the first $\sim 10^9$ years after the formation of the Moon—we could equally well say : of the solar system—and that from then (-3.8×10^9 years) on the decrease occurred significantly more slowly. The fact that there are so few meteorite craters on the Earth now requires no further explanation: the present crust of the Earth was not formed until after the supply of meteorites was exhausted. The observations on the Moon, together with the discovery that Mars, Venus (as recent radar observations show), and Mercury are covered with craters which are completely comparable to those of the Moon, suggest that in the youth of the solar system at least the neighborhood of the terrestrial planets, but probably the major part of the plane of the solar system, was rather densely filled with bodies of sizes $\lesssim$ 100 km.

When the first pictures of the far side of the Moon were obtained, some surprise was caused by the fact that the large impact craters clearly favored the hemisphere turned toward the Earth. We might conclude from that that the rotation and revolution of the Moon were already linked to one another $\sim 3.8 \times 10^9$ years ago, as they are today.

It cannot be our task here to investigate the details of lunar petrography and geology. Rather, we tackle straightaway the cardinal question: "When and how did the Earth–Moon system lose its 99.35 percent (by mass) of volatile matter (H, He, . . .)—and perhaps also huge quantities of solar matter?"

This very significant fraction of the "primeval matter" has left no kind of traces on the Moon. On the other hand, A. E. Ringwood (1960) has taken the view that the Earth's metallic nickel–iron core was formed by the construction of the Earth out of bodies less than or about 100 km in size and their immediate melting in a *reducing* atmosphere. The mass mentioned above must therefore have "vanished" in a relatively short time, even before the Moon—soon afterward—formed its first crust.

We can now try to weigh up the arguments for and against the capture theory and the fission theory of the origin of the Moon—up to now the two important rivals. To begin with, it is clear that the capture of a "finished" Moon has become unlikely as a result of the age determinations of the lunar rocks. Against the fission theory (due to G. Darwin; we need scarcely even mention the popular version, in which the Pacific Ocean is the scar left by the fleeing Moon), the objection was raised very early on that even the entire present angular momentum of the Earth–Moon system would not be sufficient to allow the primeval Earth to rotate so rapidly that at one point the centrifugal force was larger than gravity (length of day $\approx$ 2.7 hours). To this we could today reply, first of all, that with the loss of significant mass there was certainly also a related loss of angular momentum. However, we could still make no clear statement about the ratio of the two losses. As its density and composition show, the separation of the Moon must have occurred after the separation of the Earth's core and mantle had been essentially completed. A. E. Ringwood sees precisely in this fact a possible explanation for the final expulsion of the Moon from the already rather rapidly rotating Earth: as the heavy nickel–iron sank into the core of the Earth its moment of inertia became smaller and as a result (because of conservation of angular momentum) its rotation became faster. According to our present ideas, the quantitative mechanical theory of the formation of the Earth–Moon system is made extraordinarily difficult by the fact that on the one hand the system at first suffered great loss of mass and angular momentum, but on the other hand was subjected to a bombardment, at first very rapid but slower later, of small bodies, about whose original distribution of orbital elements we know as good as nothing. These are fundamentally the same considerations as were turned critically against the original, too highly schematic versions of the capture and fission theories. The synthesis of them both which is presented here was in fact already extensively hinted at in the work of H. C. Urey, T. Gold, and others.

Several investigators. such as R. A. Lyttleton and W. H. McCrea, go a step further and consider the approximate agreement of the densities of the Earth's mantle, the Moon and Mars—together with the absence or near absence of an iron core in Mars—to be an indication that originally the Earth, Moon, *and* Mars belonged together (and perhaps so did Mercury and Venus). This hypothesis appears to be possible from the point of view of celestial mechanics. The fit of the orbital radii to the geometric series (31.1)—admittedly very approximate—would then be hard to understand.

We must here leave effectively unanswered the questions of the rotation of the planets, the origin of the (actual) satellite systems of the major planets, the formation of the asteroid belt by the disruption of one or several bodies (relatively small even when gathered together), the origin of Pluto, and so on. We adopt here I. Newton's maxim: "Hypotheses non fingo."

Rather, we shall now, somewhat parochially, follow further the evolution of our own Earth. In Section 7 we have already discussed its internal structure, the sequence of geological strata, and continental drift. Furthermore, we have convinced ourselves that the Earth, together with the primeval Moon, which detached itself from the Earth, was formed together with the entire solar system in a relatively short time (a few times 10^6 or 10^7

years) about 4.5 to 4.6 × 10^9 years ago. However, the oldest rocks (W. Greenland) were formed only ~ 3.2 × 10^9 years ago, and so we are reduced to indirect conclusions about the first 10^9 years of the Earth's history.

31.4 Evolution of the Earth and of life

We turn at once to the questions of the origin of the oceans and of the atmosphere, questions which are also of particular interest in regard to the origin and evolution of life. Neither could have belonged to the original store of the Earth; we saw that in the initial stages of the Earth–Moon system (just as for the meteorites) the light elements were already expelled, at least from the inner region of the solar system. The known fact that the noble gases are very rare in the Earth's atmosphere, although on the Sun helium and neon at any rate are among the most abundant elements, shows that our atmosphere cannot be a direct remnant of the solar nebula. Also, quantitative estimates confirm the idea that the ocean and the atmosphere are of *secondary* origin,[3] from volcanic emissions, which supply H_2O, N_2, CO, CO_2, SO_2, . . . , while the most abundant argon isotope A^{40} was produced by the transformation of K^{40} in the Earth's crust and the helium by α decay of the known radioactive elements.

However, this primeval atmosphere still contained no *oxygen,* since that was completely bound up in oxides, silicates, etc., and so was absent in volcanic gases. The formation of O_2 (and with it also ozone, O_3) in the optically thin primeval atmosphere began when water vapor, H_2O, was split by ultraviolet solar radiation (photodissociation) into 2H + O. But, as H. C. Urey (1959) pointed out, this process could provide only about 10^{-3} of the oxygen now present. A denser layer of oxygen (with its associated ozone) would absorb the short-wave solar radiation so that the amount of gas would not increase further.

The additional oxygen in our atmosphere can have been formed only by photosynthesis in living organisms, that is, in connection with their evolution. An estimate made by E. I. Rabinowitch (1951) from the productive power of the existing plant world gives the interesting result that the whole of the oxygen in the atmosphere goes through the process of photosynthesis once in only 2000 years.

The history of the atmosphere is thus intimately bound up with the origin of life. As with cosmogony this problem was for long the domain of mythological concepts. After Friedrich Wöhler had in 1828 done away with the boundary between inorganic and organic matter with his synthesis of

[3]The quite different atmosphere of Venus, consisting primarily of carbon dioxide, clearly arose because the iron originally contained carbon and was partially chemically combined with oxygen. CO_2 could form in this way as a result of heating. In the atmospheres of the major planets, on the other hand, ammonia NH_3 and methane CH_4 could be formed directly out of the solar material during the chemical fractionation.

urea, more recent work in the fields of astrophysics, geology, and biochemistry has indeed still not solved the problem of the origin of life, but has nevertheless brought it further into the domain of scientific enquiry.

The complicated molecules, particularly the nucleic acids and proteins, that are characteristic of the structure of living matter, are not stable in the presence of oxygen. Therefore under existing conditions on the Earth, their formation out of inorganic substances is not possible without the cooperation of living organisms. In the "beginning" they could arise only in an oxygen-free atmosphere.

We have already seen that astonishingly complicated organic molecules are present in the interstellar medium and that the carbonaceous chondrites contain an abundant supply of amino acids and other substances which are necessary for constructing living organisms. On account of their very strong sensitivity in the short-wave ultraviolet ($\lambda < 2900$ Å) such molecules could not last long on the surface of the primitive Earth, which was not shielded by absorption due to atmospheric oxygen and ozone; however, they could collect, for instance, on the bottom of shallow lakes about 10 m deep.

Organic molecules are, nevertheless, still a long way from living organisms. How the first such organisms were created, we cannot determine. The viruses of the present time can propagate themselves only *in* higher organisms, and so we may not regard them as the primitive organisms. Possibly, however, we may look on them as *models* for the early stages of life. Even the simplest living organism forms a system in which two essential functions act together:

(1) The power of reproduction or self-replication—from complex molecules to complete living organisms—is built into the genetic material. This contains the information or the steering mechanism which—like the store of an electronic computer—takes care that an identical molecule or organism is "copied" out of suitable components.

"Errors" in this replication mechanism, which could be brought about by chemical influences or ionizing radiation, or even just by thermal motion of the particles, give rise to *mutations*. The origin of a mutation is thus purely a problem in probability. Once formed, the altered molecule or organism will then reproduce itself *exactly*.

The mutations and also the many possibilities of genetic recombination form the basis for the evolution of living creatures according to the mechanism of natural selection discovered by Charles Darwin (1859). As is only to be expected, many mutations lead to functionally inefficient forms, which therefore "vanish" again immediately. Organisms which are at least functionally efficient in themselves are at once exposed to the influence of their surroundings and (later) of other organisms. In this *struggle for existence*—to use H. Spencer's somewhat unfortunately coined cliché—the molecular complex or organism with the highest reproduction rate wins. This also gives an unambiguous measure of the selection value of a mutation.

(2) With the steering system there must be associated a mechanism for providing the requisite energy and obviously also the necessary raw material. The building up of a complicated structure with a definite form—which is therefore "improbable," and so not formed by pure chance—presupposes a certain amount of information or "know how." According to the basic rules of thermodynamics it is therefore associated with a decrease of the entropy (the entropy is proportional to the logarithm of the probability) or an increase of the negative entropy, or "negentropy," of the system. However, the latter can only be procured by an expenditure of energy, in return for which, as a consequence of the second law of thermodynamics, there is a corresponding increase of entropy in the surroundings (from which the energy was taken).

In recent decades modern biochemistry has succeeded in greatly clarifying the most important types of substances and processes which make possible the miracle of life. Here too we encounter an interplay of *information* and *function*—analogous to the legislative and executive branches of government—in a hierarchy of reaction cycles which together make possible the formation and reproduction of functionally efficient macromolecular or living systems.

The most important carriers of *functions,* for example, the "recognition" of particular (useful) substances, catalysis, the regulation of reaction rates, etc., are the *proteins*. These macromolecules are formed by linear polymerization of amino acids, that is, up to a thousand amino acid residues are joined together in a particular order by the rather rigid peptide bonds to form a polypeptide chain (Figure 31.3). It is very remarkable that in the proteins of *all* living organisms only twenty particular amino acid residues occur.

The polypeptide chains can be folded in many different ways and may in particular be tangled into a ball. The form of the ball is then uniquely determined by the sequence of amino acids (R_1, R_2, . . .) mostly by means of noncovalent bonds. The form of the molecule so determined, and the corresponding distribution of its interaction forces, enables it to specialize its activity to quite particular substances or chemical reactions. A group of proteins which are particularly important in this sense are the *enzymes*.

As can easily be calculated, so many different proteins can theoretically be produced by different orderings of twenty amino acids in chains of up to 1000 members that even a trial assortment would choke the whole universe! We can thus without more ado understand the great diversity of organic life; however, it must clearly be limited by an ordering influence.

The carriers of this *information,* of how and what should be made, are the *nucleic acids*. These are formed by linear polymerization (stringing out in a row) of *nucleotides*. These in turn are each formed out of a sugar, a phosphate residue, and a nitrogenous base.

Let us consider first the particularly important information carrier, *d*esoxyribo*n*ucleic *a*cid, or DNA for short: here desoxyribose is used as

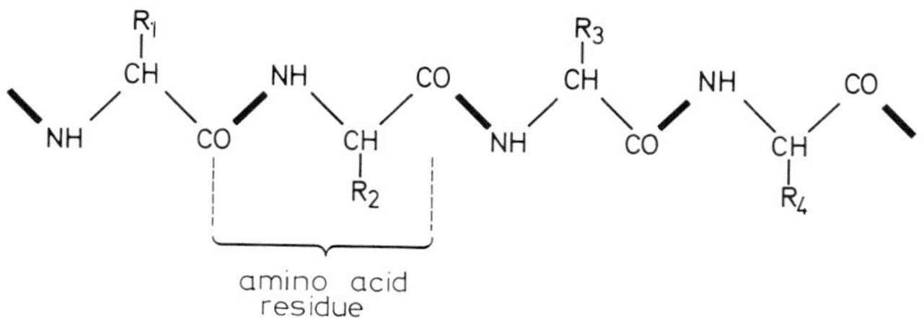

Figure 31.3. A protein molecule. This is formed by the linear connection of up to a thousand amino acid residues by means of the rather rigid peptide bonds (thick (valence) lines in figure). Each of R_1, R_2, . . . signifies an amino acid side-chain, for example, H in glycyl, CH in alanyl, and more complicated organic side chains in other amino acids. Only twenty particular amino acids appear in the proteins of all living organisms.

the sugar. However, it is more important that (obviously, in view of the limitation mentioned) only *four* different nucleotides are used, those with bases

$$\textit{A}\text{denine}, \textit{G}\text{uanine}, \textit{C}\text{ytosine}, \textit{T}\text{hymine}, \tag{31.6}$$

which are denoted briefly by their initial letters.

*R*ibo*n*ucleic *a*cid, or RNA, which is constructed according to similar principles, plays an important role in the transmission and exploitation of the information coded in the DNA (as in the punched tape of an electronic computer). However, interesting though these things are, we cannot discuss them further here and we turn at once to the decipherment of the *genetic code* by J. H. Matthaei, M. W. Nirenberg, S. Ochoa, H. G. Khorona, and others (1961), which we must without doubt count among the most brilliant achievements of recent natural science.

The "genetic alphabet" of DNA has only four "letters," namely, the four nucleotides (31.6), A, G, C, T. A sequence of any three nucleotides, a word of the genetic code so to speak, prescribes only *one* of the twenty amino acid residues. The sequence of the nucleic acid triplets along the DNA filament forms the *pattern,* so to speak, for the production of a corresponding protein with a completely determined sequence of amino acid residues. Since the genetic code could form altogether $4^3 = 64$ words, while only 20 amino acid residues are prescribed, nature "allows" some amino acid residues to be represented by several words; however, some words are used as punctuation marks or are not used at all. In order that a DNA molecule can both supply a pattern for the production of a particular protein and also replicate itself, it is formed out of two complementary strands which are arranged in the shape of a *double helix* (F. H. C. Crick and J. D. Watson, 1953). They can be separated from each other when required—like a zipper—and then made complete again (Figure 31.4).

This mechanism, which we can only hint at here, allows us to understand in principle the replication of particular systems of molecules and so

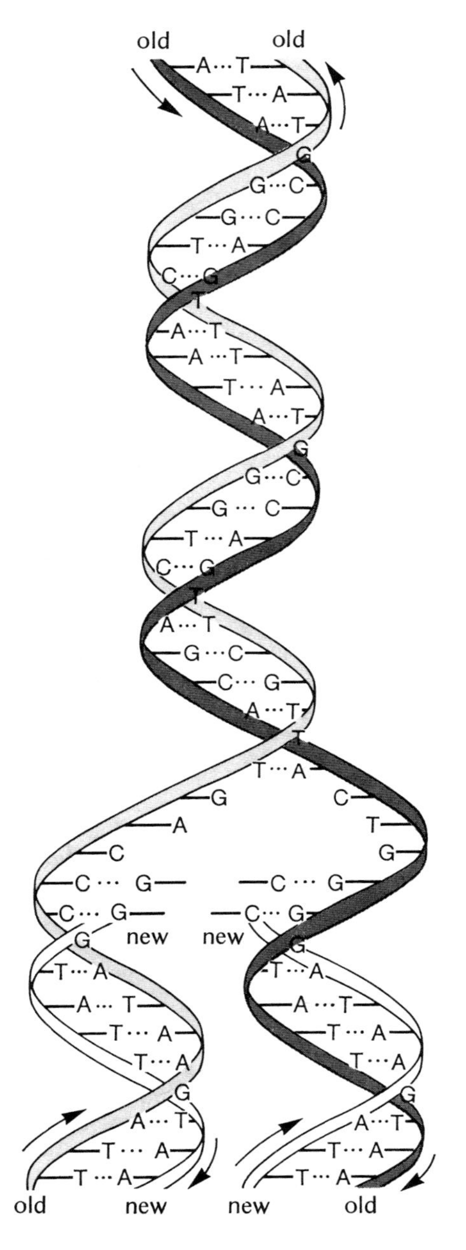

Figure 31.4
The DNA molecule is a double helix, that is, a sort of spiral staircase. The "banisters" of the staircase are formed of the phosphate and desoxyribose (sugar) groups (not indicated explicitly in the figure). The "steps" of the staircase are formed of pairs of the bases (31.6), in which specifically

*A*denine – *T*hymine
or
*G*uanine – *C*ytosine

are bonded to one another (upper part of figure). For replication, the two complementary strands of the double helix are separated like a zipper. Then to each base is attached the corresponding one according to the above scheme, and (lower part of the figure) two identical DNA double helices are formed.

of particular species. Similarly it clarifies the origin of mutations, by perturbations in word order in the DNA filament.

But how does the origin of life fare, and the evolution of "viable" structures (perhaps of the most primitive single cells) by natural selection in Charles Darwin's sense? The problem which lies behind this question, as to whether present-day physics forms a sufficient basis for biology or whether we must appeal to a mysterious "vis vitalis" or "élan vital," was, we may say, solved in principle by M. Eigen (1971). He first made it clear that a solution could not be expected on the basis of classical (equilibrium or quasi-equilibrium) thermodynamics. The interplay of proteins and nucleic acids which is required for each appearance of life, even in the most primitive form imaginable, is so complicated that its realization under the given circumstances can be described as impossible.

In fact even life is only possible with an energy supply, change of state, etc., that is, as a (quasi-) stationary *irreversible* thermodynamic process. By applying a stability principle derived for such processes by I. Prigogine and P. Glansdorff (1971) and by introducing the "selection value" of a mutation—essentially the excess of the multiplication rate over the destruction rate—M. Eigen now shows that the formation of self-organizing systems and the optimization processes of evolution are none other than the *necessary* properties of a particular reaction system which is sufficiently far from thermodynamic equilibrium. Only the *beginning*—each mutation—need be ascribed to chance; all the rest is physics.

Recently H. Kuhn (1973) has proposed a rather detailed model for the self-organization of macromolecules and the evolution of the simplest living organisms. The role of the instability mentioned above—as a "drive" for selection and evolution—is assumed in this model by a (postulated) periodicity in the conditions in the surroundings. As further important steps H. Kuhn stresses the formation of aggregates of interlocking molecules and then the step-by-step achievement of greater and greater independence of the system, particularly by the (catalytic) production of a shell. Unfortunately we must here deny ourselves any further explanation or justification of these very interesting considerations.

The uniformity of the genetic code and the use of the same building blocks for the DNA and the proteins of the most primitive bacteria and of man suggests that the origin of life was bounded in space and time.

What do we know about the oldest living organisms? About 1965 the biochemist M. Calvin succeeded in detecting in rocks, whose radioactively determined ages stretched back as far as 3.1×10^9 years ago, hydrocarbons such as pristane and phytane, which we may perhaps regard as (relatively stable and yet sufficiently complex) decay products of organisms. Although several other observations support this interpretation, there still remains above all the critical question of whether these molecules could not have been the result of a prebiotic synthesis (like that in the CI meteorites).

Then E. S. Barghoorn, J. W. Schopf, P. E. Cloud, and others ($\sim$ 1965) used microscope and electron-microscope techniques to find the remains of primitive living organisms in pre-Cambrian flints. The oldest, the so-called Fig Tree Formation in South Africa, with an age of $\sim 3 \times 10^9$ years, shows traces which may be interpreted as forerunners of the present-day bacteria and blue-green algae. The Gun Flint Formation in the USA, with an age of $\sim 2 \times 10^9$ years, already contains fossils which are attributed to primitive algae and plants.

Following P. E. Cloud (1972) and others, we try to outline a coherent picture of the early history of the Earth and its atmosphere, in connection with the evolution of life and corresponding to the present state of research in the areas of palaeontology, geology, geochronology, etc. It is obvious that for many of the details we must be satisfied with a reference to the technical literature.

As we saw, the Earth–Moon system was formed 4.6 to 4.5×10^9 years ago. However, while the rocks of the highlands or terrae of the Moon are as old as that, the oldest terrestrial rocks are significantly younger. Several places have been found which give ages of up to 3.6×10^9 years; recently (1972) gneisses have been found on the west coast of Greenland with ages of 3.70 to 3.75×10^9 years. Clearly, immediately before this large parts of the Earth's crust were melted again. We may imagine that this was connected with the end of the period of violent meteorite bombardment.

Immediately afterward volcanic exhalations (H_2O, H_2, CO, CO_2) formed the first rudiments of the ocean and the atmosphere, which however still did not contain oxygen. The rarity of the noble gases—cosmically very abundant—shows that our atmosphere does not stem from the original solar material. Such a *reducing* atmosphere was one of the preconditions for the origin of organic life; many substances of importance to life are destroyed by oxygen. However, there was therefore at that time still no ozone layer which would have protected life from the lethal ultraviolet ($\sim$ 2400–2900 Å) radiation of the Sun. The most primitive living organisms must therefore have been formed in water and in fact at depths which could still be reached by a few rays of light but not by the deadly ultraviolet.

The first organisms were certainly still heterotrophic, that is, they lived entirely on existing organic substances. An autotrophic life style, with the use of initially inorganic substances, first became possible later with the "invention" of photosynthesis by the first plants.

In the period from $\sim$ 3.2 to 2.0×10^9 years ago, it is clear—judging from the available evidence—that there were only *procaryotes,* that is, single-celled organisms without obvious nuclear membranes or chromosomes, like the present-day bacteria and blue-green algae. The small amount of oxygen which these produced remained in the hydrosphere and (according to P. E. Cloud 1968) was mostly used to transform ferrous to ferric oxide. Characteristic of this geological epoch, 2.1 to 2.0×10^9 years ago, are the *banded iron formations* BIF. According to P. E. Cloud (1973), their formation can be explained by the production in an O_2-free atmosphere of Fe^{++} solutions which are carried into the sea and there, under the action of oxygen-producing microscopic algae, periodically become Fe^{+++} compounds.

Then, 2×10^9 years ago, the first enzymes which could interact with oxygen evolved. There arose the *eucaryotes,* that is, single-celled organisms with nuclear membranes and chromosomes. The "invention" of sexuality engendered a great acceleration of biological evolution. Soon, so much oxygen was produced that it escaped into the atmosphere. Geologically, this epoch is characterized by the strongly oxidized sediments of the Red Beds and by deposits of calcium carbonate and dolomite.

Something fundamentally new happened when, 0.68 to 0.60×10^9 years ago, the oxygen content of our atmosphere reached a few percent of its present value. The oxygen was sufficient for the formation of an *ozone* layer, which absorbed the lethal ultraviolet to such an extent that now the surface of the water was inhabitable, and soon the solid ground was too.

In the *Phanerozoic* (the time of "emerging life"), about 700 to 600 million years ago, the *metazoa* appeared, that is, many-celled creatures with complicated internal structure, although at first only in soft-bodied forms. From the *Cambrian* on, that is, 600 million years ago, there appeared the multiplicity of shell-bearing metazoa. Now begins the series of geological strata already described in Table 7.2, whose fossils teach us in detail about the further evolution of the world of plants and animals.

About 420 million years ago, from the *Silurian* to the *Devonian* (Table 7.2) the first extensive forests came into existence, and in the *Carboniferous,* from whose fossil forests we still benefit today, the present level of oxygen was undoubtedly reached, if not indeed for a time surpassed.

If we may try to reduce it to its simplest elements, at the basis of the whole evolution of life there is clearly the principle, by storage and retrieval of more and more *information,* to build more complex systems which more and more successfully apply their principle of ordering (production of negentropy) in opposition to the natural tendency of the second law toward the production of statistical thermodynamic equilibrium, that is, maximum disorder (increase of entropy). This evolution leads from the fixing of genetic information already by procaryotes to the construction of instinctive, that is, programmed behavior patterns, thence further to memory and the beginnings of intelligent behavior. The development of man (a few million years ago)—compared with simpler creatures—rests essentially upon enormously improved techniques for the storage and employment of information, firstly through speech, then through writing (of every sort), and lastly through the invention of the electronic information-storage device, which may itself translate its "knowledge" into action (from computing to electronically controlled factories).

This "struggle against the entropy law," which becomes ever more gigantic, is inevitably linked with ever greater energy expenditure. An organism is in fact, as we say, in a state of flow equilibrium. That is to say, from the food—and in the case of plants from solar radiation—energy is continually extracted, which is partly necessary for building up the system, and partly goes to waste as heat, as in any heat engine. Man can realize his higher evolution according to the principle described, only if he renders natural sources of energy usable in ever greater measure. Besides the muscle-power of man and animals, the forces of wind and water were employed. The next great step was the use of fossil fuels, coal and oil, in heat engines. It is certainly no accident that our age of automation has also brought with it the release of nuclear energy.

As regards the question of the *future* of the Sun and with it life on the Earth, the theory of stellar evolution allows us to give a quite definite answer. In a few billion years the Sun will move upward to the right in the

HR diagram and will become a red giant star. Its radius will thereby grow enormously, its bolometric magnitude will increase by several magnitudes, and the temperature of the Earth's surface will rise considerably above the boiling point of water, so that the oceans will evaporate. This means without question the end of organic life on the Earth.

On the other hand, one often poses the question as to whether there is life on other astronomical bodies. At present this question is meaningful only if by life we understand the occurrence of organisms whose structure bears some resemblance to terrestial living beings. We scarcely need to say that their environmental conditions are confined to a fairly narrow range of possibilities.

In our planetary system, one can at all events think of Mars. Its rarefied atmosphere contains at any rate small amounts of the necessary gases, but it offers no ultraviolet protection; the temperature is a little lower than on the Earth. So the occurrence of extremely primitive organisms seems at least a possibility for discussion, but in our present state of knowledge not very probable.

In our Galaxy and in others there are numberless stars indistinguishable from our Sun. There is nothing against the assumption that some of these G-type stars possess planetary systems, and it seems entirely plausible that here and there in such a system a planet offers conditions on its surface similar to those on the Earth. Why should not living beings have evolved there as well?

From the study of cosmic structures and cosmic evolution we returned to problems of our planetary system, of the origin of life, and of our existence. Our historical notes showed how our deeper understanding of the universe is intimately connected with the evolution of the human mind.

What is more admirable and astonishing, all the new facts which have been brought to light or the fact that mankind has developed the ability to understand them? This is a new version of the old problem of "macrocosmos and microcosmos." Whether we penetrate into the depths of the universe or whether we search the mysteries of the human mind, on both sides we view a

New Cosmos.

Physical Constants and Astronomical Quantities[1]

1. Astronomical quantities

Astronomical unit AU $= 1.496 \times 10^{13}$ cm
(semi-major axis of Earth's orbit)
Parsec pc $= 3.085 \times 10^{18}$ cm $= 206\,265$ AU $= 3.26$ light years

A segment of length r AU (or $r/206\,265$ pc) at a distance of r pc subtends an angle of 1 second of arc at the observer; a segment of length 0.291 pc at a distance of 1 kpc subtends an angle of 1 minute of arc.

	1 day	$= 86400$ s
	Sidereal year $= 365^{\mathrm{d}}.256$	$= 3.1558 \times 10^7$ s
	Tropical year $= 365^{\mathrm{d}}.242$	$= 3.1557 \times 10^7$ s
Earth	Equatorial radius	$= 6.378 \times 10^8$ cm
	Mass	$= 5.9734 \times 10^{27}$ g
Sun	Radius $R_\odot$	$= 6.96 \times 10^{10}$ cm
	Mass $\mathfrak{M}_\odot$	$= 1.989 \times 10^{33}$ g
	Surface gravity $g_\odot$	$= 2.736 \times 10^4$ cm/s^2
	Total luminosity $L = 4\pi R^2\ \pi F$	$= 3.82 \times 10^{33}$ erg/s
	Effective temperature T_{eff}	$= 5770$ K

One magnitude corresponds to a luminosity ratio 2.512 (antilog 0.4). Absolute magnitudes refer to a distance of 10 pc.

[1]Cf. C. W. Allen, *Astrophysical Quantities*, third edition, 1973; also B. N. Taylor, W. H. Parker, and D. N. Langenberg, *Rev. Mod. Phys.*, **41**, 477ff, 1969.

2. Units

Length	1 mile = 1.609 km; 1 foot = 30.48 cm; 1 inch = 2.54 cm; 1 angstrom = 1 Å = 10^{-8} cm.
Power and energy	1 watt (MKS system) = 1 W = 1 J/s = 10^7 erg/s. 1 electronvolt (1 eV) = 1.602×10^{-12} erg corresponding to a wavenumber $\tilde{\nu}$ = 8065.5 cm^{-1} or Kayser (ky) or to a wavelength λ = 12 398 Å. 1 unit atomic weight = 1.492×10^{-3} erg = 931.48×10^6 eV.
Pressure	1 atmosphere (atm) = 760 mm Hg (torr) = 1.0132×10^6 dyn/cm^2 or microbar (μbar).
Temperature	Absolute temperature T K = 273.15 + t C. Thermal energy kT = 1 eV corresponds to T = 11 605 K.

Prefixes for powers of ten

10^3 kilo	10^{-3} milli
10^6 mega	10^{-6} micro
10^9 giga	10^{-9} nano

Mathematical constants

$\pi = 3.1416$; $e = 2.7183$; $1/M = \ln 10 = 2.3026$.

$\frac{4\pi}{3} = 4.1888$; $M = \log e = 0.4343$.

1 radian = 57°.296 and 1° = 0.017 453 rad.

3. Physical constants

Light speed	c	$= 2.997\,92456 \times 10^{10}$ cm/s
Gravitation constant	G	$= 6.673 \times 10^{-8}$ dyn cm^2/g^2
Planck's constant	h	$= 2\pi\hbar = 6.6262 \times 10^{-27}$ erg s
Electron charge	e	$= 4.8032 \times 10^{-10}$ ESU $= 1.6022 \times 10^{-19}$ coulomb (C)
Mass: Unit atomic weight	M	$= 1.6605 \times 10^{-24}$ g
Proton	M_p	$= 1.6726 \times 10^{-24}$ g
Electron	m	$= 9.1095 \times 10^{-28}$ g
Classical electron radius	e^2/mc^2	$= 2.818 \times 10^{-13}$ cm
Compton wavelength	h/mc	$= 2.426 \times 10^{-10}$ cm = 0.02426 Å
Reciprocal fine structure constant	α^{-1}	$= \frac{hc}{2\pi e^2} = 137.036$
Boltzmann constant	k	$= 1.3806 \times 10^{-16}$ erg/K
Avogadro number	N	$= 6.0221 \times 10^{23}$ particles per mole.
Gas constant	$\mathcal{R}$	$= 8.314 \times 10^7$ erg/K mol
Rydberg constant	R_∞	= 109 737.3 cm^{-1} = 1/911.27 Å
Radiation constants	σ	$= 5.67 \times 10^{-5}$ erg/cm^2 s K^4
	c_2	= 1.4388 cm K

Hydrogen atom

Atomic weight	μ	$= 1.0080$
Mass	M_H	$= 1.673 \times 10^{-24}$ g
Ionization potential	χ_H	$= 13.60$ eV
Rydberg constant	R_H	$= 109\,677.6\ \text{cm}^{-1}$
Bohr radius	a_0	$= 0.529 \times 10^{-8}$ cm or 0.529 Å.

Bibliography

The bibliography is limited to important books, journals, etc.. that should serve for the further study of particular topics and problems. References to articles on such subjects published in periodicals, observatory publications, etc., can be found in *Astronomischer Jahresbericht* or (from 1969) in *Astronomy and Astrophysics Abstracts* (see below). Historical, national, and priority considerations have to be set aside here. An asterisk * is used to indicate works that may prove suitable for the newcomer, but the distinction should not be overstressed.

[Apart from minor revisions and additions, the bulk of the bibliography is the same as in the German edition. Where an English translation of any work in another language is available, only the translation is listed here. Also a number of other books of more general scope that are not readily accessible to English readers are here omitted but, so far as possible, replaced by roughly corresponding books in English. (Translator)]

Introduction to Astronomy in General, Including Astrophysics

Abell, G.: *Exploration of the Universe*. New York: Holt, Rinehart, and Winston; 3rd edn. 1974.

*Gingerich, O. (ed.): *New Frontiers in Astronomy* (Readings from *Scientific American*). San Francisco: W. H. Freeman Co. 1975.

*Hoyle, F.: *Highlights in Astronomy*. San Francisco: W. H. Freeman Co. 1975. (Published in United Kingdom and Commonwealth by Heinemann Educational Books as *Astronomy today*.)

Lynds, B. T.: *Struve's Elementary Astronomy*. 2nd edn. Oxford University Press 1968.

Menzel, D. H., F. L. Whipple, and G. de Vaucouleurs: *Survey of the Universe*. Prentice Hall Inc. 1970.

Pecker, J. C., and E. Schatzman: *Astrophysique Générale*. Paris: Masson Cie 1959.

Smart, W. M.: *The Riddle of the Universe*. London: Longmans 1968.

An invaluable synopsis of the most important numerical quantities in astronomy and physics (in particular atomic physics and spectroscopy):

Allen, C. W.: *Astrophysical Quantities*. 3rd edn. London: Athlone Press 1973.

Charts of the Sky

Bečvár, A.: *Atlas Coeli* 1950.0. American edition: *Skalnate Pleso Atlas of the Heavens,* Field edn. Camb., Mass.: Sky Publ. Corp. 1969.

Norton, A. P.: *A Star Atlas and Reference Handbook for Students and Amateurs*. 16th edn. Edinburgh: Gall and Inglis 1973.

Histories of Astronomy

Abetti, G.: *The History of Astronomy*. London: Sidgwick and Jackson 1954.

Berry, A.: *A Short History of Astronomy*. New York: Dover 1961.

King, H. C.: *The History of the Telescope*. London: Griffin 1955.

—— *Exploration of the Universe*. London: Secker & Warburg 1964.

*Pannekoek, A.: *A History of Astronomy*. New York: Signet Science Library 1964 also London: Allen & Unwin 1961.

Struve, O., and V. Zebergs: *Astronomy of the 20th Century*. New York-London: Macmillan 1962.

Short Reference Books for the Nonspecialist

Baker, R. H., and L. W. Frederick: *An Introduction to Astronomy*. 7th edn. Princeton, N.J.: Van Nostrand 1968.

*Bizony, M. T. (ed.): *The New Space Encyclopaedia*. Horsham, Sussex: The Artemis Press 1969.

*Ernst, B., and T. E. de Vries: *Atlas of the Universe* (English translation). London: Nelson 1961.

*Weigert, A., and H. Zimmerman. Trans. J. Home Dickson: *Concise Encyclopaedia of Astronomy* (revised edition of *ABC of Astronomy*). London: Hilger 1976.

Handbooks: A Selection

Encyclopaedia of Physics (ed. S. Flügge). Vols. 50–54 (Astrophysics I–V). Berlin-Göttingen-Heidelberg: Springer-Verlag 1958–1962. With contributions in German, English, and French.

Stars and Stellar Systems (ed. G. P. Kuiper and B. M. Middlehurst). 9 volumes. The University of Chicago Press 1960 onwards.

The Solar System (ed. G. P. Kuiper and B. M. Middlehurst). 5 volumes. The University of Chicago Press 1953–1966.

Periodical publications:

Annual Review of Astronomy and Astrophysics (Vol. 1, 1963). Palo Alto: Ann. Reviews.

Advances in Astronomy and Astrophysics (Vol. 1, 1962). New York-London: Academic Press.

Transactions of the International Astronomical Union. Dordrecht: D. Reidel Publ. Co.

Annual Review of Literature

Astronomischer Jahresbericht. Berlin: W. de Gruyter. Up to 1968; continued from 1969 (vol. 1) as: *Astronomy and Astrophysics Abstracts*. Berlin-Heidelberg-New York: Springer-Verlag.

Journals: A Selection

a) *Popular magazines*

**New Scientist*. London: IPC Magazines.
**Sky and Telescope*. Cambridge, Mass.: Sky Publishing Corp.
**Scientific American*. New York.

b) *Professional journals*

The Astrophysical Journal. Chicago, Ill.: The University of Chicago Press.
The Astronomical Journal. New York: American Inst. of Physics Inc.
Astronomy and Astrophysics (A European journal). Berlin-Heidelberg-New York: Springer-Verlag (from 1969).
Monthly Notices of the Royal Astronomical Society. London: Blackwell Sci. Publ.
Soviet Astronomy = Astronomicheskii Zhurnal (Engl. translation). New York: Amer. Inst. of Physics.
Publications of the Astronomical Society of the Pacific. San Francisco, Calif.
Astronomische Nachrichten. Berlin: Akademie-Verlag.
Astrophysics and Space Science. Dordrecht: D. Reidel Publ. Co.
Space Science Reviews. Dordrecht: D. Reidel Publ. Co.
Icarus. Int. Journ. of Solar System Studies. New York and London: Academic Press.
Solar Physics. Dordrecht: D. Reidel Publ. Co.
Publications of the Astronomical Society of Japan. Tokyo: Maruzen Co. Nihonbashi.
Proceedings of the Astronomical Society of Australia. Sydney University Press.

Part I: Classical Astronomy

1. Stars and Men: Observing and Thinking. Historical Introduction to Classical Astronomy

With reference to the introductory historical Sections 1, 10, 22, as well as Sections 2, 3, see "Histories of Astronomy."

2. Celestial Sphere: Astronomical Coordinates: Geographic Latitude and Longitude

3. Motion of the Earth: Seasons and the Zodiac: Day, Year and Calendar

4. Moon: Lunar and Solar Eclipses

See "Introduction to Astronomy in General," together with:
Danjon, A.: *Astronomie Générale*. J. and R. Sennac. Paris 1952–1953.
*Kourganoff, V.: *Astronomie Fondamentale Élémentaire*. Paris: Masson 1961.
Smart, W. M.: *Text-book on Spherical Astronomy*. 5th edn. Cambridge University Press 1962.

Woolard, E. W., and G. M. Clemence: *Spherical Astronomy*. New York-London: Academic Press 1966.
The Astronomical Ephemeris. London and Washington (yearly publication), and
Explanatory Supplement to the Astronomical Ephemeris, (appearing at irregular intervals, most recently in 1961. 3rd impression, with amendments, 1974).

5. Planetary System

6. Mechanics and Theory of Gravitation

Brouwer, D., and G. M. Clemence: *Methods of Celestial Mechanics*. New York-London: Academic Press 1961.
Brown, E. W., and C. A. Shook: *Planetary Theory*. Cambridge University Press 1933, also New York: Dover 1964.
Brown, E. W.: *An Introductory Treatise on the Lunar Theory*. New York: Dover 1960.
Leech, J. W.: *Classical Mechanics*. London: Methuen Co. 1958. (A brief, clear introduction.)
Poincaré, H.: *Méthodes Nouvelles de la Mécanique Céleste*. 3 volumes. Paris 1892–1899, also New York: Dover.
—— *Leçons de Mécanique Céleste*. 3 volumes. Paris 1905–1910.
Smart, W. M.: *Celestial Mechanics*. London: Longmans 1953.

7. Physical Constitution of Planets and Satellites

Baldwin, R. B.: *The Measure of the Moon*. The University of Chicago Press 1963.
Bott, M. H. P.: *The Interior of the Earth*. London: E. Arnold 1971.
Brandt, J. C., and P. Hodge: *Solar System Astrophysics*. New York: McGraw-Hill Book Co. 1964.
Calder, N.: *Restless Earth*. London: BBC Publ. 1972.
Dollfus, A. (ed.): *Moon and Planets I, II*. Amsterdam: North Holland Publ. Co. 1967/68.
Dollfus, A. (ed.): *Surfaces and Interiors of Planets and Satellites*. New York: Academic Press 1970.
Fielder, G.: *Structure of the Moon's Surface*. Oxford-London-New York-Paris: Pergamon Press 1961.
Kopal, Z. (ed.): *Physics and Astronomy of the Moon*. New York-London: Academic Press 1961.
Kuiper, G. P. (ed.): *The Atmospheres of the Earth and Planets*. 2nd edn. The University of Chicago Press 1952.
Kuiper, G. P. and B. M. Middlehurst (eds.): *Planets and Satellites*. (Vol. III of *The Solar System*, see under "Handbooks".)
Runcorn, S. K., and H. C. Urey (eds.): *The Moon* (IAU Sympos. 47). Dordrecht: D. Reidel Publ. Co. 1972.
Sagan, S., T. C. Owen, and H. J. Smith (eds.): *Planetary Atmospheres* (IAU Sympos. 40). Dordrecht: D. Reidel Publ. Co. 1971.
Smith A. G., and T. D. Carr: *Radio Exploration of the Planetary System*. Princeton, N.J.: Van Nostrand Co. 1964.
Urey, H. C.: *The Planets, Their Origin and Development*. New Haven: Yale University Press 1952.
Wilson, J. Tuzo (ed.): *Continents Adrift*. (A Scientific American book.) W. H. Freeman and Co. 1972.

La Physique des Planètes, 8th Internat. Astrophysics Colloquium, Liège 1963.
**New Science in the Solar System* (a New Scientist special review), IPC Magazines Ltd. 1975.
**The Solar System* (a Scientific American book). W. H. Freeman and Co. 1976.

8. Comets, Meteors and Meteorites, Interplanetary Dust; Structure and Composition

Chebotarev, G. A., E. I. Kazirmichak, and B. G. Marsden (eds.): *The Motion, Evolution of Orbits and Origin of Comets* (IAU Sympos. 45). Dordrecht: D. Reidel Publ. Co. 1972.
Hawkins, G. S.: *Meteors, Comets and Meteorites.* New York-San Francisco-Toronto-London: McGraw-Hill Book Co. 1964.
Kuiper, G. P., and E. Roemer (eds.): *Comets.* Lunar and Planetary Laboratory. Tucson, Arizona 1972.
Lovell, A. C. B.: *Meteor Astronomy.* Oxford: Clarendon Press 1954.
Mason, B.: *Meteorites.* New York: J. Wiley Sons 1962.
Richter, N. B.: *The Nature of Comets* (English translation). London: Methuen 1963.
*Wood, J. A.: *Meteorites and the Origin of the Planets.* New York: McGraw-Hill Book Co. 1968.
La Physique des Comètes, Internat. Astrophysics Colloquium, Liège 1952.

9. Astronomical and Astrophysical Instruments

Stars and Stellar Systems (see under "Handbooks"), Vol. I and II.
Christiansen, W. N., and J. A. Högbom: *Radio Telescopes.* Cambridge University Press 1969.
Laustsen, S., and A. Reiz (eds.): *Auxiliary Instrumentation for Large Telescopes.* ESO/CERN Conference. Geneva: ESO 1972.
Le Galley, D. P. (ed.): *Space Science.* New York-London: Wiley 1963.
Miczaika, G. R., and W. M. Sinton: *Tools of the Astronomer.* Harvard University Press 1961.
Selwyn, E. W. H.: *Photography in Astronomy.* Rochester, N.Y.: Eastman Kodak 1950.
West, R. M. (ed.): *Large Telescope Design.* ESO/CERN Conference. Geneva: ESO 1971.
Astronomical Observations with Television-Type Sensors, Symposium held at the University of British Columbia, Vancouver 8, Canada, 1973.

Part II: Sun and Stars. Astrophysics of Individual Stars

10. Astronomy + Physics = Astrophysics. Historical Introduction

See "Histories of Astronomy."

11. Radiation Theory

Chandrasekhar, S.: *Radiative Transfer.* Oxford: Clarendon Press 1950, also New York: Dover 1960.
Unsöld, A.: *Physik der Sternatmosphären.* Mit besonderer Berücksichtigung der Sonne. 2nd edn. Berlin-Göttingen-Heidelberg: Springer-Verlag 1955.

Woolley, R. v. d. R., and D. W. N. Stibbs. *The Outer Layers of a Star*. Oxford University Press 1953.

12. The Sun

*Ellison, M. A.: *The Sun and Its Influence*. 2nd edn. London: Routledge and Kegan Paul 1959.
de Jager, C. (ed.): *The Solar Spectrum* (Sympos. Utrecht 1963). Dordrecht: D. Reidel Publ. Co. 1965.
*Kiepenheuer, K. O.: *Die Sonne*. Berlin-Göttingen-Heidelberg: Springer-Verlag 1957.
Kuiper, G. P. (ed.): *The Sun. The Solar System,* Vol. I. The University of Chicago Press 1953.
Menzel, D. H.: *Our Sun*. Philadelphia-Toronto: Harvard Books 1949.
Unsöld, A.: *Physik der Sternatmosphären* (see under 11).
Waldmeier, M.: *Ergebnisse und Probleme der Sonnenforschung*. 2nd edn. Leipzig: Akad. Verlagsgesellschaft 1955.
Zirin, H.: *The Solar Atmosphere*. Waltham-Toronto-London: Blaisdell 1966.

13. Apparent Magnitudes and Color Indices of Stars

14. Distances, Absolute Magnitudes and Radii of the Stars

15. Classification of Stellar Spectra: Hertzsprung–Russell Diagram and Color-Magnitude Diagram

In addition to the works listed under ''Introduction to Astronomy in General, Including Astrophysics'' and ''Handbooks'' are:

*Dufay, J.: *Introduction to Astrophysics: The Stars*. (English translation). London: George Newnes Ltd, and New York: Dover, 1964.
Fehrenbach, C. and B. E. Westerlund: *Spectral Classification and Multicolor Photometry*. (IAU Sympos. 50). Dordrecht: D. Reidel Publ. Co. 1973.
Hoffleit, D. (ed.): *Catalogue of Bright Stars*. Yale University Press 1965.
Morgan, W. W., P. C. Keenan, and E. Kellman: *An Atlas of Stellar Spectra*. With an outline of spectral classification. The University of Chicago Press 1942.
Russell, H. N., R. S. Dugan, and J. Q. Stewart: *Astronomy*. 2 volumes. New York: Ginn 1926 (revised 1945, 1938).
Strand, K. Aa. (ed.): *Basic Astronomical Data*. (Vol. III of *Stars and Stellar Systems,* see above.) [Also useful for Section 16.]

16. Double Stars and the Masses of the Stars

Aitken, R. G.: *The Binary Stars*. New York: Dover 1964.
Batten, A. H.: *Binary and Multiple Systems of Stars*. Pergamon 1973.
Kopal, Z.: *Close Binary Systems*. London: Chapman and Hall 1959.
Russell, H. N., and C. E. Moore: *The Masses of the Stars*. 2nd edn. Astrophys. Monographs. The University of Chicago Press 1946.

17. Spectra and Atoms: Thermal Excitation and Ionization

*Born, M.: *Atomic Physics* (Engl. translation, revised by R. J. Blin-Stoyle and J. M. Radcliffe). London: Blackie 1969.
Condon, E. U., and G. H. Shortley: *The Theory of Atomic Spectra*. Cambridge University Press 1963.

Griem, H. R.: *Plasma Spectroscopy*. New York: McGraw-Hill 1964.
Herzberg, G.: *Atomic Spectra and Atomic Structure*. New York: Dover.
Kuhn, H. G.: *Atomic Spectra*. 2nd edn. London: Longman 1971.
Marr, G. V.: *Plasma Spectroscopy*. Amsterdam-London-New York: Elsevier Publ. Co. 1968.
Moore, C. E.: *A Multiplet Table of Astrophysical Interest*, 1959; and *Atomic Energy Levels* (several volumes, 1949 onward). Washington: Nat. Bureau of Standards.

18. Stellar Atmospheres: Continuous Spectra of the Stars

19. Theory of Fraunhofer Lines: Chemical Composition of Stellar Atmospheres

Aller, L. H.: *Astrophysics. I. The Atmospheres of the Sun and Stars*. 2nd edn. *II. Nuclear Transformations. Stellar Interiors and Nebulae*. New York: Ronald 1963 and 1954.
Ambartsumian, V. A., E. R. Mustel *et al.*: *Theoretical Astrophysics*. London: Pergamon Press 1958.
Mihalas, D.: *Stellar Atmospheres*. San Francisco: W. H. Freeman Co 1970.
Thackeray, A. D.: *Astronomical Spectroscopy*. London: Eyre & Spottiswoode 1961.
Unsöld, A.: *Physik der Sternatmosphären* (see under Section 11).
Abundance Determinations in Stellar Spectra (IAU Sympos. 26). London-New York: Academic Press 1966.

20. Motions and Magnetic Fields in the Solar Atmosphere and the Solar Cycle

See also under Section 12: The Sun.
Alfvén, H., and C.-G Fälthammar: *Cosmical Electrodynamics*. 2nd edn. Oxford: Clarendon Press 1963.
Billings, D. E.: *A Guide to the Solar Corona*. New York-London: Academic Press 1966.
Brandt, J. C.: *Introduction to the Solar Wind*. San Francisco: W. H. Freeman Co. 1970.
Bray, R. J., and E. R. Loughhead: *Sunspots*. London: Chapman and Hall 1964.
—— *The Solar Granulation*. London: Chapman and Hall 1967.
—— *The Solar Chromosphere*. London: Chapman and Hall 1974.
Cowling, T. G.: *Magnetohydrodynamics*. New York-London: Interscience Publ. 1957.
Howard, R. (ed.): *Solar Magnetic Fields* (IAU Sympos. 43). Dordrecht: D. Reidel Publ. Co. 1971.
Hundhausen, A. J.: *Coronal Expansion and Solar Wind*. Berlin-Heidelberg-New York: Springer-Verlag 1972.
Kundu, M. R.: *Solar Radio Astronomy*. New York-London-Sydney: Interscience Publ. 1965.
*Smith, A. G.: *Radio Exploration of the Sun*. Princeton-Toronto-London: Van Nostrand Co. 1967.
Smith, H. J., and E. V. P. Smith: *Solar Flares*. New York: McMillan Co. 1963.
Tandberg-Hansen, E.: *Solar Activity*. Waltham-Toronto-London: Blaisdell Publ. Co. 1967.

21. Variable Stars: Motions and Magnetic Fields in Stars

Cf. references for Section 20, as well as:
*Campbell, L., and L. Jacchia: *The Story of Variable Stars*. Philadelphia-Toronto: Harvard Books 1945.
Hoffmeister, C.: *Veränderliche Sterne*. Leipzig: J. A. Barth 1970.
Payne-Gaposchkin, C.: *The Galactic Novae*. New York: Dover Publ. Inc. 1964.
Shklovsky, I. S.: *Supernovae*. London-New York-Sydney: Wiley-Interscience Publ. 1968.
Strohmeier, W.: *Variable Stars*. London: Pergamon Press 1972.

Part III: Stellar Systems
Milky Way and Galaxies; Cosmogony and Cosmology

22. Advance into the Universe. Historical Introduction to Astronomy in the Twentieth Century

See under "Histories of Astronomy."

23. Constitution and Dynamics of the Galactic System

Becker, W., and G. Contopoulos (eds.): *The Spiral Structure of Our Galaxy* (IAU Sympos. 38). Dordrecht: D. Reidel Publ. Co. 1970.
Blaauw. A., and M. Schmidt (eds.): *Galactic Structure*. (Vol. V of *Stars and Stellar Systems,* see above.)
Bok, B. J., and P. F. Bok: *The Milky Way*. 4th edn. Cambridge, Mass.: Harvard University Press 1974.
Kerr, F. J., and A. W. Rodgers (eds.): *The Galaxy and the Magellanic Clouds* (IAU Sympos. 20). Canberra 1964.
Woerden, H. van (ed.): *Radioastronomy and the Galactic System* (IAU Sympos. 31). London-New York: Academic Press 1967.
The Structure and Evolution of Galaxies, Proc. 13th Solvay Conf. London-New York-Sydney: Interscience 1965.

24. Interstellar Matter

Aller, L. H.: *Gaseous Nebulae*. London: Chapman and Hall 1956.
Dufay, J.: *Galactic Nebulae and Interstellar Matter* (English translation). London: Hutchinson 1957, and New York: Dover 1968.
Gordon, M. A., and L. E. Snyder (eds.): *Molecules in the Galactic Environment*. New York: Wiley 1973.
Kaplan, S. A., and S. B. Pikelner: *The Interstellar Medium*. Cambridge, Mass.: Harvard University Press 1970.
Menzel, D. H. (ed.): *Selected Papers on Physical Processes in Ionized Plasmas*. New York: Dover Publ. 1962.
Middlehurst, B., and L. H. Aller (eds.): *Stars and Stellar Systems,* Vol. 7: *Nebulae and Interstellar Matter*. The University of Chicago Press 1968.
Osterbrock, D. E.: *Astrophysics of Gaseous Nebulae*. W. H. Freeman Co. 1974.
Spitzer, L., Jr.: *Diffuse Matter in Space*. New York: Interscience 1968.
Van de Hulst, H. C.: *Light Scattering by Small Particles*. New York: Wiley 1957.
Wickramsinghe, N. C.: *Interstellar Grains* (The International Astrophysics Series, 9). London: Chapman and Hall 1967.

Woltjer, L. (ed.): *The Distribution and Motion of Interstellar Matter in Galaxies.* New York: W. A. Benjamin Inc. 1962.
Planetary Nebulae, IAU Symposium 34. Dordrecht: D. Reidel Publ. Co. 1968.

25. Internal Constitution and Energy Generation of Stars

Chandrasekhar, S.: *An Introduction to the Study of Stellar Structure.* The University of Chicago Press 1939, also New York: Dover 1957.
Cox, J. P. (with R. T. Giuli): *Principles of Stellar Structure. Vol. 1: Physical Principles. Vol. 2: Applications to Stars.* New York-Paris-London: Gordon and Breach 1968.
Eddington, A. S.: *The Internal Constitution of the Stars.* Cambridge University Press 1926, also New York: Dover 1959.
*Schwarzschild, M.: *Structure and Evolution of the Stars.* Princeton University Press 1958, also New York: Dover.
*Tayler, R. J.: *The Stars—Their Structure and Evolution.* London-Winchester: Wykeham Publ. (London) Ltd. 1970.
Les Processus Nucléaires dans les Astres. Col. Internat. d'Astrophysique Liège 1953.

26. Color-magnitude Diagrams of Galactic and Globular Clusters and Stellar Evolution

See Bibliography for Section 25, together with:
Burbidge, E. M., G. R. Burbidge, W. A. Fowler, and F. Hoyle: Synthesis of the Elements in Stars, *Rev. Mod. Phys.* vol. 29, p. 547, 1957.
Burbidge, G. R.: Nuclear Astrophysics, *Ann. Rev. Nuclear Science* vol. 12, p. 507, 1963.
——, F. D. Kahn, R. Ebert, S. v. Hoerner, and St. Temesvary: *Die Entstehung von Sternen durch Kondensation diffuser Materie.* Berlin-Göttingen-Heidelberg: Springer-Verlag 1960.
O'Connell, D. J. K. (ed.): *Stellar Populations.* Conf. Vatican Observ. 1958. Amsterdam: North-Holland.
Stein, R. F., and A. G. W. Cameron (eds.): *Stellar Evolution.* New York: Plenum Press 1966.
Struve, O.: *Stellar Evolution, an Exploration from the Observatory.* Princeton University Press 1950.
Tayler, R. J., (ed.): *Late Stages of Stellar Evolution* (IAU Sympos. 66). Dordrecht: D. Reidel Publ. Co. 1974.
Modèles d'Étoiles et Évolution Stellaire, Internat. Astrophysics Colloquium, Liège 1959.
Évolution Stellaire avant la Sequence Principale, 16th Internat. Astrophysics Colloquium, Liège 1970.

27. Galaxies

28. Radio Emission from Galaxies, Galactic Nuclei, and Cosmic Rays and High Energy Astronomy

29. Galactic Evolution

Baade, W.: *Evolution of Stars and Galaxies* (Ed. C. Payne-Gaposchkin). Cambridge, Mass.: Harvard University Press 1963, also MIT Press (paperback series) 1975.

Bradt, H., and R. Giacconi (eds.): *X- and Gamma-Ray Astronomy* (IAU Sympos. 55). Dordrecht: D. Reidel Publ. Co. 1973.
Burbidge, G., and E. M. Burbidge: *Quasi-Stellar Objects.* San Francisco: W. H. Freeman Co. 1967.
Evans, D. S. (ed.): *External Galaxies and Quasi-Stellar Objects* (IAU Sympos. 44). Dordrecht: D Reidel Publ. Co. 1972.
Ginzburg, V. L., and S. I. Syrovatskii: *The Origin of Cosmic Rays.* Oxford: Pergamon Press 1964.
Ginzburg, V. L.: *Elementary Processes in Cosmic Ray Astrophysics.* New York-London-Paris: Gordon and Breach 1969.
*Hey, J. S.: *The Radio Universe.* 2nd edn. Oxford: Pergamon Press 1975.
Hodge, P. W.: *Galaxies and Cosmology.* New York: McGraw-Hill 1966.
*Hubble, E. P.: *The Realm of the Nebulae.* New Haven: Yale University Press 1936, also New York: Dover 1958.
O'Connell, D. J. K. (ed.): *Nuclei of Galaxies.* Amsterdam-London: North Holland Publ. Co. 1971.
*Page, T.: *Stars and Galaxies.* Englewood Cliffs, N.J.: Prentice Hall 1962.
Palmer, H. P., R. D. Davies, and M. I. Large (eds.): *Radio Astronomy Today.* Manchester University Press 1963.
Payne-Gaposchkin, C.: *Variable Stars and Galactic Structure.* London: The Athlone Press 1954.
Piddington, J. H.: *Radio Astronomy.* London: Hutchinson 1961.
*Sandage, A.: *The Hubble Atlas of Galaxies.* Carnegie Inst. of Washington, Publ. 618, 1961.
Sandage, A., M. Sandage, and J. Kristian,: *Galaxies and the Universe* (vol. IX of *Stars and Stellar Systems,* see "Handbooks.") Chicago 1976
Sandström, A. E.: *Cosmic Ray Physics.* Amsterdam: North-Holland 1965.
Setti, G. (ed.): *The Structure and Evolution of Galaxies.* NATO Advanced Study Institute, 1974. Dordrecht: D. Reidel Publ. Co. 1975.
De Vaucouleurs, G. & A.: *Reference Catalogue of Bright Galaxies.* Texas University Press 1964.
Verschuur, G. L., and K. I. Kellermann (eds.): *Galactic and Extra-Galactic Radio Astronomy.* Berlin-Heidelberg-New York: Springer-Verlag 1974.
*Wolfendale, A. W.: *Cosmic Rays.* London: Lewnes 1963.
Woltjer, L. (ed.): *Galaxies and the Universe.* New York-London: Columbia University Press 1968.
The Formation and Dynamics of Galaxies (IAU Sympos. 58). Dordrecht: D. Reidel Publ. Co. 1974.
The Structure and Evolution of Galaxies, Proc. 13th Solvay Conf., Brussels. London-New York-Sydney: Interscience 1965.
Progress of Elementary Particle and Cosmic Ray Physics. (Annual publication since 1952.) Amsterdam: North-Holland.

30. Cosmology

*Bondi, H.: *Cosmology.* 2nd edn. Cambridge University Press 1960.
*Born, M.: *Einstein's Theory of Relativity* (Revised edition with collaboration of G. Liebfried and W. Biem). New York: Dover 1962.
Einstein, A.: *Meaning of Relativity.* 6th edn. London: Methuen 1960.
Kundt, W.: Recent Progress in Cosmology, Springer Tracts in Modern Physics, vol. 47, p. 11, 1968.

Landau, L. D., and E. M. Lifschitz: *Course of Theoretical Physics. II. The Classical Theory of Fields*. 2nd edn. Oxford: Pergamon Press 1962.
*McCrea, W. H.: *Cosmology*. Guernsey: F. Hodgson Ltd. 1969.
McVittie, G. C.: *General Relativity and Cosmology*. The University of Illinois Press 1965.
Misner, C. W., K. S. Thorne, and J. A. Wheeler: *Gravitation*. Reading: W. H. Freeman Co. 1974.
Peebles, P. J. E.: *Physical Cosmology*. Princeton University Press 1971.
Robertson, H. P.: Relativistic Cosmology, *Rev. Mod. Physics* vol. 5, p. 62, 1933.
Robertson, H. P., and T. W. Noonan: *Relativity and Cosmology*. Philadelphia-London-Toronto: W. B. Saunders Co. 1968.
Tolman, R. C.: *Relativity, Thermodynamics, and Cosmology*. Oxford: Clarendon Press 1934.
Weinberg, S.: *Gravitation and Cosmology*. New York: Wiley 1972.

31. Origin of the Solar System: Evolution of the Earth and of Life

Brancazio, P. J., and A. G. W. Cameron (eds.): *The Origin and Evolution of Atmospheres and Oceans*. New York-London-Sydney: Wiley 1963.
Eigen, M.: Self-organization of Matter and the Evolution of Biological Macromolecules, *Naturwiss*. vol. 58, p. 465, 1971. (Also available as separate publication, Springer-Verlag Berlin-Heidelberg-New York, 1971.)
Jastrow, R., and A. G. W. Cameron (eds.): *Origin of the Solar System*. New York-London: Academic Press 1963.
Hanawalt, P. C., and R. H. Haynes, (eds.): *The Chemical Basis of Life—an Introduction to Molecular and Cell Biology* (a Scientific American book). W. H. Freeman and Co. 1973.
Kuhn, H.: Entstehung des Lebens: Bildung von Molekülgesellschaften, *Forschung '74*. Frankfurt/M Fischer Taschenbuch Verlag 1974.
Kuiper, G. P.: *On the Origin of the Solar System*. In *Astrophysics*. Ed. J. A. Hynek. New York-Toronto-London: McGraw-Hill Book Co. 1951.
Monod, J.: *Chance and Necessity* (English translation). London: Collins 1972.
Reeves, H. (ed.): *On the Origin of the Solar System,* Coll. du Centre Nat. de la Recherche Scientifique, Paris 1972.
*Rutten, M. G.: *The Origin of Life*. Amsterdam-London-New York: Elsevier 1971.
*Urey, H. C.: *The Planets, Their Origin and Development*. New Haven, Conn.: Yale University Press, also London: Oxford University Press 1952.
Watson, J. D.: *Molecular Biology of the Gene*. 2nd edn. New York: Benjamin 1970.
von Weizsäcker, C. F.: *The History of Nature* (English translation). London: Routledge & Kegan-Paul 1951.

Figure Acknowledgments

The numbers are those of the figures appearing in the text.

2.3, 3.1, 3.2, 3.3, 4.1, 4.2, 4.3, 4.4, 5.1, 5.5, 7.1. Seydlitz: Part 5. *Allgemeine Erdkunde.* 7th edn. Kiel: F. Hirt, and Hannover: H. Schroedel 1961.
6.7. Unsöld, A.: *Physikal. Blätter,* vol. 5, p. 205, 1964. Mosbach: Physik-Verlag.
6.8. Schurmeier, H. M., R. L. Heacock, and A. E. Wolfe: *Scientific American,* p. 57, Jan. 1966.
6.9. Phot. NASA AS-11-40-5947.
7.2. Bott, M. H. P.: *The Interior of the Earth,* p. 203. London: E. Arnold Ltd. 1971.
7.3. Phot. Lick Observatory, *Sky and Telescope,* vol. 26, p. 342, 1963.
7.4. Phot. NASA AS-11-42-6236.
7.5. Brüche, E., and E. Dick: *Physikal. Blätter,* vol. 26, p. 351, Figure 7, 1970.
7.6. Wänke, H., and F. Wlotzka: *Universitas,* vol. 26, p. 850, 1971.
7.7. NASA and *Naturwiss.,* vol. 59, p. 395, Figure 5, 1972.
7.8. NASA and *Naturwiss.,* ol. 59, part 4, frontispiece, 1972.
7.9. Phot. B. Lyot and H. Camichel, Observatoire Pic du Midi.
7.10. Phot. H. Camichel, Observatoire Pic du Midi.
8.1. Phot. Hale Observatories.
8.2. Swings, P., and L. Haser: *Atlas of Representative Cometary Spectra,* plate IV, University of Liège 1956.
8.3. Gentner, W.: *Die Naturwissenschaften,* vol. 50, p. 192, Figure 1, 1963.
8.4. Anders, E.: *Accounts of Chem. Res.;* Oct. 1968.
9.5. Phot. Yerkes Observatory, Williams Bay, Wisc.
9.6. Phot. Mt. Wilson and Palomar Observatories.
9.7. *Das Weltall.* Time-Life International, p. 37, 1964.
9.9b. Russell–Dugan–Stewart: *Astronomy II,* Figure 254. New York: Ginn Co. 1927.
9.10. Eastman Kodak Co., Rochester, N.Y.: "*Kodak Plates and Films,*" p. 15d.
9.11. Dunham, T., Jr.: *Vistas in Astronomy II* (Ed. A. Beer), p. 1236. London and New York: Pergamon Press 1956.
9.12. Baum, W. A.: *Science,* vol. 154, p. 114, Figure 4, 1966.
9.13. Austral. Nat. Radio Astron. Observatory; 1963.
9.14. Rossi, B.: *Electromagnetic Radiation in Space,* p. 171, Figure 9. Dordrecht: D. Reidel Publ. Co. 1966.
11.1, 11.2, 11.3. Unsöld, A.: *Physik der Sternatmosphären,* 2nd edn. Berlin-Göttingen-Heidelberg: Springer-Verlag 1955.
12.1. Russell–Dugan–Stewart: *Astronomy I,* Figure 22. New York: Ginn Co. 1927.
12.2. Minnaert–Mulders–Houtgast: *Photometric Atlas of the Solar Spectrum* (section). Amsterdam: Schnabel, Kampfert & Helm 1940.
13.1. Johnson, H. L., and W. W. Morgan: *Astrophys. J.,* vol. 114, p. 523, 1951.
15.1. Morgan, W. W., P. C. Keenan and E. Kellman: *An Atlas of Stellar Spectra* (section). University of Chicago Press 1942.
15.2. Russell–Dugan–Stewart: *Astronomy II*. New York: Ginn Co. 1927.
15.3. Johnson, H. L., and W. W. Morgan: *Astrophys. J.*, vol. 117, p. 338, 1953.
15.5. Becker, W. in: *Stars and Stellar Systems III,* p. 254. The University of Chicago Press 1963.

16.1. Baker, R. H.: *Astronomy*. 6th edn. New York: Van Nostrand Co. 1955.
16.2. Unsöld, A.: *Physik der Sternatmosphären*. 2nd edn. Berlin-Göttingen-Heidelberg: Springer-Verlag 1955.
17.3. Merrill, P. W.: *Papers Mt. Wilson Observatory*, vol. IX, p. 118, 1965. Carnegie Inst. of Washington.
18.1. Unsöld, A.: *Physik der Sternatmosphären*. 2nd edn., p. 106. Berlin-Göttingen-Heidelberg: Springer-Verlag 1955.
18.4. Unsöld, A.: *Monthly Not. Roy. Astr. Soc.*, vol. 118, p. 9, 1958.
18.5. Unsöld, A.: *Physik der Sternatmosphären*. 2nd edn. Berlin-Göttingen-Heidelberg: Springer-Verlag 1955.
19.1, 19.2, 19.3. Unsöld, A.: *Angewandte Chemie*, vol. 76, pp. 281–290, 1964.
20.1. Danielson, R. E.: *Astrophys. J.*, vol. 134, p. 280, 1961.
20.2. Unsöld, A.: *Physik der Sternatmosphären*. 2nd edn. Berlin-Göttingen-Heidelberg: Springer-Verlag 1955.
20.3. Houtgast, J.: *Rech. Astron. Utrecht*, vol. 13, p. 3, Utrecht 1957.
20.4. Biesbroeck, G. van: *The Sun* (Ed. G. P. Kuiper), vol. I, p. 604, 1953. The University of Chicago Press.
20.5. *Sky and Telescope*, vol. 20, p. 254, 1960.
20.6. Royds, T.: *Monthly Not. Roy. Astron. Soc.* vol. 89, p. 255 1929.
20.7. Jager, C. de: *Handb. d. Phys.*, vol. 52, p. 136. Berlin-Göttingen-Heidelberg: Springer-Verlag 1959.
20.8. Unsöld, A.: *Physik der Sternatmosphären*. 2nd edn. Berlin-Göttingen-Heidelberg: Springer-Verlag 1955.
20.9. Cape Observatory. *Proc. Roy. Inst.*, vol. 38, no. 175, pl. I, 1961.
20.10. Palmer–Davies–Large: *Radio Astronomy Today*, p. 19. Manchester University Press 1963.
20.12. Wilcox, J. M.: *Space Science Lab. University of Calif. Berkeley*, Ser. 12, p. 53, Figure 2, 1971.
21.1. Becker, W.: *Sterne u. Sternsysteme*, p. 108. Darmstadt: Steinkopff 1950.
21.2. Minkowski, R.: *Ann. Rev. of Astronomy and Astrophysics*, vol. 2, p. 248, 1964.
23.4. Phot. Mt. Wilson and Palomar Observatories in O. Struve: *Astronomie*. Berlin: W. de Gruyter 1962, p. 326, Figure 26.3.
23.5. Duncan, J. C.: *Astronomy*, p. 408. New York: Harper 1950.
23.6. Oort, J. H.: *Stars and Stellar Systems*, vol. 5, p. 484, 1965. The University of Chicago Press.
23.7. Becker, W.: *Z. Astrophys.*, vol. 58, p. 205, 1964.
23.8. Westerhout, G.: The University of Maryland (USA).
23.11. Eggen, O. J.: *Roy. Observ. Bull.*, vol. 84, p. 114, 1964.
24.2. Oort, J. H. in: *Interstellar Matter in Galaxies* (Ed. Woltjer). New York: Benjamin 1962.
24.3. Westerhout, G.: *Bull. Astron. Inst. Netherlands*, vol. 14, p. 254, 1958.
24.4. Phot. Mt. Wilson and Palomar Observatories, in Merrill, P. W.: *Space Chemistry*, p. 122. The University of Michigan Press 1963.
24.5. Goldberg, L., and L. H. Aller: *Atoms, Stars and Nebulae*, p. 182, Philadelphia: Blackiston Co. 1946.
24.6. Phot. Harvard Observatory, in Baker, R. H.: *Astronomy*. 6th edn., p. 466. New York: Van Nostrand Co. 1955.
24.9. Mathewson, D. S., and V. L. Ford: *Mem. Roy. Astron. Soc.*, vol. 74, p. 143, 1970.
25.1. Fowler, W. A. in: *Liège Astrophys. Sympos.* 1959, p. 216.

26.1a. Johnson, H. L.: *Astrophys. J.*, vol. 116, p. 646, 1952.
26.1b. Eggen, O. J., and A. Sandage: *Astrophys. J.*, vol. 158, p. 672, 1969.
26.2. After Sandage, A., and O. J. Eggen: *Astrophys. J.*, vol. 158, p. 697, 1969.
26.3. Sandage, A.: *Astrophys. J.*, vol. 162, p. 852, Figures 13 and 4, 1970.
26.4. Sandage, A.: *Astrophys. J.*, vol. 162, p. 863, Figure 18, 1970.
26.5a and 26.5b. Kippenhahn, R., H. C. Thomas, and A. Weigert: *Z. Astrophys.*, vol. 61, p. 246, 1965.
26.6. Iben, I., Jr.: *Ann. Rev. Astron. and Astrophys.*, vol. 5, p. 585, 1967.
26.7. Iben, I., Jr.: *Astrophys. J.*, vol. 141, p. 1010, 1965.
26.8. Walker, M.: *Astrophys. J. (Suppl.)*, vol. 2, p. 376, 1956.
26.11. and cover picture. Phot. Hale Observatories.
26.12. Hogg, D. E.: *Astrophys. J.*, vol. 140, p. 992, Figure 2, 1964.
26.14. Walker, M.: *Astrophys. J.*, vol. 125, p. 651, 1957.
27.1. Phot. Mt. Wilson and Palomar Observatories, Carnegie Inst. of Washington, *The Hubble Atlas of Galaxies*, p. 18, 1961.
27.2. Hubble, E.: *Astrophys. J.*, vol. 69, p. 120, 1929.
27.3. After Hubble, E.: *The Realm of the Nebulae*, p. 45. New Haven, Conn.: Yale University Press 1936.
27.4. Reference as 27.1, p. 38.
27.5. Baade, W., and H. Swope: *Astron. J.*, vol. 66, p. 326, 1961.
27.6. Phot. Hale Observatories.
27.7. Rubin, V. C., and W. K. Ford, Jr.: *Astrophys. J.*, vol. 159, p. 390, 1970.
27.8. Zwicky, F.: *Astrophys. J.*, vol. 140, p. 1627, 1964.
27.9. Morgan, W. W., and N. U. Mayall: *Publ. Astron. Soc. Pacific*, vol. 69, p. 295, 1957.
28.2. Downes, D., A. Maxwell, and M. Meeks: *Astrophys. J.*, vol. 146, p. 657, Figure 4, 1966.
28.3. Mitton, S., and M. Ryle: *Monthly Not. Roy. Astron. Soc.*, vol. 146, p. 223, 1969.
28.4. Cooper, B. F. C., R. M. Price, and D. J. Cole: *Austral. J. Physics*, vol. 18, p. 602, 1965.
Maltby, P., et al.: *Astrophys. J.*, vol. 40, p. 44, 1964.
28.6. Sandage, A. R.: *Scientific American*, p. 39, Nov. 1964.
28.7. Lingenfelter, R. E.: *Astrophys. and Space Science*, vol. 24, p. 89, 1973.
29.1. Roberts, W. W.: *Astrophys. J.*, vol. 158, p. 132, Figure 7, 1969.
29.4. Amiet, J. P., and H. D. Zeh: *Zs. f. Physik*, vol. 217, p. 505, 1968.
30.2. Sandage, A.: *Astrophys. J.*, vol. 178, p. 34, 1972.
31.1. Taken partly from Gentner, W.: *Naturwissenschaften*, vol. 56, p. 174, Figure 3, 1969.
31.2. Acknowledgement as for Figure 8.4.
31.4. Wieland, T. and G. Pfleiderer (eds.): *Molekularbiologie*. Frankfurt/M.: Umschau-Verlag 1969, p. 46, Figure 3.

Index